Wireless Connectivity

Wireless Connectivity

An Intuitive and Fundamental Guide

Petar Popovski
Department of Electronic Systems
Aalborg University, Denmark

This edition first published 2020

Registered Offices
John Wiley & Sons, Inc., 111 River Street, Hoboken, NJ 07030, USA
John Wiley & Sons Ltd, The Atrium, Southern Gate, Chichester, West Sussex, PO19 8SQ, UK

Editorial Office
The Atrium, Southern Gate, Chichester, West Sussex, PO19 8SQ, UK

For details of our global editorial offices, customer services, and more information about Wiley products visit us at www.wiley.com.

Wiley also publishes its books in a variety of electronic formats and by print-on-demand. Some content that appears in standard print versions of this book may not be available in other formats.

Library of Congress Cataloging-in-Publication Data applied for

HB ISBN: 9780470683996

Cover Design & Image: © Peter Gregson

Set in 9.5/12.5pt STIXTwoText by SPi Global, Chennai, India
Printed and bound in Singapore by Markono Print Media Pte Ltd

10 9 8 7 6 5 4 3 2 1

In memory of my father, the unshakeable optimist.

Contents

Foreword

First: But, Why?

Why should one dare to write a relatively long book in a digital age, where everything seems to be quickly found online and watching video tutorials is used as a substitute for reading? Why read, let alone write, a long text when a single tweet results in thousands of workers getting laid off or somebody becoming a millionaire in a day? And, last, but not least, why put in so much effort, knowing that, eventually, despite all the measures, the PDF of the book will be available for illegal download at a phony website?

A book or a textbook still has its role in the digital age, but this role is significantly different from the times when books were the ultimate source of information and knowledge. It should rather be understood as a gateway to knowledge, a gadget that helps to make sense of the massive amount of online data, a hitchhiker's guide for receiving, filtering, and learning from the overwhelming waves of information. As for the illegal copies: if you are reading this on an illegal copy, then it is fair to tell you that it is getting really boring from now on, so you can stop reading.:)

Wireless became Huge and Complex

This role of a "gateway to knowledge" was one of the motivations behind writing this book. The area of wireless communications has developed immensely over the last three decades, generating a large number of concepts, ideas, articles, patents, and even myths. Identifying the crucial ideas and their interconnection becomes increasingly difficult. The area of wireless connectivity grew to be very complex, to the level where the specialists working in one part of the system, say hardware, did not know much about the functioning of the high layer protocols, and vice versa. In the extreme case, this ignorance about the concepts and functioning of the other parts of the communication system led to the "only-my-part-of-the-system-matters" attitude, sometimes resulting in disastrously sub-optimal designs.

In order to see how much wireless communication has developed in volume and complexity, here is an ultra-short overview of the generations of wireless mobile communication systems. It started with the "modest" ambition of 2G being reachable for a phone call wherever you are moving, but the unlikely hero was the short message service (SMS) that brought texting as a new social phenomenon. It was perhaps this unlikely hero that planted the obsession with the "killer application" during the development of 3G. The advent of the smartphone has shown that there was not a single killer application, but a gateway to the internet, a gadget with many apps, a hitchhiker's guide to multiple applications [sic]. Only with the data speeds offered by 4G did the smartphone start to reach its full potential, completely transforming work and play. At the time of writing, we are at the dawn of the deployment of 5G, still with a lot of predictions, enthusiasm, and skepticism. The working version of the 5G ambition is to offer truly ubiquitous and reliable coverage to the humans, but also wirelessly connect machines and physical objects.

Note that this ultra-short overview is unfair to a lot of other wireless technologies that are omnipresent and play a crucial role in daily life, such as Wi-Fi and Bluetooth. The tone in the book is leaning towards mobile wireless cellular networks, but the overall discussion is kept generic, not tied to a particular wireless system or technology.

How this Book is Structured

The book does not follow the established, linear structure in which one starts from the propagation and channels and then climbs up the protocol layers. Here the approach has been somewhat nonlinear in an attempt to follow the intuition used when one creates a new technology to solve a certain problem. With this approach, we state a problem from the real world and create a model that reflects the features of the problem. The model is a simplification, a caricature of the reality, but as every good caricature, it captures the essential features. A certain model (for example, a collision model of a wireless channel), allows the system designer to propose solutions that reside within a subspace of the space of all possible solutions to the real communication problem. By enriching the model (for example, adding a capture and interference cancellation to the collision model), the system designer can devise solutions that go beyond the boundaries of the previously mentioned subspace. Practically each new chapter brings enrichment of the models, presenting system/algorithm designs that extend the ones developed for the simpler system models.

Each chapter starts with a cartoon that carries the main message of the chapter. The narrative in the book uses characters as it facilitates the discussion about communication between different parties. This is inspired by the security literature, which deals with Alice, Bob, etc. Here I have started from the other side of the alphabet, which is populated by the letters commonly used in communication theory. The characters are Zoya, Yoshi, Xia, Walt, Victoria (which happens to be the name of my mom) and Umer, and, yes, they are more international compared to Alice and Bob. The base stations are named Basil and Bastian for obvious reasons.

Among the references in "further reading", I have inserted references to some of the research works co-authored by me. This is not to boost my citations by a self-citation (which is, righteously, often not counted in the academic record), but rather to offer a reference that shows that I, the author, have made an actual research contribution to the area. This can only increase the credibility of what is written in the chapter.

Objectives and Target Audience

When I started the book, the objective was to make it suitable for casual reading and use almost no equations. In the meantime, the number of equations increased, but still far below the number in standard textbooks on wireless communications, such that there is still hope that the reader can read it casually. Yet, the fact that there are five problems after each chapter indicates that the book can also be used as a textbook. However, for in depth reading, the reader should rely on the literature in "further reading".

The target audience is:

- Students in electronics, communication and networking. Some of the problems at the end of the chapters are actually mini-projects, which the students can do over an extended time. This is suitable for both graduate and undergraduate courses. Clearly, if used as a graduate course, then there is more reliance on external literature.
- Wireless engineers that are specialists in one area who want to know how the whole system works, without going through all the detail and math.
- Computer scientists that want to understand the fundamentals of wireless connectivity, the requirements. and, most importantly, the limitations.
- As wireless connectivity starts to play a big role in a large number of cyber-physical systems, such as smart grids, transport, logistics or similar, the engineers specializing in those areas can obtain an insight into some of the essential wireless concepts.
- As a supplement to other books on wireless connectivity that deal with the detail of analysis and design of specific technologies.

Acknowledgments

Even when a book has a single author, a large part of the authorship goes to the colleagues, friends, and family that provided inspiration, criticism, a gentle push when things looked impossible and a reminder that Sisyphus was only a mythical creature.

I am deeply grateful to Osvaldo Simeone (King's College London) for enormous support during the preparation of this book. The credit for the idea of using cartoons should go to him. He could absolutely always find the time to read the chapters that I was asking him to check, and provide prompt and rich feedback. I have been fortunate to have him, an exemplary erudite researcher, as a collaborator over many years.

Three people stood out in encouraging me throughout the long writing of the book. Jørgen Bach Andersen (Aalborg University), Angel Lozano (Pompeu Fabra University) and Hiroyuki Yomo (Kansai University). Jørgen provided me with very valuable feedback on Chapter 10 (Space in Wireless Communications). Angel removed my doubts about the usefulness of Chapter 9 (Time and Frequency in Wireless Communications). Hiroyuki decided to use this as a textbook in the early stages, when I presented him with the book concept.

I am very thankful for the feedback I got on specific chapters. Two members of my research group provided me with feedback in the early stage of writing and removed some of the doubts I had about the style. Čedomir Stefanović (Aalborg University) read the first chapters and Nuno Pratas (now with Nokia) read Chapter 6 (A Mathematical View on a Communication Channel). Anna Scaglione (Arizona State University), Emil Björnson (Linköping University) and Elisabeth de Carvalho (Aalborg University) were very kind to read Chapter 11 (Using Two, More, or a Massive Number of Antennas) and provide me with prompt and useful feedback.

A big thank you goes to the members of my research group, who had to be patient with my rants about the book throughout all these years. After one of my lectures for the master students, Rocco Di Taranto (now with Ericsson), at that time my PhD student, asked me: "Where can I read these topics explained in a way in which you did it at the lecture?". The book idea had been cooking in the background for some time, but this was perhaps the decisive push to write it. Marko Angjelichinoski (now with Duke University) was convinced that the style and the whole book project were very original and I needed to hurry up. I would like to thank Kasper Fløe Trillingsgaard (now with InCommodities) for many stimulating discussion on the information-theoretic aspects. Alexandru-Sabin Bana (Aalborg University) and Radoslaw Kotaba (Aalborg University) helped to prepare a course based on this book and spotted several errors and inconsistencies. While this book was in the final stages, I was teaching a course at Aalborg University and several students were kind to correct errors in the chapters: Andreas Engelsen Fink, Jonas Ingerslev Christensen, Taus Mortensen Raunholt, Jeppe Thiellesen and Simon Kallehauge.

The cartoons, the cover page, as well as the clipart used to make the figures, were made by Peter Gregson Studio from Novi Sad, Serbia. This is a team of immensely creative people, Jovan Trkulja, Velimir Andrejević and Milan Letić, whose ideas play a significant role in the final look of the book. I would also like to thank Aleksandar Sotirovski for making the first version of the cartoons for some of the chapters, but due to objective reasons could not continue. Thanks to Kashif Mahmood (Telenor) for suggesting Umer as a Pakistani name starting with "U". I would also like to thank the team at Wiley for being patient and supportive throughout the years, but especially in the final stage: Sandra Grayson, Louis Manoharan, Adalfin Jayasingh, and Tessa Edmonds.

My biggest support through these years came from my family: my wife Iskra, my children Andrej and Erina, as well as our extended family. Family was always there to take the blame when I was performing poorly on time management and planning of the writing. In its most severe form, that blame was ending with a threat that I was going to write something similar to the dedication written by a mathematician, who dedicates his book to his wife and children "without whom this book would have been completed two years earlier". I am obviously not doing it and, instead, I want to thank them for absolutely always being there for me. I am hoping that some of them will read the book and get to know what I am actually working with. Unfortunately, my father passed away before this book was finished. I am dedicating this book to him.

P. P.

Acronyms

ACK	Acknowledgement
AF	Amplify and forward
ARQ	Automatic retransmission request
ASK	Amplitude shift keying
AMC	Adaptive modulation and coding
AWGN	Additive white Gaussian noise
BBU	Baseband processing unit
BPSK	Binary phase shift keying
BS	Base station
BSC	Binary symmetric channel
CDMA	Code division multiple access
CoMP	Coordinated multipoint
C-RAN	Cloud radio access network
CRC	Cyclic redundancy check
CRDSA	Contention resolution diversity slotted ALOHA
CSI	Channel state information
CSIT	Channel state information at the transmitter
CSMA	Carrier sensing multiple access
D2D	Device to device
DBPSK	Differential binary phase shift keying
DoF	Degree of freedom
EGC	Equal gain combining
FDD	Frequency division duplex
FDMA	Frequency division multiple access
FEC	Forward error correction
GF	Galois field
GPS	Global positioning system
HARQ	Hybrid automatic retransmission request
IoT	Internet of Things
ISI	Intersymbol interference
LBT	Listen before talk
LDPC	Low-density parity check
LEO	Low Earth orbit

LLN	Law of large numbers
LoS	Line of sight
MAC	Medium access control; also multiple access channel
MEC	Mobile edge computing
MIMO	Multiple input multiple output
MISO	Multiple input single output
MMSE	Minimum mean squared error
mmWave	Millimeter wave
MPR	Multi-packet reception
MRC	Maximum ratio combining
NACK	Negative acknowledgement
NOMA	Non-orthogonal multiple access
OFDM	Orthogonal frequency division multiplexing
OFDMA	Orthogonal frequency division multiple access
PAM	Pulse amplitude modulation
PAPR	Peak-to-average power ratio
pdf	probability density function
PHY	Physical layer
PSK	Phase shift keying
QAM	Quadrature amplitude modulation
QPSK	Quaternary phase shift keying
RF	Radio frequency
RNC	Radio network controller
RRH	Remote radio head
SC	Selection combining
SDMA	Space division multiple access
SIC	Successive interference cancellation
SIMO	Single input multiple output
SINR	Signal-to-interference-and-noise ratio
SNR	Signal-to-noise ratio
TDD	Time division duplex
TDMA	Time division multiple access
UEP	Unequal error protection
UWB	Ultra wideband
ZF	Zero rorcing

When the teacher Walt speaks to the students, the shared wireless channel is a blessing.

This is because it is sufficient that Walt says his thing only once, instead of repeating it for each student separately.

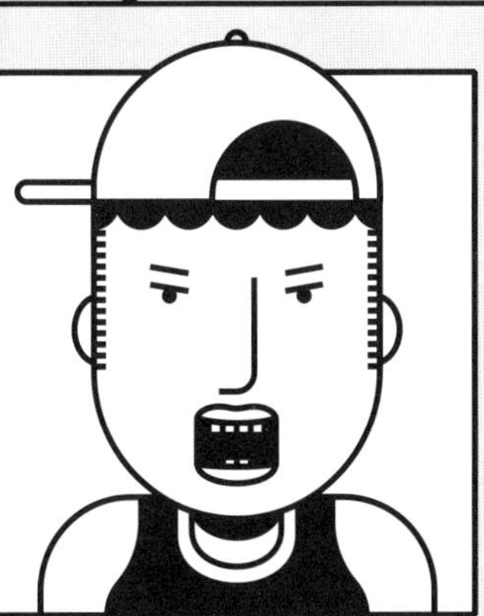

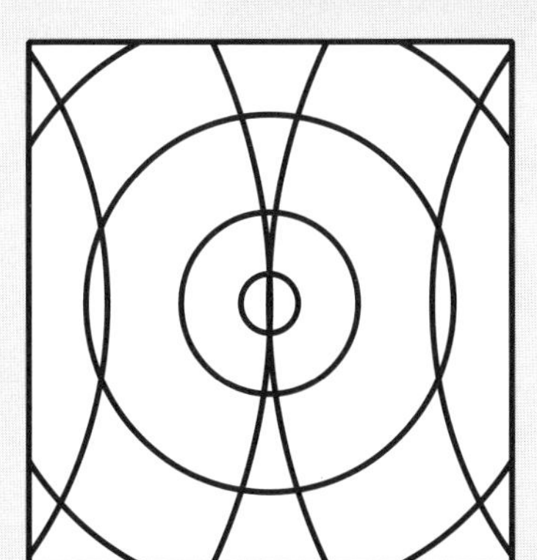

However, the shared wireless channel turns into a curse when all others try to speak to Walt at the same time.

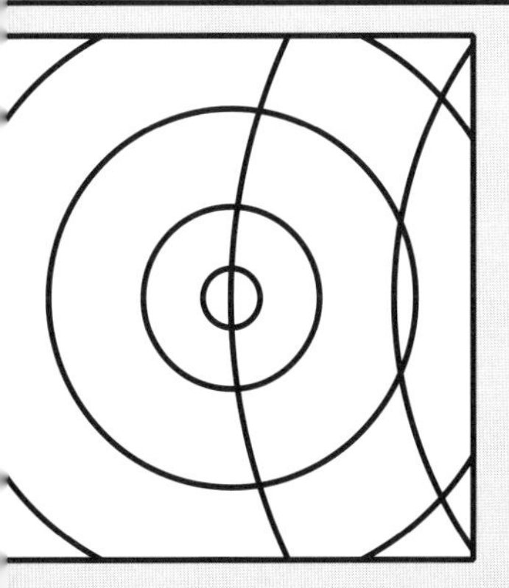

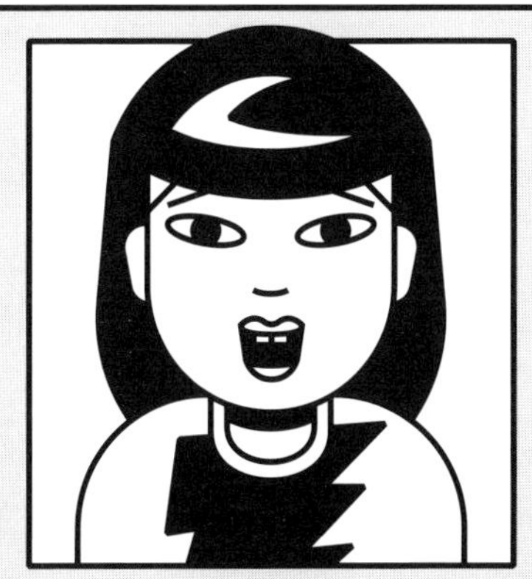

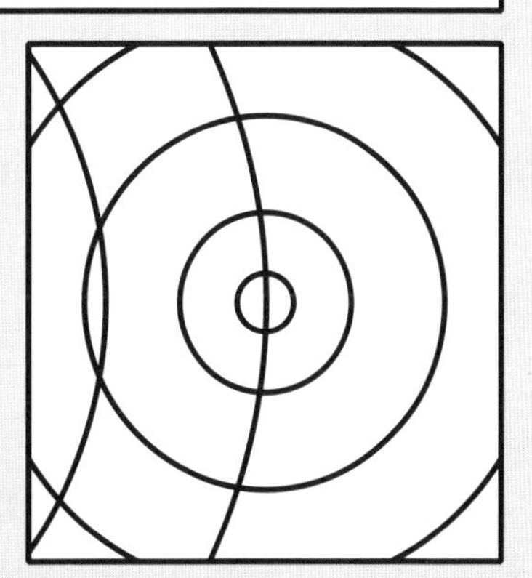

Story by Petar Popovski / Art by Peter Gregson

1

An Easy Introduction to the Shared Wireless Medium

We start by describing wireless communication through an analogy with a conversation within a group of people, named Zoya, Yoshi, and Xia. We will refer to these and some other characters throughout the book; the characters will stand for wireless devices, base stations, or similar. The *data* that they want to communicate to each other is the content of their speech, which is part of the conversation. Regardless of the speech content, the conversation can only take place if the participants follow some *conversation protocol*, such that at a given time only one person speaks while the others listen. How do they agree who gets to speak and who gets to listen? One way would be, before starting the actual conversation, to have them agree upon which conversation protocol should be followed. In that case the information exchanged in that preliminary conversation cannot be regarded as useful data, but rather as *metadata*, also called *protocol information* or *control information*. The metadata is necessary in order to enable the conversation to take place. But then, how do they agree on the protocol for exchanging the metadata?

These questions can go on to infinity, but in a normal situation the communication protocol is agreed upon by either sticking to certain rules of politeness or following visual cues and gestures that facilitate the conversation. In other words, the metadata is exchanged by using a visual communication channel that is different from the speech communication channel. However, in a commonly encountered wireless communication system there is only one communication channel through which both the data and the metadata should be sent. This is not to say that it is not possible to have one wireless communication channel for data and a separate one for metadata; even if such separation exists, then what is the protocol for exchanging the meta-metadata that is used to agree how to send the metadata?

This gets obviously complicated, but the bottom line is that we will always hit the problem of communicating over a single shared wireless channel. Now, taking the fact that there is a single channel for communicating both the data and the metadata, the key point of the analogy with the conversation is to put Zoya, Yoshi, and Xia in a dark room, such that they have only speech as a means of communicating (we exclude tactile communication) and no visual cues can be of help. In that setting, the audio channel should be used both to coordinate the conversation and to carry the actual content of the conversation.

This is the common situation in which wireless communication systems operate and will be the subject of this chapter. Here are some examples of the questions that will be discussed. If Zoya and Yoshi want to talk to each other, how do they agree who talks first and who listens first? If both Xia and Yoshi want to talk to Zoya, how should they agree who

Wireless Connectivity: An Intuitive and Fundamental Guide, First Edition. Petar Popovski.

takes a turn to speak at a given time, so that they don't all talk simultaneously? Solutions to these problems are provided by various protocols for controlling the access to the medium; hence the name *MAC (medium access control) protocols*, and they are of central importance in wireless communication systems.

1.1 How to Build a Simple Model for Wireless Communication

1.1.1 Which Features We Want from the Model

The main feature of the wireless communication medium is the fact that the medium is *shared*, in the same way in which the air through which the sound propagates is shared among the people having a conversation. MAC protocols enable multiple wireless communication users and devices to share the medium and send/receive data.

First, we must agree on how the system operates and what it takes to have a signal from one communication node received correctly at another node. In other words, we need to settle on a suitable *system model*: a set of assumptions that will allow us to talk about communication protocols and principles in a setting that is simple, but sufficient to contain the necessary properties of a shared wireless medium. We build the initial model by relying on a common sense analogy with the spoken conversation, as it captures three fundamental properties of wireless communication: *broadcast*, *interference*, and *half-duplex* operation. We illustrate these features by observing a conversation between Zoya, Yoshi, and Xia:

- *Half-duplex*: A given person, e.g. Zoya, cannot speak and listen at the same time.
- *Broadcast*: If Zoya has information to convey to Yoshi and Xia, then, provided that both Yoshi and Xia are listening, Zoya needs to say her message only once, and not repeat it individually to Yoshi and to Xia.
- *Interference*: If Yoshi and Xia speak simultaneously, Zoya will not understand either of them.

The descriptions above are arguably not always correct, but they do represent what is common sense for a conversation. Furthermore, the analogy of the communication problems with the conversation between Zoya, Yoshi, and Xia is useful, but it has its limitations, which will be pointed out when necessary.

1.1.2 Communication Channel with Collisions

For the purpose of this chapter, we define a *communication channel* to be the physical resource that is used for a wireless transmission. In that sense, in spoken communication, the channel is created by the audible vibrations that take place in air or even another sound-propagating medium. It is useful to note that the communication channel is not the *whole* physical medium with *all* the vibrations, since there are vibrations that cannot be registered by ear and thus do not carry useful audio information. Furthermore, spoken communication uses a single communication channel: one cannot switch to another channel, such as in a TV receiver, in order to listen to the desired speaker and avoid the undesired one.

As already stated above, our discussion will be limited to the case in which all nodes use a single communication channel. In reality that can be, for example, a certain frequency to which all the nodes are "tuned". Here we use the term "frequency" as it is used in a common language for, say, a TV frequency. One may argue that Zoya and Yoshi can agree to one frequency, while Xia and Walt can agree to tune to another frequency and in this way they do not need to share the channel with the link Zoya–Yoshi. This is indeed possible and we will discuss it in later chapters, when we introduce the notion of separation in frequency. On the other hand, it is also true that Zoya and Yoshi should first use some communication channel to agree upon which frequency they will use for communication. This agreement is, again, metadata or control information, such that the corresponding channel is often denoted as a *control channel* and can be shared by multiple nodes to come to an agreement about the frequency. For example, if Zoya decides to communicate with Xia, then she knows that she should try to find Xia at the control channel and, upon contacting her, use the control channel to decide which channel/frequency they should both be tuned to in order to communicate the useful data. However, the control channel is a common, shared communication channel and therefore the question of how to share that channel to send metadata remains valid.

The communication model used in this chapter is called a *collision model*. This is because the central assumption of the model is that if two or more nodes transmit simultaneously, then the interference that they cause to each other is manifested as a collision at the receiver. Upon collision, the receiver does not manage to retrieve any data successfully. Another assumption in the model, not really related to the issue of collision, is that a node operates in a half-duplex manner and cannot receive while transmitting. Most of the wireless systems that we encounter today are not full-duplex, that is, do not transmit and receive simultaneously at the same frequency channel. However, although technologically more complex, it is also possible to have full-duplex operation. Therefore, throughout the chapter we will occasionally revise the half-duplex assumption and discuss the changes that the full-duplex can bring into the design of a specific protocol or algorithm.

The communication between the wireless nodes is based on data packets. A transmitting node is capable of sending R bits per second (bps) such that a packet of duration T contains RT bits. All packets have the same duration, unless stated otherwise. In the collision model, a packet is treated as the smallest, atomic unit of information, such that either the whole packet is received correctly or it is lost. In other words, it is not possible to receive only some bits of the packet correctly. A packet sent by Zoya to Yoshi is received correctly if:

1. Yoshi is in the communication range of Zoya such that the distance between them is less than d m;
2. No other communication node that is within d m of Yoshi transmits while Yoshi is receiving the packet from Zoya.

The first condition above indicates that each transmission is *omnidirectional*. Due to the basic property of reciprocity in electromagnetic/radio propagation (see Section 10.9), each reception in our model is also omnidirectional. From this it follows that Yoshi receives a signal as long it is sent from a distance less than d m, regardless of the actual position. The ingredients of the collision model are illustrated in Figure 1.1. Specifically, Figure 1.1(a) depicts the data rate of an idealized single link as a function of the distance between two communicating nodes. An example communication scenario is depicted in Figure 1.1(b), where two nodes are connected by a line if the distance between them is less

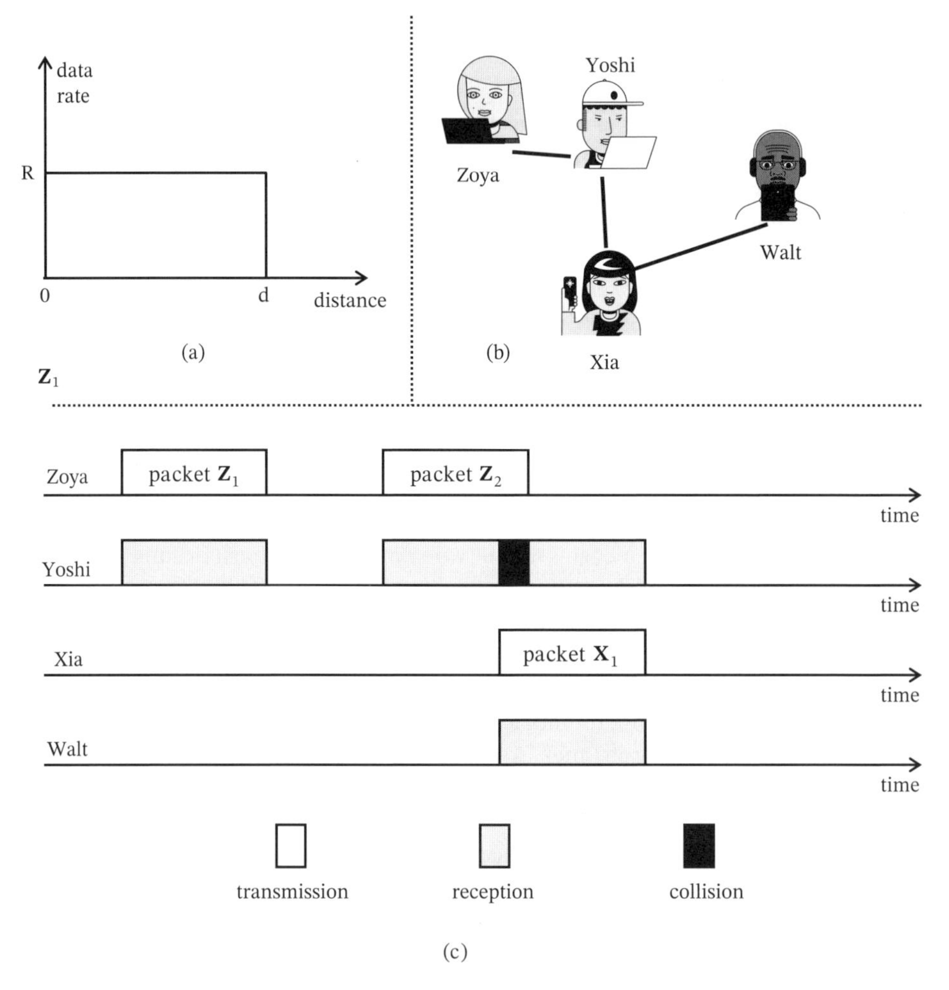

Figure 1.1 Communication model used in this chapter, referred to as a collision model. (a) Simplified dependence between the data rate and the distance, denoting a communication range *d*. (b) An example topology with possible wireless links among devices. The distance between two connected devices is at most *d*. (c) Collision model at work for the topology in (b).

than *d*, indicating the possibility of having a link between them. Figure 1.1(c) exemplifies a possible time evolution of a process of packet transmission in the framework of the collision model. The packet $\mathbf{Z}_1$ from Zoya to Yoshi is received correctly, while the packet $\mathbf{Z}_2$ is not, due to collision with the packet $\mathbf{X}_1$ sent simultaneously by Xia. Note that, by treating a packet as an atomic unit of information, even a partial overlap of $\mathbf{Z}_2$ and $\mathbf{X}_1$ causes packet loss. On the other hand, Walt is outside the range of Zoya, such that he can receive $\mathbf{X}_1$ without being interfered with by $\mathbf{Z}_2$.

The assumptions of fixed-length packets and always-destructive collisions are weakening the analogy with a conversation. If we think to relate a packet to a spoken word, then not all words have the same length and missing some letters of a word may still not destroy its comprehensibility. In fact, the collision model is rather pessimistic. In reality, one expects a certain continuity in comprehensibility/correctness of a packet: if the packets $\mathbf{Z}_2$ and $\mathbf{X}_1$ from Figure 1.1 have only a tiny overlap, then both would have to be received correctly by Yoshi. So, why are we not accounting for such a phenomenon and remain pessimistic about the collision? This is for pedagogical reasons in order to have a gradual path to system design and optimization. At the first step, make a system that works when every collision is identical and deemed destructive. In the next step, pose the question: what if not all collisions are identical? This leads to a refinement of the communication model by entering "inside the collision" and analyzing the different types of collision, which we will do in the later chapters. Notably, some types of collision will not be destructive and some collided packets can be received correctly, which sets the basis for optimizing the protocols further.

1.1.3 Trade-offs in the Collision Model

The basis of any good engineering is identification of the trade-off points that exist in a system: which benefits versus which costs are associated with given decisions on a system design. Even before discussing concrete techniques for accessing the shared medium, we can try to assess the limitations and the opportunities for protocol designs offered by the collision model. In that sense, it is at first instructive to look at the engineering trade-offs by contrasting the collision model with a model for wired communication.

For the problem of establishing and maintaining links, the obvious advantage offered in the wireless setting is that the communication is untethered and links can be established flexibly between any two nodes that come into spatial proximity. The price of this flexibility is twofold:

- Resources (time, battery) need to be consumed in order to establish the link between two nodes.
- The link is not exclusively reserved for use between the two nodes, as a third nearby node may transmit on the same channel and thus cause interference.

In contrast, in a wired model Zoya and Yoshi are connected by a dedicated cable. Precisely the lack of flexibility gives an advantage to the wired setting in certain scenarios. For example, consider the case in which Zoya and Yoshi are static devices and need to be able to reliably exchange extremely secure data, such as control data pertaining to a power plant. Then an investment in such a cable may be fully justified, despite the fact that the cable may be severely underutilized due to only occasional transmission of critical data.

The collision model captures the two essential wireless features, *broadcast* and *interference*. In Figure 1.2(a), Zoya, Yoshi, and Xia communicate with the base station (sometimes shortened to BS) named Basil. A base station can be seen as an entry point to an infrastructure through which Zoya, Yoshi, and Xia are connected to their communication peers. For example, consider the case in which Zoya wants to communicate with Walt. Zoya is in the range of the Basil, while Walt is in the range of a different base station, named Bastian. Then, Basil and Bastian are interconnected, most likely through a wired networking infrastructure, which allows transfer of data from Zoya to Walt and vice versa. In the sequel we will implicitly consider the fact that our users may want to communicate to their peers that are in the range of other base stations, but our focus will be on the communication between the users and a single base station.

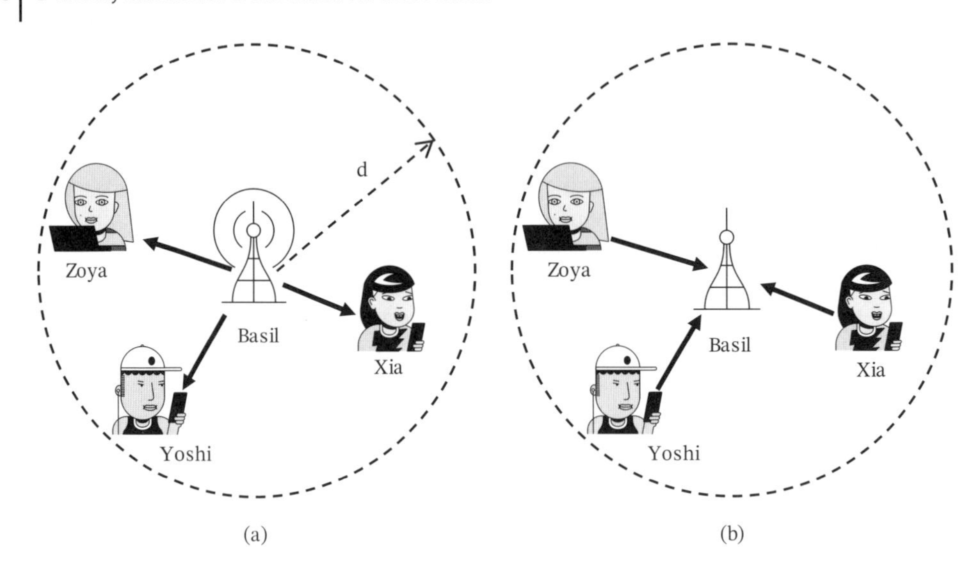

Figure 1.2 Illustration of two essential wireless features captured by the collision model. (a) Broadcast. (b) Interference.

If Basil wants to send the same information to all three devices, a single transmission would suffice, since the three devices are within a distance smaller than d. By contrast, if there were a wire between each device and Basil, then Basil should have sent each packet three times. Hence, wireless broadcast is *cheap* and this feature has been termed the *wireless broadcast advantage*. When we want to emphasize that the same message is sent to several devices, we will sometimes use the term *multicast*[1]. Clearly, if Basil has different information for each device, the broadcast advantage disappears, at least in our simple communication model.

When the communication takes part in the opposite direction, Figure 1.2(b), then the wireless broadcast advantage of the shared medium turns into a problem of interference. If the three devices transmit simultaneously, collision occurs and Basil does not receive anything useful. Therefore the devices should be coordinated in order not to transmit simultaneously and avoid collisions. This incurs certain *coordination cost*, spent on exchanging metadata. By contrast, the coordination cost is absent if each device has a dedicated wire to Basil, since he receives each signal over the wire. However, when calculating the grand total of costs, one has to account for the capital expenses incurred by installing the wires and, of course, the lack of flexibility inherent to a wired connection.

In summary, in the collision model broadcast can be an advantage, as all nodes in the range will perfectly receive the packet, while interference is always a disadvantage. In later chapters we will enhance the communication model by taking a magnifying glass and look

1 The term broadcast outside of information theory is used to denote the message that is sent to all devices within the range. In that sense, the term multicast is used when the message is sent to a subset of two or more devices within the range. Hence, in this sense a broadcast is a special case of multicast. On the other hand, broadcast in an information-theoretic sense means that the transmission of a device is received by multiple devices, regardless of who the intended receiver of the message is. Thus, we can say that a transmission over a wireless medium is a broadcast, but the actual message sent may be intended as a unicast to a specific device or multicast to a group of devices.

what happens inside a collision. This will lead to a somewhat surprising conclusion that interference can be very useful.

1.2 The First Contact

Back to the dark room analogy, we ask the question: how do two people, who have never met before, start to communicate when placed in a dark room? Reformulating this question in terms of wireless communication, we can ask: how do two wireless devices start to communicate? Who speaks first and who listens first? This is an important issue when the devices operate in a half-duplex manner, since a device cannot transmit and receive simultaneously. Before a packet from Zoya is sent to Yoshi, each of them needs to know that Zoya is about to transmit and Yoshi is about to receive. This may sound trivial and indeed it is, provided that we somehow let Zoya know in advance that she should take the transmitter role and Yoshi should take the role of a receiver. For example, if they have communicated in the past, then they may agree that, next time they are placed together in a dark room, Zoya takes the role of the one that starts to talk first. But, how do they know the roles if they have never communicated before? Let us explore this problem of first contact or rendezvous between two wireless nodes.

1.2.1 Hierarchy Helps to Establish Contact

In many cases the rendezvous problem can be solved by relying on a pre-established hierarchy or context. For example, in a conversation Basil can be the boss and Zoya an employee in a company that follows a (ridiculously) strict hierarchy. In that case, both of them know that Basil should start speaking and Zoya should listen before making any attempt to talk. Translating this idea of pre-established hierarchy into a wireless communication setting, Zoya can be a device/phone that wants to connect to the base station Basil. Then the phone can be preprogrammed to be in receiving mode and wait for an *invite* packet from a base station. Note that in this case *the context* breaks the symmetry between Zoya and Basil and thus pre-assigns the role that a device will have in trying to access the wireless medium. Basil can label the *invite* packet with his name or unique address, such that Zoya knows who sends the *invite* packet and can decide whether she wants to respond and connect to Basil.

In a more involved case, Zoya may be also in the communication range of another base station, named Bastian, see Figure 1.3(a). This second base station serves as an alternative to which Zoya can connect to in order to get access to the overall wired infrastructure and the internet. But what if Basil and Bastian send the invitation packets simultaneously? Then, following the rules of the collision model, Zoya experiences collision and she does not receive anything useful. An easy fix could be to have both Basil and Bastian to be part of a the same communication infrastructure, which makes it viable to assume that they can communicate and coordinate over a wired channel and thus agree not to send the *invite* packets simultaneously.

However, there may be cases in which Basil and Bastian cannot coordinate through the wired connection, since, for example, they belong to networks with different owners. Hence, the problem of collision over a shared medium remains. If both Basil and Bastian

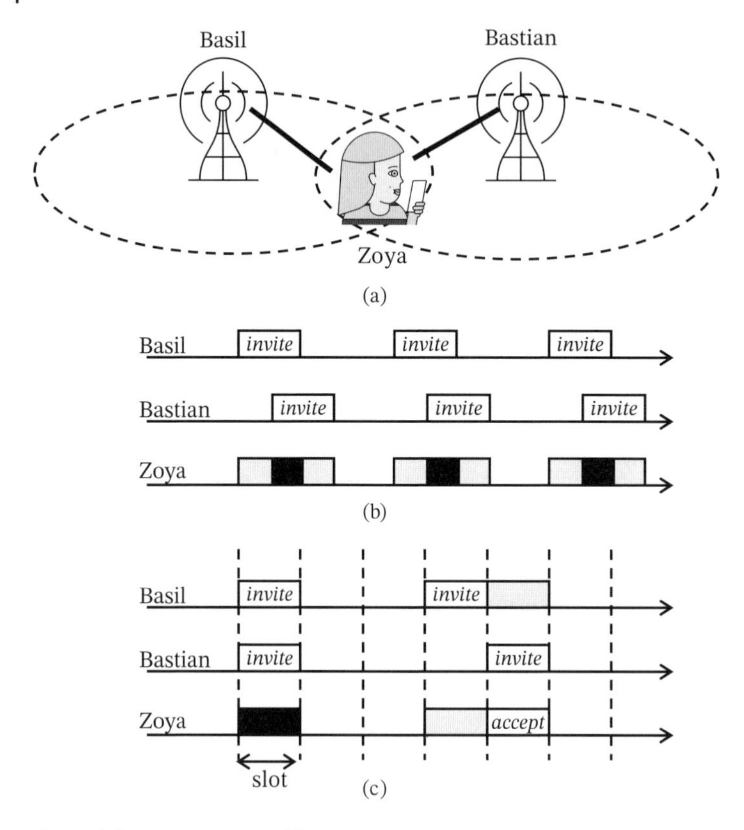

Figure 1.3 The problem of first contact when the mobile device Zoya is in the range of two base stations, Basil and Bastian. (a) Illustration of the scenario. (b) The invite packets of Basil and Bastian are persistently colliding. (c) Solution of the problem of collision between Basil and Bastian by using randomization.

go on to persistently send invitation packets over regular time intervals, then it can happen that they are synchronized in an unfortunate way. This is illustrated in Figure 1.3(b), where it can be seen that Zoya will not ever receive an invitation. One quasi-solution that Zoya might contemplate would be the following. The clocks of Basil and Bastian cannot be perfectly synchronized, so if Zoya patiently waits, at some point in the future the regular packet transmissions from Basil and Bastian will avoid overlapping. We dedicate space to this quasi-solution in order to show why is it not a usable one and thereby illustrate an important engineering point. Namely, a good algorithmic design cannot rely on randomization that is not controlled in any way by the participants in the system. Instead, the (pseudo-)random choice used in the protocol should be deliberately invoked by the participating actors in the system. In this example, imperfect synchronization is due to random deficiency in the production process of the clocks and it thus represents an uncontrolled random factor.

Let us now assume that the clocks of the two base stations are perfectly synchronized and they are both dividing the time into identical *slots*, as in Figure 1.3(c), where each slot is sufficient to send an invitation packet or receive a packet termed *invitation_accepted* from Zoya. Before the start of a slot, each base station flips a coin and decides randomly to transmit or stay silent in that slot. As shown in Figure 1.3(c), some *invite* packets will still collide, but some will be sent free of collision. Hence, the use of coin flipping leads to randomization of the transmission time between two *invite* packets. This randomization is the key to finalize the establishment of a contact, with high probability, within a reasonably short time.

1.2.2 Wireless Rendezvous without Help

Things get more complicated when the roles of the devices are not predefined. This is the situation in establishing ad hoc links between two devices that belong to the same hierarchical level, as in device-to-device (D2D) communication. For example, Zoya, Yoshi, and Xia can be three mobile phones that want to start communication, but have never communicated with each other before. The last assumption is important, since if Yoshi and Xia have already communicated in the past, they may have agreed who should be the one sending the *invite* packet next time they need to communicate. In the absence of such a context, it is impossible to predefine the roles. For example, if Zoya and Yoshi are predefined to be the ones sending *invite* packets, while only Xia is waiting to receive them, then Zoya and Yoshi cannot establish a link between them. The problem of first contact when devices are symmetric is exacerbated by the half-duplex nature of devices: if Zoya and Yoshi are continuously sending invitations to each other, then neither of them is able to receive the invitation from the other one.

Coin flipping again helps to resolve this situation. Let us assume that Zoya and Yoshi have a common time reference for a slotted channel, as in the case with Basil and Bastian. Achieving the required synchrony is harder to justify in this case, compared to the example in which a device establishes connection with a base station. Nevertheless, for the purpose of this example, one can assume that both Zoya and Yoshi have GPS receivers that can be used for clock synchronization. Let both Zoya and Yoshi be quiet in the $(t-1)$th slot, see Figure 1.4. Before the start of the slot t, Zoya and Yoshi flip a coin and randomly decide to either send or listen for an *invite* packet. In the example in Figure 1.4, Zoya decides to transmit. Note that in the $(t+1)$th slot Zoya must be in the "listen" state, since she should receive a *invite_accepted* packet in case Yoshi received Zoya's *invite* packet.

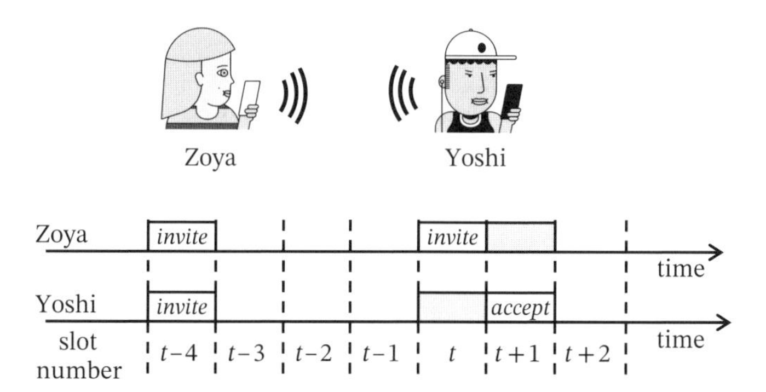

Figure 1.4 Rendezvous protocol for Zoya and Yoshi where both of them use half-duplex devices and their roles in the protocol are not predefined.

But what if Zoya and Yoshi are not synchronized in using the slotted channel; can that help or make things even more difficult? In fact, the synchronous case can be seen as the worst case: if both Zoya and Yoshi decide to transmit in a certain slot, then their packets are completely overlapping. On the other hand, the lack of synchronism can be helpful in breaking the symmetry between Zoya and Yoshi. For example, we can have the situation in which both Zoya and Yoshi decide to transmit, but the transmission slot of Yoshi starts slightly later than Zoya's slot. Yoshi starts to receive the *invite* packet and postpones his transmission in order to complete the reception of Zoya's packet. In this example, Yoshi adjusts its clock to Zoya and, after the *invite* packet is received, Yoshi sends back a reply using Zoya's slot timing. It can be seen that, when the devices are not in synchrony, the symmetry can be broken without using randomization, as some form of randomization is already embedded in the asynchronism. However, recalling our discussion on a proper protocol design and controlled randomness, asynchronism can facilitate the rendezvous protocol, but it should not be the definitive solution; the protocol should always have the opportunity to rely on an intentional randomized choice.

Through the problem of first contact and link establishment, we have introduced randomization as the key idea used in breaking the symmetry among different wireless devices that want to access the same shared medium. In the next chapter we will see that randomization can also be useful for efficient medium access after the links have been initially established.

1.2.3 Rendezvous with Full-Duplex Devices

The problem of first contact becomes easier if the mobile devices are equipped with full-duplex capability. If Zoya and Yoshi both transmit simultaneously an *invite* packet to each other, then each of them will simultaneously receive the *invite* packet from the other device. Furthermore, in the next slot both Zoya and Yoshi transmit an *invitation_accepted* packet and, again, Zoya receives this packet from Yoshi and vice versa. With this, the link can be considered as being established. Therefore, full-duplex avoids the need for randomized assignment of transmit/receive roles and thus speeds up the procedure of link establishment. This may appear to be one of the most important advantages of full-duplex technology in scenarios in which fast link establishment and device discovery is of high importance.

We note that, even with full-duplex, there is still the problem of interference due to collision. Hence, full-duplex cannot help to solve the problem of colliding *invite* packets in Figure 1.3, as Zoya still does not get either of the two *invite* packets.

1.3 Multiple Access with Centralized Control

Referring again to Figure 1.2, let us assume that each of the mobile devices has established a link with Basil and we turn to the problem of actual data communication. Since in some examples we have to refer to a set of K mobile devices or terminals connected to the same base station, we will also use depersonalized identification of the mobile devices $MD_1, MD_2, \dots MD_K$ interchangeably. Here it will be understood that Zoya uses MD_1, Yoshi uses MD_2, etc. The objective of this section is to bring the reader to a simple, but functional design of a wireless system that can establish/disconnect links and exchange data between the devices and a base station Basil that acts as a central controller.

1.3.1 A Frame for Time Division

Let us first consider downlink traffic, such that Basil has data to transmit to each of the devices. Then a straightforward idea is the one in which Basil divides the time into slots, where each slot has a duration of T, which corresponds to the duration of the transmission of a single packet. In each slot, Basil can send a single packet to a particular device. For the moment we are ignoring the possibility that Basil can broadcast a common packet to all devices, using the shared property of the wireless medium. Keep in mind that we are still assuming that all packets have the same size. This mode of communication is a simple variant of *time division multiple access (TDMA)*: at a given time, the whole shared medium is allocated to a single user, that is, the terminal to which Basil transmits. The simplest TDMA scheme with K users is depicted in Figure 1.5(a), where each device gets an equal fraction $\frac{1}{K}$ of the shared channel. It is convenient to define a *TDMA frame*, which is periodically repeated, and see how one can define a *logical channel* by using the physical, shared communication channel. For the example in Figure 1.5(a), the logical communication channel between Basil and Zoya is the first slot in each TDMA frame, as depicted in Figure 1.5(b). The data rate of the logical channel is easily calculated as follows. The data rate of each packet is R, such that in each frame a total of $R \cdot T$ bits are sent to Zoya. If F frames are observed, Zoya receives in total $F \cdot R \cdot T$ bits over a duration of $F \cdot K \cdot T$ s, which makes the equivalent data rate for Zoya

$$R_c = \frac{F \cdot R \cdot T}{F \cdot K \cdot T} = \frac{R}{K} \qquad \text{(bps)}. \tag{1.1}$$

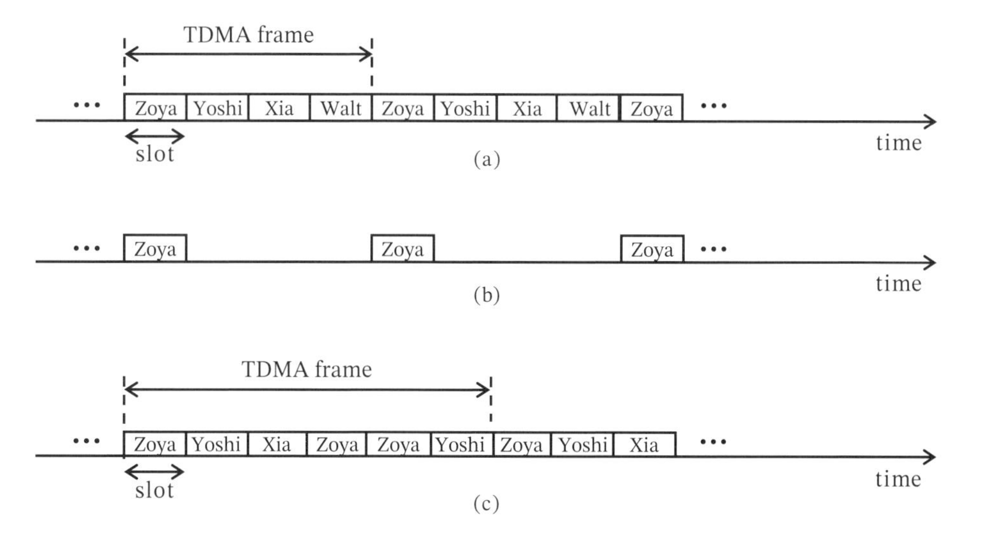

Figure 1.5 Downlink time division multiple access (TDMA). (a) Periodic equal allocation to all terminals. (b) The downlink communication channel from the perspective of Zoya. (c) Periodic allocation, but unequal across the terminals.

This very simple fact is, surprisingly, often neglected in practice. For example, a wireless system may have a nominal speed of, e.g., $R = 50$ (Mbps, megabits per second), but this data speed is rather instantaneous and valid during the times when the user receives the data. However, if a lot of data needs to be sent by the user over a long period, then the average data rate is decreased because there are time intervals in which no data is sent to the user and thus their data rate in those time intervals is strictly zero. This leads to the conclusion that, for a consistent definition, we should always keep in mind that there is a particular time interval T_R over which the data rate is calculated:

$$R(T_R) = \frac{< \text{data bits sent over period } T_R >}{T_R}. \tag{1.2}$$

Most of the time T_R is implicitly clear from the context.

If Basil is certain in advance that sending $R \cdot T$ bits each $K \cdot T$ s is sufficient for each terminal, then the simple TDMA is an easy and, in fact, perfect solution. In addition, the frame structure can be established once and kept indefinitely, as long as the devices stay synchronized to the frame defined by Basil. This is the essence of a *circuit-switched connection* in communications, where the usage of a certain communication resource is agreed in advance for a long period of time. In this particular example, Basil may have agreed with the mobile devices in the past that Zoya will use the first slot in a frame, Yoshi the second, etc. This means that whenever the time slot allocated to Zoya comes, Basil does not need to use a part of the $R \cdot T$ bits to send signaling information and thus label the packet "*This packet is for Zoya*". Instead, it is sufficient to send pure data to Zoya as every other device in the network knows that what is being sent in that slot is data for Zoya.

Circuit-switched operation is useful in minimizing signaling whenever it is known in advance which resources are required over a certain period. In practice, TDMA allocation can be more complex than what is described above. Let the demanded data rates be: $\frac{R}{2}$ for Zoya, $\frac{R}{3}$ for Yoshi and $\frac{R}{6}$ for Xia. An example of a TDMA allocation that can satisfy these data rate demands is depicted in Figure 1.5(c).

In real systems, even in the case of static, circuit-switched allocation, it is unrealistic to assume that the logical channels and frames will stay ideally allocated for an indefinite period. For example, there might be a period of time in which Basil has no data to send to Zoya. If Zoya does not receive anything within several consecutive frames, she might easily get out of synchronization with Basil, which would result in irrecoverable errors. If the internal clocks of Zoya and Yoshi have a large relative drift, then Zoya might start to receive the data for Yoshi, not knowing that it is not intended for her. This cannot be prevented in the described simple TDMA scheme, since no resources are spent in sending control information after the initial, circuit-switched allocation. This control information would be used to describe what kind of data is sent in a particular slot. Therefore the periodic frame structure with fixed allocations to the users is only an approximation, as there must be flexibility to change the allocation in the frame when new devices are coming in the system, as well as to release slots when some devices are leaving the system.

1.3.2 Frame Header for Flexible Time Division

The conclusion from the previous subsection is that a robust system operation requires some of the physical communication resources to be invested into transmission of metadata

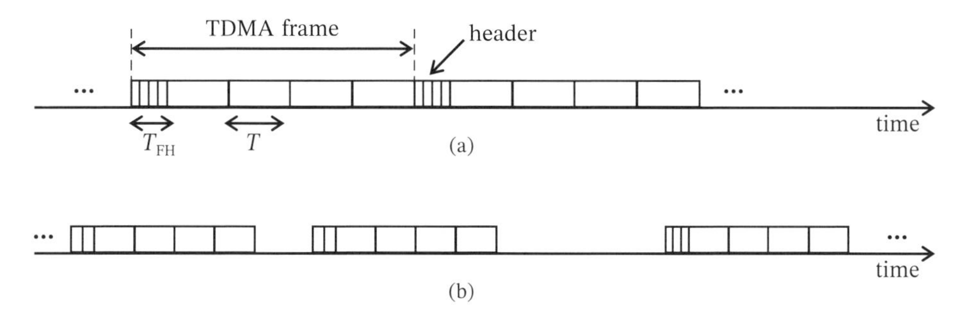

Figure 1.6 Introduction of a header in the TDMA frame. (a) Periodic TDMA system with full occupancy of the channel. (b) The header brings flexibility: the central controller (Basil) can decide to start the frame at an arbitrary time after the previous frame is finished.

or control information. In this way, the base station can regularly inform the mobile devices about the actual allocation of the data in the logical channels.

A step towards achieving such flexibility is the introduction of a *frame header* of duration T_{FH}, as illustrated in Figure 1.6(a). In the simplest form of circuit-switched operation, the channel allocation in each TDMA frame remains fixed, such that the frame header is only required to mark the start of the frame, and not carry information about the allocation. Even so, the introduction of a frame header that carries only information "*This is the frame start*", introduces additional flexibility, as depicted in Figure 1.6(b). In the example, there are $K = 4$ served in a frame, each of them getting an equal share of the resources, such that the total frame duration is $T_{FH} + 4T$. The allocation is still circuit-switched, but now a given terminal does not locate its communication resource (slot) in terms of absolute time, but rather the relative time, measured with respect to the frame header: for example, Yoshi receives the information sent in the second slot after the frame header. Basil can now decide *when* to start a frame and actually leave some blank inter-frame space. This is very important, since shared communication channel in the inter-frame space can be used for other purposes, such as link establishment, as it will be readily seen.

The frame header can further be enhanced in order to support time division between downlink and uplink traffic. For example, the frame headers in Figure 1.6 can contain information "*This is a downlink frame*", such that Zoya knows that she should receive her data during the slot allocated to her. The reasoning for the uplink is analogous. Besides marking the frame start, now the header contains an additional, single bit of information to announce whether the frame is intended for downlink or uplink, respectively. Based on that, Zoya knows whether to receive or transmit during the slot that is allocated to her. Now the system can flexibly allocate resource for communication in both directions (uplink/downlink), such that the system operates with a flexible *time-division duplex (TDD)* mode.

1.3.3 A Simple Two-Way System that Works Under the Collision Model

Using the ideas described so far, we can create a rudimentary, but fully functional scheme for medium access control (MAC) that allows using the shared channel between Basil and a group of terminals in his range. This scheme works provided that the communication channel behaves according to the collision model adopted in this chapter. Each frame header has a duration of T_{FH}. We define four frame types, each one associated with its respective frame header. A frame type can be represented by two bits, which is sufficient to encode four frame types. The frame header should include these two bits in order to identify the type of frame that follows the header. The frame headers are denoted $H_{00}, H_{01}, H_{10}, H_{11}$, and the meaning of each header is specified as follows:

- H_{00} : link establishment frame
- H_{01} : start of a link termination
- H_{10} : this frame contains K slots for downlink transmission
- H_{11} : this frame contains K slots for uplink transmission.

It should be noted that the number of users K is a predefined value, not conveyed through the header, such that we must assume it is known by Basil and the devices. Basil acts as a central controller and each header can be treated as a command transmitted from Basil to the devices. By default, each device is in a receive state (recall the hierarchy!) in order to detect the header and it subsequently takes action as instructed by the header.

We first describe how the headers H_{10} and H_{11} work. A frame that starts with either of these two headers has a total duration of $T_{FH} + KT$ s. For the example in Figure 1.7(a), it is assumed that $K = 4$. During the link establishment process, a terminal is allocated a number between 1 and 4. Zoya is allocated the slot number 2 and if Zoya detects a header H_{10}, she expects a downlink packet (of RT bits) in slot 2 after that header. If Zoya detects H_{11}, she is allowed to transmit an uplink packet in slot 2. It should be noted that, by system design, after H_{10} or H_{11} is sent, then no new header from Basil can arrive within the next KT s.

The header H_{00} is treated as an *invite* packet. If Basil has already K established links and cannot accommodate a new terminal, then he simply just does not send a frame with the header H_{00}. If Xia has already established a link with Basil, then she ignores this header. Otherwise, if Xia is still not connected to Basil, then one of the sequences depicted in Figure 1.7(b) occurs, where Xia sends an *invite_accept* packet to Basil within the slot that follows the header. In the second slot, following the one in which she transmits, Xia enters the receive state. Basil sends back a *link_established* packet, which also contains the number of the slot that is allocated to Xia for uplink/downlink transmission. Hence, the frame starting with H_{00} has a duration $T_{FH} + 2T$ s. Unlike the other frame types, this frame can feature both downlink and uplink data transmission. A device that has not yet established link with Basil ignores the headers H_{01}, H_{10}, H_{11}.

If more than one device, say Xia and Walt, respond simultaneously to the same H_{00}, then Basil observes a collision and does not send back a *link_established* packet. Receiving no packet from Basil in the second slot after H_{00}, both Xia and Walt conclude that there has been a collision; this is because the simple collision model does not contain other sources of error. To deal with this situation, a randomization mechanism should be used, similar to the one introduced in Section 1.2: when the next H_{00} comes, Xia and Walt should flip a coin in order to decide whether to send *invite_accept* to the next H_{00} that they will receive.

Figure 1.7(c) depicts a worked-out example, including all of the system elements introduced so far.

Next, let us look at the issue of link disconnection. When a terminal receives H_{01}, then it waits in the next slot for another packet from Basil. This packet contains the name of

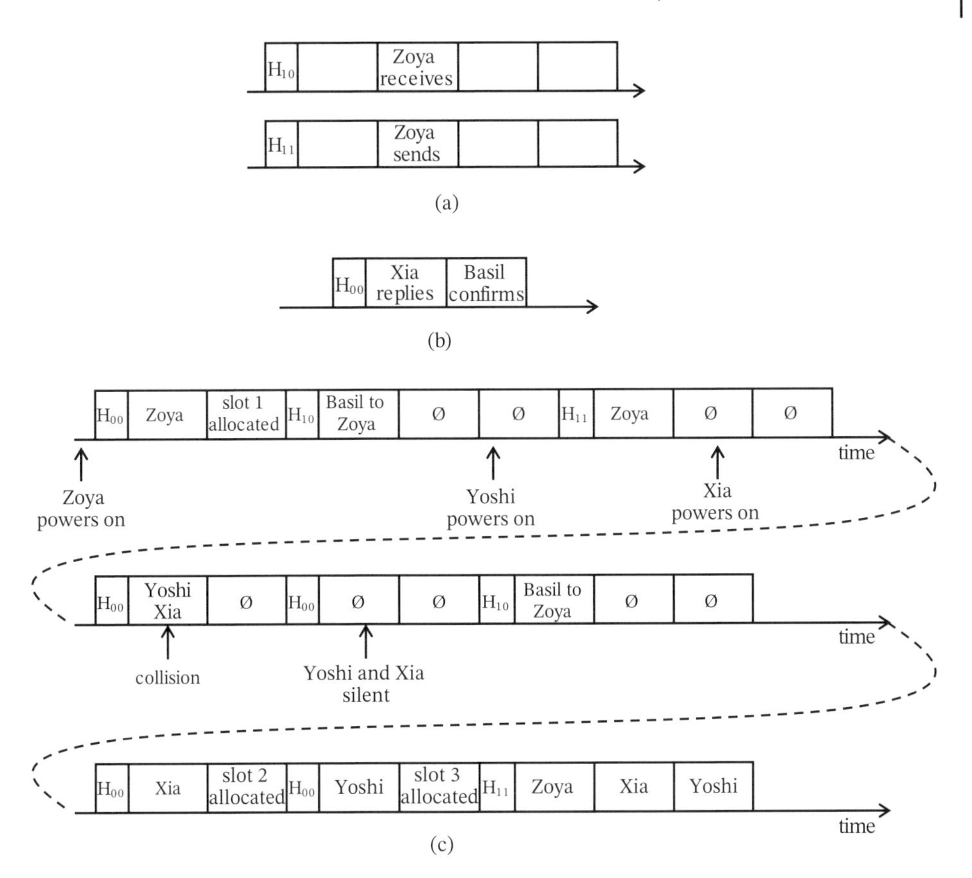

Figure 1.7 Illustration of several ingredients required in a simple wireless TDMA system with centralized control. (a) Using H_{10}/H_{11} for downlink/uplink allocation to/from Zoya, where $K = 4$. (b) Link establishment between Basil and Xia using H_{00}. (c) Example of the system in action with $K = 3$ terminals.

the terminal that will be disconnected from the network. All terminals receive this packet from Basil and the terminal that reads its own address considers itself detached from the network, such that if it needs to communicate with Basil, it has to go again through the link establishment procedure. A frame that starts with H_{01} has a single data slot, such that its total duration is $T_{\text{FH}} + T$ seconds.

In general, $T_{\text{FH}} \neq T$ and, in fact, in many practical systems the objective is to have $T_{\text{FH}} \ll T$, in order to maximize the time during which the wireless channel is loaded with useful data. However, in that case, our concept of a time slot, as defined above, is ruined, as there is no single, atomic, time duration that can be used as a time slot. In order to still have a common time reference that describes the events important for the MAC protocol, we can define a slot structure as follows. Assume that the ratio of T and T_{FH} can be represented as a ratio of two integers:

$$\frac{T}{T_{\text{FH}}} = \frac{k}{l}. \tag{1.3}$$

Then the duration of a time slot can be defined to be $T_S = \frac{T}{k} = \frac{T_{\mathrm{FH}}}{l}$. From the perspective of a MAC protocol, this is indeed an atomic time unit that can be used to describe all the important time instants. Note that now one TDMA slot corresponds to k time slots, while a TDMA frame consists of $l + k \cdot K$ slots.

And, that is it! We have specified a very simple, but functional system with centralized MAC in which a single node (Basil) decides who will enter or leave the network, or when a device should receive or transmit. For the sake of simplicity, we have made several design choices that are sub-optimal. Each data packet contains RT bits, which can be questionable in the case of a frames with H_{00}, H_{01}. In practice it takes many fewer bits to identify a node (in this case the disconnected device) compared to the number of bits in an average data packet used in a broadband communication system. It is noted that this can be violated in systems that operate with very short data packets, such as wireless sensors or other devices under the umbrella of the Internet of Things (IoT), but in this introductory discussion we mostly refer to packets where the amount of data bits in a packet is large.

The circuit-switched allocation has an additional degree of flexibility, since Basil decides when to set the transmission timing by sending the headers at appropriate instants. Even more, the circuit-switched allocation is "realistic" in a sense that it does not start in the infinite past, but instead it starts with a link establishment and ends with a link termination.

The reader can easily extend the system design to address the case of multiple base stations that have overlapping communication areas, as for the example depicted in Figure 1.3(a). In this case, a frame header should also contain an address that identifies the base station sending the header. There is again the issue of collision between the frames, observed at a device that is in the communication range of both base stations. As discussed in the problem of link establishment, this issue can be solved by randomization.

1.3.4 Still Not a Practical TDMA System

The system design presented above is not a proper one that can thrive in real life, but rather a sketch of a system that works reasonably well under the assumed collision model. Let us first define Basil's *cell* to be the area around the base station Basil in which a terminal is in the communication range of Basil. Using our simple collision model, the cell is a circular area.

We can now state that a condition for correct operation of the described TDMA scheme is that any terminal connected to Basil should remain in Basil's cell until Basil decides to terminate the connection. This condition is somewhat strange with respect to the way a link is terminated, since it does not consider the wishes of the terminal. In other words, Basil may decide to terminate the connection to Walt, although Walt may have more data to send in the uplink. However, this is not critical for the overall system operation, as Basil can continue to use the same TDMA structure, substituting Walt with another terminal. What is critical is the case in which Walt has an active link with Basil and Walt walks out of Basil's cell. With the protocol specified above, this leads to a rather fatal system error: the slot allocated to Walt will remain unused forever, as Basil has not terminated the connection and the slot for Walt stays reserved, potentially forever. A practical fix to this situation could be to introduce a certain *timeout mechanism*: if Basil does not hear from Walt for a certain time and several consecutive time slots allocated for uplink transmission to Walt are silent, then Basil considers Walt to be out of the network and makes Walt's logical channel available to another terminal.

This is still not sufficient to ensure a system design that is robust in practice. Take the following situation: Walt walks temporarily out of reach of Basil but he is back after the timeout has passed. Now Walt does not know that his slot has been allocated to someone else, which may lead to collision in the uplink transmissions made by Walt. The system design can be further patched in different ways in order to deal with this challenge. One solution is that Walt also uses a timeout mechanism, such that if Walt does not send anything to Basil for a time longer than the timeout, then both Basil and Walt claim the link to be non-existent. With this, Walt now knows that he needs to go again through the link establishment procedure.

Alas, this patch is still not sufficient. Recall that the collision model is only a model of reality, but does not fully grasp the practical conditions. One such practical condition is that, even in the absence of collision, the packet is not always received correctly by a receiver that is in the communication range. For example, several consecutive transmissions of Walt may be received incorrectly by Basil due to random noise. In such a situation, Basil starts the timeout for deciding link termination, but Walt does not. This can lead to inconsistent perception of the link between them, since Basil thinks the link is terminated, but Walt thinks the opposite. Yet another patch to the system design can be to use a mechanism based on two-way transmissions between Basil and Walt to check if the link is alive.

We could largely broaden this discussion by spotting other practical deficiencies and finding out suitable patches to the system design. The objective here is not to make a full real-life protocol, but rather illustrate how a simple protocol specification can operate under certain assumptions. However, this protocol needs to be enriched in order to be robust to other practical issues, even for ones that have a very low probability of occurrence.

1.4 Making TDMA Dynamic

1.4.1 Circuit-Switched versus Packet-Switched Operation

The introduction of a frame header for the MAC protocol relaxes, to some extent, the strict constraints of the circuit-switched operation. Note that the usage of the communication channel as a system resource is not deterministic between the moment the link is established and the moment it is terminated. This is because Basil has the freedom to determine when a certain frame should start, after which Zoya reads from the frame header which of the slots that follow that frame header are allocated to her. Zoya does not have a predefined, absolute, time instant to use a communication resource and needs to receive some form of command from Basil associated with a particular physical slot in which she will send or receive data. Therefore, the transmissions of Basil should contain control information or control signaling, which we have also termed *metadata*. This type of operation can be characterized as *packet-switched operation*, which stands in contrast to a circuit-switched operation.

In the described rudimentary system, one can already spot the main trade-off between circuit-switched and packet-switched operation. In circuit-switched operation, signaling is minimized at the expense of losing flexibility. On the other hand, frequent transmission of control signaling or metadata in packet-switched operation introduces *overhead*, which can be considered a waste, since it does not represent data that is of use to the end user. However, the metadata can be used to describe changes in the operation mode, such as a new allocation of the resources. This enables Basil to adapt the allocation of the slots to the current traffic demand from the users and thus offers flexibility advantage over the circuit-switched operation. Note that here we speak about circuit- and packet-switched operation in order to describe the possible ways in which the MAC protocol can use the shared communication channel. Nevertheless, the concepts of circuit-switched, which stands for inflexible but low-overhead operation, and packet-swtiched operation, which stands for flexible, but high-overhead operation, is universal and applicable to all communication protocols.

Strictly speaking, the usage of a frame header in the MAC protocol of the previous section is not packet-switched, since there is no separate control signaling sent for each packet, but the header is valid for all the packets transmitted in a frame. This is useful to note, because it illustrates, in a simple way, the main principle that can be used to get the desired trade-off of signaling and flexibility. Namely, sending a common frame header is a kind of a hybrid design, by which a portion of control signaling is used as a metadata for several packets. On the one hand, the signaling overhead is decreased by increasing the number of packets associated with a signaling information. On the other hand, more flexible allocation requires control signaling to be used more frequently, which would increase the overall signaling overhead.

1.4.2 Dynamic Allocation of Resources to Users

The portion of the TDMA channel allocated to a device does not need to be constant from the instant of link establishment and until the disconnection of the device. Furthermore, there is no need to divide the total amount of communication resources equally among the users. The latter one has already been illustrated in Figure 1.5(c), where Zoya, Yoshi, and Xia get different amounts of resources in a frame. It needs to be noted that there is no header in Figure 1.5(c), but, based on the previous discussion, the reader can easily add it. The common feature of these examples is that, once established, the allocation of communication resources is fixed for all the frames.

At this stage is it clear that, in order to be able to flexibly allocate resources to a user, the frame header should be enriched with more bits used for control signaling. Specifically, these bits should describe how TDMA slots are allocated in the actual frame, preceded by that frame header. For simplicity, let us stick to the fact that each packet has a fixed duration T, equal to a TDMA slot, and the frame is composed of header and F TDMA slots. Using the information in the frame header, Basil should be able to allocate any TDMA slot in the frame to any user. Thus, potentially, all slots of a given frame may be allocated to Zoya.

Let there be in total K devices that have established links with Basil. In order to get a proper design of the frame header we need to address the following question: how many bits need to be added to the frame header in order for Basil to be able to make any possible flexible allocation of the F slots to K users? Assuming that no slot remains unallocated, there are K^F ways to allocate the slots. In order to be able to describe any of those allocations, the frame header should contain at least

$$\lceil \log_2 K^F \rceil \leq F\lceil \log_2 K \rceil \leq F\lceil \log_2 K_{\max} \rceil \tag{1.4}$$

additional bits, where $K_{\max}$ is the maximal possible number of active users in the system. The number $K_{\max}$ is known in advance and Basil is never expected to admit more than $K_{\max}$ users. This assumption may look limiting, but is applied in practically all existing wireless systems. For example, some of the existing wireless standards impose that each device has a unique MAC identifier consisting of 48 bits, which means that the system design assumes $K_{\max} = 2^{48}$.

In order to get to the expression $F\lceil \log_2 K \rceil$ in (1.4), let us assume that the identity of each device that is active in the system is described by a fixed number of bits. Since there are K devices, the minimal number of bits to identify a device uniquely is $\lceil \log_2 K \rceil$. If the user allocation is described by specifying the address of the device to whom the slot is allocated, then the header needs to use $F\lceil \log_2 K \rceil$ bits in total. The reader may object to the latter observation, since the number of connected users K is variable, such that each time K is changed, Basil needs to adjust the number of bits he uses to address the active devices. On the other hand, choosing $K_{\max}$ to define the length of the address may be very conservative and wasteful. However, in that case no additional signaling would be needed when the actual number K of users is changed.

The most important implication from the previous discussion is: *any flexibility in the allocation of the communication resources corresponds to additional* ***signaling information*** *or* ***metadata*** *that needs to be communicated between the base station and the devices.* Thus the flexibility can offer better use of resources, but then the overall correct operation of the protocols becomes more vulnerable to the loss or errors in the metadata.

In order to make the most of the flexibility offered by the additional signaling bits, Basil should somehow know what is the most appropriate way to allocate the slots to the users in a given frame. For example, in an ideal case, Basil should allocate two slots to Xia only if he is certain that both slots will be filled with data to/from Xia. This is not a problem for downlink traffic from Basil to Xia, as Basil precisely knows how much data there is to send to Xia and can allocate the appropriate number of slots. More precisely, the data that can be allocated in this way should have arrived to the transmit queue of Basil before the header of the actual frame has started, such that the allocation can be announced in the header.

However, making the right allocation is not that simple in the case of uplink transmission. Unless the packets of Xia arrive in a perfectly predictable way, Basil cannot always know a priori how much data Xia has to send during the upcoming uplink frame and therefore Basil has either to guess it or learn it. Consider the case in which Basil allocates two slots to Xia in the uplink frame. Then the frame header sent by Basil can be understood as a *polling* or an invitation to Xia to send. If Xia has only one packet to send, then the second allocated slot remains empty. Intuitively, polling should use cautious allocation of resources and minimize the number of empty responses. However, if this allocation is overly cautious, then the devices end up being inhibited in sending all their data. This can be based on a knowledge or prediction about the current demand for uplink traffic across the population of terminals that have active connections to Basil.

But, how is this knowledge obtained by Basil? Going back to the analogy with speech, one can think of a conference scenario, in which the chairman (Basil) gives word to the individuals from the audience (devices or terminals). The first difference with the TDMA communication scenario from above is that whenever an individual speaks, not only the chairman but all the people in the audience receive the data. On the contrary, in our setting a mobile device does not communicate directly with another device. Furthermore, there is another difference with the TDMA operation described previously, which is essential for protocol operation. This is the way in which the individuals signal to the chairman whether they have an "uplink traffic", which can be done by raising a hand or pressing a button. As already mentioned several times, a raised hand or a pressed button represent an additional communication channel. So, we stick to the dark room analogy and ask: how can we imitate the raising of a hand in a dark room in which the only way to communicate is to speak? This question leads to the idea of *reservation packets*.

1.4.3 Short Control Packets and the Idea of Reservation

Instead of directly allocating an uplink slot to send data, each device is given the opportunity to send a short control packet, termed a *reservation packet*, which is used to inform Basil how many slots for sending data it will need in the TDMA frame. We need to make a distinction between reservation and data packets. The reservation packet should only carry a few bits in order to indicate how many data slots in the frame it can use. For example, if the TDMA frame has a fixed number of F data slots, then the reservation packet should carry at least $\lceil \log_2(F+1) \rceil$ bits, as it should be able to describe numbers from 0 to F[2].

Figure 1.8(a) describes a possible way to use reservation packets. Basil sends the frame header H to indicate that it allows uplink reservation transmissions to start. This is followed

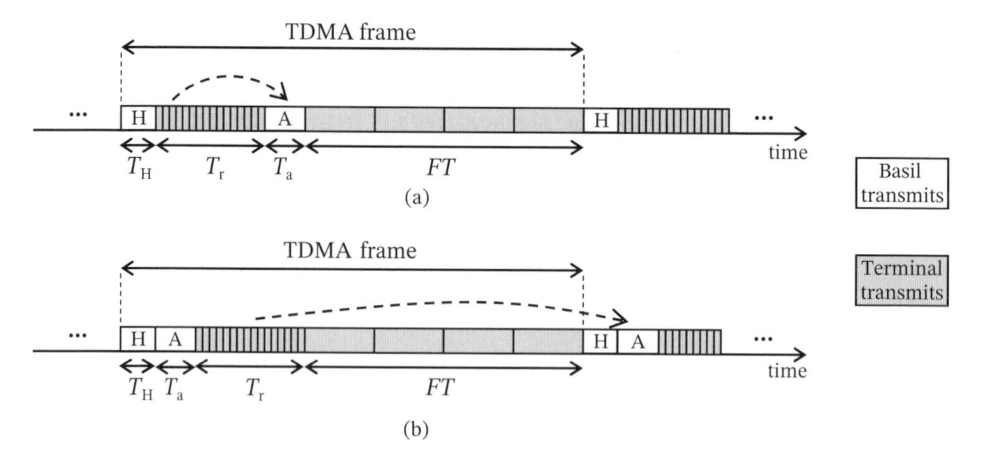

Figure 1.8 Uplink transmission with a reservation frame. (a) Case when the allocation is done based on the reservation outcome in the same TDMA frame. (b) Case when the allocation is done based on the reservation outcome in the previous TDMA frame.

2 Here we conservatively assume that a device sends a reservation packet even if it has no data to send.

by K reservation slots, that have a total duration to T_r. Here K is the total number of devices that have established a link with Basil. Each of the K devices is pre-allocated a unique reservation slot in which a user is allowed to send and it is guaranteed that there will no be a collision with another user. Due to this unique association between a device and a reservation slot, the reservation packets sent by the device do not need to carry the address of the device that transmits it. After the uplink reservation slots, Basil sends a short *allocation packet* A that announces how the data slots are allocated to different users. Considering the total number of possible allocations, the allocation packet A should contain $F\lceil \log_2 K \rceil$ bits. It should be noted that, through the reservation slots, the total amount of resources required by the devices can be larger than F. In this case Basil uses a certain *scheduling policy* to decide how to allocate the F data resources to the devices. The requests that are not met in that frame can be scheduled in future frames.

Figure 1.8(a) also illustrates the cost introduced by the reservation packets. If Basil somehow knows to whom to allocate the data slots in the current uplink frame, then the reservation slots should be omitted and the allocation packet A becomes a part of the frame header. We now proceed to evaluate how much the reservation slots are affecting the performance in terms of useful data rate experienced by the users. The total duration of the frame is $T_H + T_r + T_a + FT$, where T_r is the duration of the reservation frame, T_a is the duration of the allocation frame, such that the data rate for a device that uses a single uplink slot is

$$\overline{R} = \frac{RT}{T_H + T_r + T_a + FT} \tag{1.5}$$

which indicates the average data rate, observed in a period of a frame and by a terminal that has a single slot allocated to it.

While it is clear that the rate (1.5) is lower than $\frac{R}{F}$, we still lack an illustration of how short the reservation packet should be in order to justify its role. To do that, we need to enrich our system model with an additional assumption: the duration of a single bit is always equal to T_b, regardless of whether it is a bit that belongs to a data packet or a bit describing a signaling information (headers/reservation/allocation). Denote by $D = RT$ the number of bits in a data packet, such that $T_b = \frac{T}{D} = \frac{1}{R}$. Thus, we can express the durations T_H, T_r, T_a as fractions of T:

$$\begin{aligned} T_H &= 2T_b = 2\frac{T}{D} = \frac{2}{D}T \\ T_r &= K_{\max}\frac{\lceil \log_2(F+1) \rceil}{D}T = \frac{r}{D}T \\ T_a &= \frac{F\lceil \log_2 K_{\max} \rceil}{D}T = \frac{a}{D}T \end{aligned} \tag{1.6}$$

where it is assumed that the header has only two bits, a reservation packet contains $r = \lceil \log_2(F+1) \rceil$ bits, while the allocation packet has $a = F\lceil \log_2 K_{\max} \rceil$ bits. Using (1.5) we get:

$$\overline{R} = \frac{R}{\frac{2+r+a}{D} + F}. \tag{1.7}$$

The effect of the overhead caused by sending signaling information is more clearly seen if, instead of (1.7), we look at the total useful data sent in a frame, irrespective of which user is sending it. This is often called *sum rate* or *system goodput* and in this particular case

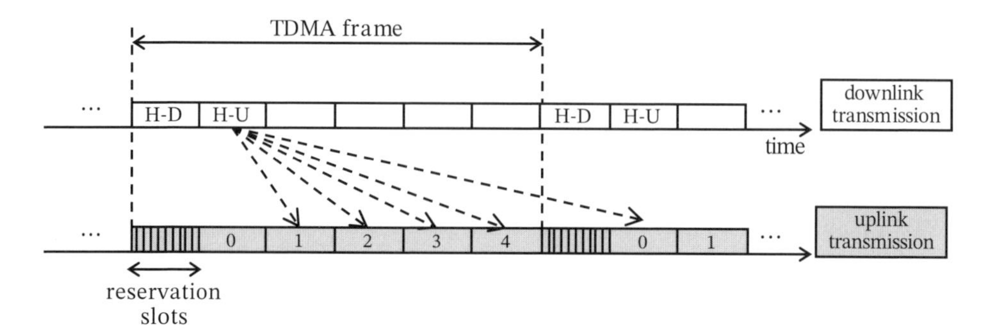

Figure 1.9 Overlapped downlink and uplink frame for a full-duplex base station and devices. H-D is the header for the downlink, H-U for the uplink. The allocation made by H-U is decided based on the reservation slots that occur during H-D. The slots allocated by H-U are slots 1, 2, 3, 4 from the same frame and slot 0 from the next frame.

can be expressed as:

$$\overline{R}_s = \frac{FR}{\frac{2+r+a}{D} + F} \tag{1.8}$$

This is clearly showing that having large data packets, $D \gg r$ and $D \gg a$, improves the overall efficiency of the system.

One might object to the fact that the number of reservation slots is equal to the number of users K, since the number of users might change. This is not the same as having fixed number of F data slots in a frame, since Basil can use a scheduling policy to decide how the F slots are used, as informed through the allocation packet A. However, each user should have the opportunity to voice her requirements in a frame; otherwise there should be another, pre-reservation frame to pre-reserve the reservation slots, etc. Expressions (1.6) use a rather conservative estimate of the required resources, as it is always assumed that the number of users is maximal possible. This is clearly inefficient if there are only few active users. A quick fix would be to replace $K_{\max}$ with K, the number of currently active users. However, in that case Basil needs to correctly signal this to all active users, along with the allocation of the reservation slots. There can be other ways to make a better use of the reservation slots, based on the partial knowledge that Basil may have about the traffic demands. For example, Basil may use some machine learning methods to estimate which group of users is likely to have data to send in a certain period. We will not go into detail for all possible techniques, as our purpose has been to illustrate only the most important principles. The general class of protocols that addresses the case in which the number of active user is (much) lower than the total number of users is the one of *random access protocols*, which is the subject of the next chapter.

1.4.4 Half-Duplex versus Full-Duplex in TDMA

Our half-duplex model assumes ideal switching between transmit and receive state, without causing any additional inefficiency in the system. As mentioned before, achieving two-way communication by switching in time between a transmit and a receive state is known as TDD. In practical TDD wireless systems, switching between transmission and reception is not without cost, as it takes a certain turnaround time. Hence, one should avoid having protocols that require from a half-duplex device to switch frequently between transmission and reception. For our discussion, this implies that it is not desirable to have the situation from Figure 1.8(a), where a device needs to switch multiple times from receiving the header, transmitting the reservation packet, receiving the allocation and, potentially, be the first transmitter in the data slots. A possible solution is shown in Figure 1.8(b), where the reservation slot is used to reserve data slot in the next frame or another frame in the future, instead of the current one, thus achieving some form of pipelining.

If the wireless devices have the capability for full-duplex transmission/reception, then there is no need to use TDD and separate the uplink and the downlink frame in time. Figure 1.9 shows a possible configuration of a TDMA frame in which the downlink and the uplink transmission take place simultaneously. The gain, usually claimed from full-duplex, is the doubling of the overall data rate or the throughput, as the same communication resource can be simultaneously used twice. This figure illustrates the gain that full duplex brings in terms of *latency*, which is the time since the packet in a terminal is ready to be transmitted until the time the terminal gets the opportunity to actually transmit it over the air. The header H-U announces the allocation of the uplink slots that follow it, such that the user allocated to slot 1 can start to send immediately after receiving H-U. By contrast, in the previous example of TDD with half-duplex transceivers, the first slot available for transmission is slot 0 in the next frame.

1.5 Chapter Summary

This chapter has dealt with the problem of sharing a single wireless communication channel among multiple communication links. We have used the simplest possible communication model that captures important features of the shared wireless medium, such as broadcast and interference, where the latter is modeled as a collision. We have adopted a packet to be the atomic unit of transmission, meaning that either the whole packet is received correctly or it is completely lost. The objective has been to introduce the main ideas for sharing the channel, such as TDD and TDMA and sketch the elements of a protocol that closely approximates the practical protocols. Where relevant, we have also discussed how the full-duplex capability of the wireless devices can contribute to the design of protocols that are more efficient compared to the case of half-duplex devices.

1.6 Further Reading

A classical book that introduces elements of data networking, along with rigorous models is the one by Bertsekas and Gallager [1992]. For rendezvous and link establishment procedures, the reader is referred to the operation of various standards, such as 4G LTE in Dahlman et al. [2013] (chapter 14, Access Procedures), or 802.11 Wi-Fi networks, both in ad hoc and infrastructure mode, see Standards [2016]. Besides Bertsekas and Gallager [1992], another book that offers insights into models for communication over a shared channel as well as stochastic modeling of communication traffic is Rom and Sidi [2012].

1.7 Problems and Reflections

1. *State machine for a TDMA system.* Describe a possible state machine through which the devices and the base station implement the protocol from Figure 1.7.
2. *An even more practical state machine.* Extend the state machine from the previous model in order to make the protocol practical, such as introduction of timeout mechanisms, dealing with device mobility, etc.
3. *More than one rendezvous channel.* Assume that there are L different rendezvous channels. Two nodes can establish a link only if one of them sends and the other one receives on the same channel. At a given time, a device can use only one channel. Devise a strategy for establishing a link between two devices and try to compare its performance to the case when there is a single $L = 1$ rendezvous channel. When and why would it be useful to have $L > 1$ channels for rendezvous?
4. *Unequal slots.* Consider a generalization of a frame, which consists of a header, followed by communication resources. However, now assume that the communication resources are not organized into F equal slots and instead the frame can contain slots with different lengths. Discuss how does this affect the type and amount of signaling bits used in the header.
5. *Reservation with variable number of data slots.* The analysis in Section 1.4.3 is done for the case in which the number F of data slots following the reservation slots is fixed. Let us now consider the case in which F is variable and adapted to the actual number of resources required in the reservation slots. Assume that each of the K devices can request up to L resources through the reservation packets.
   (a) Find the number of bits that are required in the reservation packets and the allocation packet.
   (b) Using the assumptions for bit duration from Section 1.4.3, find the maximal throughput that can be offered in a given frame. NB: the maximal throughput depends on the amount of resources requested by the devices.
   (c) In practical systems, the allocation packet A may not be received by some of the devices due to errors caused by noise or interference. When is the impact of not receiving A worse, when F is fixed or when F is variable? How do you suggest to design the system to be more robust to this type of error?

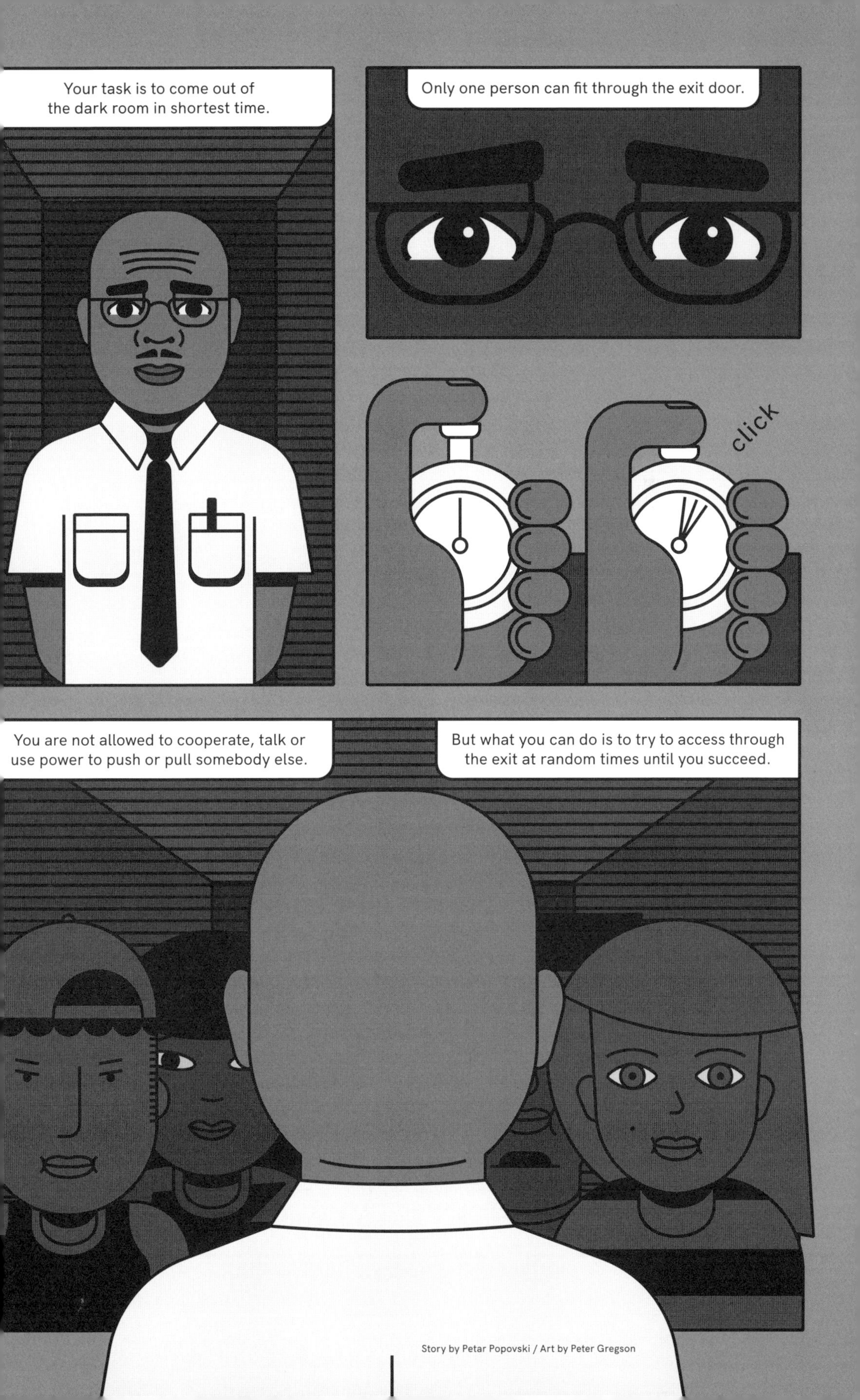
Your task is to come out of the dark room in shortest time.
Only one person can fit through the exit door.
click
You are not allowed to cooperate, talk or use power to push or pull somebody else.
But what you can do is to try to access through the exit at random times until you succeed.
Story by Petar Popovski / Art by Peter Gregson

2

Random Access: How to Talk in Crowded Dark Room

In the previous chapter, the dark room analogy was used to introduce the problem addressed by rendezvous protocols. Thinking about the same analogy, let us assume that Basil is in a dark room and some of the other people in the room want to talk to him. Basil cannot use visual cues, such as a raised hand, in order to schedule who should speak at him at a given time. Furthermore, the room is crowded, there are many other people in it, but only a few of them are active, in the sense that they want to say something to Basil at a given time instant. With this in mind, it is clearly not efficient to ask the people one by one if they have something to say, as most of them will be just silent and thus most of the time will be spent inefficiently. This observation paves the way for *random access protocols*, in which the reservation slots or data transmission slots are not exclusively pre-allocated to a device. The attribute "random" comes from the fact that the decision to transmit is randomized. The randomness can be caused by random packet arrival to the device. Alternatively, when the packet is already in the buffer of the device, the device can make a deliberate randomized choice to transmit or not. The dark room reflects the fact that we need to use the same wireless medium both to obtain the right to transmit data, which is a form of *metadata*, and to send the actual data.

Similar to the rendezvous protocol, random access is an indispensable solution when the devices need to perform an *initial access*. The objective of the initial access is to connect a device to the base station Basil, potentially going through a process of authentication, allocation of a temporary short address, etc. Clearly, in the case of initial access there are many, potentially infinite, number of devices that can connect, but at a given time only one or very few of them want to do that. After the initial access, the communication can either proceed as a scheduled one or, if the activity of the device is sporadic, rely again on a random access. The latter is typical for scenarios in which a massive number of small Internet of Things (IoT) devices are connected to a base station Basil. However, at a given time only a small subset of them is active; this subset is random and unknown to Basil.

The canonical scenario for random access is depicted in Figure 2.1. In this scenario, a number of uncoordinated mobile devices attempt to transmit data or control information to a common receiver.

In presenting the algorithms for random access we will keep the assumptions on the collision model established in the previous chapter: a packet is an atomic transmission unit and any overlap between two or more packets leads to collision and incorrect reception of

Wireless Connectivity: An Intuitive and Fundamental Guide, First Edition. Petar Popovski.

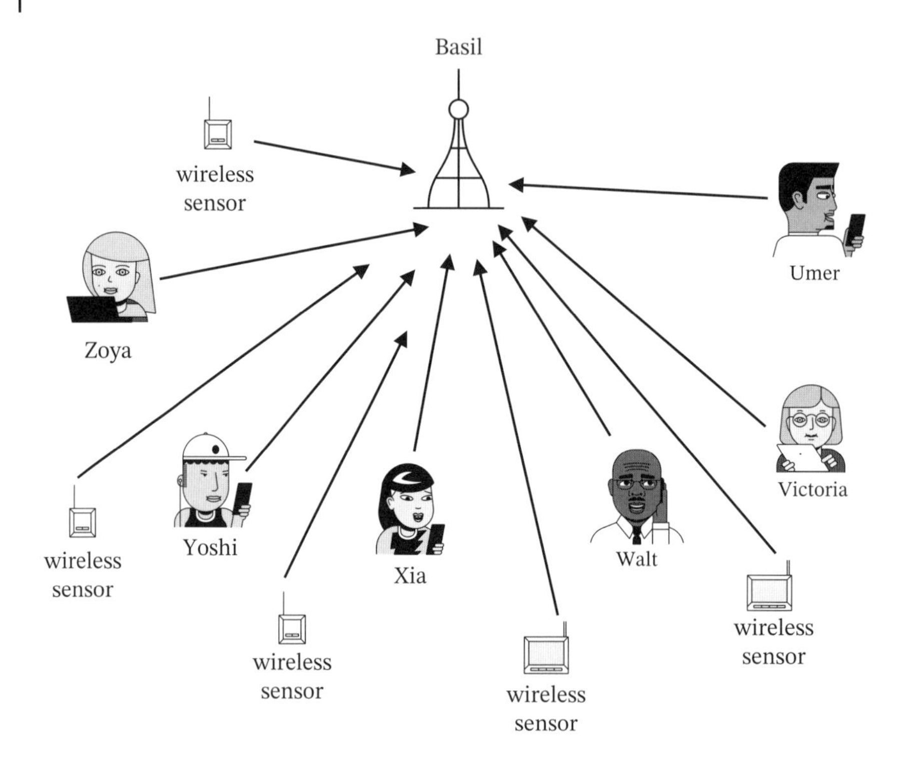

Figure 2.1 Canonical scenario for random access protocols, where a number of uncoordinated devices attempt to transmit in the uplink to the same receiver, which here is the base station Basil.

all collided packets. We will consider two classes of random access protocols: ALOHA type and probing (also known as tree type protocols). The variant of ALOHA considered here is not the original ALOHA, but rather framed ALOHA, a variant that logically extends the reservation schemes discussed in the previous chapter.

2.1 Framed ALOHA

The context for this discussion is in Section 1.4.3, where we introduced the reservation slots. Looking only at the expression (1.8), we can try to understand in which situation the usage of reservation slots may not lead to an efficient operation. For example, let us take the scenario in which the terminals connected to Basil are not phones, but sensors that monitor certain physical phenomena and only occasionally have data to send.

The parameters of this scenario can be described as follows. The total number of sensors K connected to Basil is large, while the number of users that have some data to send at a given frame is small. In other words, the probability that a particular sensor has data to send in a particular frame is very low. The amount of data D in a packet of each sensor is also small. Let us assume that the number F of data slots that follow the reservation slots is small and, furthermore, all F slots are full with data. In that case, using the equation (1.8), one can see that $\frac{2+r+a}{D}$ starts to dominate in the denominator, leading to decrease in the goodput. This is an example of a case in which the resources consumed by the metadata, which include reservation slots and all the other auxiliary packets, become comparable to the amount of data that needs to be sent and the overall system efficiency drops.

In the extreme case there is only $F = 1$ sensor transmitting. Let us fix $F = 1$ and assume that, in a given frame, the probability that a particular sensor has data to send is $q = \frac{1}{K}$. This means that, on average, only $qK = \frac{K}{K} = 1$ packet comes in a frame from the total population of K sensors. Let then the number of reservation slots be $S_R = 2$. Recall that, in the previous chapter we had $S_R = K$, such that each reservation slot was deterministically and exclusively allocated to a single device (sensor). Here we have $S_R < K$, such that an exclusive allocation is not possible. Let us assume that, at the start of the frame, each sensor that has data to send picks randomly one of the $S_R = 2$ reservation slots and sends a reservation packet. Note that, unlike the case with deterministic allocation of reservation slots from the previous chapter, here Basil cannot know who is the sender unless its address is included in the reservation packet. Although in our example the expected number of sensors with data is $qK = 1$, it can happen, with a significant probability, that two or more sensors have data to send in the same frame. If exactly two out of the K sensors, Zoya and Yoshi, have data to send in the same frame, then the following outcomes are possible:

Case 1. Each sensor picks a different reservation slot. Then Basil receives both reservation packets and decides to allocate the $F = 1$ data slot to, for example, Zoya. Yoshi tries again to send its reservation packet in a future frame.

Case 2. Both sensors pick the same reservation slot and end up in a collision. Then Basil cannot allocate the data slot to any of the two sensors, leaving the data slot empty.

Each of the two outcomes occurs with probability $\frac{1}{2}$ and in both cases it becomes apparent that having a fixed $F = 1$ is not efficient. In the first case, the successful reservation of Yoshi is wasted[1]. In the second case, the data transmission slot is empty and wasted, as none of the two sensors can use it for transmission.

This leads us to think of a more efficient solution: F does not need to be fixed, but it would be the best if the value of F can be adapted to be equal to the number of successful outcomes, denoted by S, where $S \leq S_R$ in the reservation frame of size S_R. Basil needs to dynamically set $F = S$, since in each new frame S is a random number. Recalling the discussion from the previous chapter, this flexibility demands additional signaling information, as Basil needs to decide the value of F after the reservation phase is finished and then communicate the value F to the terminals. Since there can be at most S_R successful reservations, the number of data slots $F = S$ for a frame can range from 0 to S_R and this number can be specified in the allocation packet, along with the addresses of the devices to which the slots are allocated.

The essence of the described scheme is to allow all the users to randomly access the S_R reservation slots. This method of random access is known in the literature as *framed ALOHA*, as it is a variant of the basic ALOHA protocol. The next question is: how do we choose the number of reservation slots S_R? We will carry out a quick, non-rigorous analysis, in order to get an insight into the design choices for the described type of system.

1 One may object to this by noting that the successful reservation by Yoshi can be memorized for a future frame, where Basil allocates the data slot to Yoshi regardless of the actual outcome of the reservation slots in that particular future frame.

2.1.1 Randomization that Maximizes the ALOHA Throughput

The question of choosing the optimal S_R cannot be answered without providing additional elements of the model in which random access is used. To start with, we have not specified the random process that describes the way the sensors attempt to send their packets. The way to model this is to assume a random process that describes whether a sensor device has something to transmit in a given frame. In order to shed light on these issues, we can formulate a simpler problem that is still relevant for making the optimized choice of S_R. Let there be K_A active sensors with data to send among the population of K sensors. Each active sensor is trying to send a reservation packet in one of the S_R slots. It is important to state that the sensors are not mutually coordinated in any way before starting the random access process.

Having said that, there is a certain (dark room) symmetry in the problem: all the sensors look equal to the receiver and each of the S_R reservation slots looks equal to each sensor. This means that, if a particular sensor Zoya needs to pick a single reservation slot, then each of the reservation slots should have an equal chance to be picked, with probability $\frac{1}{S_R}$. Considering this, the probability that Zoya will have a successful transmission of her reservation packet in a particular slot is

$$\frac{1}{S_R}\left(1-\frac{1}{S_R}\right)^{K_A-1} \tag{2.1}$$

which is the probability that Zoya sends in that slot and that none of the other $K_A - 1$ sensors chose it for transmission.

The probability that there is a successful transmission in that slot by *any* of the K_A sensors is:

$$\binom{K_A}{1}\frac{1}{S_R}\left(1-\frac{1}{S_R}\right)^{K_A-1} = \frac{K_A}{S_R}\left(1-\frac{1}{S_R}\right)^{K_A-1}. \tag{2.2}$$

It can be shown that the latter expression is maximized when $S_R = K_A$. Hence, the best way is to choose the number of reservation slots to be equal to the number of active sensors that are contending via random access (framed ALOHA), such that the probability of successful reception in a given slot is:

$$P(K_A) = \left(1-\frac{1}{K_A}\right)^{K_A-1}. \tag{2.3}$$

Clearly, this requires knowledge of the number K_A of active sensors in the total population of K sensors.

Let us see the implications that this has on our system. Before the frame starts, Basil knows that there will be K_A sensors that will be contending with each other in order to request access, but he does not know the identities of these sensors. Note that this is the crucial assumption in the problem setup for random access, since if Basil knows which K_A sensors will require access, then there is no need for randomized contention: namely, Basil can simply set $F = K_A$ and allocate one data slot to each sensor. Although this observation seems trivial, it is very often overlooked when random access is considered in practical systems. Basil sets $S_R = K_A$, but the number S of sensors that successfully send reservation requests is random. This implies that the number of allocated slots F is also random. The *expected* value of F is $K_A P(K_A)$, and one can use (1.8) to calculate the expected goodput in a frame.

Another observation is that $P(K_A)$ decreases with K_A and it reaches $e^{-1} = 0.3679$ as the number of users goes to infinity. The engineering insight from $P(K_A)$ is that, when the users are contending in smaller groups, then the probability of successful transmission experienced by an individual user is higher.

The assumption that Basil knows the exact value of K_A is rather artificial. On the other hand, Basil may know some statistics about the random process according to which the sensors send reservation requests. In that case, it can be reasonable to conclude that Basil knows the expected value of K_A. Although not mathematically rigorous, Basil can work with the expected value as if it is the exact value and use the following approach. At the start of L–th frame the expected number of sensors that require access, denoted by $\overline{K}_A$, is given by:

$$\overline{K}_A = \overline{K}_N + \overline{K}_C \tag{2.4}$$

where $\overline{K}_N$ is the expected number of new requests generated from new sensors in the previous, $(L-1)$th frame. $\overline{K}_C$ is the expected number of sensors that tried to send request in the previous frames, but did not succeed due to collision. If a frame becomes sufficiently long, then we can apply the law of large numbers, by which the expected values can be approximated as the exact values. For this to be true, the random arrival process of the requests from the sensors should satisfy certain conditions, which we will not discuss in detail here. It suffices to say that, for example, Poisson arrivals of requests over a sufficiently long interval would work. Going back to (2.4), we remove the averaging bar and recast the same equation as $K_A = K_N + K_C$. Using the previous analysis on the probability of successful transmission of a request, we can express $K_C = K_A(1 - P(K_A))$, but since K_A is large, we can write $K_C \approx K_A(1 - e^{-1})$, which leads to:

$$K_A = K_N \cdot e. \tag{2.5}$$

Hence, if Basil uses long frames and applies the law of large numbers, then he can have a good guess at the number of contending sensors and practically choose the reservation frame size K_A in an optimal way.

The equation (2.5) can give us further very important insights into the random access protocols. Let us, for a moment, put aside the frame structure considered until now, in which Basil first lets the sensors contend using short reservation frames and then allocates data slots to the successful contenders. Instead, consider the following situation. A very large population of sensors is synchronized to Basil. A periodic frame of K_A data slots and duration of $K_A T$ is used, without any additional overhead at the frame start, since all sensors and Basil are assumed to be perfectly synchronized and thus have a perfect knowledge about the moment at which a frame starts. Each sensor that got data to send before the start of the Lth frame, chooses a random number j between 1 and K_A and sends its data in the jth slot of the Lth frame. At the end of the frame, all sensors that sent data successfully receive feedback from Basil. This feedback is assumed to be sent extremely quickly, taking practically zero time. The sensors that did not send the data successfully, treat their data packet as a newly arrived one during the Lth frame and try again in the $(L+1)$th frame. Looking again at the equation (2.5), we can interpret it as follows: if the number of newly arrived requests during each frame of duration $K_A T$ is $K_N = K_A \cdot e^{-1}$, then this number is equal to the number of successfully sent requests in a frame. Hence, the system is in equilibrium in the sense that each arrived request eventually gets served. Therefore the *throughput* of this system is, calculated in number of requests (packets) per unit time, is:

$$G = \frac{K_N}{K_A T} = \frac{K_A e^{-1}}{K_A T} = \frac{e^{-1}}{T} \quad \text{(packet/s)}. \tag{2.6}$$

Note that, due to the absence of overhead, here the throughput is equal to the goodput. If we take $T = 1$, then the throughput is conveniently expressed in packets per slot and we arrive at the well known formula for maximal throughput of a slotted ALOHA system equal to $e^{-1} = 0.368$ packets per slot.

However, what does this theoretical value of the ALOHA throughput mean for a practical system? The randomized protocol coordinates the sensor transmissions, such that each sensor eventually transmits its request successfully. The presented analysis captures the following extreme case: the total population of sensors K is very large, practically infinite, and each new request comes from a new sensor, which also means that each sensor has only one request. Such a hypothetical scenario represents the most difficult case for coordination among the sensors. In the following we provide the reasoning behind the choice of the infinite-size sensor population.

Instead of K_A active sensors, each with a single request, we consider $K_x \ll K_A$ sensors, where each sensor has D_y packets. The total number of packets to be sent in the system is $K_A = K_x \cdot D_y$, which makes the overall traffic load equal to the case with K_A single-packet users. The following protocol is run by each sensor. The sensor Zoya applies the framed ALOHA protocol until it successfully sends her *first* request. After succeeding, Zoya records (a) the number of sensors $(K_Z - 1)$ that sent their first requests successfully before Zoya, which she learns from Basil's feedback; (b) puts on hold her access until the remaining $(K_A - K_Z)$ sensors have sent their first requests successfully. Note that, after this randomized contention is finalized, Zoya has a unique number K_Z, where $1 \leq K_Z \leq K_x$. Since every sensor applies the same protocol, each sensor has a unique *token*, which is a number between 1 and K_x. After contending to send the first request and obtaining the token, the K_x sensors no longer need to contend, but they are served through a TDMA frame with K_x slots, where, for example, the slot number K_Z is allocated to Zoya. This is reminiscent of the use of random access as a technique for initial access, after which the transmissions are coordinated and scheduled.

When there are K sensors with a single request each and K goes to infinity, the system throughput is $e^{-1} = 0.368$ packets per slot, since the sensors need to contend indefinitely. Let now the number of packets in the system K go to infinity in a different way: the number of sensors K_x is kept finite, but the number of packets per sensor D_y goes to infinity, while having $K_A = K_x \cdot D_y$. Then the sensors waste some *finite* time to coordinate and obtain tokens, but after that they are served in TDMA frames ad infinitum, where each frame has a duration of K_x slots and has K_x successful transmissions. Thus the throughput of this latter system is, asymptotically, 1 packet per slot, much better than 0.368.

The infinite population assumption contains an inherent paradox. When the total sensor population K goes to infinity, then the size of the address of each sensor, requiring $\log_2 K$ bits, also goes to infinity. This implies that the size of each slot should increase to infinity in order to accommodate the address. Hence, if we insist that each sensor provides its address in the request, then the slot size cannot be fixed. The paradox stems from the fact that the analysis of ALOHA looks at the packet as a single, atomic unit of communication, not taking into account its internal structure. It is thus clear that, when the model for access protocol is enriched to reflect the internal packet structure, one cannot straightforwardly use the infinite population assumption.

One way to circumvent the paradox of the infinite population was devised by Polyanskiy in 2017. Consider an application that includes a large number of sensors in a certain field and the base station needs to compute certain functions where the sensor identification is irrelevant. This could be, for example, the average value of the sensor readings in the field. This means that the packet size does not need to grow with the population size K and one can work with the assumptions of an infinite population.

2.2 Probing

In trying to optimize the frame size for framed ALOHA, we have used some favorable statistical assumptions that allow us to approximate the actual number with the expected number of contending sensors. However, these statistical assumptions may not always be valid. Consider, for example, a large set of K wireless sensors, in which upon occurrence of some event, an unknown number K_A of sensors gets activated and each activated sensor attempts to send a data packet. Such an event might occur very rarely and the number of activated sensors may vary significantly. It would not be feasible to wait for a long time in order to make the number of activated sensors statistically predictable, since this would incur intolerable delay. We therefore need to find a method that can deal with the situation in which an unknown number K_A of sensors are trying simultaneously to send a packet/request to Basil.

We make a slight digression to our dark room metaphor to illustrate the problem addressed by probing. The reader can refer to the cartoon from the beginning of this chapter. There are K_A people in a dark waiting room that has a door through which only a single person can pass at a given time. The people do not know each other from before (otherwise some hierarchy could have been established) and therefore do not talk to each other. Walt stands outside the waiting room and he should ensure that each of the persons in the waiting room should eventually come out of the door. Which strategy should Walt use? Clearly, if Walt knows the names of the people that are in the room, the problem is trivial as he would call each of them by name and get them one by one out of the waiting room. If Walt does not know them, then he should start by saying “one of you come out of the room”; however, the people in the dark room cannot make a mutual agreement who should be that one and the arbitration is left to Walt. Hence, this is an another instance of a multiple access problem that requires some form of random access.

Going back to our original multiple access problem, let us assume that, at a certain time instant, the base station Basil starts a multiple access frame by sending a packet that invites reservation requests. In the single reservation slot that follows all K_A active sensors transmit their requests. For this discussion, it will be useful to think that the packet sent by Basil is a polling packet, containing the address *ALL*. This is not an address of a particular sensor, but it should be interpreted as if Basil is inviting any sensor that has a data to send, to transmit in the slot that follows. This is certainly different from the framed ALOHA, where Basil assumes that there are $K_A > 1$ active sensors and pre-emptively asks them to try to avoid the collisions by selecting randomly one of the multiple slots in a frame. If there is no sensor to transmit and $K_A = 0$, then Basil perceives an idle slot and he does not need to take action, except sending again a frame header at a future point in time. If $K_A = 1$ then there is a single transmitting sensor, Basil receives the request successfully and sends acknowledgement (ACK). If $K_A > 1$, then Basil observes a collision. More importantly, Basil now *knows* that there are two or more sensors attempting to communicate with him, since in our model we assume that a collision is detected perfectly. The objective of Basil is to resolve this collision by running a protocol that will enable each of the sensors involved in this initial collision to send the data packet successfully at some later point in time.

Let us continue the example by taking $K_A = 8$ and refer to Figure 2.2. For this example it is more convenient to denote the sensor devices as $s_1, s_2, \cdots s_8$. After observing the initial collision, Basil sends again a packet to invite transmissions from the sensors. However, if Basil addresses the packet that invites all sensors to respond, the same collision will occur again. Therefore, randomization is needed and here is an idea of how it can be introduced. Basil now performs *probing* by sending a poll packet addressed to a sensor 0. Initially there is no sensor with address 0, entitled to respond to this packet. However, upon receiving the 0–addressed polling packet from Basil, each of the eight sensors randomly tosses a coin, getting an outcome of 0 or 1, each with probability 0.5. For the example in Figure 2.2, the sensors s_1, s_2, s_3 obtained 0, while the rest obtained 1 as an outcome from the coin tossing. An interpretation of this random outcome is that each of the sensors s_1, s_2, s_3 randomly assigned to itself the address 0, while each of the other sensors randomly self-assigned address 1. In slot 2 in Figure 2.2 only the sensors with address 0 are transmitting. Basil again observes collision, not knowing how many sensors are transmitting. For example, it could have happened that all eight of them toss coin 0, leading to a situation that is identical to that in slot 1. Yet, statistically speaking, with the first polling packet Basil manages to split the initial group of sensors into two approximately equal groups, such that the problem of resolving one large collision is divided into resolving two smaller collisions. This principle of using coin tossing to split a group of collided sensors into smaller groups can continue in a recursive fashion, until each group contains a single sensor (successful transmission) or no sensor (idle slot).

We continue with the example from Figure 2.2. Note that Basil's packet with address 0 is not exactly a poll packet, since there is no certainty that a sensor with self-assigned address 0 exists; this happens when all sensors have tossed 1. It is rather a *probing packet* or a *probe*, aiming to explore/learn about the set of contending sensors rather than letting the sensors transmit over a large interval in order to avoid collisions. The next probe sent by Basil has address 00, such that the sensors that have already generated address 0, and a coin toss is used in order to decide if the next bit of the address is 0 or 1. Now only the sensor s_1 has the address 00, such that it is the only one responding to the probe addressed with 00. The reader can continue to follow the full example in Figure 2.2. For example, there is no sensor that has a sequence of random outcomes 110, such that when the probe 110 is used in slot 12, Basil receives no response. The tree representation in Figure 2.2 can be used to track the random outcomes based on the coin tossing. This is why these algorithms are sometimes referred to as splitting tree algorithms or simply tree algorithms.

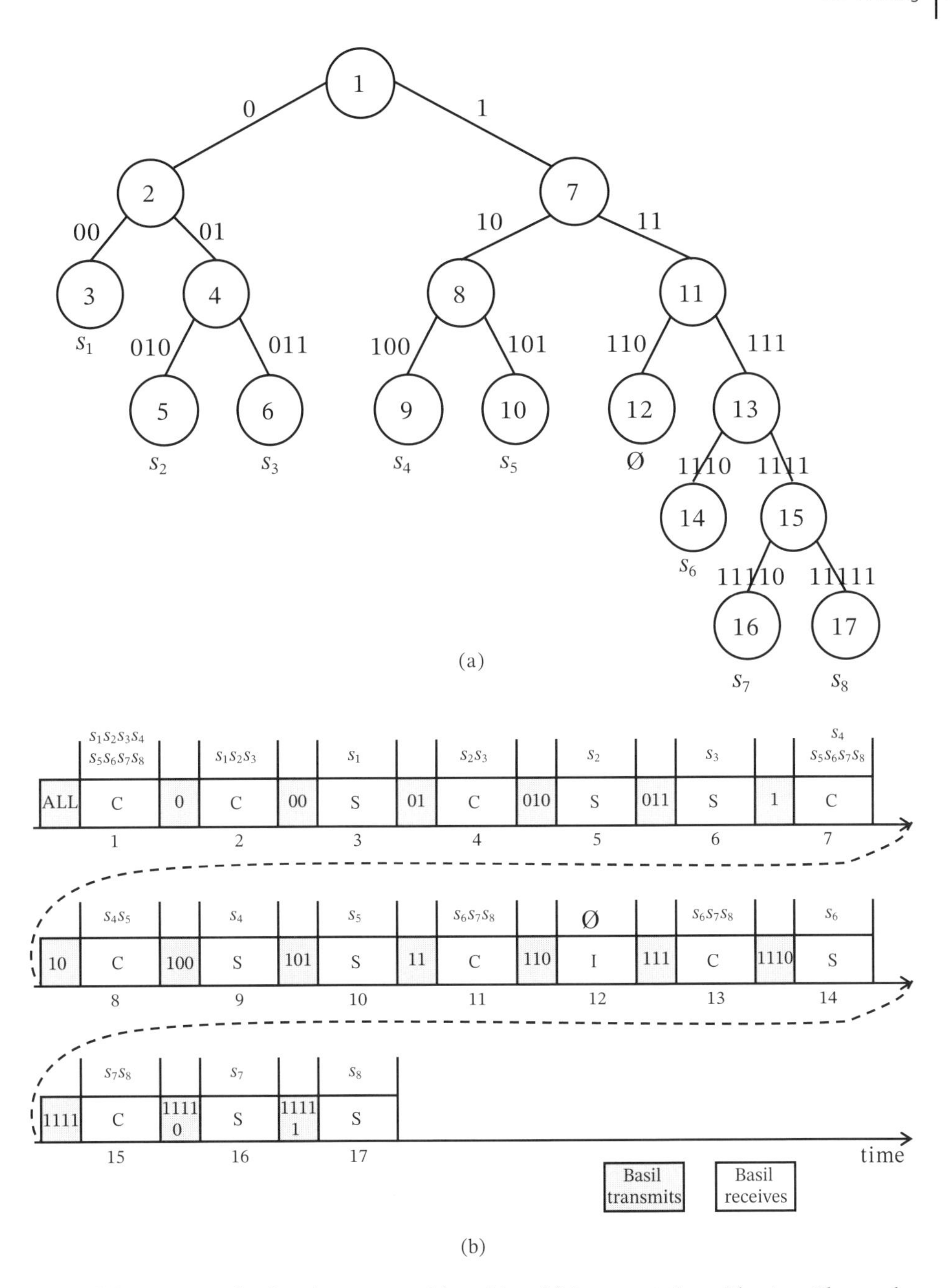

Figure 2.2 An example of random access with probing. (a) Representation with a tree. The number in each circle is the time slot in which that tree node is visited. Each edge is labeled with the probe address that enables the next tree in the node. (b) Representation of the tree in time. Basil can receive three different outcomes: C (collision), S (single), I (idle). For each slot in which Basil receives, the set of transmitting sensors is put at the top. For example, in slot 11 the sensors s_6, s_7, s_8 transmit and Basil receives collision (C).

The concept of probing does not need to be implemented by random coin tosses, it can also be run by using the actual addresses of the contending nodes, provided each node has a unique address. To see how this works, let us assume that each of the eight sensors from the example in Figure 2.2 has a unique address and Basil knows that fact. Such an address, for example, can be hard-wired into the sensor and consists of a unique pattern of b bits. Clearly, b needs to be at least 3 for this example to work. The probes sent by Basil have the same content; however, now the sensor actions and the probe interpretation is different. For example, a probe with address 00 means "any sensor that has a unique address starting with 00 should transmit".

An alternative view could be that the probing process creates temporary short addresses by which the nodes can be identified/polled within the communication process. For example, let each of the sensors s_i in Figure 2.2 have a unique, worldwide address that consists of 48 bits. Furthermore, let the sensors be in a sleep mode most of the time, such that the probing process is used to establish the initial contact with the sensors that have just woken up. In other words, the initial probe packet can be interpreted as "has anyone out there woken up"? After a sensor is awake, it may have multiple data exchanges within a short period. It should be noted that during the probing process, the sensors are allocated unique addresses: s_4 has the address 100, while s_7 has the address 11110. After the probing process, the base station can make the data exchange with the sensors $s_1, \dots s_8$ by using these temporary addresses, which are much shorter than the unique 48 bit addresses.

We now look closer at the packets that need to be sent by Basil in order to run the splitting tree algorithm. In Figure 2.2, Basil uses the full probe address each time, which is redundant. After initiating the process in slot 1, Basil only needs to send the last generated bit of the probe address, not the full address. This is because each sensor can track the outcomes and therefore the sensor knows what is its current position in the tree. For example, after the collision in slot 8, Basil sends only 0. The sensors know that the last received probe was addressed to 10, such that they append 0 and get the current probe address 100. After each receiving slot, Basil can send a feedback message to inform the sensors about the outcome in the *previous* slot, which can be collision (C), single (S), or idle (I). This is different from Figure 2.2, apart from the initial probe sent to all, where the probe sent by Basil tells who is eligible to transmit in the slot *after* the probe. It can be seen that usage of feedback instead of a probe leads to an equivalent result. For example, after the first slot, the feedback C denotes that there has been a collision and the next probe address is 0. As another example, the feedback messages received after the first four slots are C, C, S, C; this uniquely determines the next probe address to be 010.

The ideas behind random access with probing have led to the most efficient random access protocols that operate with the collision model, attaining throughput of more than 0.487 packets per slot. In practice, a limiting factor can be the feedback packet sent by Basil. For the explanation above we have assumed that the feedback is instantaneous, and it therefore does not affect the time efficiency/throughput of the random access protocol. On the other hand, probing algorithms use feedback messages very extensively and this must be taken into account when designing a practical random access/reservation protocol. Another remark is that each participant in a probing protocol should often switch between transmit and receive state. As mentioned in Chapter 1, such an operation may not be desirable, for example due to the fact that the switching takes time and therefore affects the throughput.

2.2.1 Combining ALOHA and Probing

One can easily think of creating new random access algorithms by combining the principles of ALOHA and probing. An advantage of framed ALOHA is avoidance of collisions and economic usage of feedback messages. ALOHA uses *memoryless* randomization: the random choice of a slot made to access at a given time is independent of the random choice made at another time. Contrary to this, probing or splitting tree algorithms create memory in the randomization process and utilize the history of collisions to resolve the set of colliding sensors more efficiently. Here is an example of what a hybrid ALOHA probing algorithm could look like. Basil sends a probe-to-all packet, which, instead of a single slot, is followed by a frame that has K_A slots. Each sensor selects randomly a slot j to send its packet. We can now interpret j as a self-assigned random address, as we did in the probing algorithm. If Basil observes a collision in slot j, then he proceeds to resolve this collision recursively. At first, he sends a probe addressed to the sensors that have collided in the jth slot or, equivalently, all sensors that have self-assigned the address j. This probe is followed by a frame with K'_A slots. Hence, the generalization is that a probe can be followed by more than a single slot. This would require generalization of the tree structure used to describe the basic probing algorithm.

2.3 Carrier Sensing

2.3.1 Randomization and Spectrum Sharing

Randomization is the key ingredient for solving the problem of uncoordinated access to a shared communication medium. We have seen its use in the rendezvous problem, where randomization helps to assign roles to two half-duplex devices that would like to talk to each other. In the random access algorithms from the previous section, randomization is used to resolve the conflict among devices that are transmitting to the same receiver.

The third case that requires use of randomized access over a shared wireless medium is depicted in Figure 2.3, where two independent links are in spatial proximity. Specifically, Zoya wants to transmit to Yoshi and Xia wants to transmit to Walt. This is often referred to as *spectrum sharing*, as it is necessary when the two collocated links use the same communication channel. Another common term to denote this situation is that both link operate at the same “frequency” or use the same “frequency channel”. The quotation marks are due to the fact that we have not yet introduced the concept of frequency, which will be done much later in Chapter 9. However, at this point it is sufficient to say that if the collocated links are using different frequency channels, then no channel is shared and there is no interference between the links. Contrary to this, the assumption for the scenario of Figure 2.3 is that all nodes used the same frequency channel and are in communication/interference range of each other. Furthermore, the scenario shows that communication happens in a specific direction, Zoya to Yoshi or Xia to Walt, while there is is no cross-communication between the devices that belong to different links, say Zoya and Xia. In Figure 2.3 there is no interfering line between Yoshi and Walt, although they are in range; the reason is that none of them acts as a transmitter. The bidirectional interference between Zoya and Xia does not mean that they simultaneously interfere with each other, but it shows the possible interference

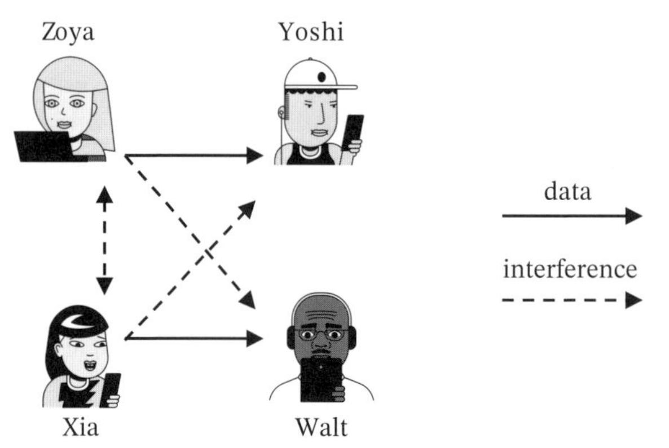

Figure 2.3 A simple scenario for sharing the wireless spectrum between two collocated links: Zoya–Yoshi and Xia–Walt. The assumption is that only Zoya and Xia can be transmitters.

that can occur when one of them is transmitting and the other listening or receiving. If Zoya transmits, but Xia does not and is in listening mode, then Xia can overhear the signal and detect that there is a transmission. By reciprocity, the same happens when Xia transmits and Zoya listens.

The only way in which the two links Zoya–Yoshi and Xia–Walt can impact each other is through interference. In the framework of the collision model, if Zoya and Xia transmit simultaneously, then both Yoshi and Walt detect collisions and neither of them receives the desired packet correctly. The latter assumption is rather subtle, as if Zoya transmits and Xia is quiet, then Walt is able to overhear and receive the packet of Zoya correctly. Then why not use this overhearing to communicate coordinating messages between the two links? This is indeed possible, but in practice there can be collocated devices that are not *logically connected* and/or are not part of the same administrative network domain. It is thus viable to assume that data can be communicated only with a limited group of collocated nodes to which the observed node is *networked*, while all the other transmissions can be detected, but not used to extract data from them. However, it is important to note that if Walt is connected to Xia, but he overhears and detects packets from any arbitrary network nearby, then he may spend a substantial amount of battery energy to detect or receive packets he does not care about.

Figure 2.4(a) depicts a possible method for randomized spectrum sharing between the links Zoya–Yoshi and Xia–Walt, respectively, by assuming that both systems are synchronized at the slot level. Specifically, Figure 2.4(a) illustrates an example of execution of a transmission protocol for a given pattern of arrival of packets to Zoya, denoted by $\mathbf{Z}_1, \mathbf{Z}_2, \ldots$, and Xia, $(\mathbf{X}_1, \mathbf{X}_2, \ldots)$. Upon experiencing a collision, such as $\mathbf{Z}_2$ and $\mathbf{X}_1$ in the third slot, both Zoya and Xia apply randomization in order to decide in which slot each of them will retransmit the packet $\mathbf{Z}_2$ and $\mathbf{X}_1$, respectively. In this example, Zoya stays idle for two slots and only after that retransmits $\mathbf{Z}_2$.

We can detect at least two inefficiencies created by the slotted structure:

- If a packet arrives after the start of the slot, then the node needs to postpone the first transmission of the packet until the start of the next slot, even if the previous slot is idle.
- Consider the collision between $\mathbf{Z}_2$ and $\mathbf{X}_1$. Although the packets have not arrived at the same time, both Zoya and and Xia need to wait until the start of the next slot. Thus, the slotted structure forces them to be sent at exactly the same time, which leads inevitably to a collision.

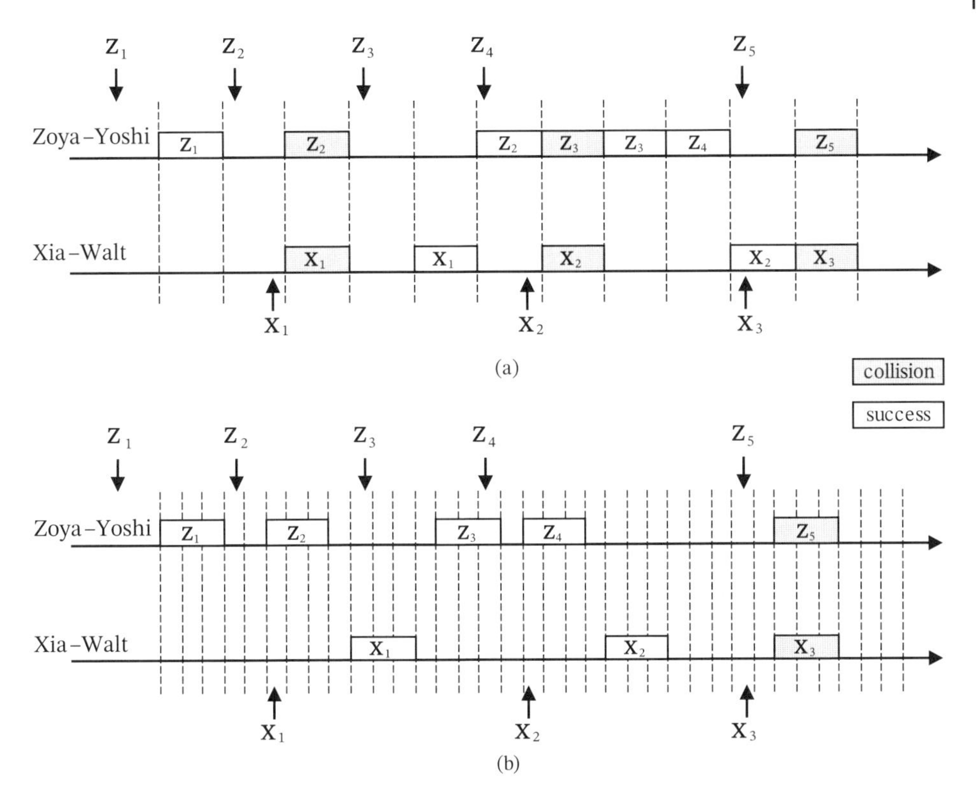

Figure 2.4 Illustration of randomized spectrum sharing between two interfering links, Zoya–Yoshi and Xia–Walt, respectively. Zoya and Xia are transmitters. The pattern of packet arrivals to Zoya ($\mathbf{Z}_i$) and Xia ($\mathbf{X}_j$) is identical for both cases. (a) Slotted channel. (b) Carrier sensing solution, with a minislot duration being a third of the slot duration.

2.3.2 An Idle Slot is Cheap

The key idea in overcoming these inefficiencies is to shorten the duration of the idle slot and introduce *carrier sensing multiple access (CSMA)*. Figure 2.4(b) shows an example in which the idle slot duration T_I is one third of the slot duration T. Let us use the term *minislot* to denote the time unit corresponding to an idle slot. After $\mathbf{Z}_2$ arrives and Zoya has it available for transmission, she waits until the start of the next minislot and then starts to listen during that minislot. If the minislot is idle, Zoya starts transmission after the minislot ends. The packet $\mathbf{X}_1$ arrives at Xia during a minislot in which Xia is not in a transmission state. Since Zoya transmits $\mathbf{Z}_2$ during that minislot, Xia detects a *carrier* and concludes that the medium is busy from a transmission carried out by somebody else. After the transmission of $\mathbf{Z}_2$ is finished, Xia waits for one idle minislot to verify that the transmission of Zoya has finished. In other words, Xia cannot learn that the transmission of Zoya will finish during the minislot in which it actually finishes and she has to verify the end of Zoya's transmission by detecting the subsequent idle minislot. After that idle minislot, Xia transmits $\mathbf{X}_1$.

Note that the introduction of a minislot has drastically reduced the number of collisions. This is because the minislot is cheap in terms of time duration, offers a better time resolution, and the system operation can benefit from the asynchronism among the arrivals of the packets across the different devices. However, the minimal resolution is brought to the level of a minislot, such that if two packets arrive at the same minislot, then collision cannot be avoided. This is illustrated by the packets $\mathbf{Z}_5$ and $\mathbf{X}_3$. Upon collision, Zoya picks a random number of minislots (not slots!) to wait until the next transmission and Xia does the same. In this way, they create the required asynchronism and likely avoid another collision between the packets of Zoya and Xia.

While the idea of minislots and CSMA is introduced here in the context of spectrum sharing, the same mechanism can also be used to design a random access protocol. We can reuse Figure 1.1(b) and think of a system in which Zoya, Yoshi, and Xia use random access to transmit to Basil. Recall that, when we were using the same setting to describe random access, the devices received signals only from the base station Basil and it was not relevant to consider the fact that a device can detect the transmission of another device. By contrast, the new requirement in CSMA is that a device should listen to find out whether the medium has been taken by a transmission from another device.

Figure 2.4(b) presents a rather basic version of CSMA. For example, there can be a variant in which, upon detecting that the medium is not busy, a node waits for a random number of time slots before starting the transmission. The rationale is that, while the medium is busy, there could have been multiple packet arrivals at different transmitters, and if all of them wait only for a single idle slot then a collision occurs. A similar argument is valid for the following feature that is used in practical systems, such as Wi-Fi. Assume that Zoya experienced a collision and decided to wait for 10 minislots. While waiting, Zoya detects that the medium has been busy for 15 minislots. If Zoya counts down the waiting minislots when the medium is busy, then she finishes the countdown while the medium is still busy and transmits after the idle minislot that follows the busy period. Again, the main problem is that many other nodes could have done the same and thus they get synchronized towards a collision. An elegant solution to this is to stop the counter while the medium is busy, thus removing the synchronizing effect that the busy medium may have on the waiting nodes.

We note that, as the minislot becomes the basic time reference of the protocol, then this removes the need to assume that all packets are of the same length. The example in Figure 2.4(b) can be easily reworked by assuming that each of the packets $\mathbf{Z}_i$ or $\mathbf{X}_j$ has a different length, expressed as an integer number of minislots.

The gains of carrier sensing improve when the minislot is shorter. Ideally, it should be equal to zero. However, there are practical constraints that put a lower bound on the minislot duration. While it is not part of our collision model, in practice there is always a *propagation delay* in the wireless signals. This means that, when Zoya starts to transmit $\mathbf{Z}_2$ in Figure 2.4(b), Xia does not immediately detect that there is a transmission, and only detects one after a certain time that is required for the wireless signal to travel from Zoya's transmitter to Xia's receiver. The duration of the minislot should be set to be equal to the maximal propagation delay that a carrier signal can experience. In our communication model, depicted in Figure 1.1(a), we work with the (very artificial) assumption that no signal propagates beyond a distance of d; then the minislot duration can be determined as:

$$T_\mathrm{I} = \frac{d}{c} \tag{2.7}$$

where c is the propagation speed, equal to the speed of light. We can soften the propagation assumption and enrich our collision model by assuming the following: the communication range is d, but the range within which the carrier can be sensed, called the sensing range, is larger, say $2d$. The rationale can be drawn from the analogy with speech: as Yoshi goes away from Zoya, after some distance he can hear that she is speaking, but cannot understand the words. If Zoya and Yoshi are terminals that are situated at the opposite ends of Basil's cell, then the distance between them is $2d$ and the duration of the minislot should be chosen to be $\frac{2d}{c}$. In practice, it gets more complicated as not all signals travel over a straight line, as some of the waves that carry wireless information are reflected and arrive later.

We remark that carrier sensing is well suited for ALOHA type protocols, where collisions are avoided. On the other hand, the random access protocols based on splitting tree are good at resolving collisions once they occur, such that the gains that CSMA introduces in splitting tree type protocols are rather modest, as the main effect of CSMA is to decrease the probability of occurrence of a collision.

2.3.3 Feedback to the Transmitter

The way we have described the system operation in Figure 2.4 assumes that a transmitter, Zoya or Xia, knows perfectly if their packet has been received successfully or was subject to collision. On the other hand, collision or success is a phenomenon that occurs at the receiver, such that it is the receiver that needs to inform the transmitter about the outcome. In addition, Zoya's transmitter is half-duplex and she cannot detect the collision with Xia while transmitting, although Xia is within the communication/interfering range and her signal reaches Zoya. In fact, due to the use of half-duplex transmission, after sending the packet, Zoya should go into receiving mode. In this mode she waits for a packet from Yoshi that carries feedback to inform her whether the packet reception outcome was successful or not. This feedback packet from Yoshi, denoted by $\mathbf{Y}_f$, can, in principle, carry a single bit of information, either acknowledging (ACK) the packet reception or sending a negative ACK (NACK). The latter is sent in the case there is a collision and thus an incorrect reception. The feedback packet is, generally, much shorter than the data packet.

Figure 2.5(a) shows the transmission of the feedback packet $\mathbf{Y}_f$ from Yoshi for the case when the channel is slotted. Differently from the slotted model used before, where the slot duration was equal to the packet duration, here the slot duration is longer than the duration of Zoya's packet. The remainder of the slot, after the end of Zoya's packet, is reserved for Yoshi to transmit the ACK message $\mathbf{Y}_f$, while Zoya is in a listening mode. Referring

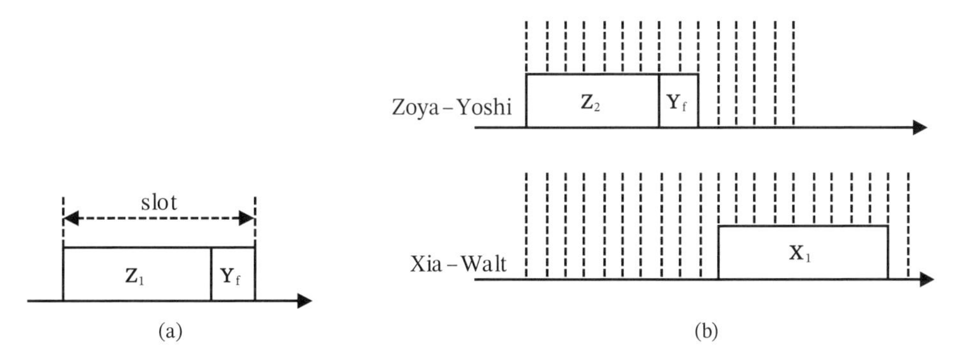

Figure 2.5 Transmission of a feedback packet $\mathbf{Y}_f$ from Yoshi to Zoya for (a) a slotted channel, (b) carrier sensing.

to Figure 2.4(a), Yoshi receives $\mathbf{Z}_2$ successfully and sends $\mathbf{Y}_\mathrm{f}$ =ACK. The packet $\mathbf{Z}_2$ is not received correctly and Yoshi sends $\mathbf{Y}_\mathrm{f}$ =NACK. If we assume that all nodes are in each other's communication range, then at the same time when Walt sends NACK to Xia, collision occurs at the receiver of Zoya, as well as that of Xia. As a result, Zoya does not receive the NACK. To avoid this type of situation, one can use randomized contention for sending the feedback packets, which is a contention that occurs in addition to the contention used for data packets. Nevertheless, in our communication model, Zoya can receive NACK also *implicitly*, through the absence of ACK. In other words, the mere existence or absence of $\mathbf{Y}_\mathrm{f}$ can be understood as a single bit of information sent from Yoshi to Zoya. If Yoshi does not receive the data packet correctly, he "sends" NACK by not transmitting $\mathbf{Y}_\mathrm{f}$.

Figure 2.5(b) shows how the transmission of a feedback packet would operate in a CSMA setting. As in CSMA the basic time unit is a minislot and not a slot; we cannot say that the end of the slot is reserved for transmission of a feedback packet. Note that Yoshi, after receiving $\mathbf{Z}_2$, does not wait for an idle minislot and he immediately sends $\mathbf{Y}_\mathrm{f}$ to Zoya. Xia, and potentially other contending transmitters, will detect that the medium is busy and postpone her transmission to start after $\mathbf{Y}_\mathrm{f}$ is followed by an idle slot. We can interpret this as if the feedback packet has a higher priority over a data packet sent by another node.

This observation reveals the inherent capability of the CSMA mechanism to introduce different priority classes. In the simplest case, there can be two classes of traffic: high and low importance, respectively. Then the protocol can be designed such that a high-priority data uses a single minislot for carrier sensing, while low-priority data uses two minislots for carrier sensing. Let, for example, Zoya send a high-priority packet that corresponds to some data for critical control system managed by Yoshi, while Xia sends to Walt data from, for example, an entertainment service. After the busy medium is released, then Zoya will always be the first to start a transmission after a single idle minislot, while Xia will defer her transmission. On another note, this property of CSMA can be misused by a malicious user. For example, the CSMA protocol can be specified such that the minimum idle time before the transmission is T_I. However, a malicious user can set his device to wait for a time less than T_I and in this way always gain an advantage in accessing the shared medium.

Transmission of feedback has a special role if we assume that the devices have the full-duplex capability. Namely, full-duplex enables the use of CSMA with collision detection (CD), which was widely used in the early days of wired ethernet. Consider the collision of $\mathbf{Z}_5$ and $\mathbf{X}_3$ from Figure 2.4(b) and let us assume that a single minislot is sufficient for the receiver to determine whether the received signal is a single transmission or collision among multiple packets. Then after the first minislot in which $\mathbf{Z}_5$ and $\mathbf{X}_3$ are overlapping, Yoshi and Walt start to transmit a signal termed *busy tone*. Since Zoya and Xia can receive while transmitting, they will both detect a busy tone. To be precise, each of them will detect a collision of the two busy tones sent by Yoshi and Walt. The introduction of busy tone is another enrichment of our communication model, as it is a special signal that only tells that the medium is busy and does not carry any additional information. This enables us to assume that a collision of busy tones is a special case of collision that does not lead to error, but it can be again treated as a busy tone signal. Therefore, at the end of the first minislot during which $\mathbf{Z}_5$ and $\mathbf{X}_3$ are colliding, both Zoya and Xia detect a busy tone and stop their transmission. This saves two minislots that are otherwise wasted in collision in Figure 2.4(b); instead, Zoya and Xia can already, after receiving the busy tone, make a random choice on which minislots to retransmit their packets. It can thus be concluded that the capability of collision detection, unleashed by full-duplex, enables early termination of the collisions and thus brings performance gain in addition to the gain brought by carrier sensing.

2.4 Random Access and Multiple Hops

The scenario from Figure 2.4 in which all nodes are in range of each other is rather limiting. Carrier sensing works well in that scenario because each transmitter and the corresponding receiver are able to receive the same signals from external transmitters; for example, both Zoya and Yoshi are in range of Xia. However, if the positions of the communication nodes are changed, while still applying the simple model based on a strict definition of a communication range d, then we arrive at a completely new setting from a system viewpoint. Some of the possible communication/interference configurations that can be obtained in this way are depicted in Figure 2.6. There are other possible configurations that are not depicted in

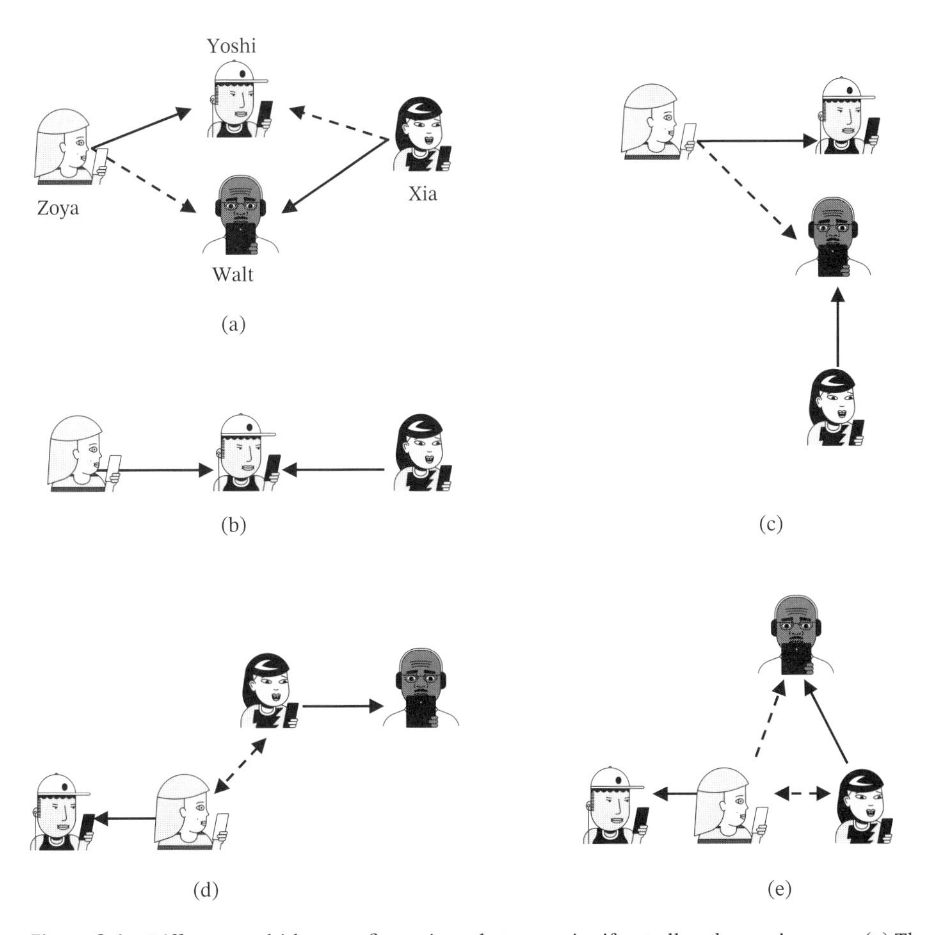

Figure 2.6 Different multi-hop configurations that can arise if not all nodes are in range. (a) The hidden terminal problem for two interfering links. (b) The classical hidden terminal problem. (c) Zoya and Xia are hidden from each other, but only Walt can experience collisions due to the hidden terminal problem. (d) The exposed terminal problem. (e) Multi-hop structure in which carrier sensing still works well.

the figure, but, due to symmetry, they can be analyzed in an analogous manner. Compared to the all-in-range scenario, also known as *single-hop*, in each of the scenarios in Figure 2.6 there are at least two nodes that are outside the communication range, such as Zoya and Xia in Figure 2.6(a). The name *multi-hop* attached to the scenarios in Figure 2.6 comes from the following. If Zoya would like to communicate with Xia, Zoya sends first her packet to Walt and then Walt sends it to Xia. If we say that one link corresponds to one transmission hop, then Zoya can reach Xia through multiple hops.

In the discussion that follows we stick to the communication model for which the communication range and the interfering range are the same. Figure 2.6(a) depicts the basic difficulty posed in carrier sensing in a multi-hop spectrum sharing setting, termed the *hidden terminal* problem. Zoya and Xia are not in range and thus they cannot sense each other's carriers. If Zoya starts a transmission to Yoshi, Xia thinks that the medium is still idle and starts transmission to Walt, collisions occur at Yoshi and Walt, and both packets are lost. Carrier sensing does not help here. Figure 2.6(b) depicts the standard defining scenario for a hidden terminal, where both Zoya and Xia try to send to the same receiver Yoshi, but they are hidden from each other. In Figure 2.6(c), Zoya and Xia are hidden from each other, and Yoshi is outside the range of Xia. Zoya and Xia cannot apply carrier sensing with respect to each other and can thus end in a situation where they transmit simultaneously. However, in that case only Walt experiences a collision, but not Yoshi.

One can extend the communication model by assuming that the carrier sensing range is larger than the communication range, as already mentioned before when the carrier sensing was introduced. In a model where the carrier sensing range is larger than the communication range, the hidden terminal problem can be mitigated. Furthermore, if the carrier sensing range is sufficiently larger than the communication range, then it can happen that even if the terminals are in a multi-hop setting with respect to the communication range, they are still in a single-hop setting with respect to carrier sensing.

Nevertheless, it is not only the absence of carrier sensing that can cause problems in a multi-hop setting. Figure 2.6(d) illustrates what is known as the *exposed terminal problem*. Zoya and Xia are in range, such that they can inhibit each other by using carrier sensing. Let us look at the example in Figure 2.4(b): when $\mathbf{X}_2$ arrives, Zoya is already transmitting $\mathbf{Z}_4$, such that Xia defers her transmission until one idle slot after the transmission of $\mathbf{Z}_4$ is over. However, if the physical positions of the nodes are the ones depicted in Figure 2.6(d), then Xia can start transmitting while Zoya sends $\mathbf{Z}_4$ and both $\mathbf{Z}_4$ and $\mathbf{X}_2$ will be received correctly. In short, the exposed terminal problem is manifested by unnecessary inhibition of a terminal that can, in fact, transmit.

Not all multi-hop settings exhibit problems with carrier sensing; the reader can verify that it works correctly for Figure 2.6(e), although Yoshi cannot be interfered by Xia.

The discussion so far leads to the observation that the gain that carrier sensing brings in a multi-hop setting is not straightforward. This is because collisions can still occur and each collision is expensive, as the medium is wasted for at least one packet duration. In order to see the worst case, consider the hidden terminal problem in Figure 2.6(a). Let Zoya start to send to Yoshi and, just before that transmission ends, Xia starts to send to Walt. According to the collision model, both packets are lost and the time for which the wireless medium has been wasted corresponds to a time that can go up to the sum of the duration of the two packets.

2.4.1 Use of Reservation Packets in Multi-Hop

In order to make the collisions less expensive, we can reuse the concept of a reservation packet from Section 1.4.2. The main idea is to constrain the collisions to occur only for packets that are short. First we look at the simplest setting for hidden terminals from Figure 2.6(b). If Zoya has a data packet for Yoshi, then she sends a short *request-to-send (RTS)* packet. The RTS packet should contain information about the originator (Zoya) and how long the transmission from Zoya will last. Assume that Xia does not transmit while Zoya transmits an RTS. Then Yoshi receives the RTS from Zoya correctly, and he acknowledges it by sending a short *clear-to-send (CTS)* packet to Zoya. The CTS should contain the address of Zoya, but it also repeats the information about how long Zoya will need to send her packet to Yoshi. The latter information from the CTS is intended for the terminals that are hidden from Zoya, such as Xia, and the CTS blocks their transmissions while Zoya transmits. If Zoya receives the CTS, then she starts to transmit her data and, at the end of the transmission, she receives an ACK from Yoshi. Note that Xia is in the range of Yoshi, such that she gets inhibited by the carrier sensing mechanism in the case when Yoshi has something to send to Zoya, be it a CTS or an ACK. Recalling from the previous section the use of idle slots for packet prioritization, both control packets CTS and ACK that are sent as responses to other packets can use an idle slot that is shorter compared to an RTS packet. With this, when Yoshi sends a CTS packet, Xia senses a busy medium and is thus prevented from sending an RTS packet.

To see the other effects of the RTS/CTS mechanism, consider the exposed terminal problem in Figure 2.6(d). Zoya sends an RTS, Yoshi sends a CTS; Xia receives the RTS, but not the corresponding CTS. This is an indication for Xia that the intended receiver of Zoya is outside Xia's range and Xia can freely initiate a transmission to Walt. The only problem is that, after Yoshi receives Zoya's packet and sends an ACK to Zoya, Xia may be still transmitting and Zoya will not receive the ACK. Therefore, in this simple form, an RTS/CTS does not completely solve the exposed terminal problem and the reader is encouraged to think what other amendments can be done to the protocol in order to address this problem. We should also note that the utility of an RTS/CTS decreases if the range for carrier sensing is larger than the communication range.

2.4.2 Multiple Hops and Full-Duplex

We look briefly into the changes required when full-duplex devices operate in a multi-hop setting. Recall from the previous section that the receiver, upon detecting collision, can send a busy tone to the transmitters, such that they can interrupt their transmissions and shorten the time wasted in a collision. Consider the hidden terminals Zoya and Xia from Figure 2.6(c) and assume there is no RTS/CTS mechanism in place. With half-duplex devices, the time that can be consumed by collision can go up to two packet durations; the extreme case is when Xia starts transmission just before Zoya's packet ends and both packets are wasted. With full-duplex devices, the time consumed by a collision is at most a single packet duration, as Zoya and Xia interrupt transmission immediately after a busy tone is sent to them. If the RTS packet is much shorter than the data packet, then it is still useful to limit the maximal duration of the collision to be short, even if the devices have full-duplex capability. Full-duplex operation can also be beneficial to address the exposed terminal problem, Figure 2.6(d). As discussed above, Xia can start to transmit after receiving an RTS from Zoya and no CTS from Yoshi. When Zoya has finished transmitting, she sends a signal to issue a command to Xia to temporarily switch off her transmission, until Zoya receives an ACK from Yoshi. If Xia has a full-duplex device, then while still transmitting to Walt, Xia can detect the command sent by Zoya, suspend her transmission while Zoya receives an ACK, and continue transmitting afterwards.

2.5 Chapter Summary

In this chapter we have used the dark room analogy to depict a situation in which the same wireless channel needs to be used for coordination and control of transmissions, as well as for the transmission of the actual data. This problem is addressed through the broad class of random access protocols. Two different paradigms for random access have been presented: protocols based on ALOHA and the tree-splitting protocols based on probing. By extending the communication model to introduce minislots, we have introduced the widely used mechanism of carrier sensing. Finally, the chapter presented some challenges and possible solutions to random access problems applied in a wireless multi-hop setting.

2.6 Further Reading

The history of random access protocols is very rich, but also surprisingly vital in identifying new models, aspects and associated problems, for example related to the recent developments in massive communication for the IoT. It has started with the paper on ALOHA Abramson [1970], while the paradigm based on probing and splitting tree was introduced later on in Hayes [1978], Tsybakov and Mikhailov [1978] and Capetanakis [1979]. Detailed analysis of random access protocols can be found in Bertsekas and Gallager [1992] and Rom and Sidi [2012]. A beautiful example of modeling and analysis of random access protocols can be found in Bianchi [2000].

2.7 Problems and Reflections

1. *Random access over multiple channels.* Let us consider a scenario in which a number of devices attempt to communicate with the base station Basil through random access. Assume that there are F available communication channels. At a given instant a device or Basil can be active (transmit or receive) on only one channel. All devices and Basil are half-duplex. Propose a design of random access protocols for the following two cases:
   (a) All F channels are used for data transmission.
   (b) Part of the channels are reserved for random access and coordination of the devices, while the remaining channels are used for data transmission.
2. *The room is not dark.* In problem 1(b) it seems that we are departing from the dark room analogy, as there is a dedicated channel for reservation/signaling. Compare this to a classroom in which the students reserve a speech channel by raising a hand through

the visual channel. Hence, this classroom scenario has $F = 2$ different channels. Explain how the model from assignment 1. should be changed in order to represent correctly the communication model in the classroom.

3. *Detecting packet multiplicity*. Consider an ALOHA type protocol with a single channel, but let us upgrade the communication model by assuming that, when more than one device transmits simultaneously and there is a collision, Basil can perfectly detect how many packets are present in the collision, but he cannot decode the packets. Propose a random access protocol that can utilize this upgraded model to improve the overall throughput when:
   (a) Basil knows only the number of the packets involved in the collision, but not the identities of the devices that transmitted the packets.
   (b) For each collision, Basil knows the identities of the devices that have transmitted the packets that constitute the observed collision.
4. *Errors beyond collisions*. In order to make the collision model more realistic, let us assume that even when Zoya is the single device that transmits to Basil, her packet can be received with errors due to, e.g., noise, such that she needs to resend the packet. To make the things more challenging (and even closer to reality), assume that Basil cannot distinguish between a collision and a single packet that is in error due to noise. Analyze how the introduction of error in single packets affects:
   (a) ALOHA type protocols.
   (b) Probing and splitting-tree protocols.

   For both cases suggest a suitable re-design of the protocols.
5. *Longer sensing range*. Consider the cases of multi-hop communication depicted in Figure 2.6. How would their operation change if we assume that the sensing range is:
   (a) Two times longer than the communication range.
   (b) Three times longer than the communication range.

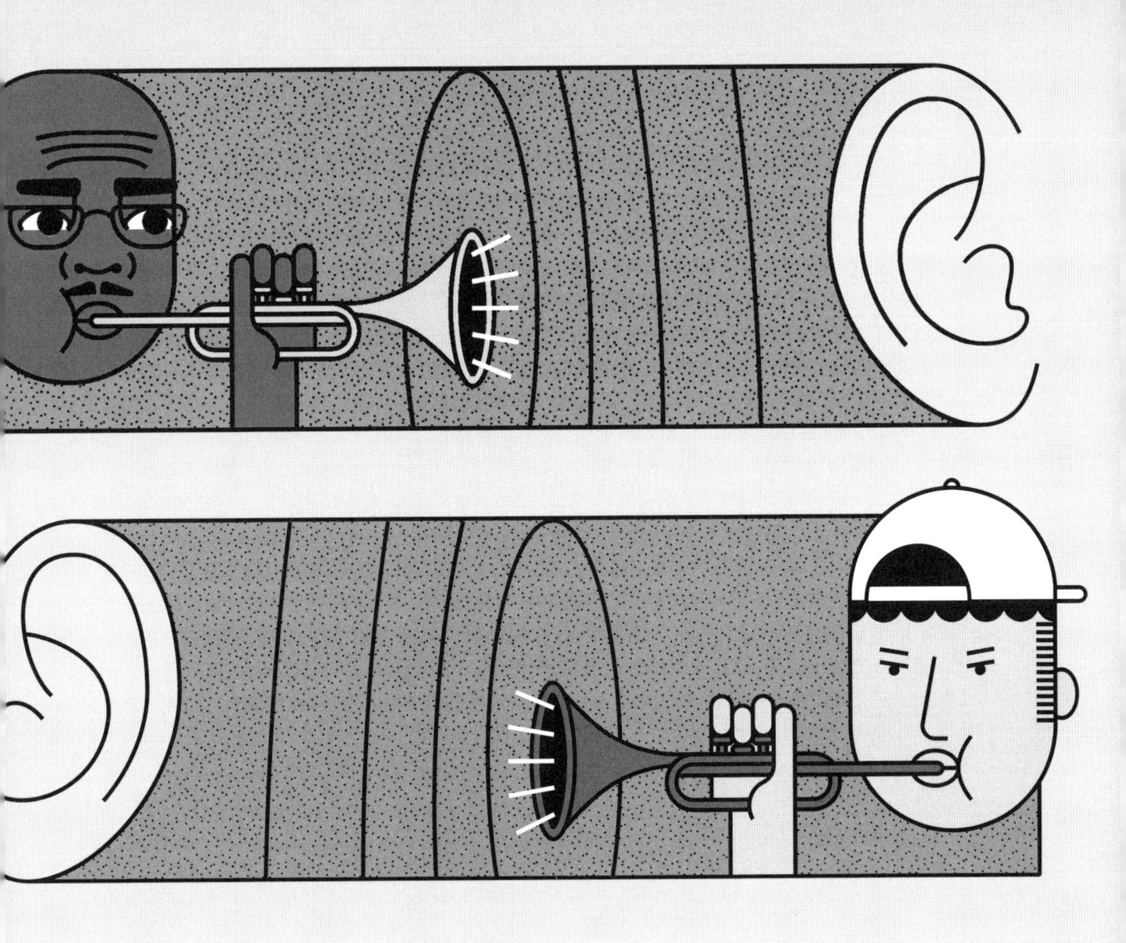

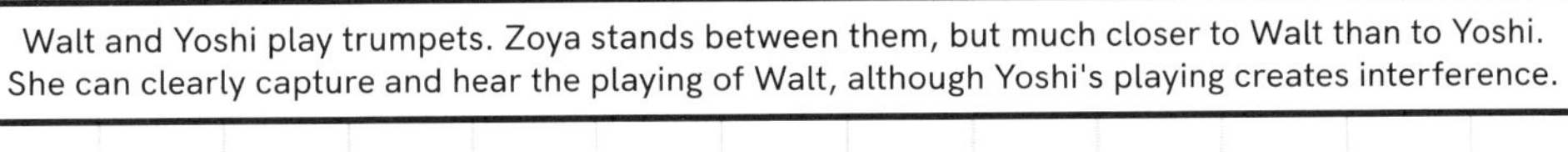

Walt and Yoshi play trumpets. Zoya stands between them, but much closer to Walt than to Yoshi.
She can clearly capture and hear the playing of Walt, although Yoshi's playing creates interference.

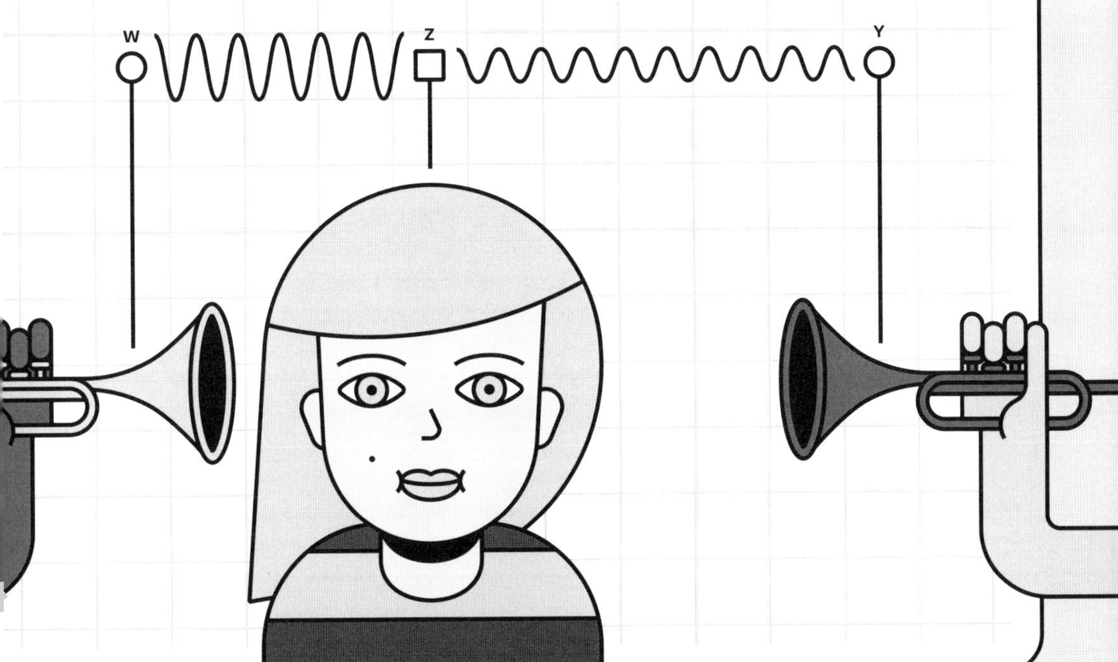

Story by Petar Popovski / Art by Peter Gregson

3
Access Beyond the Collision Model

The collision model is very useful for introducing the basic ways in which a wireless medium can be accessed and shared among several communicating nodes. Yet, recalling the speech analogy, one would expect that when a single node is transmitting and there are no collisions, the quality of the communication gradually decreases as the distance between the transmitter and the receiver increases. In the cartoon that illustrates this chapter, Walt and Yoshi are both playing a trumpet simultaneously, but the listener Zoya is much closer to Walt. Then there is, practically, no collision as the music played by Walt suppresses the weak tones that come from Yoshi, such that Zoya captures only the sound of Walt. This type of dependence on distance is not captured by the collision model.

Another idealized assumption adopted in the collision model is that each packet is an atomic, indivisible unit of information. This results in a rather pessimistic interference model, in which any time overlap of the packets destroys the involved packets and the receiver cannot recover any of them. This disagrees with the intuition obtained from the speech analogy. Namely, even when a part of the word or a sentence is interfered by another speaker or sound, the listener may still be able to recover the word or the sentence.

In this chapter we deal with models that take a look inside the collisions and the packets. The premise of these models is that not every collision is a waste and/or that the packet is not an atomic unit for carrying information. Following the intuition gained from the distance effects in the conversation analogy, the collision model will be enriched to bring distance into the picture. We will also look into its internal structure and see how this can lead to new operation modes that are not attainable under the simple collision model.

3.1 Distance Gets into the Model

3.1.1 Communication Degrades as the Distance Increases

The dependence of the data rate on the distance, depicted in Figure 1.1(a), is a very crude approximation of the physical reality. In revising this model, we start by still keeping the packet with $D = RT$ bits as the elementary, atomic unit of information that can be sent or received over the wireless medium. This means that we still keep the assumption that no part of the packet can be treated as a separate entity. The objective is to to build a model in which the communication from Zoya to Yoshi gracefully degrades as the distance between them increases.

Wireless Connectivity: An Intuitive and Fundamental Guide, First Edition. Petar Popovski.

Since the packet can either be received correctly or not, there is no possibility of receiving correctly only some of the bits within the packet. The data throughput that Yoshi receives during a given transmission of duration T is either R, if the packet is received correctly, or 0 otherwise. In order to get around this rather limiting situation, instead of observing a single transmission slot, one can observe a long sequence consisting of L packet transmissions sent by Zoya. The reception of Yoshi can now be treated as a *random event*: in a given transmission slot, Yoshi does not receive the packet correctly with probability p_e or receives it correctly with probability $(1 - p_e)$. If the observed packet sequence is sufficiently long and L is large, then approximately $L(1 - p_e)$ packets are received correctly. Then the *long-term average* throughput measured by Yoshi during LT seconds is

$$\text{Th}(p_e) = \frac{L(1 - p_e)RT}{LT} = (1 - p_e)R. \tag{3.1}$$

The probability p_e is a parameter that can be made dependent on the distance d between Zoya and Yoshi. In that case we can write it as a function of the distance $p_e(d)$ and this probability should increase as the distance d increases. This is a simple way of enriching the communication model in order to bring in the required feature of *graceful degradation* of the quality of the link between Zoya and Yoshi. In analogy to a conversation, graceful degradation corresponds to the decrease in speech comprehensibility as the speaker and the listener become more and more separated.

We can choose the function $p_e(d)$ in various ways, respecting the fact that it should increase monotonically from 0 to 1 as the distance d increases. However, electromagnetic radio waves propagate, in theory, until infinity and there is always "something" from Zoya's signal that arrives at Yoshi. This may lead us to conclude that $p_e(d)$ can attain the value 1, that is, certain error, only when the distance d between them is infinitely large. This counters the idea that the communication range is always finite and we need to look again at the meaning of the statement *"Yoshi is in the communication range of Zoya"*.

It should be noted that the transmission model from Figure 1.1(a) actually uses a particular form of the increasing function $p_e(d)$, defined for a communication range d_c. Here $p_e(d) = 0$ when Zoya and Yoshi are in communication range $d \leq d_c$, while $p_e(d) = 1$ when they are out of range, determined by $d > d_c$. In order to have a meaningful definition of a communication range, there must be a finite distance d_c such that packet errors occur with certainty when $d > d_c$. In practice such distance is determined by the minimal signal strength that is required for the receiver Yoshi to be able to detect the signal sent by Zoya. Wireless receivers are made such that they start to decode the received data only if the received power[1] is above some threshold P_{th}. This threshold is determined in a way to distinguish the signal from the always present noise. Hence, at distances d for which the signal strength is much lower than the noise, we can confidently state that $p_e(d) = 1$. However, this definition assumes that the communication range depends on the noise strength at the receiver. If we do not adopt the usual assumption the noise strength is uniform in space, then, in principle, the communication range d_c can depend on the actual geographical location of Zoya and Yoshi. We will not introduce that complication in this chapter, but we would like to illustrate that the model enrichment requires additional assumptions to

---

1 For the moment we do not have a working definition of signal and noise power (see Chapter 5), so the reader needs to rely on her or his intuitive, everyday understanding.

retain consistency. In fact, in this chapter we do not assume any specific model for signal propagation and power decrease with the distance, we only state the rather obvious fact that in practice the power must decrease with the distance. When the transmitting power is P, then there is a certain distance at which the received signal power drops to the value P_{th}. In the following, we use d to refer to the communication range in the sense described above.

3.1.2 How to Make the Result of a Collision Dependent on the Distance

Even in the previous chapter our discussion was not completely faithful to the collision model with a strict communication range. Recall that, while introducing the ideas of carrier sensing in Section 2.3.2, we referred to a model in which the carrier sensing range is larger than the communication range. In other words, the interference range is larger than the communication range. This feature is consistently supported by defining the communication range d_c to be the distance at which the received power is equal to P_{th}. Any signal received beyond the distance d_c has a power that is too weak to be useful for communication, but it may be sufficiently strong to cause interference and disturb another communication signal. However, considering that a signal carries energy even to very large distances, the use of the strict collision model would set the limit to have, at a given time, only one transmitter in the whole universe in order to have a successful packet reception. This, of course, does not make much sense and henceforth we need to amend the collision model.

We will use Figure 3.1, where Zoya and Xia transmit simultaneously packets to Basil. The distances of Zoya and Xia to Basil are d_{ZB} and d_{XB}, respectively. The receiving outcome at Basil is random and the probability of a particular outcome depends both on r_{ZB} and r_{XB}. One of the following outcomes is possible:

(a) The packets cause collision with each other and what Basil receives is completely incomprehensible. This is a "real" collision, in a sense defined in the previous chapter, as both transmissions are wasted.
(b) Both packets are too weak to be received by Basil.
(c) One of the packets is received correctly, either Zoya's or Xia's. This is the *capture effect*. Note that this case embeds two possible sub-outcomes, one when Zoya's packet is received and one when Xia's packet is received.
(d) Both packets from Zoya and Xia are received correctly!

The left side of Figure 3.1 illustrates four different distance scenarios (a)–(d), where each scenario will *most likely*, rather than certainly, result in a specific outcome from the four described above. The fifth possible outcome is symmetric to Figure 3.1(c), by exchanging the roles of Zoya and Xia and bringing Xia closer to Basil. The cases (a)–(c) can be easily explained using only the speech analogy. However, this is not the case for the surprising outcome (d). We therefore turn to another analogy, described in the right side of Figure 3.1 where it is assumed that there is a piece of paper in which both Zoya and Xia want to write a message to Basil. Zoya writes "Loremipsum" and Xia writes "Pantareirei".

For the cases (a) and (b) we use both analogies, speech and writing. In (a), Zoya and Xia are both in range of Basil and each of them at approximately the same distance. Their voices

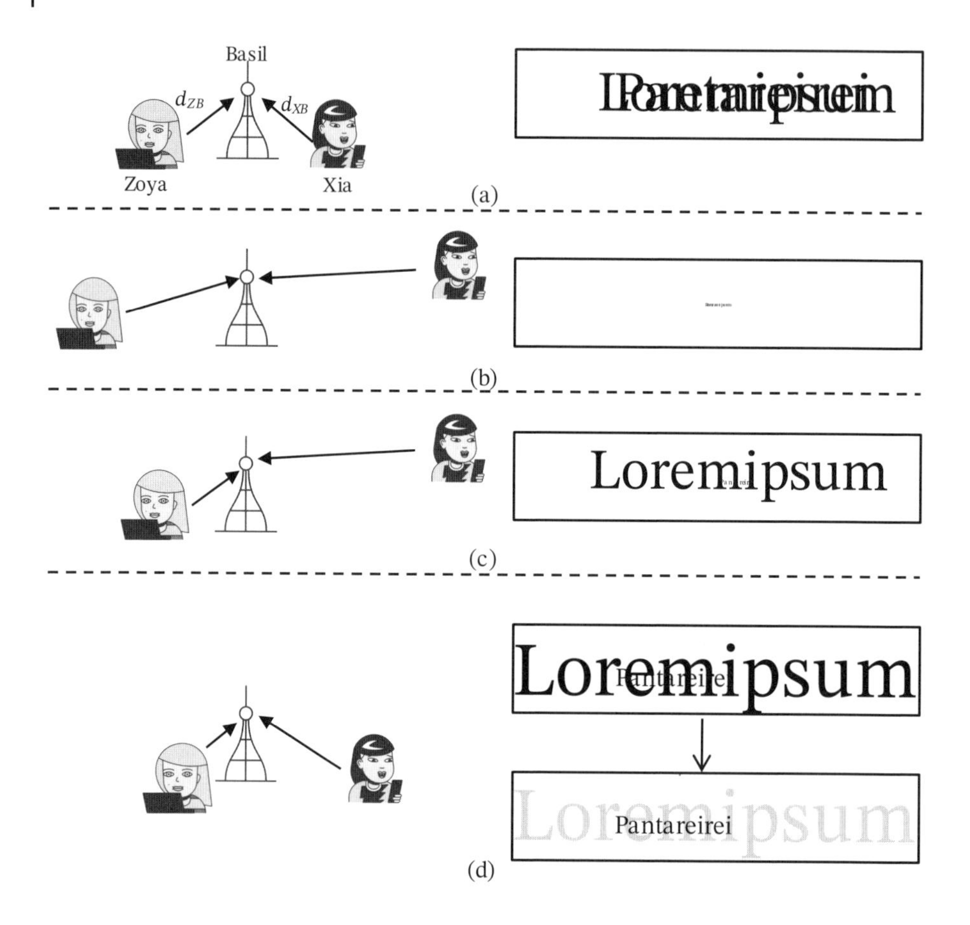

Figure 3.1 Zoya and Xia communicate simultaneously with Basil. The left side describes four distance scenarios, while the right side describes the most likely outcome via the analogy of Zoya and Xia writing the messages *Loremipsum* and *Pantareirei*, respectively, on the same piece of paper. (a) Collision. (b) Idle. (c) Capture effect. (d) Successive Interference Cancellation (SIC).

are equally strong and Basil receives them as garbled nonsense. Using the writing analogy on the right side of Figure 3.1(a), Zoya and Xia both use a font size of approximately the same size; the size is chosen such that the text is readable when either Zoya or Xia writes alone on the paper, but Basil cannot recover any of the texts when both of them have written on the paper. In Figure 3.1(b), Zoya and Xia are far away from Basil and he can hardly, if at all, hear their voices. This corresponds to small font sizes and unreadable messages. The messages would stay unreadable even if the texts are not overlapping. The capture effect is illustrated in Figure 3.1(c), where the message "Loremipsum" of Zoya is readable and can be "captured" by Basil, undisturbed by the tiny message written by Xia. This is the situation in which Zoya is standing next to Basil and talking, while Xia is talking from far away. Another similar outcome occurs when Xia and Zoya exchange their roles.

In Figure 3.1(d), both Zoya and Xia use a font of a readable size, but the font used by Zoya is much larger. If Zoya and Xia talk to Basil, but not simultaneously as on Figure 3.1(d), then Basil can get both messages, hearing Zoya with a really loud voice. Basil can then easily capture or read the message of Zoya and, afterwards, erase this message from the paper. After erasing it, even non-ideally, Basil can read the "Pantareirei" message of Xia. This illustrates the important idea of *successive interference cancellation (SIC)*, by which two or more messages transmitted simultaneously can be received correctly. Since in this case Basil receives two packets simultaneously, this is also known as *multi-packet reception (MPR)* capability.

The illustration in Figure 3.1 shows the most probable outcome for each scenario, which does not imply that the other outcomes are impossible. For example, in case (c) it may happen that both packets are received correctly. In order to build the desired model, for each possible pair of distances (d_{ZB}, d_{XB}) we need to specify five probabilities p_1, p_2, p_3, p_4, p_5, corresponding to the five possible outcomes[2]. Furthermore, to write it compactly, we can define a two-dimensional vector of distances $\mathbf{d} = (d_{ZB}, d_{XB})$ and define each probability $p_i(\mathbf{d})$ for $i = 1 \dots 5$ as a function of $\mathbf{d}$, similar to the way we defined $p_e(d)$ for a single transmission.

3.2 Simplified Distance Dependence: A Double Disk Model

In the context of the simple collision model, two important parameters for a communication protocol are the range and the data rate. Recall that in the simplest collision model, the communication range coincides with the interference range. The price paid for using the enriched wireless model, which incorporates distance effects, is seen in the significant increase in the complexity for specification of the model and the associated communication schemes. In the enriched model one needs to specify at least the following probabilities as functions of the device distances:

- Probability of reception when only Zoya transmits to Basil from a distance d.
- Five probabilities of reception outcomes when both Zoya and Xia transmit to Basil from distances given by $\mathbf{d} = (d_{ZB}, d_{XB})$.

If, in addition to Zoya and Xia, there are more nodes in the wireless network, then it cannot be guaranteed that at most two nodes will transmit simultaneously. Therefore, the model needs to be further specified through additional probability values. Let there be K mobile devices $MD_1, MD_2, \cdots MD_K$ connected to Basil and let their respective distances to Basil be given by $d_1, d_2, \cdots d_K$. It is possible to have any subset of k nodes transmitting simultaneously, where $1 \leq k \leq K$. The remaining $K - k$ nodes are silent. An important question for specifying the model is the following: given the transmissions of k active devices $MD_1, MD_2, \cdots MD_k$, how many different outcomes can be observed by Basil? There are at least 2^k different outcomes, corresponding to the fact that the packet sent by the ith mobile device MD_i is either received successfully or not. However, the packets that are not decoded correctly can be perceived either as collision or as idle, where the latter happens for weak signals. Considering the exponential number of outcomes for given k and the fact that k can vary between 0 and K, the number of probabilities that needs to be specified becomes prohibitively high.

2 To be precise, only four probabilities p_1, p_2, p_3, p_4 need to be specified; p_5 is found as $p_5 = 1 - p_1 - p_2 - p_3 - p_4$.

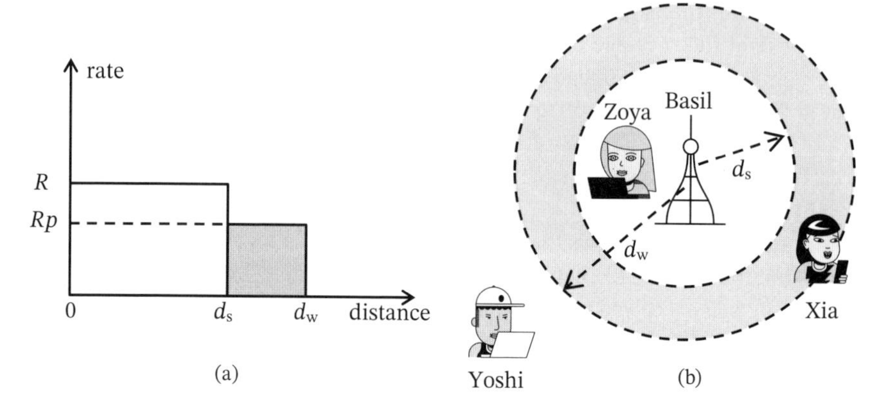

Figure 3.2 Enhanced communication model with a strong and weak region. (a) Dependence of the data rate on the distance. If the transmitter is placed in the weak region (gray shade), the receiver Basil gets the packet successfully with probability p. (b) The coverage area of Basil represented as a double disk. Basil receives strong signal from Zoya, weak signal from Xia, and no signal from Yoshi.

This clearly indicates that the introduction of distance dependence in the communication model comes with significant increase in complexity. It is possible to create models that have a lower complexity, but also contain a lower level of detail. This model can still be useful and capable of reflecting the physical reality better than the simplest model from Figure 1.1(a). Here is an example how such a model could be created. Instead of using a single disk to represent the communication and the interfering range, as in Figure 3.2(a), one can define a *strong reception region*, for distances up to d_s, and a *weak reception region*, for distances between d_s and d_w; this is reminiscent of the communication range and sensing range in the previous chapter. Then the wireless coverage area around Basil looks like a double disk, depicted in Figure 3.2(b). We use d_{ZB} to denote the distance between Zoya and Basil. We can consistently say that Basil receives a strong signal from Zoya if $d_{ZB} \leq d_s$, a weak signal from Zoya if $d_s < d_{ZB} \leq d_w$ and no signal if $d_{ZB} \geq d_w$. This model is *symmetric*: if Zoya receives a strong (weak) signal from Basil, then Basil will also receive a strong (weak) signal from Zoya when the transmitter/receiver roles are exchanged.

Using this terminology, the model can be specified by Table 3.1. The way to read the table is the following: for example, if the received signal from Zoya is strong, there is at least one weak interfering signal and no strong interfering signal, then Basil receives correctly Zoya's packet with probability p. One can easily argue that this model is not realistic. For example, the probability parameter p can vary depending on the actual distance and also different p can be used in the different events described in the table. Yet, even with this rather simple model, we obtain new insights and principles in designing the protocols, as the next section shows.

3.3 Downlink Communication with the Double Disk Model

In this section we reconsider the ideas and the design principles for downlink communication presented in the previous chapter by enriching the communication model. We adopt

Table 3.1 Specification of the reception outcomes at Yoshi when Zoya and possibly other devices transmit using the double disk model.

Other signals → Zoya ↓	None	≥ 1 weak, no strong	1 strong	≥ 2 strong
Strong	**Received, 1**	**Received, p**	**Collision, 1**	**Collision, 1**
Weak	Received, p	Idle, 1	Received, p if the strong is received otherwise 0.	Collision, 1

NB: Each table entry is a pair: "outcome, probability". For example, "received, p" means that Basil received the packet from Zoya successfully with probability p.

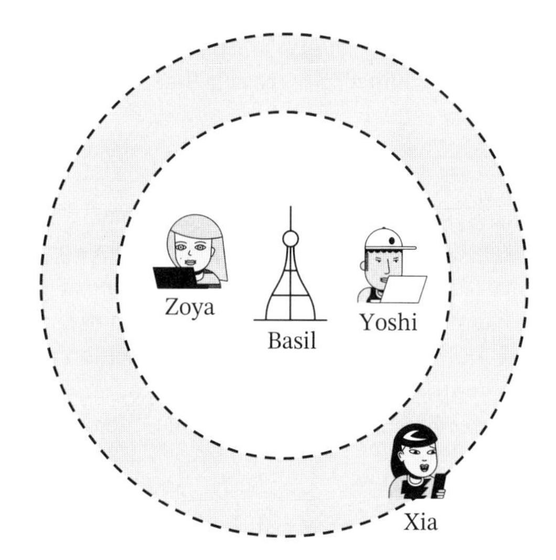

Figure 3.3 Communication system with a strong/weak region around the base station Basil.

the model of weak/strong reception region discussed above and we apply it to a communication system, in which a base station Basil communicates in the uplink or downlink with several devices, which we will interchangeably call Zoya, Yoshi, Xia or $MD_1, MD_2, \ldots MD_K$ for easier notation. This leads to the communication scenario depicted in Figure 3.3.

Consider downlink transmissions from Basil to the terminals. As illustrated in Figure 3.3, Zoya and Yoshi receive strong signals, while Xia receives a weak signal. In the downlink there are no collisions and Basil serves the terminals by using, for example, the frame structure from Figure 1.6. Any transmitted packet from Basil is perfectly received by Zoya/Yoshi, but Xia receives a packet from Basil with probability p. In practice, some packets to Zoya/Yoshi can also be lost, but the probability for that happening is much lower than the probability that Xia loses a packet, which is equal to $1 - p$. We can thus adopt the simplification that the communication to Zoya and Yoshi is free of errors. The same observation should also be valid for the frame header, not only the data. Indeed, so far we have emphasized only the fact that the header has different content compared to the data packets, but otherwise it is still only a specific kind of packet sent over the wireless medium and the communication model should be applicable to it as well. If Xia does not receive the frame header, then she ignores the data content of the frame, since she cannot interpret properly whether there is a data packet intended for her.

These considerations lead to the conclusion that the information in the frame header is, in a way, more important than the actual data. Clearly, the end user is interested in the actual data, but in order to attain an acceptable performance for the overall communication protocol, one has to invest an effort in the control information or metadata, carried in the frame header. It is therefore of interest to find a way to send the frame header more reliably compared to the data, such that the terminals in the weak region can also receive it.

This is further clarified through the following example. Let the frame contain a header and three data slots, all of them allocated to Xia. If Xia receives the frame header correctly, then she receives each data packet with probability p, such that the expected number of received packets in that frame is $3p$. However, if Xia does not receive the frame header correctly, then all the data packets are discarded. This simple example shows that the expected number L_{p} of packets that Xia receives in a frame in which 3 packets are addressed to her is

$$L_{\mathrm{p}} = p \cdot 3p + (1-p) \cdot 0 = 3p^2. \tag{3.2}$$

A simple improvement would be to have a frame structure in which it is predefined that the header is repeated two (or more) times at the frame start. In this case, it would be sufficient for a terminal to receive at least one of the headers in order to be able to interpret the content of the data packets transmitted in that frame. If the header is repeated twice, then the probability that the header information is received by Xia is $1-(1-p)^2$, as Xia fails to receive the header only if she does not receive both copies of the header. Now the expected number of packets that Xia receives increases to:

$$L'_{\mathrm{p}} = (1-(1-p)^2) \cdot 3p + (1-p) \cdot 0 = 3p^2(2-p). \tag{3.3}$$

This is always higher than L_{p} from equation (3.2), since the multiplier $2-p \geq 1$.

However, the redundant header transmission brings an additional cost as it increases the metadata related overhead in the system. Using the notation from Figure 1.6, we see that the time spent in sending control information is raised from T_{FH} to $2T_{\mathrm{FH}}$, which is a time interval that could otherwise be used to send data. The expected goodput of a frame that carries three packets to Xia with a single header, denoted by Gp, and two frame headers, denoted by Gp′, is given by:

$$\begin{aligned} \mathrm{Gp} &= \frac{L_{\mathrm{p}} D}{T_{\mathrm{FH}} + 3T} = \frac{3p^2 D}{T_{\mathrm{FH}} + 3T} = \frac{3p^2 R}{\tau + 1} \\ \mathrm{Gp}' &= \frac{L'_{p} D}{2T_{\mathrm{FH}} + 3T} = \frac{3p^2(2-p)D}{2T_{\mathrm{FH}} + 3T} = \frac{3p^2(2-p)R}{2\tau + 1} \end{aligned} \tag{3.4}$$

where each packet carries D bits, the nominal data rate of a packet is $R = \frac{D}{T}$ and $\tau = \frac{T_{\mathrm{FH}}}{T}$ is a fraction that measures the overhead with respect to the useful data. When the fraction of the overhead τ is small, the loss in goodput is less significant compared to the gain in reliability. In this example we have repeated the header only two times, but for increased reliability the header can be repeated multiple times, thus inducing a respectively larger penalty in terms of goodput.

3.3.1 A Cautious Example of a Design that Reaches the Limits of the Model

The observations from the previous section are used to construct a cautious example of protocol optimization. The example is cautious as it illustrates that making optimization by chaining several "nice small ideas" may lead to significant, but unintended, consequences for the overall system.

We define two types of frames: one having a single header, while the other has two repetitions of the same header. Let us call the latter one the "repeated header". In a frame with a single header, downlink data is only transmitted to the devices or terminals that have strong signals, or, in short, strong terminals. Conversely, in a frame with a repeated header, the data is always targeted to one or more weak terminals. This leads to an optimized system goodput since the penalty due to the repeated header occurs only when the frame needs to be received by the weak terminals. However, one has to be careful in proposing such an optimization, as it turns out that it can bring only an insignificant benefit when time consumed by the frame header is negligible and thus the overhead $\tau = \frac{T_{\mathrm{FH}}}{T}$ is low, while the overall protocol operation can be fatally flawed, as explained next.

At first, we need to specify who needs to know what in order to enable optimized operation. On the one hand, Basil should somehow know that Zoya and Yoshi have a strong signal, while Xia has a weak signal. On the other hand, each of the terminals Zoya, Yoshi, and Xia needs to know whether his or her signal is strong or weak. However, in our model, the only way by which a terminal can know if it is weak or strong is to receive many packets from Basil and obtain a statistics about his or her link. Based on that, a device can make a simple estimation: if almost all of them are received correctly, then this device classifies itself as a strong terminal; if a fraction p of all the packets are received correctly, then it is a weak terminal. Hence, in practice, the system should always start to work with a repeated (robust) header in order to ensure that even the weak terminals can get it and thus estimate whether their channel is weak or strong.

After a sufficiently long time, and assuming that no terminal changed its position from the zone with the strong signal to the zone with the weak signal (or vice versa), then for the example in Figure 3.3, Zoya and Yoshi know that they are strong terminals, while Xia knows that she is a weak terminal. Furthermore, we assume that, after that long estimation period, Basil knows this as well[3]. After Basil decides to use optimized frames, he needs to inform the devices that he will start to use both single and repeated header frames. This needs to be done reliably, such that all involved communication nodes, base station and devices, have a common knowledge that an optimized frame is being applied for system operation.

Now the important question is: how does a terminal differentiate between a single and a repeated header? The key mechanism that enables that type of operation is the common

3 It is interesting to note that this knowledge can be obtained by Basil only by having feedback transmission from the devices to Basil through uplink transmissions. In other words, we cannot speak strictly only about downlink transmissions, which is the focus of this section.

knowledge of Basil and the devices. For example, Basil knows that Zoya has a strong signal and Zoya knows that Basil knows it (sounds convoluted, but it is indeed required!). Then, the terminals operate as follows:

- *Strong terminal.* Zoya decodes the first header of the frame. If the header does not indicate Zoya as a destination, the rest of the frame is ignored. The same holds in the rare situation in which the header is not received by a strong terminal.
- *Weak terminal.* Xia tries to decode the first header. The following outcomes are possible:
  1. Header decoded, but not addressed to Xia. The rest of the frame is ignored.
  2. Header decoded and addressed to Xia. Xia knows that the header is repeated and starts to decode the data that comes after the second header.
  3. Header not received correctly. Xia tries to decode the second header; if the second header is targeted at Xia, then she proceeds to decode the data. For any other outcome, Xia ignores the rest of the frame.

Problems can potentially arise in step 3 for a weak terminal. One can think about the following situation: Basil sends a single header frame with data to a terminal with a strong signal (say, Zoya) and Xia does not decode the header. Then Basil proceeds to send data to Zoya, while Xia tries to decode the second header. The system needs to be designed in a way that data intended for Zoya cannot produce something that to Xia looks like a valid header. Indeed, if the data targeted at Zoya can be any combination of bits, then there is a specific combination of bits that is equal to the content of the second frame header that would indicate that there is data for Xia that follows after the second header. Due to this, there should be restrictions in the ways the bits are coded into packets, but our current model is unable to deal with it, as our atomic unit is still the packet, without looking into how the content in the packet is encoded. Evidently, at this point we are reaching the limit of our model and any further discussion of the packet content needs to be put in a context of an enriched model, which further unwraps the packets and considers the actual bits and bytes that are used to build the packet.

Besides reaching the limit of our model, this optimization example brings other valuable lessons. First, an intuitively good idea for system improvement can lead to many additional requirements in the system operation, such that one has to judge if the benefit obtained from the optimized operation is worth such an effort. Second, we were focusing on optimizing the downlink transmission, but that required multiple reliable uplink–downlink transmissions and careful considerations about who knows what. This fact should not be overlooked when practical protocols are designed.

3.4 Uplink Communication with the Double Disk Model

The change of the communication model has a more significant impact on the uplink transmissions. This is because, besides the dependence on the range, the transmissions of the terminals may be uncoordinated and carried out through a random access protocol. This opens up the possibility of causing collisions, such that the generalizations introduced in treating collisions start to have an effect. Note that we have not used those generalizations for the downlink transmission. For the purpose of this discussion, we will work with the model described in Table 3.1.

We first look at the case without collisions, where Basil allocates the transmission rights to the terminals at the beginning of the frame, for example using the frame header H_{11} described in Section 1.3.3 and Figure 1.7. Similar to the downlink case, Xia has a weak signal and thus she might not receive the frame header that is granting her rights to transmit in one or more slots of the current frame. This reiterates the need to make the headers more reliable for the weak terminals. The difference is that here the header that Xia receives in the downlink controls her behavior as a transmitter in the uplink, not her reception of the downlink data further on in the frame. Xia transmits only if she receives the frame header correctly *and* she has been allocated a slot in that frame. Basil can interpret the lack of response from Xia in two different ways:

1. Xia did not receive the frame header correctly, or
2. Xia has no uplink data to send.

In order to differentiate between the two cases, there could be a dummy packet with a predefined content that Xia can use to indicate that she does not have data to send. The differentiation between cases 1 and 2 above is important for efficient system operation. For example, if Basil knows that Xia does not have data to send, then he could predict that she does not have data to send in the upcoming frame as well, thus avoiding unnecessary resource waste.

Things get complicated by the fact that the dummy packet could be erroneously received by Basil such that Basil cannot be sure what has exactly happened in his communication with Xia. One can also think of the dummy packet as a response from Xia to a ping message from Basil, indicating that that she is "alive" and still in Basil's communication range, i.e. within his coverage area. If a predefined maximal number L_{max} of pings are not responded to by Xia, then Basil assumes that Xia is no longer within his coverage area. There is a trade-off in choosing the value of L_{max}. If L_{max} is too low, then there is an increased probability of false disassociation, where Basil concludes that Xia is not in range, while she still is. The cost is that, after being disassociated, Xia needs to spend communication resources on establishing the connection again. On the other hand, if L_{max} is is large, then it would be very improbable to miss all L_{max} pings even if Xia had a weak signal. The cost is that, if Xia goes outside Basil's cell, Basil still wastes a lot of resources to ping Xia. To summarize, whenever the protocol needs to deal with mobile terminals, such as Xia, there needs to be a finite L_{max} used to decide if the mobile terminal is still connected or not.

Another important issue that needs to be addressed occurs when Xia receives a packet with errors. The questions that arise in that case are: (i) what does Xia observe when she receives an erroneous packet and (ii) how does that compare to the case in which no packet was received at all. In short, if Xia is capable of differentiating between "something, but with error" and "nothing", then this could help in the following way. If Xia is in the weak signal region and receives the frame header correctly, then she responds in the allocated slot with data or a dummy packet. If Basil receives "something, but with error" rather than "nothing", then he still observes that something was received and knows that Xia is still in the communication range.

3.4.1 Uplink that Uses Multi-Packet Reception

The introduction of multi-packet reception (MPR) capability opens up new possibilities for protocol design. In the collision model, Basil is always trying to allocate the uplink resources in order to ensure collision-free transmissions. Using the model from Table 3.1, an idea would be to schedule the transmission of a strong and a weak terminal jointly, in the same data slot. If Basil asks Zoya and Xia from Figure 3.3 to transmit in the same slot, then the following events are possible:

- Zoya's packet is not received correctly and thus Xia's packet is not received. The probability that no packet is received is $1 - p$.
- Zoya's packet is received correctly, Basil applies SIC, but Xia's packet is not received correctly. The probability that only one (Zoya's) packet is received is $p(1 - p)$.
- Zoya's packet is received correctly and Xia's packet is received correctly after a SIC. Thus, both packets are received with probability p^2.

It can be calculated that the expected number of data packets that are decoded when both Zoya and Xia are sending in the same slot is:

$$0 \cdot (1 - p) + 1 \cdot p(1 - p) + 2p^2 = p(1 + p)$$

which is greater than 1 whenever the probability of successful reception p is greater than 0.618. Thus, the average goodput can be improved if we take an advantage of the collisions.

This line of reasoning can carry over to the random access protocols as well. Under the simple collision model, the objective of the random access algorithms, such as ALOHA or probing, is to maximize the probability of getting a transmission from a single terminal. However, with the enriched model that allows multi-packet reception, this objective should be changed. Consider the example where the expected number of received packets from a collision in which two nodes are involved is higher than 1. Let there be K contending nodes; then the size of the ALOHA frame should not be chosen to be equal K, but lower in order to increase the probability of collisions, such that each collision will, with certain probability, produce one or more successfully decoded packets for the receiver.

3.4.2 Buffered Collisions for Future Use

The concept of SIC can have implications even when the protocol operates under a model that can be obtained with only a mild alteration of the collision model. This is done as follows. Zoya and Xia are both in range of Basil, they transmit simultaneously, and cause collision. The received collision is not dropped, but is rather *buffered* by Basil. The fact that we allow this type of buffering is another type of enrichment of the collision model. Referring to Figure 3.4(a), the collision happens in slot 1. In slot 2, Zoya transmits again, Xia stays quiet and Basil receives packet **Z** from Zoya correctly.

Recall that in the SIC mechanism described in relation to Figure 3.1(d), the first signal is decoded successfully from the collision, then it is canceled from the *same* collision and finally the second signal is decoded. In order for such SIC to take place, one of the colliding signals needs to be sufficiently stronger than the other signal, as explained in relation to Figure 3.1(d). If they are of approximately the same strength, then it is impossible for SIC to take place. Let us call this type of SIC an *intra-collision* SIC, where one signal is decoded successfully from the collision and then canceled from that same collision. However, there

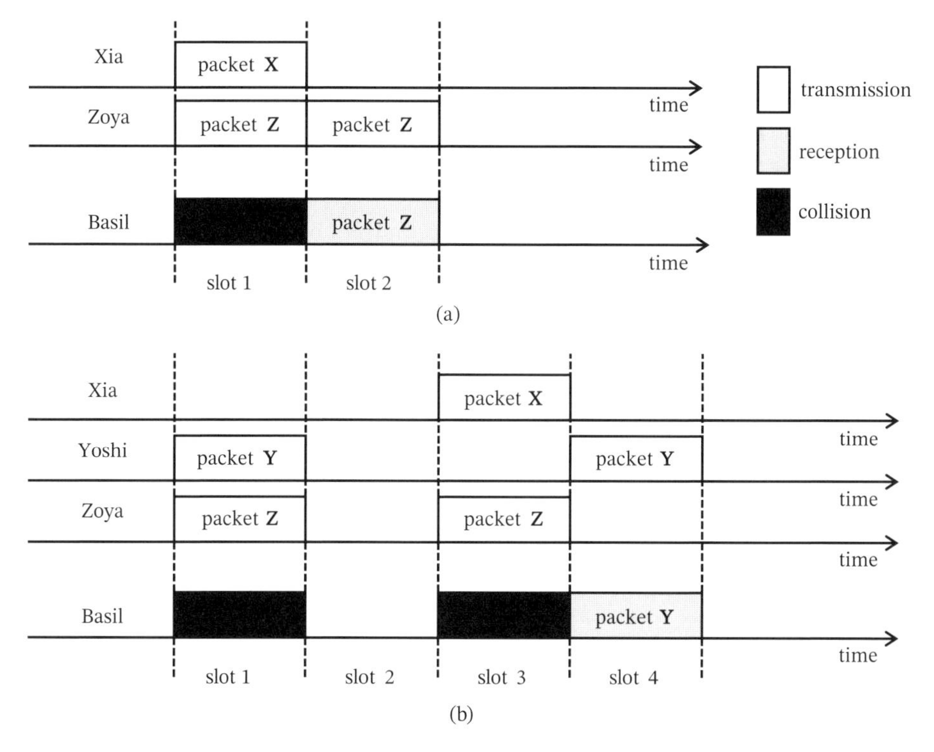

Figure 3.4 Illustration of the use of SIC in the collision model. (a) Collision of two packets; the throughput with SIC is 1. (b) Collision among multiple packets. Decoding and SIC starts in slot 4, using the buffered collisions from slots 1 and 3, resulting in throughput of $\frac{3}{4}$.

is an alternative way; let us call it *inter-collision* SIC. Consider now the collision of two strong signals in Figure 3.1(a), which is received and buffered by Basil. Let us assume that, after that collision is received and saved, Basil somehow learns the packet **Z** send by Zoya; then he can go back to the buffered collision, cancel the packet **Z** and then try to decode the packet **X** of Xia without being disturbed by any interference. Hence, the collision is buffered in order to be used in the future, in relation to another received packet. This is the basis for a scheme originally proposed under the name *contention resolution diversity slotted ALOHA (CRDSA).*

This is precisely what happens after Basil decodes the packet **Z** in slot 2 in Figure 3.4(a): he can use this packet to remove the interference from the packet **Z** from the collision buffered in slot 1. However, it should be noted that after decoding the packet **Z** in slot 2, Basil also needs to know that the packet **Z** had been previously involved in the collision buffered in slot 1. This issue can be solved by assuming that each transmitted packet carries a pointer towards the other slots in which this packet appears[4]. As a result, Basil receives both packets

4 Strictly speaking, if the copy of the packet **Z** sent in slot 2 contains a pointer to slot 1 and the copy of the packet **Z** from slot 1 contains a pointer to slot 2, then the replicas of the packet **Z** from slots 1 and 2 are not completely identical. One way to address this inconsistency is to assume that each packet contains pointer to all the slots where a replica of the packet **Z** is sent. In this case, the packet **Z** contains pointers to slot 1 and slot 2. Note that this is only one possible way of implementation and there are other ways as well.

successfully after two slots, resulting in a throughput of 1, which is the same as if Zoya and Xia were coordinated to send the packets one after another!

Of course, there is a price to be paid, as Zoya transmits twice instead of once. In general, in order to take advantage of the described SIC mechanisms, at least some of the nodes need to repeat the packet multiple times, which is different from ALOHA, where in each attempt only one packet transmission is made. Figure 3.4(b) shows a more general setting where the idea of SIC is useful. Zoya, Yoshi, and Xia transmit in a time period of four slots. Zoya and Yoshi transmit two replicas of their packet and Xia only one. Basil receives collisions in slots 1 and 3 and decodes the packet **Y** of Yoshi in slot 4. If no SIC is used, then the throughput achieved over the four slots is $\frac{1}{4}$. On the other hand, if SIC is applied, the operation of Basil proceeds as follows. After decoding packet B from slot 4, he removes this packet from the buffered collision in slot 1 and decodes the packet **Z**. Finally, he removes the packet **Z** from the buffered collision from slot 3 and decodes the packet **X**. As a result, the achieved throughput is $\frac{3}{4}$, three times higher for this particular instance.

The use of SIC in the collision model is very suitable for an access protocol based on probing. To see this, consider the following example:

- Slot 1: Zoya and Yoshi transmit, Basil memorizes the collision and probes the address 0. Assume that Zoya flips a coin and gets 0, while Yoshi gets 1.
- Slot 2: Zoya transmit successfully. Basil uses the packet of Zoya, cancels it from the buffered collision and extracts the packet **Y** of Yoshi.

In order to see SIC in action in an actual instance of a probing protocol, we can go back to the example in Figure 2.2 and see how the idea of using SIC can reduce the total number of slots required to resolve all eight sensors. After getting a successful reply from s_2 in slot 5, the probe 011 does not need to be sent as the packet from s_3 can be recovered from the collision buffered in slot 4. Continuing in a similar fashion, the reader can find out that the other tree nodes that can be skipped are the ones labeled with 10 and 17, such that the collision is resolved in 14 instead of 17 slots.

Both in the cases of ALOHA and probing, the fact that Basil has a more advanced receiver can at least partially compensate for the losses in efficiency due to collisions. The main difference between using SIC in ALOHA and probing can be explained as follows. The probing procedure goes inside a collision and tries to resolve it through a sequence of recursive collisions. Such an operation makes it natural for the probing to use buffered collisions and SIC. On the other hand, in the ALOHA setting, the transmitters need to send multiple replicas of their packets in order to make the buffered collisions useful and enable a SIC process.

3.4.3 Protocols that Use Packet Fractions

The central feature of the collision model is the all-or-nothing approach to the packet reception: even if two colliding packets are not completely overlapping in time, all data from both packets are lost. Consider now the situation in Figure 3.5(a), where Zoya and Yoshi transmit to Basil. Zoya repeats her packet twice. The packet of Yoshi has the same duration as the packet of Zoya and it overlaps with both packets sent by Zoya. The collision caused by Yoshi's packet **Y** divides the packet **Z** of Zoya into two parts, denoted by **Z**1 and **Z**2, respectively. This division is identical for both copies of the packet **Z** sent by Zoya. However, the

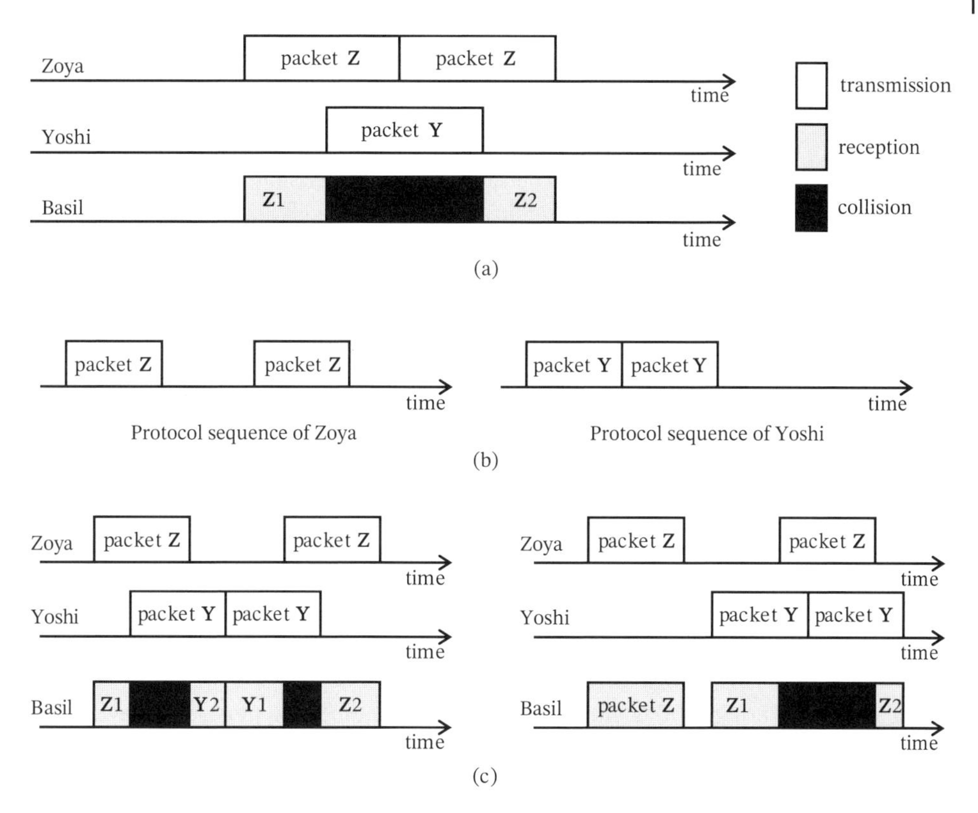

Figure 3.5 Strategies for communication in a collision channel without feedback that use fractions of the packets that are not destroyed by collisions. (a) Recovering the packet of Zoya from two non-collided packet fractions **Z**1 and **Z**2. (b) Protocol sequences for Zoya 1010 and Yoshi 1100. (c) Two examples of possible collisions between Zoya and Yoshi at the receiver of Basil that still allow recovery of both packets **Z** and **Y**.

part **Z**2 of the first transmission of **Z** is garbled due to the collision, while the part **Z** of this first replica remains unaffected. The situation is opposite for the second transmission of **Z**. This leads to the following idea: the receiver can take **Z**1 from the first packet and **Z**2 from the second packet, concatenate them together and get a single correct copy of the packet **Z**. The only additional requirement is that the receiver is able to differentiate between the collided and the healthy, non-collided parts of the packets.

To see where this feature can be useful, we take the example of a scenario where Zoya and Yoshi are two simple sensors that have transmitters but not receivers, such that they cannot coordinate between themselves. Another thing that aggravates the communication scenario is that the clocks of Zoya and Yoshi are not completely synchronized and they are drifting over time. This is known as a *collision channel without feedback*, since Basil cannot use feedback to indicate collision/success. How can one then ensure that the collisions between Zoya and Yoshi will still result in decodable packets? We can use the concept of packet fractions and assign suitable *protocol sequences* to Zoya and Yoshi. The protocol sequence of Zoya is 1010, where 1 means that a packet is sent, while 0 means that Zoya stays silent (pause) for a time equal to a packet duration. When using transmission in a given protocol sequence, for each 1 in the protocol sequence, Zoya sends a copy of the same packet **Z**. The protocol sequence of Yoshi is 1100. Figure 3.5(b) illustrates the transmissions sent with these protocol sequences. Due to the lack of synchronization, the start of Zoya's protocol sequence can be arbitrarily positioned with respect to the start of Yoshi's sequence. Figure 3.5(c) shows two different relative positions: in both cases the receiver is able to get healthy packet fractions in order to assemble full packets for both Zoya and Yoshi.

The idea of using protocol sequences can be generalized to the case when there are more than two nodes in the system. The essential assumption that makes this idea work is that the packet length is equal to the pause length and they are identical for all the nodes.

The concept of SIC can bring further improvement to the idea of using packet fractions. In Figure 3.5(a) it can be noted that the receiver can recover the packet of Zoya from the healthy parts, then it can use these parts to subtract **Z**1 and **Z**2 from the collided parts and thus recover the packet of Yoshi. In this case Yoshi does not need to send two replicas of his packet, only one would be sufficient. The usage of SIC enables, in principle, designing protocol sequences that have different length for different nodes, as long as we can guarantee that from any possible collision all the packets can be recovered.

3.5 Unwrapping the Packets

In this section we bring another enrichment of the communication model. We unwrap the packets to reveal their internal structure in terms of constituent bits and their content. Note that the consideration of packet fractions does not need to care about the actual packet structure; for example, in Figure 3.5(a) one does not need to know the content of **Z**1 and **Z**2, as long as it is known that they are identical for both packet transmissions of Zoya. In other words, even if packet fractions are considered, the protocols considered so far do not require going to the packet structure at the bit level.

In the model that we have used, a packet of duration T sent at a data rate R is said to have $D = RT$ bits. What we want to show now is that, in order to send D information bits, the number of physically represented bits *must be larger than D*. Let us denote the array of data bits by **D**. Assume we need to send a single byte of information, such that the array **D** can have $2^8 = 256$ different values. The reception of this data is subject to errors, manifested by changing the value of a bit from 0 to 1 or vice versa. When Yoshi receives a packet, he extracts an array of $D = 8$ bits. However, since all 256 possibilities represent valid information, Yoshi cannot know if some of the bits have been flipped due to random errors in the communication channel.

This brings us to the concept of an *integrity check* for packets, which enables the receiver to decide if the received packet represents a valid chunk of data or not. This can only be achieved by making some of the received bit patterns invalid, such that if Yoshi receives such a bit pattern, then he knows that an error has occurred. Alternatively, we can keep all D information bits and implement an integrity check by adding C *redundant bits*, also called *check bits*. Let us denote the array of check bits by **C**. The check bits do not carry information, but C is determined as a function of the data bits **D**. The function that is used to create C from **D** is an *error detection code*. A packet is created by appending the C bits to the D bits and getting the resulting array [**DC**]. In fact, relating to to the previous notation, we can write that the packet sent by Zoya is $\mathbf{Z} = [\mathbf{D}_Z\mathbf{C}_Z]$. We use the brackets to denote a packet: the array of bits put in $[\cdot]$ does have an integrity check. For example, if $D = 8$ and $C = 2$, the total number of bits sent is 10. A group of 10 bits can have $2^{10} = 1024$ different values, but we now know that only 256 values are valid packets and represent valid information. If Yoshi receives a packet that does not represent valid information, then he knows that an error has occurred and the packet is dropped or perhaps used for future processing.

The check bits introduce a certain redundancy or overhead, such that the data rate R_C at which the packet $[\mathbf{DC}]$ is transmitted must be higher than the useful data rate (or goodput) obtained in the system:

$$R_C = \frac{D+C}{T} > \frac{D}{T} = R.$$

The efficiency of the check bits is better when R is closer to R_C and it can be represented by the ratio:

$$\frac{D}{C+D} \tag{3.5}$$

which increases as D becomes much larger than C. In practice, the size of the integrity check is in the order of several bytes. This is not significant when the payload is large, for example thousands of bytes, but it does represent a significant overhead when the payloads are of a size comparable to the data size D. The latter is relevant in a number of Internet of Things (IoT) applications where the communicating devices are sensors and actuators that transmit small data payloads.

The errors usually occur randomly and, if Zoya sends the packet $[\mathbf{DC}]$ there is always a possibility of being unlucky and getting a pattern of errors in both $\mathbf{D}$ and $\mathbf{C}$ that produces a valid packet $[\mathbf{D}'\mathbf{C}']$. In other words, the error pattern converts the correct packet into an incorrect packet $[\mathbf{D}'\mathbf{C}']$, but the error checking code is not capable of detecting it and will pass the validity check. Then Yoshi will conclude that the information sent by Zoya is $\mathbf{D}'$, which is not correct as $\mathbf{D}' \neq \mathbf{D}$. This event is called an *undetected error*. It cannot be fully avoided and it is therefore not possible to have perfect error detection. The fact that the error check is often considered perfect by researchers as well as engineers is that the error checking code has an extremely low probability for an undetected error; for example less than once in 10^{12} packets. The key point in achieving a low probability of undetected errors is the choice of a good error detection code. The design of of error detection codes has matured over the past decades and *cyclic redundancy check (CRC)* codes are commonly used to carry out an integrity check. There is a variety of CRC codes. For example, the International Telecommunication Union (ITU) standardizes several CRC codes and, for example, the CRC-8 code produces $C = 8$ check bits for data sequence $\mathbf{D}$ of any size.

3.6 Chapter Summary

The main underlying idea of this chapter has been to show how the introduction of a more complex, enriched model can give rise to algorithms and protocols with superior performance. Our way into building a more complex model consists of two steps. First, we enrich the model in a way that a collision is not necessarily treated as a waste, but it is possible to extract one or even several packets from the collision. The mechanisms that are put to work when the collision is generalized are the capture effect and SIC. The possibility for implementing SIC motivates redesign of the random access protocols by sending multiple packet replicas that facilitate the use of SIC. The second step towards generalization of the collision model has been to abandon the assumption that a packet is an atomic unit. This leads to access protocols that are designed to use packet fractions. We have also established the fact that not every arbitrary array of bits can be considered to be a packet, but only an array of bits that has a check of its integrity. The integrity check has a price, expressed through the amount of redundant check bits that need to supplement the data bits and thus together constitute a packet.

3.7 Further Reading

As mentioned at the end of the previous chapter, the area of random access is still vital and relevant in the context IoT connectivity. This chapter has illustrated the fact that extensions of the collision model lead to rich possibilities for protocol design. The model of capture was introduced very early on in Roberts [1975]. Working with packet fractions was one of the key ideas behind the work on collision channel without feedback, presented in Massey and Mathys [1985]. The use of SIC was introduced in splitting-tree type protocols in Yu and Giannakis [2007] and in ALOHA type protocols in Casini et al. [2007]. The latter led to a new line of research in coded random access; see, for example, Paolini et al. [2015]. The recent interest in massive IoT connectivity has resulted in novel information–theoretic models for uncoordinated access, see Polyanskiy [2017] and Chen et al. [2017]. The main premise in these latest works is that the packet abstraction used from a perspective of random access protocols cannot be decoupled from the content/code of the packet, as has commonly been done in random access protocols (and we have also done it in this and the previous chapter).

3.8 Problems and Reflections

1. *Generalized probabilistic parametrization.* Table 3.1 provides a very simplified probabilistic parametrization of the double disk model. For example, one of the simplifications is that the same parameter p is used both in the reception of the weak and the strong signal. Propose a different parametrization of the model and argue the choice of the parameters.
2. *Reception with strong and weak regions.* Section 3.3 discussed the possibility of frame header repetition in order to ensure that it is received more reliably by the terminals in in the weak region. Let the number of terminals in the strong region be K_S and in the weak region K_W. The probability that a device receives a packet, regardless whether it is a frame header or a data packet, is p. Analyze how should Basil choose the number of frame header repetitions as a function of p, K_S, and K_W in order to maximize the throughput.
3. *Full-duplex base station.* Consider a scenario in which the base station Basil has full-duplex capability, while all the devices connected to him are half-duplex. Describe a possible parametrization of the model, similar to Table 3.1, considering that Basil can decide to transmit or not while he is receiving.

4. *Combination of intra- and inter-collision SIC*. The capture effect with intra-collision SIC is a mechanism that is different from SIC used with inter-collisions. Discuss how these mechanisms can be combined in the design of a random access protocol.
5. *Integrity check for multiple packets*. The use of an integrity check and CRC introduces overhead. A way of decreasing this overhead would be to use a single integrity check applied to multiple packets. For example, when Basil transmits a downlink frame with packets for Zoya, Yoshi, and Xia, he can use single CRC for an integrity check of all the data packets in the frame. Analyze the pros and cons of this choice.

lass production of layered cakes...
ach cake layer is
repared in isolation
om the others...
... and has a specific role in the cake.
Cakes are served by slicing.

Story by Petar Popovski / Art by Peter Gregson

4

The Networking Cake: Layering and Slicing

Protocols are distributed algorithms that run at different nodes, such as devices or base stations. In the previous chapters we have been concerned with describing how communication algorithms and protocols work under different communication models. In this chapter we will deal with the following question: *which components should a wireless node have in order to run a given communication protocol?* Our approach is to start from the simplest possible system architecture that can support a specified communication protocol and show how it should gradually become more complex as the communication model becomes richer and the communication algorithms become more sophisticated. This will eventually lead us to a *layered system design*, omnipresent in digital communication systems, where each layer has well defined functionality. Layered system design is a prime example of a good architectural design that can lead to proliferation and support of various connectivity types and services. The way different services can use the resources of the layered architecture is through *network slicing*, which cuts through the layers to use the resources in a way that is customized for each specific service.

4.1 Layering for a One-Way Link

4.1.1 Modules and their Interconnection

The simplest place to start is the one-way connection, illustrated in Figure 4.1(a). The transmitting device Zoya has two modules: *DataSender* and *TXmodule*, where the latter stands for a transmitting module. The receiving device Yoshi has also two modules: *DataCollector* and *RXmodule* (receiving module). Each module has to perform a certain task and it has to communicate with the other modules in order to get the task done. We assume that new data is created locally in the DataSender of Zoya and the task of the DataSender is to transfer this data to the data receiver of Yoshi. The task of the TXmodule is to get data bits from the DataSender, prepare data packets, convert them into physical signals, and send them over the air. The RXmodule receives the data through wireless signals from the air, converts the physical signals into data bits and provides these bits to the DataCollector. The task of Yoshi's DataCollector is to get data from the RXmodule and use it in a meaningful way, as required by some applications.

Wireless Connectivity: An Intuitive and Fundamental Guide, First Edition. Petar Popovski.

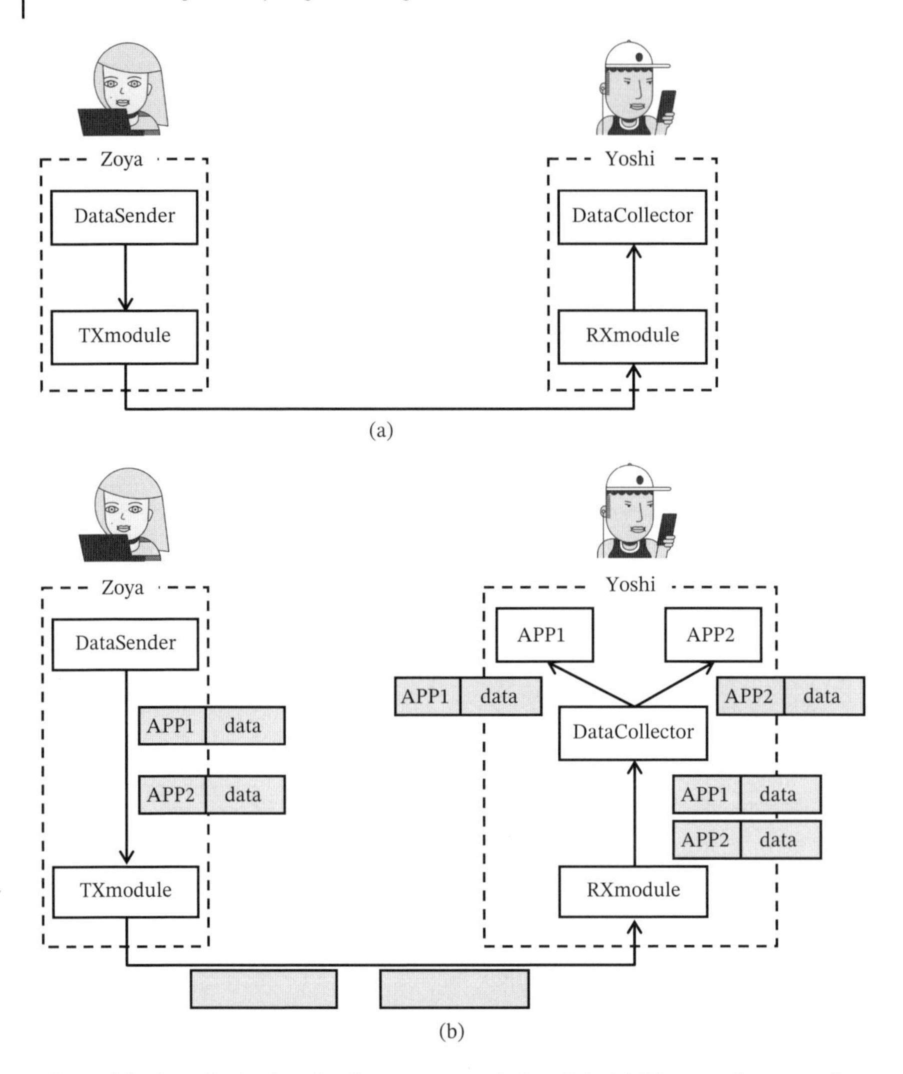

Figure 4.1 Introducing layering for a one-way wireless link. (a) Diagram of a system for one-way connection from Zoya to Yoshi. (b) Flow of packets in the system. The metadata intended for the two applications at Yoshi's side, APP1 and APP2, is not visible in the communication between the TXmodule and RXmodule.

The description in the previous paragraph is quite simple and reasonable, but at a second look it contains some imprecise and unclear statements. Note that we have considered the DataSender as a single module and treat it as a *black box*, neglecting the fact that it has an internal structure and sub-modules, which we have not described. These sub-modules (and their sub-sub-modules, etc.) can be defined in a large variety of ways, but we will introduce only the minimal level of detail in order to support our discussion. For example, in order to create new data, the DataSender must have a sub-module that represents a certain application used by Zoya. One way to carry out the communication task is to organize the new data in several packets and pass this data to the TXmodule. Nevertheless, passing the data to the TXmodule requires, in principle, *another communication link*, represented by the arrow between the DataSender and the TXmodule.

At this point it seems that we are entering into a recursion, since in order to describe a wireless communication link, we have to describe another communication link, required to support the wireless link. That is, the DataSender must have some transmitting sub-module and the TXmodule must have some receiving sub-module in order to get the packets from the DataSender. One can think of these different communication types as Russian dolls that keep appearing as we dig deeper into the structure of the system/modules. In order to avoid our description entering into a vicious recursive spiral, we need to make some assumptions about the local, inter-module communication link. We can think of this link as being a wire that connects the modules and every packet on the wire must come from the DataSender module and go to the TXmodule. Hence, there is no need for link establishment protocols and the packets going through the wire do not need to contain any source/destination address, since the source and the destination are predefined. The wire is perfectly reliable, such that the TXmodule can detect when a new packet starts and so receive all the bits of this packet successfully. In practice, it means that the reliability of the inter-module intra-device links is several order of magnitudes higher than the reliability of an inter-device link. And let us not forget that the objective of the whole modular, layered design is, in the first place, the functionality of the inter-device links. Having such a component of the system with pre-defined, almost perfect communication is not only a theoretical abstraction, but rather a necessary element for building more complex systems. We will return to this issue in Section 4.5.

If Zoya's TXmodule and Yoshi's RXmodule are predefined as a source/destination, then the wirelessly transmitted packets traveling through the air from Zoya to Yoshi do not need to carry any identification, since Yoshi's RXmodule knows that it needs to receive them all and hand them to the DataCollector of Yoshi. Furthermore, if no transmission errors are occurring, then the packets passed from the DataSender to the TXmodule will be delivered to the RXmodule in the same order, such that the DataCollector can perfectly reconstruct the original data created at Zoya's side. Under such ideal conditions, it is sufficient to have one-way connections only, as depicted in Figure 4.1(a).

4.1.2 Three Important Concepts in Layering

The packet on the wire between the sender and the TXmodule is different from the corresponding packet sent over the air, although both packets are carrying the same data bits. When there is a danger of ambiguity, we will use the term *data frame* to denote the packet sent wirelessly over the air, between the two devices. The difference is in the physical representation of the packets, such as voltage levels over the wire versus modulated radio waves, but also in the way the data bits are interpreted.

To illustrate this point, we take the example in Figure 4.1(b). Assume that Zoya and Yoshi are mobile devices that can run two different applications. The DataSender can generate two different data types, each of them corresponding to a different application. Each data type should be received by the corresponding application at Yoshi's side. Therefore, in addition to the actual data, the packets sent from the DataSender to the TXmodule need to contain metadata, put into a suitable header. This metadata is intended to be read and interpreted by the DataCollector of Yoshi, such that it can forward the received data bits to the appropriate application. However, the metadata inserted by the DataSender does not need to be interpreted by the RXmodule.

This simple example illustrates three important concepts in layering:

1. *Module reuse*. The same RXmodule can be used to carry data for different applications and different data types. One can think of each application as a different user of the RXmodule. In order to enable sharing of the RXmodule, the data bits need to be supplemented with metadata that describes which application is sending the current data bits through the RXmodule.
2. *Information hiding*. The RXmodule does not need to be aware of which bits represent metadata and which ones are used to represent data. In fact, the RXmodule does not need to interpret the bits, but it should simply "obey" and pass on the received bits to the DataCollector. Then it is the task of the DataCollector to interpret the metadata and deliver the actual data to the appropriate application. Hence, the metadata is only visible to the module that needs to use it and it remains hidden for the RXmodule.
3. *Service through a black box*. DataSender and DataCollector do not need to know how the actual communication between the TXmodule and the RXmodule is done, as long as the TXmodule and the RXmodule provide the data transfer service that they are asked for. Hence, the TXmodule and the RXmodule are seen as being in a (distributed) black box, where the DataSender uses the input of the black box and the data collector uses the output of the black box. The only thing that needs to be specified is the *interface* of the black box: how to provide the data bits to it at the input and how to read them at the output.

4.1.3 An Example of a Two-Layer System

Before going to a more general description of the layer functionality, let us look at an analogy for the system from Figure 4.1(a) in the old telegraphy system. The telegraphic system has two layers, a lower and a higher layer, respectively. The lower layer is represented by the two telegraphists at the two ends of the physical connection, the higher layer by two persons who want to exchange telegrams. The telegraphists are focused on the letters/symbols and their correct transmission/reception, while they do not (or, more correctly, should not), care about the words and the actual message content. To take the analogy further, let us consider the setup in which Yoshi dictates a message to the transmitting telegraphist. At the side of the receiving telegraphist, there are two different recipients, Zoya and Xia. Furthermore, let us assume that the receiving telegraphist does not know which of the persons is Zoya and which one is Xia. The received telegram contains the messages: "Zoya buy. Xia sell". The data are "buy" and "sell", while the metadata are "Zoya" and "Xia", respectively. Clearly, the Zoya and Xia can interpret the data correctly, while the telegraphist cannot.

This analogy contains all the three ingredients:

1. Module reuse: the telegraphist is used by two different connections: Yoshi-Zoya and Yoshi-Xia.
2. Information hiding: the telegraphists read letters, not words.
3. Black box service: Zoya, Yoshi, and Xia do not care if the telegraphist is right- or left-handed, whether the telegraphs are connected via a cable or a radio connection.

The system in Figure 4.1 is a general communication system for one-way connections that contains two layers. The lower layer contains the TXmodule and the RXmodule, which are *peers* with respect to each other. On the one hand, this means that they can communicate with each other. On the other hand, both modules are at the same level of hierarchy in the system. Allowing slight imprecision, we can informally say that the peer modules belong to the same black box, speak the same language, and can understand the bits in the same way. The higher layer contains two other peers, the DataSender and the DataCollector. The DataSender can communicate with the TXmodule, but they are not at the same hierarchical position in the system, as the DataSender can interpret some of the bits as metadata, while the TXmodule is unable to do so. The modules at the higher layer do not communicate directly, but through the connection that is established within the black box that resides at the lower layer.

Layering and layers are not specific to wireless networks, but rather an underlying paradigm for all communication networks. As we will see in the following text, the concept of layers stacked on top of each other, black boxes with well-defined interfaces and peers that communicate within each layer, is very powerful and essential in building communication protocols for large systems.

4.2 Layers and Cross-Layer

The simple models in Figure 4.1 can support one-way connection in the absence of errors and various other surprises that can emerge from the underlying black box. Here by "surprises" we refer to behaviors that deviate from the expected, normal behavior of the black box that resides in the lower layer. The black box at the low layer will be referred to as LoLaBBox.

In order to illustrate the impact of those surprises, let us take the following example. Let the DataSender of Zoya work under the assumption that the packets that are delivered in a certain order to the input of the LoLaBBox are delivered in the same order at the output of the LoLaBBox. Then, clearly, the "surprise" for the DataSender/DataCollector occurs when the LoLaBBox does not behave in this way and does not deliver the packets in the same order. This can happen, for example, if we allow a larger freedom in implementation of the black box.

Let us take the example in which the data packets used by the DataSender can vary in size. The TXmodule of Zoya receives the packets from the DataSender and buffers them for later transmission. This kind of buffering may be necessary if there is temporarily no connection between the TXmodule of Zoya and the RXmodule of Yoshi. This can occur, for example, due to the fact that Zoya and Yoshi are not in range. Assume further that the internal implementation of the TXmodule is such that the longest packets in the buffer are sent first. This makes sense from the viewpoint of the TXmodule, if it wants to manage its buffer in a way that it always keep maximal possible space for new packets. Consider the case in Figure 4.2(a): the DataSender delivers first a shorter packet, then a longer packet, but they are delivered to the DataCollector in the opposite order. The DataSender can deal with this surprise using the LoLaBBox by adding a sequential number to each packet it sends to the TXmodule. However, this also brings forward a fundamental property: the price for ensuring correct operation in the system is an additional control overhead.

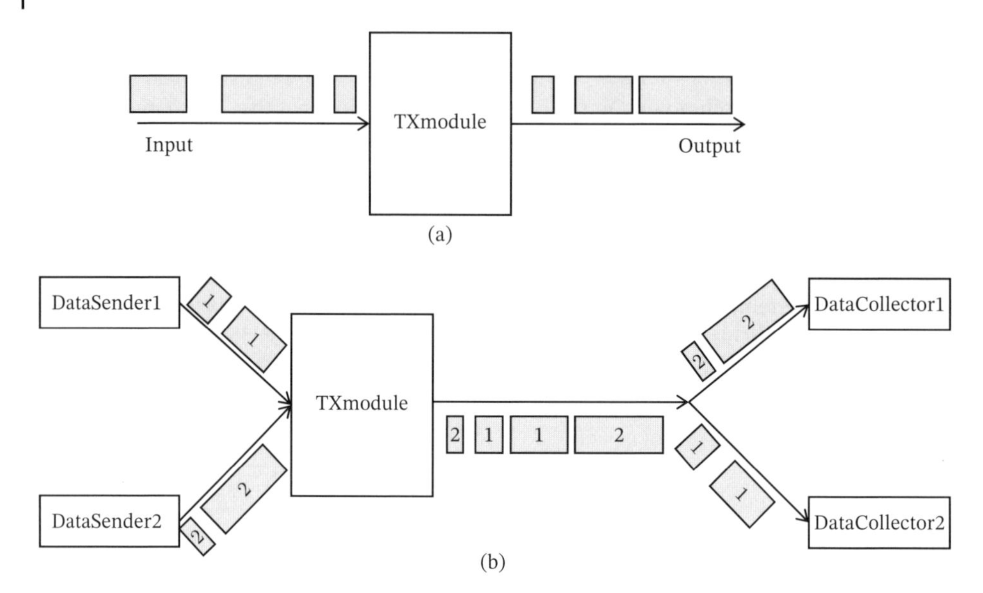

Figure 4.2 Illustration of how layering deals with packets that can get out of order. (a) TXmodule with internal scheduling policy to send out the largest packet first. (b) If the users of the TXmodule (DataSenders) know the internal scheduling policy, then they can ensure that the packets arrive in order without using a sequential number. The packets sent out by DataSenders 1 and 2 are labeled 1 and 2, respectively.

This example raises several issues. The buffering policy applied by the TXmodule is advantageous whenever multiple, uncoordinated, DataSenders are supplying packets to the TXmodule, because keeping the buffer as empty as possible will allow acceptance of new packets in the TXmodule. Nevertheless, if there is a single DataSender that always uses the TXmodule, then the operation of the TXmodule does not seem optimal, as it unnecessarily permutes the order of the packets from the same communication flow. In fact, if the DataSender is aware of the internal buffering/transmission policy of the TXmodule, the use of sequential numbers can be avoided in the following way. The DataSender can segment the data into L packets, starting from the first packet as the longest one, decreasing the length of each of the subsequent packets and ending up with the Lth packet, which is the shortest one. Then the DataSender supplies these L packets to the TXmodule. When the TXmodule informs the DataSender that all its packets from the buffer have been delivered, and the DataSender supplies L new packets. This would work well even if multiple DataSenders are concurrently using the TXmodule and its buffer, as the example in 4.2(b) shows.

The discussion above illustrates the important principle of *cross-layer optimization*. Namely, if the only thing that a DataSender knows about the LoLaBBox is that the packet supplied to the TXmodule will, at some point in time, appear at the output of the RXmodule, then it is necessary to use sequential numbers. However, if the DataSender knows more about the operation of the TXmodule, then it can optimize the system operation by avoiding additional overhead and choosing appropriate packet sizes to be supplied to the TXmodule. Hence, the entity residing at the higher layer adapts its operation to an entity residing at the lower layer in order to get better overall system operation. This optimization is not done at a single layer, but couples the parameters selected at different layers; therefore the name cross-layer optimization.

4.3 Reliable and Unreliable Service from a Layer

In the example above we have quietly introduced the need for communication in the opposite direction to the communication flow (TXmodule to the DataSender). In fact, two-way communication is necessary if the DataSender needs to be sure that the appropriate application at the DataCollector has received the data. A system model that supports two-way communication between Zoya and Yoshi is described in Figure 4.3. Since here both Zoya and Yoshi can have senders and receivers, we prefix the name of the component by the name of the device, such as ZoyaDataSender or YoshiTXmodule. The signaling links within a device can be, in general, bidirectional, while the wireless, over-the-air, transmission of signaling information is unidirectional. However, there can still be a bidirectional communication of metadata between any component belonging to Zoya and its peer at Yoshi. For example, the transmission from the ZoyaDataSender to the YoshiDataCollector uses the links already described above. Consider the case in which YoshiDataCollector has to notify Zoya not to send more data. Then Yoshi has to pass this control information to the module that is delivering data to it, which is YoshiRXmodule. The whole flow of the control information is YoshiRXmodule → YoshiTXmodule → ZoyaRXmodule → ZoyaTXmodule and, finally, to ZoyaDataSender, which stops sending data.

Errors can occur in absolutely any practical communication link, including the links over the wires between the modules residing in the same devices. However, errors can occur

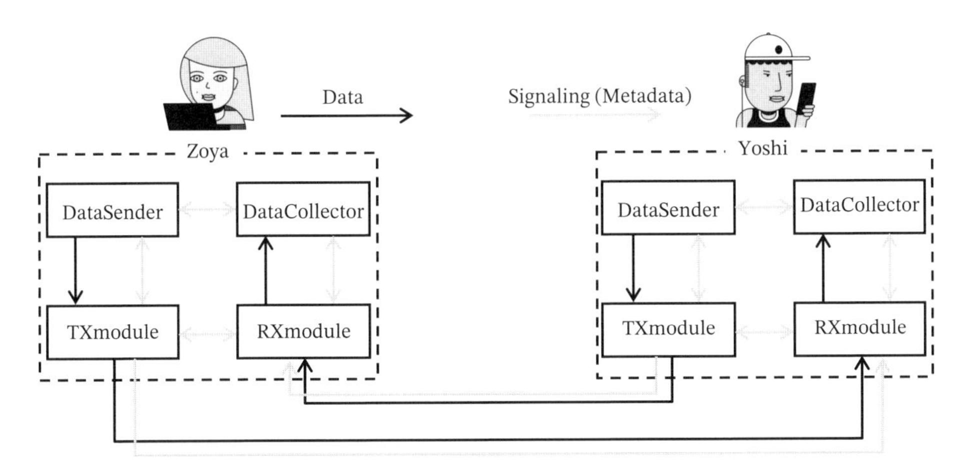

Figure 4.3 A simple layered model for two-way communication. Note that the arrows are bidirectional within a device, but unidirectional in the wireless (over-the-air) part.

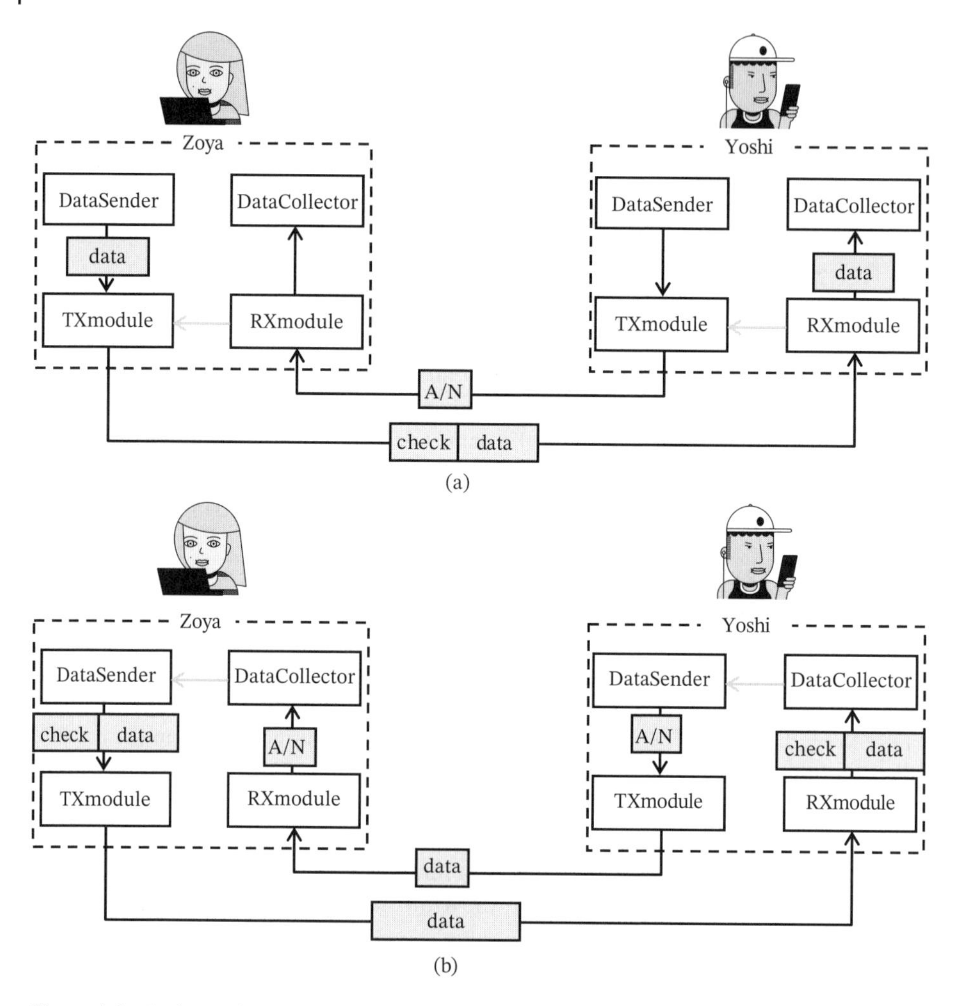

Figure 4.4 Packet exchange when the black box of the lower layer provides (a) reliable service and (b) unreliable service. The packet with content A/N stands for ACK or NACK. In both cases an ARQ protocol is run, but in case (b) the lower layers is unaware of it. to preserve clarity, only the relevant signaling links have been marked.

within the other elements of the system as well, such as in the data that is stored in a memory. Arguably, the most serious source of errors is the wireless link between the TXmodule and the RXmodule, since the probability of error in the wireless (over-the-air) part is usually magnitudes of order higher than the probabilities of the other errors mentioned above. Considering the possible occurrence of errors, LoLaBBox can be designed to provide two different service types, reliable and unreliable, respectively. These are illustrated in Figure 4.4.

If LoLaBBox is designed to provide *reliable service* to the higher layer, then, once the packet is supplied to the TXmodule, the TXmodule is obliged to inform the DataSender whether the packet has been delivered or not to the RXmodule. In order to be able to do that, the TXmodule treats the packet from the DataSender as a component carrying pure data and adds an integrity check, as discussed in Section 3.5. This is described in Figure 4.4(a). The integrity check is used by the RXmodule to verify the correctness of the received packet.

Using the integrity check, the LoLaBBox can implement protocol for an automatic retransmission request, in short called *ARQ protocol*, that can operate as follows. After receiving the packet, the YoshiRXmodule verifies the correctness of the received packet and, depending on the outcome, it prepares one bit of information to be sent back to the ZoyaTXmodule: “1” if the packet has been correctly received and “0” otherwise. As a convention, the short packet that carries the bit “1” is called the ACK (acknowledgement) packet, while the other is called the NACK (negative acknowledgement) packet. From the example in Figure 4.4(a), it can be noticed that the YoshiRXmodule of Yoshi supplies the ACK/NACK packet back to Zoya through the TXmodule of Yoshi. After receiving the ACK through the ZoyaRXmodule, the ZoyaTXmodule can inform the ZoyaDataSender that the packet has been delivered. On the other hand, if it receives a NACK, the ZoyaTXmodule can try to resend the packet several times until it receives ACK. If NACK is repeatedly received, say, L times, then the ZoyaTXmodule informs the ZoyaDataSender that the packet delivery has been unsuccessful. This is to avoid the situation in which Zoya keeps transmitting packets while Yoshi is unavailable for communication due to, for example, being out of range or experiencing failure. Then it is up to the higher layer module, the ZoyaDataSender, to decide whether it can again supply the same packet to be transmitted through the LoLaBBox.

By contrast, when the LoLaBBox provides *unreliable service*, then the ZoyaTXmodule guarantees the ZoyaDataSender that it will send the packet to the ZoyaRXmodule at least once, but will not give information on whether the packet has been received or not. Thus, reliability in the transmission should be introduced through a suitable ARQ protocol that runs at the higher layer, between the peers ZoyaDataSender and YoshiDataCollector. This means that ZoyaDataSender should add the integrity check to the packet and the ACK/NACK is a regular packet supplied by YoshiDataCollector to YoshiTXmodule. This is illustrated in Figure 4.4(b). Note that the communication between the TXmodule and RXmodule does not have additional integrity check on the data received from the upper layers. Also, the ACK/NACK sent by the DataSender is treated as data, not as a special signaling packet. By contrasts with the reliable service in Figure 4.4(a), here the ACK/NACK packet (metadata) is not passed on from the YoshiRXmodule to the YoshiTXmodule, since they do not understand it. Instead, YoshiDataCollector performs an integrity check and passes on the result (ACK/NACK) to the YoshiDataSender, which then supplies the packet to the YoshiTXmodule as if it is ordinary data.

Although it seems unnecessary to introduce error control at a higher layer if the lower layer is already providing a reliable service, today's networks rely on the end-to-end principle, which basically states that the highest layer in the system, which represents the end user of the communication system, should ensure that the data packet arrives reliably at the peer. The reason is that in complex systems there are many components where errors can occur and, since the integrity check is not ideal, sometimes an error can be disguised as a correct operation. Therefore, a final check can be advantageous, despite the extra redundancy.

4.4 Black Box Functionality for Different Communication Models

Here we illustrate how the system components, as well as layering, become related to the communication models and protocols discussed in the previous chapters.

We start with the collision model and treat the packet as an atomic unit of information. Let there be multiple terminals that send data to the base station Basil through a random access protocol. Let us first make an attempt to address the problem by using the layered structure in Figure 4.5. Note that, for example, the TXmodule and the RXmodule of each terminal are collapsed into the same block, representing the module of the device denoted by PHY. This representation is more compact than the one in Figure 4.3 and it allows us to assume wireless two-way communication, while keeping in mind that it consists of two one-way links that go in opposite directions. The layer functionalities are:

- PHY: *physical layer*. Enables physical transmission of data over the air.
- MAC: *medium access control (MAC) layer*. Ensures that each terminal gets an opportunity to send a packet to Basil without experiencing collision.
- DL: *data link layer*. Ensures reliable transfer by doing an error check and sending ACK/NACK.

A possible way of operating under this layered structure can be described as follows. The PHY module of Basil, denoted by BasilPHY, receives whatever comes over the wireless channel in a given slot. The information that BasilPHY needs to pass on to BasilMAC is whether the observed slot has been idle (I), in collision (C) or contains a single transmission (S). In addition, when S occurs, BasilPHY passes on the received packet to BasilMAC. The role of the BasilMAC module is to gather the information on the slots (I, S, C) and send appropriate feedback to support the desired operation of the random access protocol. In addition, the BasilMAC module should store all the packets received from the S-slots and,

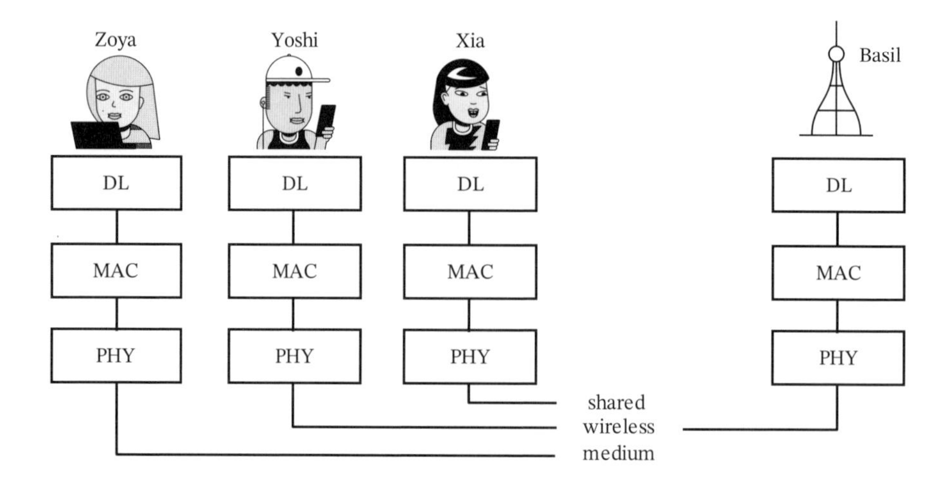

Figure 4.5 An example of a layered structure for multiple terminals sending data to a common base station Basil.

when it deems that the random access process is finished, passes on these packets to the BasilDL module. BasilDL runs an integrity check for each of the packets, prepares individual ACK packets and sends them to the terminals that succeeded in sending their packets correctly. If a terminal that has transmitted previously does not receive an ACK, then it re-enters the random access procedure. Note that, in general, Basil cannot send individual NACKs to the terminals that did not succeed, since an erroneous packets reception prevents Basil from knowing who had been the sender of that packet.

This protocol does not guarantee that each terminal will succeed in sending its packet to BasilMAC in a collision-free manner. For example, if there is a capture effect, then if multiple terminals transmit in a slot it can happen that the strongest signal is decoded, such that the slot is considered to be in state S and the signals of the other terminals in the slot are ignored. Under such an effect, BasilMAC cannot reliably detect if the random access protocol has successfully terminated. This problem can be addressed with the described layered structure; however, in general, it cannot be solved with certainty. One solution is that BasilMAC runs several successive instances of the random access protocol until he observes a predefined number of idle slots, which is treated as an indication that the MAC protocol has terminated.

Using a protocol that strictly follows the layered structure may result in an inefficient operation. Let us, for example, take the extreme case in which all the packets stored in BasilMAC and passed on to BasilDL are received with errors. Then all the terminals that do not receive ACKs have to start the random access procedure all over again. This is inefficient, since the first run of the random access protocol can be treated as an investment to enable each terminal to get the opportunity to send a collision-free packet. This investment is lost if the random access procedure is restarted without retaining some memory.

In order to avoid this problem, a cross-layer design can again be helpful. Let us now consider a structure in which the functionality of DL and MAC are merged into a single module. Now, when S is received, the packet is immediately checked for errors. If the packet is correct, then the ACK is sent in the following way: Basil sends feedback S, along with the terminal address of the correctly received packet. If the packet is incorrect, then one option is to interrupt the usual random access protocol and broadcast a NACK packet. This NACK packet is not addressed to any terminal in particular, as Basil does not know who sent the incorrect single packet, but it is understood as a command "the terminal that transmitted last, repeat your packet". This can be understood as a cross-layer optimization of the ARQ protocol with the MAC protocol, which can potentially lead to a more efficient operation.

Despite the successful re-use of the investment made by the random access protocol and the potential performance benefit, this, and similar tricks of cross-layer optimization, should be used with caution. In fact, the success of a cross-layer optimization technique depends critically on the capability of the PHY module to distinguish S from collision (C)[1]. Note that, in many practical systems, the receiver only differentiates between the states "no transmission" (I) and "transmission present" (S or C). Collision is not explicitly detected, but the receiver tries to run an integrity check whenever transmission is present and if it does not check, then erroneous packet can mean collision or channel error. If the reason

1 This capability depends on even lower layers, below the level of packet and at a level of a transmitted symbol, as discussed in Chapter 5.

for error is collision, then the idea of integrating ARQ and MAC, described above, would not work, as the terminals that caused the collision will keep colliding upon retransmitting their packets. Therefore, in the absence of reliable collision detection, ARQ and MAC may still be integrated, but the protocol needs to be optimized based on the likelihood that the observed error in the packet integrity check is due to transmission error or due to collision. This type of knowledge can only come from the lower, physical layer and it can be based, for example, on the power level of the received signals.

The layered operation for protocols that apply SIC and can operate with packet fractions is even more involved. The PHY module needs to be able to identify healthy packet fractions and merge them into a single packet that is likely correct. For example, in this case it is feasible to push the error check down to the PHY module, as it needs to frequently apply it when trying to construct packets out of the received combination of collided and collision-free packets.

4.5 Standard Layering Models

The simple examples so far have been solely focused on direct communication between two nodes that are in communication range, also known as a single-hop connection. Let us look at the scenario in Figure 4.6, where Zoya wants to communicate with Xia, but they are not in communication range. On the other hand, there is another node, Yoshi, which is in range of both Zoya and Xia. Now Zoya can create a multi-hop connection, by sending the packets to Yoshi and then Yoshi forwards (or relays or routes) the packets to Xia. A multi-hop connection uses the building blocks defined for the single-hop connections and we can again apply the paradigm of layering and black boxes. In the example, the *network*

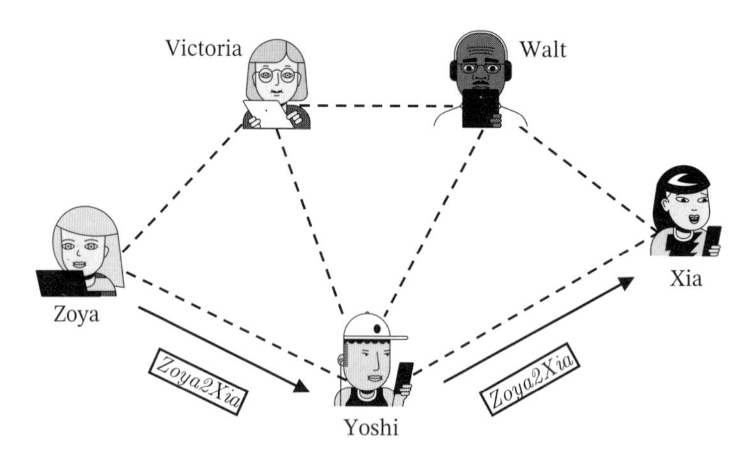

Figure 4.6 Scenario in which Zoya can establish a multi-hop connection to Xia. A dashed line between two nodes indicates that they are in communication range. First Zoya sends the packet *Zoya2Xia* to Yoshi, who then forwards the packet to Xia. Another option can be the multi-hop connection Zoya–Victoria–Walt–Xia or Zoya–Victoria–Yoshi–Xia.

OSI model	TCP /IP model
7-Application layer	Application
6-Presentation layer	
5-Session layer	
4-Transport layer	Transport
3-Network layer	Internet (IP)
2-Data link (DL) layer	Network interface
1-Physical layer (PHY)	

Figure 4.7 Protocol stacks: OSI versus TCP/IP.

layer module of Zoya passes the packet *Zoya2Xia* to its data-link (DL) module, instructing it to send it to Yoshi. The DL module is a black box that gets the packet reliably over a single hop to Yoshi and does not need to be aware that this same packet is to be forwarded further from Yoshi to Xia. On the other hand, the network layer module of Yoshi knows this and supplies the packet *Zoya2Xia* to the DL module of Yoshi to be sent further to Xia.

What we have described starts to look like the well known seven-layer OSI (Open Systems Interconnection) model or a *protocol stack*. It is a canonical model for layering put forward by the International Organization for Standardization for describing and standardizing the functions and modules of a communication system. This is not the only layering system in use, as the internet uses the TCP/IP model. Although the two systems are different, there is a significant functional similarity between them. Both layering systems are depicted in Figure 4.7, in such a way to point out the functional correspondence of different layers in the two models. In each layer a specific protocol runs between the peers, using the services of the protocols running at the lower layers and providing services for the protocols running at the higher layers. Before we describe the layer functionalities, we need to explain the notions of connection-oriented and connectionless protocols.

4.5.1 Connection versus Connectionless

In *connection-oriented* protocols, the two communicating parties first establish a communication session through a *handshake*, such that they are both aware that they are present and communicating. After that is established, data can flow between them and, in principle, once this connection is established, the packets of this data flow need to carry only limited metadata, as the source and the destination are known. Conversely, in a *connectionless* protocol, a message between the two nodes can be sent without making any prior setup or handshake and without even being certain that the other node is ready to receive the message. Clearly, any communication required to make a handshake and establish a session must be connectionless. Such is, for example, the rendezvous protocol, described in Chapter 1[2]. A random access protocol is essentially connectionless, since if the receiver has had agreements with the nodes about who should send when, then there would be no collisions. On the other hand, a reservation protocol based on random access can be seen as a collective handshake procedure of the base station with multiple terminals. After this procedure is finalized, then the communication between the base station and each individual terminal can be treated as connection-oriented.

As a general rule, a connection-oriented protocol that runs at a given layer can be composed by using a connectionless protocol provided by the black box of the lower layer. A connection-oriented protocol is useful when two nodes need to exchange many packets, which can be shown as follows. Assume that Zoya and Yoshi are devices that have unique 48 bit addresses, such that there can be at most 2^{48} devices of that type worldwide. If Zoya sends a packet to Yoshi in a connectionless manner, then the metadata of this packet should include at least 96 bits in order to uniquely identify the source and the destination of the packet. However, if Zoya and Yoshi first run a handshake procedure in order to establish a session and obtain the same context, then they can agree upon, for example, a 10 bit identifier of the connection. The 10 bits could be selected randomly, which could work well if Zoya (or Yoshi) has very few simultaneous connections, such that it is unlikely that any two connections will generate the same random number. Hence, each packet exchanged with a connection-oriented protocol will carry 86 bits less in overhead compared to the connectionless-type of communication. On the other hand, connectionless transmission can be useful when sending multicast messages, for which it would be unfeasible or unnecessary to run handshake protocols with multiple terminals.

4.5.2 Functionality of the Standard Layers

This book is almost entirely dedicated to wireless networking issues at the *physical (PHY)* and the *data link (DL)* layer from the OSI model. The combination of both layers correspond to a single layer in the TCP/IP model, termed *network interface* layer. Referring to Figure 4.5, it can be seen that the medium access protocols, which we have extensively discussed so far, can be thought of as if they lie at the intersection of the physical and the data link layer. The physical layer corresponds to the LoLaBBox with unreliable service, discussed in relation to Figure 4.4. The functionality of the DL layer has been discussed in relation to the model in Figure 4.5. In OSI terminology, the DL layer works with packets, while the PHY layer works with the bits from the packets supplied from/to the DL layer.

The *network layer* in OSI and the internet layer in TCP/IP are responsible for discovering a path between the source and the destination and composing a multi-hop connection based on single-hop connections. Another important function of the network layer is to ensure

2 Taking the notions of connection-oriented and connectionless protocol to a very precise description may be problematic. For example, even if two devices are running a rendezvous protocol to establish the initial link, both devices have some assumptions, such as the assumption that the other device is running a compatible protocol. In other words, the handshake is used to establish the same context for the devices in the short term, but the long-term context is embedded in the protocol design itself. In that sense, the ultimate example of connectionless transmission are the golden records carried by the Voyager spacecrafts towards extraterrestrial civilizations.

that each node in an interconnected network has a unique address. For example, such as the IP (internet protocol) address in the in the TCP/IP model. The network layer protocol in TCP/IP is connectionless, since a packet with an IP address traverses the network without assuming a session or a "context" at the receiver. It is the upper layer, the *transport* layer that can use the connectionless transmission of IP packets in order to create a connection-oriented protocol. The name of the connection-oriented protocol in the TCP/IP stack is TCP (transmission control protocol). The transport layer of the TCP/IP model also supports a connectionless protocol, termed UDP (user datagram protocol). Regarding the remaining layers, we first note that the correspondence in Figure 4.7 is not entirely precise, since some of the functionality of the *session layer* in OSI is done by the transport layer in the TCP/IP. More precisely, the session layer is responsible to open, manage, and close a communication session between two end nodes. Such a session could be established over a single or over multiple hops. On the other hand, the same responsibility in the case of the TCP/IP protocol is delegated to the TCP at the transport layer. Finally, the upper layers (application, presentation) specify the interfaces and data formats that are available to the applications for accessing networking services.

4.5.3 A Very Brief Look at the Network Layer

We take a quick look at the issues and protocols at the network layer, as this is the intermediate layer between the lower wireless layers and the upper layers, where the latter are not treated in this book. Coming back to Figure 4.6, the establishment of a connection between Zoya and Xia as a multi-hop connection via Yoshi seems like a natural choice, as it is the shortest path between Zoya and Xia. A bit less natural would seem to use a multi-hop connection Zoya–Victoria–Yoshi–Xia. One can think of a scenario in which there is some obstacle between Zoya and Yoshi, such that the link Zoya–Yoshi experiences excessive errors. Under such an assumption, it becomes justified to circumvent the bad link and find an alternative path.

In the previous example, Zoya makes a preference of the path Zoya–Victoria–Yoshi–Xia over the shorter path Zoya–Yoshi–Xia based on the knowledge that the link Zoya–Yoshi is bad. If such knowledge is not available, but Zoya still wants to increase the probability that the packet *Zoya2Xia* will reach Xia, then an idea can be to use *multi-path diversity*: send the same packet over two (or more) different paths (routes): Zoya–Yoshi–Xia and Zoya–Victoria–Walt–Xia. The concept of *diversity* has an important role in communication engineering and we will encounter it in different forms in the following chapters.

In general, we use the term diversity if, for the sake of improved reliability, the same information is encoded and sent through two or more communication resources. It is desirable that these resources are statistically independent, in a sense that, if one of them is bad (or good), then this does not imply that the other resource is necessarily bad (or good). It should be noted that that the concept of diversity should be used when there is no or very limited information about the goodness of the resources. By contrast, if a resource is known to be bad, then it may not have to be used in the first place. In the example of routing, the communication resources are the paths from the source to the destination. The path Zoya–Yoshi–Xia is not independent of the path Zoya–Victoria–Yoshi–Xia, as they are both sharing the link Yoshi–Xia; but it is independent of the path Zoya–Victoria–Walt–Xia.

An important ingredient of the network-layer protocols is *route discovery*: deciding which single-hop links can be used to compose a multi-hop connection between the source and the destination. Discovering a route is associated with a cost in terms of overhead, since the nodes should exchange messages, which do not carry data but only metadata. If the nodes are not moving and the links between any two nodes have constant quality, then the communication topology is static. For a static topology, in principle all the routes should be discovered only once and then memorized, theoretically until infinity, such that the time that the network spends to send route–discovery–metadata is negligible compared to the total time for which the network is used to send actual data.

A perfectly static topology is only a theoretical construct and, in practice, all communication topologies are dynamic, as the connection between two wireless nodes can change due to node movement, node failure, and sudden excessive interference, for example, from a microwave oven next to the node. Note that even a wired connection between two nodes can change the state due to node failure. For a dynamic topology it is important to differentiate if the changes are *slow*, where the topology will likely stay static while the route discovery procedure is being executed, or *fast*, where the topology changes unpredictably even during the route discovery. For slow topology dynamics, a viable approach is to have periodic updates among the nodes about the availability of links and routes. For example, Yoshi in Figure 4.6 can periodically check if the link to Zoya is operational and, if it is, also inform Zoya that Yoshi can provide a route to Xia.

However, if the topology dynamics is fast, then the route information is likely to be outdated at the time when it should be used. In that case it becomes feasible to send the data by *flooding*. In its simple variant, flooding is implemented by asking each node to forward the packet and thus, hopefully, the packet will eventually reach its destination. Hence, flooding can be seen as a connectionless transmission that combines route discovery and data transmission in a single operation.

Next, we summarize the engineering trade-offs associated with route discovery. For a topology with slow dynamics, the route information is periodically refreshed, such that, when needed to send data, there is a minimal delay. In addition, the selected route is well directed in a sense that each transmission by an intermediate or relaying node brings the data closer to the destination. On the other hand, when the topology changes rapidly, discovering a route is not feasible and flooding is used. Besides the larger delay, the problem with flooding is that some of the forwarded packets are necessarily useless. For the example in Figure 4.6, if Zoya floods the packet towards Xia, then even if Xia gets the packet through Yoshi, the same packet will still be broadcast by Victoria and Walt.

Knowing these elementary trade-offs can help in building hybrid solutions, optimized for certain scenarios. For example, an idea for a hybrid solution could be to use *clustering* and this is illustrated in Figure 4.8. The nodes are grouped into clusters and one node within that cluster has the role of a clusterhead. Each node in the cluster, except the clusterhead, refreshes only the route to its clusterhead, while the clusterhead connects to other clusterheads. If Umer in Figure 4.8 wants to connect to Walt who is outside of his cluster, then the clusterhead (Umer's car) runs a route discovery in order to locate the cluster in which the destination node resides and then forwards the packet to the clusterhead (Walt's car) to which the destination node is connected to. In the example where the clusters move significantly with respect to each other, as with the case where each cluster is a car or a bus,

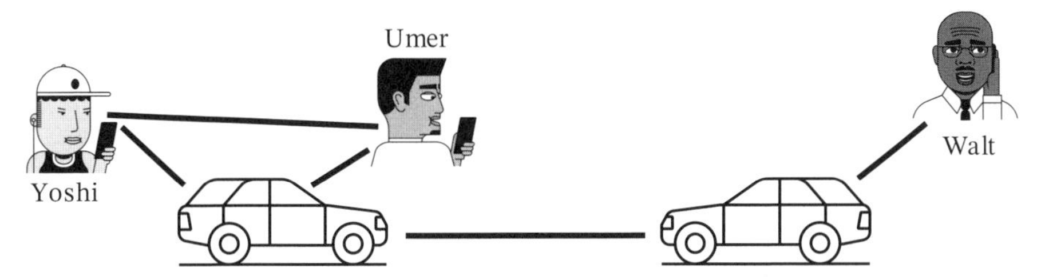

Figure 4.8 Example of clustered wireless networks. Each car represents a cluster to which the users sitting in that car are connected. Reaching a user that sits in another car requires inter-cluster routing. NB: none of the depicted people (Yoshi, Walt, Umer) are using the wireless device while driving.

then flooding may be used among the clusters only, but a specific route within the cluster; therefore the solution is a hybrid one.

4.6 An Alternative Wireless Layering

The definitions of the OSI or TCP/IP layers are biased towards wired networks. They are centered on the concept of *link*: reliable communication is achieved within a link, a route is a sequence of links, etc. The notion of link is somehow understood as a wire or cable: each link corresponds to a separate transmission effort employed by the transmitter. On the other hand, we have seen in Chapter 1 that the broadcast feature of the wireless medium enables a single transmission to reach multiple nodes. Hence a relevant question would be: what kind of layering corresponds to the wireless broadcast feature?

We take the scenario from Figure 4.6, where Zoya wants to communicate with Xia. The packet broadcast by Zoya is received by both Yoshi and Victoria. The objective of Zoya's transmission can now be formulated as follows: the data sent by Zoya should be received correctly by the *union* of Yoshi and Victoria. The requirement formulated for the union of Yoshi and Victoria is less strict compared to the requirement that at least one among Yoshi and Victoria gets the whole packet correctly. For example, if the packet has 100 bits and Yoshi receives the first 37 bits correctly, while Victoria has the correct values of the last 63 bits, then the union of Yoshi and Victoria has received the packet. It should be noted that the probability that the union of Yoshi and Victoria receives a packet correctly is higher compared to the probability that either Yoshi or Victoria receives the packet correctly; this is the basic motivation behind the multi-path diversity transmission. Once the union of Yoshi and Victoria has the packet sent by Zoya, then Zoya should cease to transmit, as it has already succeeded to move the data closer to Xia. In the next step, the union of Yoshi and Victoria transmits the packet to the union of Walt and Xia. If the union of Walt and Xia receives the packet correctly, then, if necessary, Walt can forward the last piece of data required by Xia, so that she can reconstruct the original packet sent by Zoya.

The previous paragraph contains several statements and concepts that may seem rather superficial and thus need further elaboration. The described transmission scheme, along with the notion of union of devices, abandons the traditional link concept and adapts to the broadcast nature of the wireless medium. One may think that we are reusing the same layering concept, only that now a union of devices acts as a *super-device*. Indeed, suppose for a moment that Yoshi and Victoria are connected by an optical fiber with very high data rate, such that for all purposes Yoshi and Victoria can be seen as a single device, or a super-device. Suppose the same for Walt and Xia. Then there is an equivalent two-hop transmission Zoya→ {Yoshi, Victoria} → {Walt, Xia}, since, due to the ideal fiber, the last transmission from Walt to Xia is unnecessary.

This two-hop transmission with super-devices can be described through the traditional layering paradigm. The problem is that there is no fiber within a union of devices and, in principle, there should be a dedicated wireless communication protocol within a union of devices through which the union of devices will try to approximate the operation of a super-device. After Zoya sends the packet, Yoshi and Victoria need to have some form of coordinative talk to each other to see if they, as a union, are able to reconstruct the complete data sent by Zoya. Furthermore, they also need to coordinate how to carry out the next transmission, otherwise Yoshi and Victoria may cause a collision for Walt's receiver. The communication between Yoshi and Victoria takes place over the same shared wireless medium, such that the coordination between them consumes wireless resources, which can easily make the whole transmission concept inefficient.

The idea of cooperative communication is discussed further in Chapter 12. For now, we can conclude that if the communication scheme that involves union of devices is operated through ordinary layering and emulation of super-devices, then the scheme may turn out to be much less efficient compared to the usual multi-hop communication with single or multiple paths. Therefore, the task of sending data between Zoya and Xia can be decomposed in a different way, suited to operate with a union of devices, which would lead to an alternative layering. We will not pursue the discussion in the direction of how exactly to do such alternative layering, but we will outline its two most important features:

1. The alternative layering should provide better integration between the actual data communication with the task of coordination within a union of devices.
2. The alternative layering should give rise to *scalable* protocols applicable to universal scenarios. It cannot be overemphasized that the success of the traditional layering is in the fact that it provides an architecture for creating connections in practically any imaginable scenario where such connection is physically possible. The example above might be tempting to devise layers and protocols that can provide superior performance in some scenarios. Nevertheless, the goodness of an architecture or protocol should not only be judged based on its superiority in certain very specific scenarios. A good architectural design aims to offer an acceptable performance in a large majority of scenarios.

4.7 Cross-Layer Design for Multiple Hops

In this section we abandon the ambitious goal of devising an alternative layering architecture and we stick with the traditional layering. We stay with the scenario in Figure 4.6 and explore the how the cross-layer optimization, or enhanced exchange and use of information across different layers, can bring benefits in terms of communication performance.

Let us first assume that single-path routing is used, Zoya→Victoria→Walt→Xia. We consider the case when there is a communication flow of multiple packets that need to get from Zoya to Xia and these packets have sequential numbers 1, 2, 3, ... At a particular instant, Walt has the packet 7 to be delivered to Xia, while Zoya should send the next packet 8 to Victoria. At the MAC layer, Zoya and Walt see each other as usual competing nodes that are trying to reserve transmission time on the wireless medium. Since Victoria is in the range of Walt, if Walt and Zoya transmit simultaneously, Victoria observes a collision and has a lower chance of decoding packet 8 sent by Zoya. One cross-layer solution would be to use the network layer information: since they are on the same path, Walt and Zoya know that their transmissions can collide at Victoria and the MAC layer protocol can be designed to avoid this type of collision. Yet, the MAC protocol should still be able to deal with other, uncoordinated transmissions: for example, Yoshi may have his own packet for Xia, which can cause collision at Victoria (with Zoya or Walt) or Xia (with Walt).

Recalling some advanced interference cancellation techniques from Chapter 3, another, more efficient cross-layer solution comes to mind. Since Victoria has already forwarded packet 7 to Walt, she knows the content of that packet. Therefore, if Zoya and Walt transmit simultaneously, Victoria receives a collision, but she cancels the known packet 7, sent by Walt, and decodes successfully packet 8 sent by Zoya. At the same time, Xia receives successfully packet 7 from Walt. The cross-layer optimization here consists of the fact that Victoria can participate in the medium access reservation procedure for both Zoya and Walt in order to, somewhat paradoxically, ensure that they will transmit at the same time and cause a collision at Victoria. Simultaneous transmission is advantageous, since it maximizes the overall efficiency: within the time of one packet duration, two packets are delivered in the network, packet 8 to Victoria and packet 7 to Xia. This is desirable from a system point of view since the total time for which a given path reserves the wireless medium is minimized, and other communication flows, such as new data flow from Yoshi to Xia, can use the wireless medium.

As a final example, let us assume that multi-path diversity is used and the packets are sent through two different routes: Zoya→Yoshi→Xia and Zoya→Victoria→Walt→Xia. For simplicity, let us assume that the probability that any specific link is in outage is constant and equal to p_0. Note that Yoshi is in the range of Walt and Victoria. If Walt learns that Yoshi has received the packet number 1, then he can remove this packet from his transmission queue. Walt can learn this information by overhearing the ACK packets sent by Yoshi. The fact that Walt ceases to transmit a packet that is already known by Yoshi will remove the possibility of having two copies of packet 1 colliding at Xia. However, this loses the advantage of multi-path diversity, since the packet delivery depends only on the link Yoshi–Xia and the probability that the packet will not be delivered is equal to the link outage probability p_0.

Alternatively, a similar optimization trick can be used by Victoria: if she overhears the ACK from Yoshi, then she removes the packet 1 from her queue. However, the situation is now different, since Walt has not received packet 1 before and, when Yoshi sends it to Xia, Walt can overhear it. If Walt receives the packet and does not overhear an ACK from Xia, then he concludes that the transmission through the link Yoshi–Xia has not been successful and he transmits the packet 1 to Xia. In this case the probability of outage is calculated differently: outage occurs if the link Yoshi–Xia is in outage AND if the *route* Yoshi–Walt–Xia is in outage. The route Yoshi–Walt–Xia is in outage when at least one of the links Yoshi–Walt and Walt–Xia is in outage. The probability that the route Yoshi–Walt–Xia is *not* in outage is thus $(1 - p_o)^2$ and thus the probability that the route is in outage is $1 - (1 - p_o)^2$. Hence, the probability that the packet will not be delivered from Yoshi to Xia is:

$$p_o(1 - (1 - p_o)^2) < p_o \tag{4.1}$$

since $1 - (1 - p_o)^2 < 1$. In this case some part of the multi-path diversity has been retained although the link Victoria–Walt is not used.

4.8 Slicing of the Wireless Communication Resources

Layered architecture enables proliferation of systems and devices: its black box principle allows innovation and implementation to occur within each module (layer), while ensuring compatibility between modules through well defined interfaces. In this section we focus on another feature of the layered architecture: its universal support for diverse connectivity types and services.

4.8.1 Analog, Digital, Sliced

Let us start from analog communication systems and consider two types of services, say, voice communication and TV picture broadcast. Each analog communication service is treated as a system that requires its own architecture, adjusted to the specific type of information carried by the system. The point where different analog systems may start to *converge* is at the level of analog radio signals, as illustrated in Figure 4.9(a). This means, for example, that both voice signals and TV broadcast signals can use the same type of power amplifier or their receiving filter can be built based on the same principles, possibly using different parameters. The two services can cause interference to each other whenever they simultaneously use the same frequency channels. The way to avoid interference is to ensure physical sharing of the connectivity resources: for example, either different services are allocated different frequency channels or they are time sharing the same channels.

In contrast to analog systems, the starting premise of digital communication systems is that information can be measured and represented in a unified manner, regardless of the information source. In other words, a bit is a universal currency of information[3]. From this it can be concluded that an obvious point of convergence in digital systems is at the level of a bit, which corresponds to the physical layer where all bits can have identical representation, regardless of the actual service to which a specific group of bits belongs. This is illustrated in Figure 4.9(b).

Nevertheless, the key feature of the modular layered architecture is that it offers multiple points at which different services can converge. Thus, instead of using commonly represented bits, the services can rely on commonly represented packets. For example, the underlying idea of the internet is to use IP packets as basic units for transferring data, such that any connectivity service can be mapped onto IP packets. Following the previous example from analog systems, voice communication and TV picture streaming may

3 See the next chapter for a definition of information, entropy, and bit.

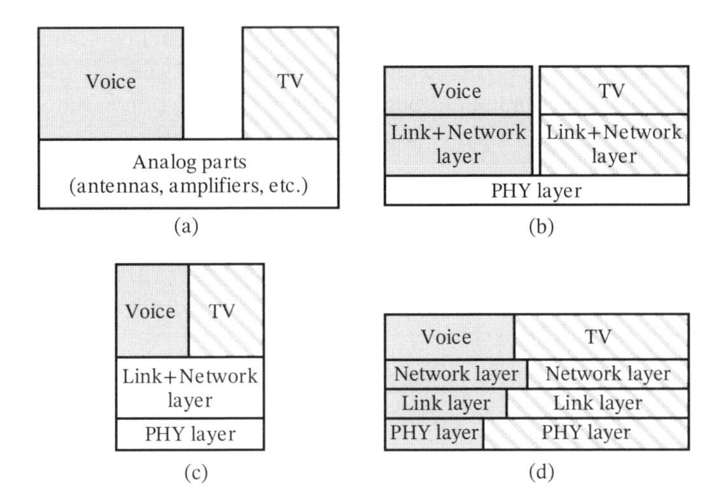

Figure 4.9 Illustration of the points of convergence in different types of systems for supporting two connectivity services: voice communication and TV picture transmission. (a) Analog communication system, where the only possible convergence point is at the analog hardware parts. (b) Layered digital communication system where the two services converge at the bit level, within the physical layer. (c) Layered digital communication system where the two services converge at the packet level, within the transport layer. (d) Slicing of the layered architecture for flexible allocation of resources to support the two services; the point of convergence can be flexibly adjusted.

be designed in such a way that they use a common transport layer protocol, as illustrated in Figure 4.9(c). In that case, the layers below the transport layer cannot recognize the service to which a particular packet belongs; packet differentiation happens at the transport layer and above.

The requirements of different services may vary very significantly. For example, the data rate of the TV streaming service is much higher than the data rate required by voice communication; however, voice communication is based on interaction and therefore it puts forward certain requirements on latency. In other words, these two services have different demands for quality of service (QoS) and/or quality of experience (QoE) that is delivered to the end users. For each of these services, the layered architecture should use the available resources in a different way.

In general, the services can share three types of resources: computation, data storage, and connectivity. Here "connectivity" resources refers to the resources (bits, packets, parts of packets, signaling information) that are eventually transferred through the physical layer toward other devices. The term *network slicing* refers to the allocation of these resources to different services by meeting the requirements of each service. In that sense, a *slice* is a set of resources allocated for a specific type of service (e.g. voice communication) and dimensioned to support a certain number of instances of that service (e.g. a certain number of voice connections). This is illustrated in Figure 4.9(d). The illustration reflects the fact that different services run over a common layered architecture, but the layers are parametrized in a way to flexibly support the demands of different slices.

Another interpretation is that the point of convergence can be flexibly adjusted throughout the layers. For example, a service requiring low latency and high reliability should be differentiated from a broadband service that aims for maximization of the data rate. This can happen at the physical layer, even below the bit level, as each of the services can use a different transmitted waveform. At the same time, both services may use common functions and structures at the higher layers; for example transmission of metadata or security related features.

We have now arrived to the cake analogy depicted in the cartoon at the beginning of this chapter. Efficient "production" of communication protocols and communication nodes can happen by relying on layering; this is reminiscent of the mass production of cakes. Once the cake is produced, it should get sliced and offered for consumption; each slice can be customized in size or some special ingredient can be added to a slice if required. Although not completely, this gets close enough to the layering and slicing of the "networking cake".

4.8.2 A Primer on Wireless Slicing

In order to illustrate the performance trade-offs for services with heterogeneous requirements, let us consider the case in which Zoya and Yoshi are transmitting to Basil in the uplink, as shown in Figure 4.10(a). We consider the following two connection types:

- *Broadband connectivity*. Zoya wants to *continously* transmit to Basil, for example a stream from a video camera, at the highest possible data rate that can ensure the highest quality of the received video stream.
- *Low-latency reliable control*. Yoshi wants to transmit to Basil short control messages, but each of these messages occurs *intermittently* and, possibly, randomly. When a message arrives at Yoshi, he needs to send it with the lowest possible latency. If Yoshi does not have such a message, then he stays idle and does not send anything.

We will refer briefly to these services as broadband and low latency, respectively. For the purpose of this illustration we have selected uplink, rather than downlink, transmissions. The reason is that in the uplink Zoya and Yoshi are competing for resources without coordinating with each other. In contrast to this, in a downlink scenario Basil would be the central point that allocates resources to Zoya and Yoshi, which is less challenging compared to the uplink case.

4.8.2.1 Orthogonal Wireless Slicing

One obvious way in which the requirements of the different services can be met is to allocate exclusive communication resources to each service. For example, one frequency channel[4] can be allocated to Zoya and another to Yoshi, which would avoid any interference and thus the two services would be ideally isolated in terms of performance guarantees. On the one hand, this system offers ideal low latency to Yoshi, as he can just send the message whenever it arrives, without waiting. On the other hand, if the packet arrivals to Yoshi are not frequent, then the overall system suffers from inefficiency, since the channel allocated to Yoshi will stay idle most of the time.

Another way to allocate resources for these two services would be to allow Zoya and Yoshi to time share the same channel. This is illustrated in Figure 4.10(b). For simplicity, the time

4 Here we still use the intuitive notion of a frequency channel: signals transmitted simultaneously at two different frequency channels do not cause interference to each other.

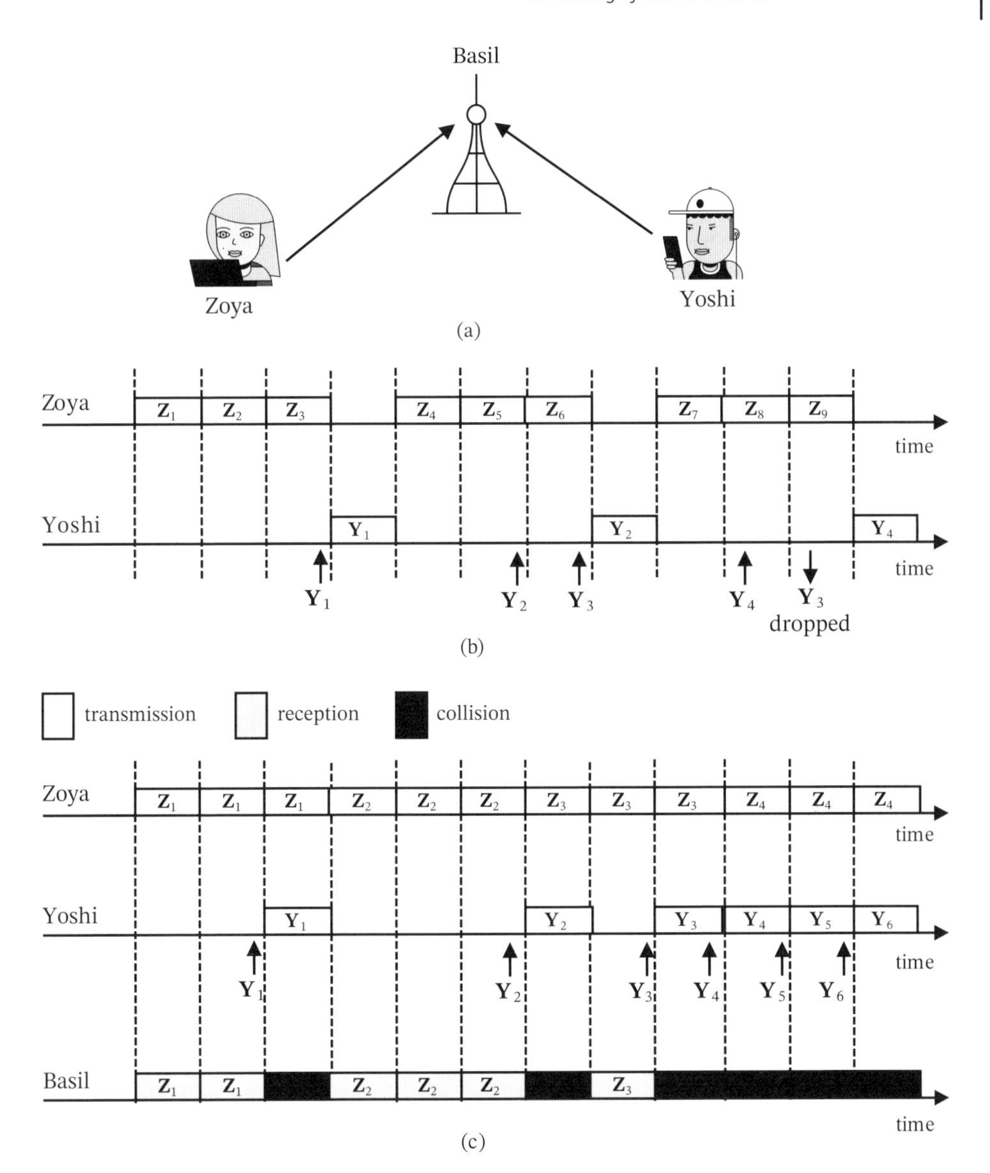

Figure 4.10 A primer on wireless slicing for two services, broadband and low latency, respectively. (a) Uplink scenario in which Zoya sends broadband traffic and Yoshi low-latency traffic to Basil. (b) Orthogonal slicing where one out of L slots is allocated to Yoshi, while the other $L - 1$ slots are allocated to Zoya. (c) Non-orthogonal slicing, in which Yoshi can transmit in any slot, while Zoya repeats each of her packets L times.

is divided into equal slots. We have ignored the existence of frame headers, i.e. we have assumed that they take an insignificant amount of time compared to the time slots used for data transmission. Similar to the previous case, this is another example of *orthogonal* resource allocation, which ensures that the wireless signals of Zoya and Yoshi are not interfering. However, here the performances of the two services do have an impact on each other and there is a trade-off between the rate achieved for Zoya and the minimal latency available to Yoshi. Let us assume that one out of L slots is allocated to Yoshi for low-latency transmission, while the rest of the slots are allocated to Zoya for broadband transmission. For the example in Figure 4.10(b), $L = 4$. In each allocated slot Zoya transmits at a rate R (bps); hence, the average goodput available to Zoya is:

$$\mathrm{G} = \frac{L-1}{L}R \qquad \text{(bps).} \tag{4.2}$$

The performance available to Yoshi depends on the process that describes the arrival of the messages at Yoshi's transmitter. In the simplest case, Yoshi's messages arrive periodically, one of them at each Lth slot. If the slot allocated to Yoshi is timed to start just after the arrival of the message, then the latency is zero. In addition, all messages sent by Yoshi are received perfectly by Basil.

Now consider the situation in which the message arrivals for Yoshi are randomized. To keep the things simple, assume that just before the beginning of each slot, an arrival occurs at Yoshi with probability a, while there is no arrival with probability $(1 - a)$. Yoshi has the possibility to queue the messages and send a message whenever there is a transmission opportunity. A basic and intuitive result in queuing theory says that if $a < \frac{1}{L}$ then each of Yoshi's messages will eventually be delivered to Basil within a finite time. In contrast, if the message arrivals to Yoshi are more frequent and $a \geq \frac{1}{L}$, then the queuing delay tends to infinity. In that case, a possible strategy applied by Yoshi could be to drop some of the messages that are already excessively delayed, beyond the low-latency value required by Yoshi's service. This is what happens with the packet $\mathbf{Y}_3$ in Figure 4.10(b). In general, when the message arrivals are random, then latency experienced by a particular message becomes a random variable. If packets start to get dropped, some messages never get delivered; in other words, the probability or message delivery becomes less than one. Furthermore, if the orthogonal allocation is fixed, then the randomized arrivals of Yoshi's messages affect only the latency and delivery probability of the Yoshi's low-latency service, while the broadband service offered to Zoya remains intact. Indeed, all packets of Zoya are delivered perfectly and the goodput stays at the value (4.2).

Let us still keep the message arrivals for Yoshi random, with an arrival probability of a, but now let us assume that Yoshi's packets should have zero latency and be delivered immediately in the slot that starts after the message arrival. If we keep the allocation orthogonal, then this requirement can only be satisfied by having $L = 1$, such that each slot is allocated to Yoshi only and Zoya has a goodput of zero. This is an extreme situation in which all packets of Yoshi are delivered perfectly and with zero latency. Nevertheless, the inefficiency problem remains since, when a is small, most of the slots will stay unused.

4.8.2.2 Non-Orthogonal Wireless Slicing

We have thus hit the limits of orthogonal allocation, as we cannot offer zero latency to Yoshi while offering non-zero data rate to Zoya. Instead, we can use the wireless channel models developed in the previous chapter and create schemes for *non-orthogonal allocation*, i.e. *non-orthogonal wireless slicing*. Specifically, we show how one can use the technique for inter-collision successive interference cancellation (SIC) in order to offer a non-zero rate to Zoya while ensuring that every delivered packet from Yoshi has zero latency. In order to make this possible, Zoya has to relax her requirement and allow that some of her packets are not delivered with probability ϵ_Z. Furthermore, let us assume that Yoshi allows the same type of flexibility and he is satisfied if the probability of undelivered low latency message is ϵ_Y.

A possible scheme is illustrated in Figure 4.10(c). Zoya repeats each of her packets L times; $L = 3$ for the example in Figure 4.10(c). Yoshi transmits the messages as they arrive. The receiver Basil works according to the collision model: he receives successfully a packet if only one transmitted (Zoya or Yoshi) is active in a slot. If both of them transmit in a given slot, the received collision is buffered for potential use in an inter-collision SIC. Consider the first group of $L = 3$ repetitions of $\mathbf{Z}_1$ in Figure 4.10(c). Basil receives $\mathbf{Z}_1$ already in the first slot and uses it in the third slot to cancel $\mathbf{Z}_1$ from the collision and thus decodes $\mathbf{Y}_1$. Receiving only one of the L repetitions of Zoya's packet is sufficient to use it for SIC and recover the packets of Yoshi that ended up in a collision. However, if all L replicas sent by Zoya end up in a collision, as is the example with $\mathbf{Z}_4$, then neither Zoya's packet nor the L packets sent by Yoshi are received correctly by Basil. Hence, the probability that a particular packet of Zoya is not received correctly is a^L.

Given the activation probability a and the target error probability ϵ_Z, we can choose the number of repetitions L to satisfy:

$$a^L \leq \epsilon_Z. \tag{4.3}$$

The average goodput for Zoya is:

$$\mathrm{G} = \frac{R}{L}(1 - a^L) \qquad \text{(bps)}. \tag{4.4}$$

For a given target ϵ_Z, a larger activity probability a requires larger L, but this decreases the goodput factor $\frac{R}{L}$.

The probability that a packet sent by Yoshi is not received is a^{L-1}, which is larger than a^L. The reason for this is that the observed packet, for which we calculate the probability of error, is already active and occupying one of the L slots; error occurs when Yoshi has transmissions for the other $L - 1$ slots. For a target probability of error ϵ_Y, the value of L should be chosen such that:

$$a^{L-1} \leq \epsilon_Y. \tag{4.5}$$

Finally, we should choose L that meets both requirements (4.3) and (4.5).

This primer on slicing the communication resources shows that more involved communication models can lead to non-trivial schemes for sharing of communication resources between services. The reader may object that, in the non-orthogonal slicing the performance of the services is not isolated. However, as long as each service meets its target probability of packet delivery, then they can be treated as isolated in terms of performance. As a last remark, the crucial element that makes the non-orthogonal slicing work is the knowledge of the activity probability a.

4.9 Chapter Summary

Segmenting the design of a communication system into functional modules that reside at different layers is an idea that appeared early in the era of digital communication; it can be recognized in the seminal paper by Shannon from 1948 that laid the foundations of information theory. In this chapter we have described the main ideas and principle behind the layered protocol design. Layered protocol architecture leads to proliferation of network nodes and devices, as it allows each module from each layer to be developed independently, as long as it conforms to the requirements of the well defined interfaces towards the other modules. We have also shed light on the alternative, largely speculative, forms of layering that can be defined by taking into account the specifics of the wireless communication medium. In the last part we have considered the design possibilities and trade-offs in the communication architecture when considering connectivity services with heterogeneous, sometimes vastly different service requirements. It can be seen that one can take advantage of the more advanced wireless communication models in order to come up with non-trivial schemes for efficient sharing of the wireless resources among different services. This has completed the analogy to the “network cake”: produced efficiently through layers, consumed adaptively through slicing.

4.10 Further Reading

Layering and separation of functionalities in a communication systems was one of the several ideas raised in the groundbreaking paper by Shannon [1948]. However, Shannon does not mention explicitly the term “layering”, but shows how the source coding and channel coding functionalities can be separated into respective modules. A pedagogical approach to layering can be found in Wetherall and Tanenbaum [2013] and Bertsekas and Gallager [1992]. Cross-layer design and revision of layering from a wireless perspective can be found, for example, in Kawadia and Kumar [2005] and Scaglione et al. [2006]. The notion of slicing emerged in relation to a 5G wireless system; a primer on modeling for wireless network slicing can be found in Popovski et al. [2018].

4.11 Problems and Reflections

1. *Pseudocode for the TDMA system.* Consider the system from Chapter 1, depicted in Figure 1.7. Describe a layered architecture for the base station and the devices, as well as pseudocode for each of the modules in the layered architecture that implements the described protocol.
2. *Retransmission protocols.* The protocols for controlling the transmission/retransmission of packets according to the positive/negative acknowledgments are called ARQ protocols. In relation to Figure 4.4, it has been explained that a given layer can provide either a reliable service (with its own flow control, which is a mechanism for providing reliable communication) or an unreliable service for the higher layer. Nevertheless, in some cases both layers apply a flow control protocol; that is, even if the lower layer provides a

reliable service, the upper layer also adds its own mechanism for acknowledgement of the received packets.

(a) Discuss what could be the possible interaction among the flow control protocols from the two layers in terms of queuing, timeout mechanism (how long a node waits for ACK/NACK until it resends), etc.

(b) Consider a multi-hop connection Zoya→Yoshi→Xia→Walt. The transport layer sees a connection between Zoya and Walt and applies its own flow control. In addition, each link, say Zoya–Yoshi, has its own flow control. Discuss the interaction of the two flow control protocols by considering the properties of the shared wireless medium and assuming different types of access protocols.

3. *Cross-layer and SIC*. Following the description of the layered architecture, along with cross-layer considerations in Section 4.4, provide the same layered design considerations for MAC protocols that work with SIC and packet fractions (see the previous chapter).
4. *Alternative layering*. Describe yet-another-alternative layering for wireless networks taking into account the properties of the shared wireless medium. Note that the notion of a link is not as solid and symmetric as with a wired connection. For example, Walt can increase his transmit power to be able to reach Yoshi. This may result in an asymmetric situation, in which Walt uses a higher transmit power compared to Yoshi, such that the signal of Walt can reach Yoshi, but not vice versa. Include these factors in your description of the alternative layering.
5. *Wireless slicing and latency-rate trade-offs*. In the primer on wireless slicing from Figure 4.10 we have shown how the decrease in Zoya's rate as well as a sacrifice in terms of reliability of packet delivery (both for Zoya and Yoshi) can be used in non-orthogonal allocation of resources to ensure low latency for Yoshi. However, in this scheme the latency of Yoshi stays strictly at zero. Propose a scheme in which Yoshi's latency can be increased, while simultaneously offering a higher rate for Zoya and/or reliability for both Zoya and Yoshi.

wo people use the beach sand to send
message. One of them uses only his own
ower and hand tools.

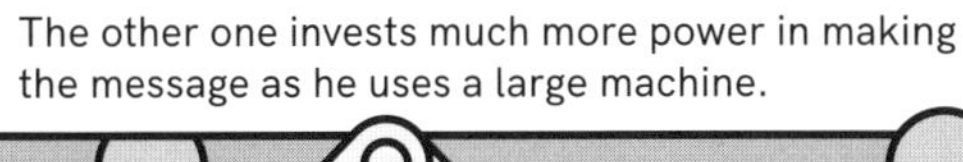

The other one invests much more power in making
the message as he uses a large machine.

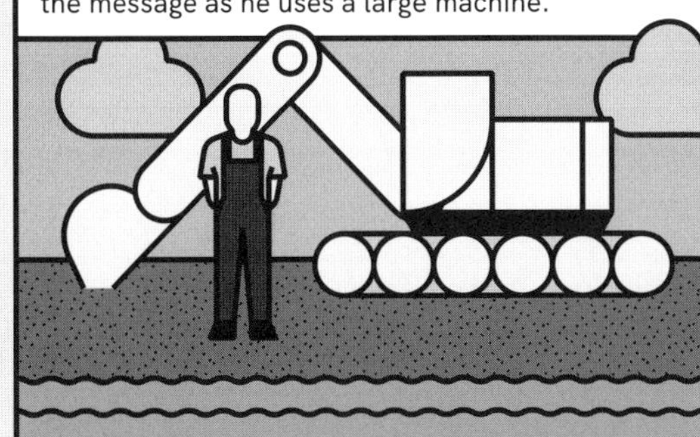

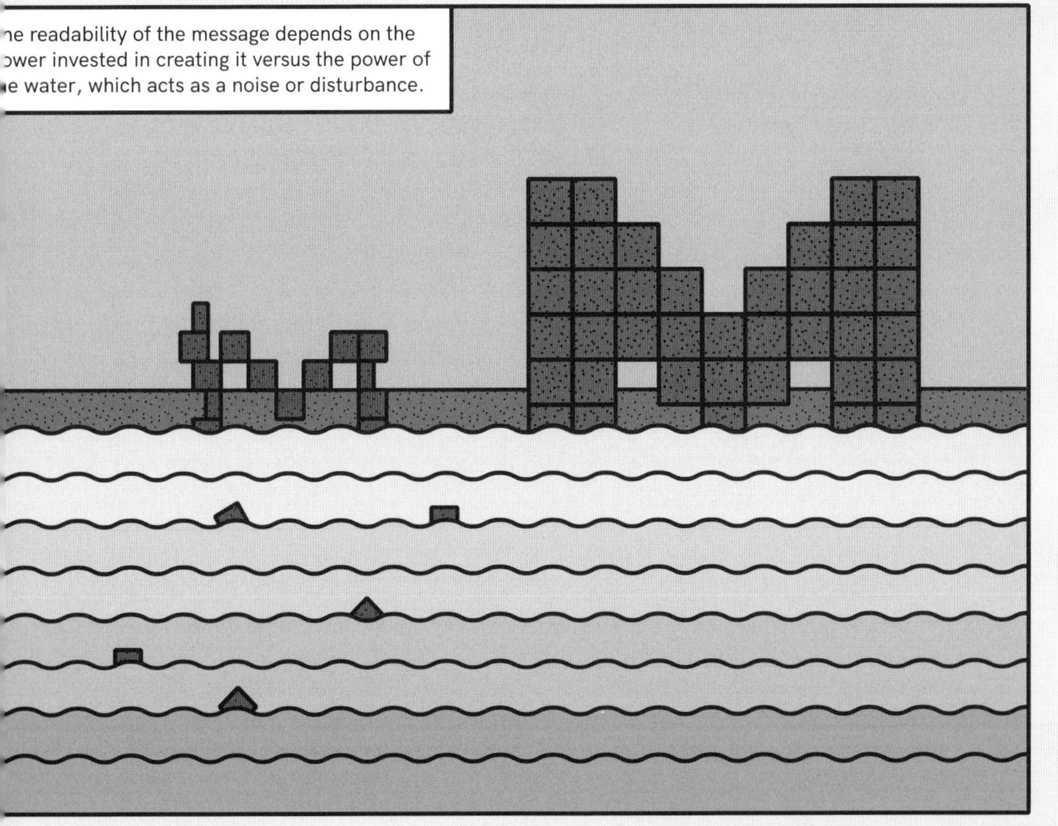

e readability of the message depends on the
ower invested in creating it versus the power of
e water, which acts as a noise or disturbance.

Story by Petar Popovski / Art by Peter Gregson

5

Packets Under the Looking Glass: Symbols and Noise

The information content of a packet can always be described by 0s and 1s. On the other hand, the representation of a data packet may change depending on the layer at which the packet is observed. For example, the packet can be represented by 0/1 at the link layer, while it can be represented via specific voltage levels at the physical layer or as a specific form of a radio wave when sent wirelessly over the air. We use the term *modulation* to denote the way in which information bits are represented in a particular (sub-)module of the communication system.

We refer again to Figure 4.1, describing the modules and layers for a simple one-way link. Once the 1/0 data packet is passed on to the TXmodule, it can undergo several physical transformations. In other words, the layer that contains the TXmodule and the RXmodule consists of several sub-layers and each sub-layer may use a different physical representation of the data. Our objective in this chapter is not to describe all those different physical representations, but rather to discuss the first sub-layer, termed the *baseband layer*. In the baseband layer, the data represented by 0/1 is mapped into complex numbers, which serve as carriers of information. A special case of a baseband layer is the one that uses only real numbers.

In the previous chapters we have dealt with packets and sometimes with packet fractions, but in this chapter we take the looking glass and examine the internal structure of a packet. What we will find there are baseband signals, which are sequences of complex (or real) numbers that are used to represent the information contained in a data packet. So far, we have used the notions of bit, data, and information in a rather informal manner. However, now that we need to describe how information is modulated to symbols, we need to look at the fundamentals of information theory and make these notions more precise.

5.1 Compression, Entropy, and Bit

Using the layering model, it can be stated that the useful data, contained into a *message*, is created at the highest, application layers. This message is then passed on to the lower layers and, finally, it is sent as a wireless, most often radio, transmission. In this context, an important question is the one related to the initial digitalization: how does the system obtain the data bits that constitute a message?

Wireless Connectivity: An Intuitive and Fundamental Guide, First Edition. Petar Popovski.

5.1.1 Obtaining Digital Messages by Compression

While any information can be represented in a binary form using 0s and 1s, the binary representation is not the one that is naturally associated with many information sources, such as picture or audio signal. Information can be understood, for example, as a description of the state of an object. It is intuitively clear that the more states the object can have the more symbols (characters, bits, signs) we need to have in order to describe the state unambiguously. In fact, if Zoya can have 2^B different states, then using B bits she can represent all her possible states in an unambiguous, *lossless* manner. Hence, Zoya can describe her current state to Yoshi by providing him with the B bit values that uniquely describe that state.

However, there is an omnipresent idea in nature and many aspects of human life: using shortcuts for things that are used more commonly or states that occur more frequently. Indeed, much before information theory was conceived, Morse code was built by using the idea that one does not need to always use an equal number of symbols to represent all the letters. Intuitively, fewer bits should be used to describe the states that occur more frequently and, vice versa, more bits are used to describe the states that occur rarely. The motivation is that, when we observe a long sequence of states, the average number of symbols sent will decrease compared to the case in which each state is encoded with a constant number of symbols.

To make this clearer, let us consider a simple example, in which Zoya can observe a system with three possible states, S_1, S_2, S_3 that occur with probabilities $p_1 = 0.5, p_2 = 0.25$, $p_3 = 0.25$, respectively. In order to communicate the state to Yoshi, Zoya can encode it into bits using the following mapping:

$$S_1 \rightarrow 0 \quad S_2 \rightarrow 10 \quad S_3 \rightarrow 11. \tag{5.1}$$

The *average* number of bits used to describe the three states of the system is:

$$\overline{B} = 0.5 \cdot 1 + 0.25 \cdot 2 + 0.25 \cdot 2 = 1.5 \quad \text{(bits/state)}. \tag{5.2}$$

This essentially puts forward *information as a probabilistic concept* or, more precisely, as the average number of bits required to describe the current state of a system, where the state occurrence follows a certain probability distribution.

As a more general case, let the system have S different states that occur with probabilities $p_1, p_2, \dots p_S$. Shannon proved that the minimal average number of bits $\overline{B}$ required to describe this system cannot be lower than the *entropy* of the distribution $p_1, p_2, \dots p_S$, defined as follows:

$$H(p_1, p_2, \cdots p_S) = -\sum_{s=1}^{S} p_s \log_2 p_s \tag{5.3}$$

and is always a non-negative quantity. The entropy for the example used above is:

$$H(0.5, 0.25, 0.25) = 1.5$$

such that the encoding that we have devised in (5.1) is optimal.

5.1.2 A Bit of Information

There is another important feature of the optimally compressed sequences, such as the one that can be produced using (5.1). Let us consider the following array of bits that Zoya sends to Yoshi in order to describe a sequence of states:

$$0010110101001010010 0100 \ldots . \quad (5.4)$$

Encoding is done in such a way that Yoshi can uniquely determine the state sequence as $S_1 S_1 S_2 S_3 S_1 \ldots$. Let us pick a random position in the sequence and ask the following: what is the probability that the bit at that position has the value 0/1? The selected position must be one of the following three: (1) the single bit representing S_1, (2) the first bit representing S_2 or S_3, (3) the second bit representing S_2 or S_3. Assume that the observed sequence of states is of length L, where L is a very large number. Then the total number of bits used to represent the sequence L is approximately $B = 1.5L$. Within the sequence of L states, S_1 occurs approximately $0.5L$ times, while S_2 and S_3 occur, each of them, $0.25L$ times. The total number of zeros within the B bits representing the sequence is a sum of:

- $0.5L$ zero bits corresponding to the single bits that represent occurrences of S_1
- $0.25L$ zero bits corresponding to the second bit that represents the occurrences of S_2.

The probability that a randomly picked position b from (5.4) carries a zero bit value is

$$P(b = 0) = \frac{0.5L + 0.25L}{L} = \frac{0.75L}{1.5L} = 0.5. \quad (5.5)$$

In fact, this is a general feature, stating that if the sequence of bits represents an optimally compressed sequence of states, then the bit values 0 and 1 are equiprobable. Furthermore, it can be shown (which we are not doing here) that the binary digits from the sequence (5.4) are independent in the sense that the sequence looks as if it has been generated by flipping a fair coin that results in 0 or 1 with probability 0.5.

This brings us to a meaningful *definition of a bit* as a binary variable that gets value 0 or 1 with probability 0.5. Indeed, one can calculate that the entropy of such a binary random variable is 1. The operation of encoding the source information into a compressed sequence of bits is referred to as *source coding*. In the layered system in Figure 4.1, the source coding operation takes part within the DataSender. Since the focus in this book is what happens with the data bits after the DataSender delivers them to the lower layers, our discussion always assumes that the information sources are compressed optimally and the data bits are equiprobable, such that one data bit carries one bit of information.

It can thus be stated that one bit of information corresponds to the level of uncertainty that one has about the outcome of the random experiment with two possible outcomes, both equally probable. As stated above, when compression is ideal, the messages produced by the source encoder contain "pure" bits, such that each bit carries new information that is in no way correlated to the information contained in another bit of the optimally compressed message.

5.2 Baseband Modules of the Communication System

Figure 5.1 describes the decomposition of the TXmodule of Zoya. In this chapter we will focus on the functionality of the TXbaseband and RXbaseband. Zoya's TXbaseband accepts bits and outputs a complex number, denoted z and called *baseband symbol* or simply *symbol*, which is passed on to the TXRFmodule. After various physical transformations and

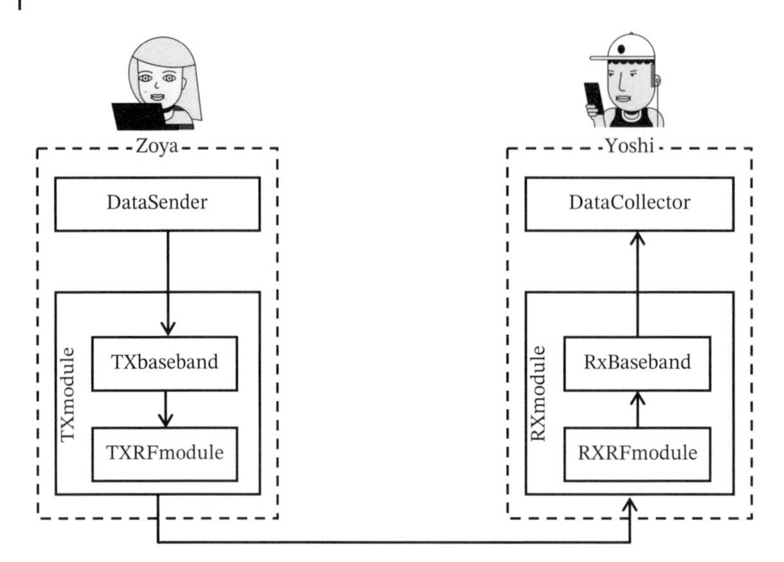

Figure 5.1 A look inside the black boxes of the TXmodule and RXmodule. TXbaseband receives bits from the sender and outputs complex numbers to TXRFmodule, transforms them and send them to the RXRFmodule, which outputs complex numbers to the RXbaseband.

external impacts, the symbol sent from the TXRFmodule arrives at the RXRFmodule of Yoshi. The RXRFmodule passes on a transformed version of z to the RXbaseband module. Yoshi received a complex number, denoted by y, and given by

$$y = hz + n \tag{5.6}$$

where h is the *channel coefficient* and n is the *noise*. Both h and z represent the external factors that affect the communication system. The expression (5.6) is the simplest representative of the *baseband model* or *symbol-level model*[1] of a wireless communication system. More precisely, (5.6) models a single-user system, as there is only one transmitter, but we will see that interference among multiple users is naturally integrated in the model. The TXmodule can send one symbol z each T_s s. It should be noted that the choice of T_s is due to factors that are outside the scope of the baseband model and will be discussed in Chapter 9.

5.2.1 Mapping Bits to Baseband Symbols under Simplifying Assumptions

Let us assume the setup in which $T_s = 1$ (μs, microsecond) and $h = 1$. In addition, let us adopt for a moment the assumption, highly unrealistic, that there is no noise and $n = 0$. In the simplest case, z can have only two complex values, say -1 and $1 + j$. We use -1 to represent the bit value 0 and $1 + j$ to represent the bit value 1. Under these assumptions, Yoshi receives the symbols of Zoya perfectly $y = z$, such that we can say that the baseband layer offers data transfer at a rate of 1 (Mbps, megabits per second).

We can do better than this, by picking four different complex values, say $1, j, -1, -j$ and associate each of them with a 2 bit value. A possible mapping is the one where the bits 00

1 We will use the terms baseband and symbol-level interchangeably, unless explicitly stated otherwise.

are represented by 1, written in short as $00 \to 1$; the remaining three bit combinations are mapped as $01 \to j$, $10 \to -1$, and $11 \to j$. By using this quaternary mapping, each symbol transmission carries two bits, thereby increasing the data rate to 2 (Mbps).

Following this line of reasoning, one can pick M different complex values, such that each baseband symbol carries $\log_2 M$ bits, leading to a data rate of $\log_2 M$ (Mbps). The condition to get this data rate is that each single symbol is received correctly, thereby implying that *all* $\log_2 M$ bits that are mapped to that symbol are received correctly. However, this involves several strong assumptions:

1. In order to receive useful data bits, Zoya and Yoshi must have agreed beforehand that the communication is about to take place.
2. Yoshi knows the time instants, spaced T_S s apart, at which Yoshi should receive the complex symbols sent by Zoya. In a realistic setting, the symbol sent by Zoya does not instantaneously appear at Yoshi's side, but rather after some delay ΔT.
3. The values of the channel coefficient h and the noise n are perfectly known *in advance* for each received symbol. Note that we have assumed that $n = 0$, but assuming that n is known in advance is equally good, as Yoshi can perfectly recover Zoya's transmitted symbol as $z = \frac{y-n}{h}$.

5.2.2 Challenging the Simplifying Assumptions about the Baseband

These assumptions are too optimistic for any practical setting. The first assumption may look trivial at first glance, but it becomes less so when one asks: how it is possible to have the very initial communication, the one used to exchange the control information or *metadata* and establish the channel? To understand the significance of this problem, consider a binary sequence that Zoya sends e.g. 00101101010010100100100 .... It is legitimate to ask: what had Zoya been sending before the first 0 of this sequence? Obviously there has to be a way to tell Yoshi where the data sequence begins. The problem becomes intricate if all that Zoya can send to Yoshi is 0s and 1s, but there is no special start symbol that is different from 0 and 1. We will return to this problem in Chapter 6, which treats the problem of how a communication channel is defined.

By far the strongest of the three assumptions is the one that the noise n is known. In fact, in the role of n in (5.6) is to model the ignorance about all the disturbances that can affect the reception of symbols at Yoshi's RXmodule. Such an ignorance can be modeled by *random noise*, such that for each symbol the value that is added to hz is a random complex number. Another common assumption is that the random noise for different symbols is *uncorrelated*: the random complex number that represents the noise that affects the ith received symbol is independent of the random complex numbers that represent noise elements added to the other received symbols. The assumption that the noise samples are not correlated is, in a way, the worst-case assumption for the receiver. This is because in the case in which the noise samples for different symbols are correlated the receiver can, in principle, use the knowledge derived from the noise that affects one symbol in order to remove at least part of the noise in another symbol.

An important observation is that, in the process of receiving the desired data from Zoya, Yoshi needs to receive other, auxiliary information that is a pre-condition for correctly receiving the desired data. For example, Yoshi needs to synchronize to the transmissions of Zoya, such that he knows the time instants at which the symbols sent by Zoya arrive. Furthermore, Yoshi needs to perform channel estimation in order to learn h. In practice there is even more auxiliary information that needs to be acquired, such as frequency estimation, but since we have not treated the concept of frequency in detail, we assume that frequency estimation is done perfectly. Recalling the discussion in Section 1.2 on the initial contact and link establishment, it can be stated that such type of initial communication should, in principle, be established by assuming no synchronization and no shared knowledge between Zoya and Yoshi. Hence, the *invite* packet should fulfill this role and be used to gain synchronization at the receiver; let Yoshi learn the channel coefficient and prepare for the subsequent communication.

The communication is further challenged by the fact that Yoshi should occasionally repeat the operation that enabled him to gather auxiliary information. For example, the delay ΔT at which the symbols sent from Zoya arrive at Yoshi may change over time and, if these changes are significant with respect to the value of T_s, Yoshi needs to account for them and perform a re-synchronization. Unless stated otherwise, in our discussion we will always assume that the synchronization is perfect.

The channel coefficient may also change in time due to, for example, movements by Yoshi and/or Zoya, such that Yoshi needs to repeatedly invest resources to estimate the channel. In the quest to receive the desired information, there is always a cost paid in terms of resources that need to be invested to acquire auxiliary information. A common way to acquire the value of h is for Zoya to use several symbols as *pilots* and send dummy data, known to Yoshi in advance, such that Yoshi can extract the value of h.

It is usually considered that the resources spent on pilots are negligible with respect to the resources used for the desired information. However, this is not always the case as the following idealized counter-example shows. Let us assume that h can take only two possible values, $h = -1$ or $h = 1$ and assume that this value changes randomly after each second symbol. If the random change is such that $h = -1$ and $h = 1$ occur with equal probability, then this means that Yoshi needs to receive one bit of auxiliary information every second symbol. On the other hand, if Zoya uses modulation that sends one data bit per symbol, then this means that the auxiliary information is 50% of the useful information that should be received by Yoshi, which is substantial. This example is rather extreme, but puts forward a caution that one cannot always ignore the resources that need to be used to acquire auxiliary information. As we will see later on, the knowledge of h has a decisive impact on how the communication channel is defined and the data rates that can be achieved when communicating over that channel.

5.3 Signal Constellations and Noise

5.3.1 Constellation Points and Noise Clouds

We start with a simple non-trivial case in which $h = 1$ and it is assumed that this fact is known to the transmitter Zoya and the receiver Yoshi. We further assume that the receiver perfectly knows the time instants at which the received symbols arrive. A packet that consists of B data bits is represented at the baseband layer through a sequence of

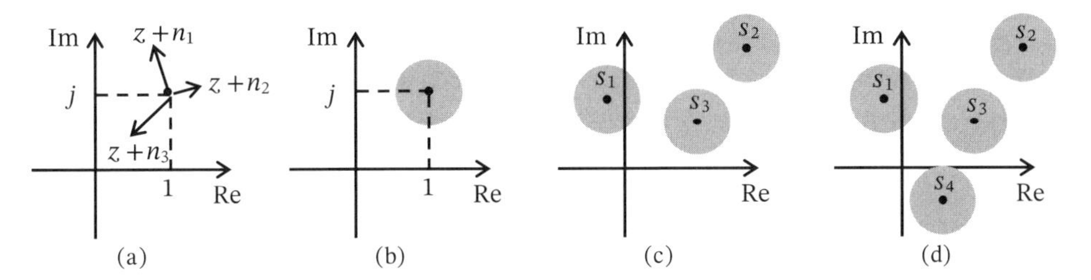

Figure 5.2 Effect of the additive noise complex baseband symbols. (a) Three different received symbols for the same transmitter symbol z. (b) The noise cloud is represented by the circle around z in which noise lies with very high probability p_c, e.g. 0.999. (c) Three example constellation points A, B, C; if binary modulation is used, then A and B should be chosen to represent 0 and 1 (or vice versa). (d) Constellation with $M = 4$ points.

L symbols. The relation between B and L will become clear further in the text. We will use the notation

$$y_i = hz_i + n_i = z_i + n_i \tag{5.7}$$

where $h = 1, i = 1, 2, \dots L$ and z_i is the ith baseband symbol that constitutes the packet sent by Zoya. The noise n_i is additive and it is independent of the value of the transmitted symbol z_i. We will assume that the noise can be described by a stochastic process that is *stationary*, such that for each received symbol the noise sample is drawn from the same probability distribution. It is further assumed that the noise has a mean value 0. This assumption is not limiting, since if the mean value of the noise is some non-zero value μ, then noise would have a constant bias of μ which, over a long period, could be estimated quite correctly and subtracted, leading again to the equivalent case of noise with zero mean value.

We take $1 + j$ as one complex value that is used to represent a specific sequence of bit values. Consider the case in which $z_i = 1 + j$ is sent three times $i = 1, 2, 3$. Since each n_i is a complex number, it is added to the signal, as shown in Figure 5.2(a), resulting in three non-identical received points. Intuitively, noise introduces the highest uncertainty if it is equally likely to move the received point y_i away from z_i in any direction on the complex plane. In other words, the noise introduces the highest uncertainty when it is *circularly symmetric*. Another commonly assumed property of the noise is that occurrence of noise with low magnitude is more likely than the occurrence of noise with large magnitude. This is certainly true for the widely used model of Gaussian noise. Therefore, the impact of the noise can be conveniently represented by a circular *noise cloud* around the correct signal z_i. Given that z_i has been sent, such a cloud represents the area in which the received signal y_i lies with very high probability, for example, $p_c = 0.999$; this is depicted in Figure 5.2(b). The higher the probability p_c, the larger the cloud size.

One of the key issues related to the baseband module is the design of the *constellation* of complex points that are transmitted from the TXbaseband. Each constellation point represents a certain combination of data bits, such that the important question is how to design the bit-to-symbol mapping, which tells us which bit combination should be represented by each particular constellation point. In the simplest case of *binary modulation*, the constellation consists of two points, representing 0 and 1, respectively. Figure 5.2(c) shows three candidate constellation points, s_1, s_2, and s_3. The best pair is s_1 and s_2, since they are the two points that have the largest *Euclidean distance* between them. The reason is the following: since low-magnitude values of the noise are more likely, it follows that a large Euclidean distance decreases the probability that the received symbol will come close to s_1 when s_2 is sent, and vice versa.

If each symbol can have only two complex values, then this symbol can be used to represent a single bit of information. However, there are many, uncountably infinite, complex values out there, so one may opt to choose four constellation points and use them to represent two bits of information, since the number of different states that can be represented by two bits is $2^2 = 4$. We can carry on and say that when we have a packet of B bits we can select a complex constellation that consists of 2^B complex points such that the whole packet can be transmitted with a single baseband symbol.

Where is the catch in this line of thinking? In order to explain this, we need to specify the exact way in which the receiver operates. With a slight abuse of the terminology, let us use the term *transmitted noise cloud* for the noise cloud around the constellation point that has been transmitted. This corresponds to the received noise cloud with channel coefficient $h = 1$. For each received noisy symbol, the receiver determines whether it belongs to any of the noise clouds such that the point that represents the received symbol lies in one of the shaded areas in Figure 5.2. If it does, then the receiver decides that the transmitted symbol that had been sent is the one that is at the center of that noise cloud. If the received symbol does not belong to any noise cloud, the receiver treats that reception as an error.

Let us fix the size of the noise cloud based on the desired probability p_c of the event that the received baseband symbol lies inside the transmitted noise cloud. When adding a new constellation point, we want its noise cloud not to overlap with the noise cloud of another constellation point. If the noise clouds are separated, then one can guarantee that the probability of successful reception of the symbol will correspond to the probability p_c. This is clearly not the case when the noise clouds of two or more constellation points are overlapping. As the constellation points become more separated, the performance of the communication link can be improved in two ways:

1. If the noise clouds are kept fixed and therefore the probability of success p_c is constant, then larger separation implies that more points can be added to the constellation.
2. If the noise clouds are increased and the number of constellation points is kept constant, then the probability of success p_c increases.

The only way to keep the newly added constellation points whose noise clouds do not overlap is to choose complex points z for which the magnitudes $|z|^2$ are large. This increases the average magnitude of the transmitted symbols, which is directly related to the *transmission cost*, something that we have not yet introduced in our communication model. The most common way to introduce transmission cost is to put a limit on the *transmission power* that can be used by the TXmodule. The physical power of a transmitted complex symbol z_i is proportional to $|z_i|^2$. The limit on the physical power can always be converted into a mathematical power limit imposed on the baseband complex symbols, such that we can say that $|z_i|^2$ represents the power of the symbol z_i. Once we take into account the transmission cost, it becomes clear that one cannot use constellation points with arbitrary magnitudes, which puts a limit on the number of different constellation points whose noise clouds are not overlapping.

5.3.2 Constellations with Limited Average Power

For analytical tractability and pedagogical approach to the constellation design, it is usually assumed that the *average power* is limited. However, in practice, the *maximal* power of the transmitter is always limited.

In order to illustrate the constraint imposed by the average power, let us assume that the TXmodule uses four constellation points, which is known as quaternary modulation. Furthermore, let us assume that the constellation be chosen to be as it is shown in Figure 5.2(d). Since each symbol represents two bits of data, the probability of sending any of the symbols s_1, s_2, s_3, s_4 is equal to $\frac{1}{4}$. Then the average power used by the TXbaseband is

$$P_{\mathrm{T}} = \sum_{i=1}^{4} \frac{1}{4} |s_i|^2. \tag{5.8}$$

Before we proceed to show why the quaternary constellation in Figure 5.2(d) is not particularly good, we make a slight digression about the properties of the noise. An important noise parameter is its average power P_{N}, which is assumed to be limited. For fixed P_{N}, it turns out that the noise distribution that leads to the highest uncertainty at the receiver is the Gaussian distribution. Specifically, each noise sample n_i is a complex number drawn independently from a Gaussian distribution with mean value 0 and variance P_{N}. The most common baseband communication channel is the one in which the input is z_i, the coefficient h is fixed throughout the whole transmission and known by both communicating parties, and the noise n_i has a Gaussian distribution. This channel plays the central role in designing and analyzing communication systems and is termed the *additive white Gaussian noise (AWGN)* channel. We postpone the explanation of the attribute "white" to Chapter 9, in which we discuss the frequency characteristics of the signals.

The complex Gaussian distribution is circularly symmetric and it can move the received symbol away from the transmitted symbol uniformly in any direction. This resonates well with the model of a noise cloud, in which no direction is preferred. The noise cloud is only an approximation of the Gaussian noise and its radius is proportional to $\sqrt{P_{\mathrm{N}}}$. Let us fix the probability $p_{\mathrm{e}} = 1 - p_{\mathrm{c}}$ that the received symbol falls out of the transmitted constellation point. A larger noise power P_{N} results in a larger noise cloud. For example, if the radius of the noise cloud is taken to be $3\sqrt{P_{\mathrm{N}}}$, then the probability that the noise falls in the cloud is around 0.95.[2] The Gaussian distribution can potentially produce infinitely high values and there will be always a non-zero probability that the received symbol falls outside of the transmitted noise cloud in Figure 5.2(b). Further, with the Gaussian noise it is always possible that the received symbol falls out of the transmitted noise cloud and lies in the noise cloud of another constellation points such that the receiver can even not detect that an error has occurred.

Based on this, the objective of the constellation design should be: *for given transmit power P_{T}, noise power P_{N}, and a given number of constellation points, pick the points in the complex plane such that they are separated as much as possible, thereby minimizing the probability that the noise causes confusion of one symbol with another*. In that sense, the constellation

2 Note that this is due to the complex, two-dimensional Gaussian random variable. For a real Gaussian random variable, the probability that the noise falls within the interval $[-3\sqrt{P_{\mathrm{N}}}, 3\sqrt{P_{\mathrm{N}}}]$ is 0.997, the well known "three sigma" number.

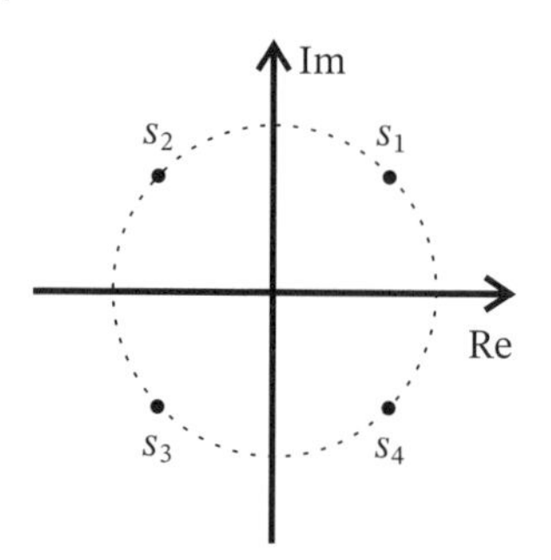

Figure 5.3 The QPSK constellation. All constellation points lie on a circle such that each of them has the same power. The decision region A_i for $i = 1, 2, 3, 4$ is the quadrant that contains the point s_i.

in Figure 5.2(d) is not optimal, as the four points can be separated in a better way, while keeping P_T unchanged.

The best way to separate the four points under a given power constraint is by using the QPSK (quadrature phase shift keying) constellation, depicted in Figure 5.3. Note that any rotation of this constellation works equally well. The decision criterion that has been created by using the noise clouds includes error detection, that is, it announces an error if the received symbol does not fall in any of the possible noise cloud.

It is also possible to make a decision criterion without error detection, which we illustrate as follows. We partition the complex plane into *decision regions* $A_1, A_2, \dots A_M$, where the number of decision regions M correspond to the number of different constellation points. For QPSK we have $M = 4$. The decision regions can be defined by specifying a function $g(\cdot)$:

$$\hat{z}_i = g(y_i) \tag{5.9}$$

that for a given received symbol y_i outputs the decided symbol $\hat{z}_i$, which is not necessarily equal to the transmitted symbol z_i. In this way, each possible received point y_i is mapped to a transmitted constellation point.

A *symbol error* occurs whenever the decided symbol $\hat{z}_i$ is not equal to the symbol z_i that had been originally transmitted. The decision regions are defined in such a way that the probability for symbol error is minimized. According to the common noise property, the received symbol y_i will stay, with high probability, close to the transmitted point z_i. It therefore makes sense to define $g(\cdot)$ such that for a received symbol y_i it is decided that the transmitted symbol $\hat{z}_i$ is the one that has a minimal Euclidean distance to y_i. For a general value of h, $g(y_i) = \hat{z}_i$ such that the distance between y_i and $h\hat{z}_i$ is minimized. For the AWGN channel, assuming that a decision is made for each symbol and that each of the possible symbols occurs with equal probability of $\frac{1}{4}$, the described per symbol decision rule is optimal. For example, for each of the four decision regions for the QPSK constellation, Figure 5.3 corresponds to one quadrant of the complex plane: A_1 is the quadrant that contains s_1, A_2 is the quadrant that contains s_2, etc.

5.3.3 Beyond the Simple Setup for Symbol Detection

Three observations are in order to supplement our simplistic treatment of constellation design, noise, and symbol detection.

Let us first take the case in which the transmitted symbols are not equiprobable. This can happen if the source of information is not encoded into an optimal, compressed sequence of data bits, since in that case each bit would get value 0 or 1 with equal probability. Specifically, let the information source be encoded in such a way that the bit sequence 00, represented by s_1, occurs with probability 0.7. Furthermore, assume that each of the remaining bit sequences $01, 10, 11$ and its respective symbol s_2, s_3, s_4 occurs with probability 0.1. Then the constellation should be designed so that s_1 is well separated from the whole group s_2, s_3, s_4, thereby decreasing the probability of error when s_1 is sent. On the other hand, s_2, s_3, s_4 are clustered together such that the probability of error when either of them is sent is higher compared to the probability of error when s_1 is sent. This is an example of *unequal error protection (UEP)*, in which the symbol s_1 is better protected compared to the other symbols. Despite the high probability of error when one of the symbols s_2, s_3, s_4 is sent, the average probability of error can be kept low, as the symbol s_1 occurs more frequently.

The second observation is related to the optimality of per symbol decision. If all the transmitted symbols are independent of each other, then knowing z_i cannot tell us anything about another symbol z_j and vice versa. Since the correspondent noise samples n_i and n_j are independent, then it follows that y_i and y_j are independent, such that it is optimal to make an individual per symbol decision: decide on z_i from y_i and on z_j from y_j. However, this is not true when z_i and z_j are dependent. Let us take the simple case in which each symbol is repeated twice such that $z_2 = z_1, z_4 = z_3, \cdots z_{2i} = z_{2i-1}$, and this is in advance agreed by the transmitter Zoya and the receiver Yoshi. Then the receiver should make a decision by *jointly* considering y_{2i-1}, y_{2i} instead of deciding separately on y_{2i-1} and y_{2i}. Assuming that $z_1 = z_2 = z$ and $h = 1$, it is optimal to combine the received symbols in the following way:

$$y' = y_1 + y_2 = z + n_1 + z + n_2 = 2z + n_1 + n_2. \tag{5.10}$$

Note that the desired symbols have added up coherently: z_1 and z_2 can be seen as two-dimensional vectors, both pointing in the same direction, i.e. the direction of z. On the other hand, the noise samples n_1 and n_2 are uncorrelated and point in random directions with respect to each other. Following the properties of the Gaussian distribution, the resulting noise $n_1 + n_2$ is again Gaussian noise with power $2P_{\text{N}}$, where P_{N} is the power of n_1 and n_2. The new decision function made for y' should output the closest point from a modified constellation: since $2z$ belongs to the set $\{2s_1, 2s_2, 2s_3, 2s_4\}$, the decision function should select one of those four points. Hence the distance between the constellation points doubled, while the radius of the noise cloud increased only $\sqrt{2}$ times to $\sqrt{2P_{\text{N}}}$. Therefore, the decision based on y' will have a lower probability of error compared to any of the per symbol decision for y_1 or y_2 and is the essence of the widely used techniques of *maximum ratio combining (MRC)* and *Chase combining*.

However, one may object to this statement as follows. It may happen that, for a particular instance of y_1 and y_2, the receiver does not make an error if it decides purely to use y_1, but it does make an error if it makes a decision to use y'. This could happen, for example, if the noise sample n_1 is close to zero, while n_2 has a very high magnitude in a wrong direction, i.e. bringing the received sample close to an incorrect received symbol. However, if we consider many received symbols, then the *average* number of errors when the receiver makes a decision by using y' is lower than the number of errors made when the receiver uses only one of the outputs y_1 or y_2.

Finally, the third observation is related to the issue of correlated noise. At first assume that n_1 and n_2 are perfectly correlated $n_1 = n_2 = n$. Then the noise in y' adds up coherently and thus has a power of $4P_\text{N}$, i.e. the noise cloud has the power of $2\sqrt{P_\text{N}}$. Seemingly, the correlation of the noise does not help in any way. However, if the sender Zoya knows in advance that the noise is correlated, she could send $z_1 = z$ as the first symbol and $z_2 = -z$ as the second symbol such that Yoshi could create the value $y_1 - y_2 = 2z$ and decode a perfect noiseless version of the transmitted symbol. In general, if the transmitter knows how the noise at the receiver is correlated, it can apply communication strategies that utilize this correlation and achieve a lower error probability.

5.3.4 Signal-to-Noise Ratio (SNR)

In the previous examples we have used $h = 1$ for simplicity. Assuming a fixed probability of error, which corresponds to a fixed probability that the received signal falls outside the transmitted noise cloud, then the following conclusions could be derived.

For a fixed noise power P_N, and therefore fixed size of the noise cloud with radius $\sqrt{P_\text{N}}$, the link can accommodate more constellation points as the transmit power P_T increases. Alternatively, for fixed P_T, the link can accommodate more constellation points as the power P_N of the noise, and thereby the radius of the noise clouds decreases. Therefore, the number of separated constellation points depends on how large P_T is relative to P_N, rather than the value of P_T in an absolute sense. This leads us to define the *signal-to-noise ratio (SNR)*, a quantity measured at the receiver that serves as an indicator of how many constellation points can be used and is defined as

$$\gamma = \frac{\text{received signal power}}{\text{received noise power}} = \frac{P_\text{T}}{P_\text{N}}. \tag{5.11}$$

The SNR is a dimensionless, scalar quantity, but it is convenient to express it in *decibels (dB)*. If the scalar value of the SNR is γ, then the dB value is:

$$\gamma_\text{dB} = 10\log_{10}\gamma \quad \text{(dB)}. \tag{5.12}$$

In general, the received signal is given by (5.6), with a channel coefficient h. Assuming that the receiver Yoshi knows h, he can use the complex conjugate value of the channel coefficient, denoted by h^* and create the quantity

$$y' = h^* y = |h|^2 z + h^* n. \tag{5.13}$$

This is an elementary example of *matched filtering* applied by the receiver. The useful signal $|h|^2 z$ is multiplied by the real quantity (phase-0 complex number) $|h|^2$ and therefore has identical structure as the originally transmitted signal z, except for an amplification factor $|h|^2$. The noise n is not correlated with h, such that h can be treated as a constant when finding the average power of $h^* n$, which in (5.13) is equal to $|h|^2 P_\text{N}$. The SNR of the received signal is calculated through:

$$\gamma = \frac{\text{received signal power}}{\text{received noise power}} = \frac{|h|^4 P_\text{T}}{|h|^2 P_\text{N}} = \frac{|h|^2 P_\text{T}}{P_\text{N}}. \tag{5.14}$$

Note that in the case of a scalar channel (5.6), one can create the following quantity:

$$y' = \frac{y}{h} = z + \frac{n}{h} = z + n' \tag{5.15}$$

and work with the equivalent channel that has $h = 1$ and noise power $\frac{P_N}{|h|^2}$, which is the variance of the Gaussian random variable n'. In this way the SNR remains as $\frac{|h|^2 P_T}{P_N}$.

5.4 From Bits to Symbols

5.4.1 Binary Phase Shift Keying (BPSK)

BPSK is a modulation method in which one bit is mapped to one baseband symbol. Figure 5.4(a) illustrates the BPSK constellation $S_B = \{-\sqrt{2}, \sqrt{2}\}$; the average power of this constellation is 2. There are only two ways to map the bits into symbols, the one shown in Figure 5.4(a) and the opposite one, where 0 is represented by $-\sqrt{2}$. Both ways work equally well since, due to the symmetry of the Gaussian noise, they have identical probability of error, irrespective of whether 0 or 1 is sent. In the case of BPSK the probability of *symbol error* is identical to the probability of *bit error*. The probability of error is a function of the SNR, denoted by γ defined in (5.14), and is given by:

$$P_B(\gamma) = Q(\sqrt{2\gamma}) \tag{5.16}$$

where the Q-function is defined as

$$Q(x) = \frac{1}{\sqrt{2\pi}} \int_x^{\infty} e^{-\frac{x^2}{2}} dx \tag{5.17}$$

and it decreases with x. More precisely, the error probability $P_B(\gamma)$ goes to 0 as the SNR γ goes to infinity, since in that case the size of the noise cloud relative to the Euclidean distance between the two constellation points goes to zero.

There is one subtlety regarding how SNR is defined in the case of BPSK and its impact on the probability of error. Recall that noise is modeled as a complex random variable that is circularly symmetric and the SNR in (5.14) is defined with respect to the total power P_N. Consider the matched filter in equation (5.13). If BPSK with constellation set $S_B = \{-\sqrt{2}, \sqrt{2}\}$ is used, then the information-bearing signal that is obtained after the receiver applies matched filter is either $-|h|^2\sqrt{2}$ or $|h|^2\sqrt{2}$. The SNR of the received signal is $\gamma = \frac{2|h|^2}{P_N}$. The noise n can be represented as

$$n = n_{Re} + jn_{Im}. \tag{5.18}$$

It is circularly symmetric, meaning that is has half of its power in the real dimension and half of its power in the imaginary dimension. In other words, n_{Re} is a Gaussian random variable with power $\frac{P_N}{2}$ and is independent of the Gaussian random variable n_{Im} that has also a power of $\frac{P_N}{2}$. Now let us consider h^*n that represents the noise that is relevant and directly affects the decision made by the receiver. The receiver applies a matched filter and, since h^* is a constant complex coefficient that is known by both transmitter and the receiver, it follows that the noise that affects the decision is:

$$n_1 = h^*n. \tag{5.19}$$

Since h is a constant value, n_1 is also a Gaussian random variable that is circularly symmetric, but it has a variance of $|h|^2 P_N$. Following the explanation from above, it follows that n_1

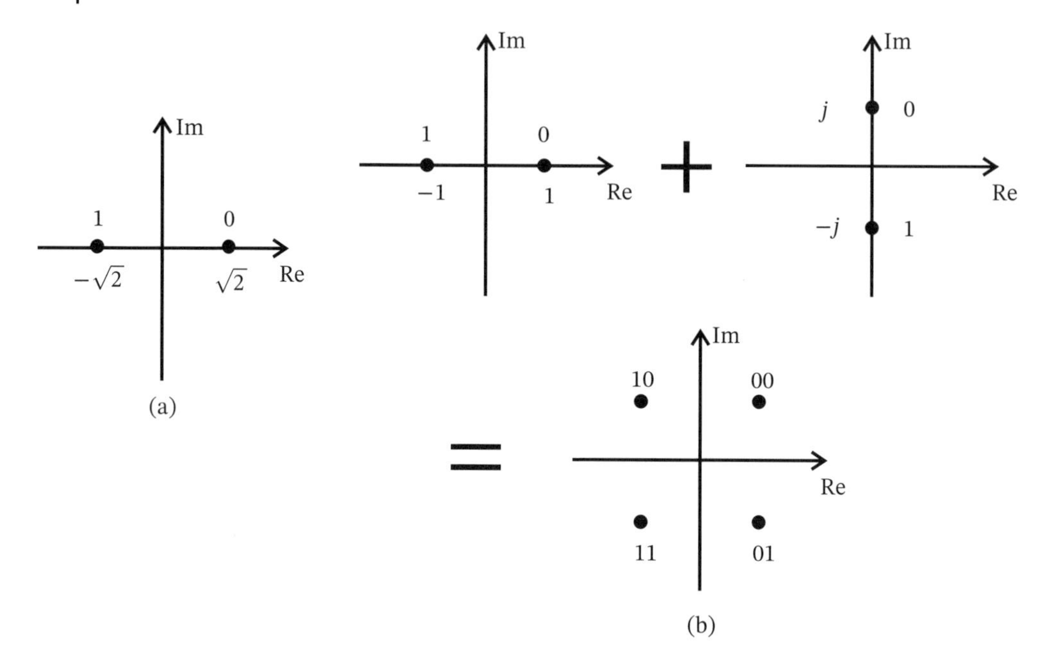

Figure 5.4 Bit-to-symbol mapping for two constellations of power 2. (a) BPSK. (b) QPSK with Gray mapping as a superposition of two orthogonal BPSK signals.

can also be decomposed into real and imaginary components:

$$n_1 = n_{1,\mathrm{Re}} + jn_{1,\mathrm{Im}} \tag{5.20}$$

where $n_{1,\mathrm{Re}}$ and $n_{1,\mathrm{Im}}$ are independent real Gaussian random variables with zero mean and each of them has variance $\frac{|h|^2 P_\mathrm{N}}{2}$. Note that for the BPSK constellation depicted in Figure 5.4(a) only the real noise component $n_{1,\mathrm{Re}}$ can cause error as it can bring the received signal close to the other constellation point. In fact, the noise cloud in this case is *one-dimensional*. The imaginary component $n_{1,\mathrm{Im}}$ is irrelevant, since it always moves the received signal within the same decision region. Therefore only half of the power of the complex noise is relevant for calculating the error performance of a BPSK signal.

5.4.2 Quaternary Phase Shift Keying (QPSK)

We have thus established the fact that the noise in the real dimension does not affect the transmission in the imaginary dimension and vice versa. Therefore, the imaginary component can be used to *multiplex* one more BPSK signal on the same complex symbol, thus leading to a QPSK modulation. Figure 5.4(b) shows how to synthesize a QPSK symbol from two independent BPSK symbols. The obtained QPSK constellation is:

$$\mathcal{S}_\mathrm{Q} = \{s_1, s_2, s_3, s_4\} = \{1+j, -1+j, -1-j, 1-j\}. \tag{5.21}$$

Therefore, the data rate of a QPSK system is equivalent to a double-speed BPSK system, which is a BPSK system whose symbol time is half of the QPSK symbol time and thus there are two times more BPSK symbols per unit time. The probability of each individual bit for given γ remains $P_B(\gamma)$, as given by (5.16). On the other hand, *symbol error* in QPSK is not identical with a bit error. A symbol is received correctly only when both constituent bits are received correctly, such that the probability of symbol error in QPSK at given SNR of γ is:

$$P_Q(\gamma) = 1 - (1 - P_B(\gamma))^2. \tag{5.22}$$

We have obtained the bit-to-symbol mapping in QPSK by synthesizing two independent BPSK symbols. Note that there are $4! = 24$ ways in which two bits can be mapped to four symbols. Due to the noise symmetry all these mappings fall into two different mapping classes. The way the two bits are mapped to a QPSK symbol in Figure 5.4(b) is known as *Gray mapping*. Besides allowing the symbol to send two bits independently in QPSK, Gray mapping has another desirable property. In order to explain it, we need to make a digression and define the concept of *Hamming distance*.

Let $\mathbf{b}_1$ and $\mathbf{b}_2$ be two arrays of bits or bit vectors, each consisting of L bits. The Hamming distance between $\mathbf{b}_1$ and $\mathbf{b}_2$ is the number of bit positions in which these two packets differ. For example, if $L = 3$ and $\mathbf{b}_1 = (010)$, $\mathbf{b}_2 = (100)$, then the Hamming distance is

$$d_H(\mathbf{b}_1, \mathbf{b}_2) = 2$$

Hence, if the bit sequence $\mathbf{b}_1$ was sent, but the bit sequence $\mathbf{b}_2$ was received, then the Hamming distance $d_H(\mathbf{b}_1, \mathbf{b}_2)$ is equal to the number of erroneous bits in the received packet. Following the idea that the small values of the noise are more likely than the large ones, then it is more likely that the noise causes less rather than more bit errors. Under such assumption, it is understandable that the number of bits in error should be minimized, as in that case the received data is closest, in terms of Hamming distance, to the data that has been sent. Furthermore, as it will be seen in the later chapters, there are error correction codes that can be used to correct the errors in the received packets and those codes cope better with fewer errors rather than a larger number of errors.

If QPSK modulation is considered, the bit array that is mapped to a single QPSK symbol consists of two bits. In Gray mapping, the Hamming distance between the bit vectors that correspond to neighboring symbols in the constellation is 1. Recall that the probability that symbol s_i has been received when symbol s_j has been sent, under influence from Gaussian noise, is inversely proportional to the Euclidean distance between the two symbols. In the particular QPSK constellation from Figure 5.4(b), if $1 + j$ is sent, then it is more likely to receive $1 - j$ than $-1 - j$ and, if Gray mapping is used, it is thus more likely to have one rather than two bits in error.

In the case of QPSK, one can say that the Gray mapping makes the Hamming distance between the bit vectors proportional to the Euclidean distance between the corresponding symbols and minimizes the average number of bit errors upon transmission of a single QPSK symbol.

5.4.3 Constellations of Higher Order

Each complex symbol consists of two dimensions, real and imaginary, and each dimension can carry information bits independently of the other dimension. In other words, each complex symbol contains *two degrees of freedom* for modulating information. In the case of BPSK and QPSK, each dimension contains only one bit of information. Note that the noise,

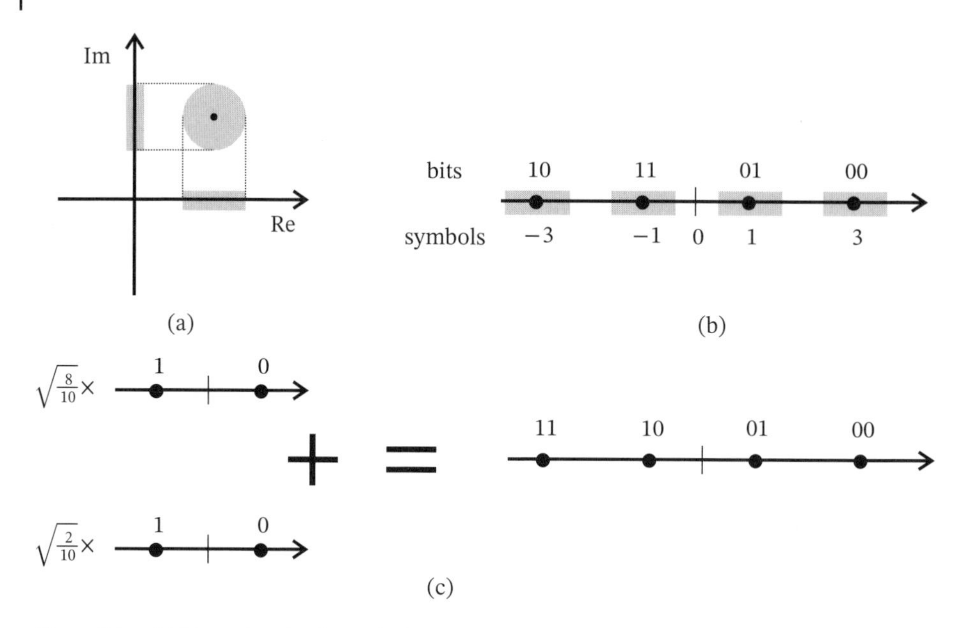

Figure 5.5 (a) The noise cloud is a circle in two dimensions and becomes an interval in one dimension. (b) Pulse amplitude modulation (PAM) with Gray mapping; the noise intervals are marked. (c) PAM as a superposition of two BPSK modulated signals with identical phase.

which is represented by a circular cloud in a two-dimensional complex plane, is projected on an *interval* when a single dimension is considered, see Figure 5.5(a). Recall also that the noise cloud in the case of BPSK is one-dimensional. This interval represents the region that will contain the noise-contaminated received signal in, say, 99.99% of the cases.

We can now use the idea to put more constellation points on a single dimension, which leads to *pulse amplitude modulation (PAM)*. If the noise intervals that are around each constellation point are non-intersecting, then it can be guaranteed that when a symbol is sent, error does not occur in 99.99% of the cases. Similar to the complex case, one cannot extend the constellation points arbitrarily over a single real dimension, as they need to be confined to a finite interval in order to satisfy the power constraint of the transmitter.

Figure 5.5(b) illustrates a 4-PAM constellation that uses Gray mapping of the bit vectors. The average power of this particular constellation is $P = \frac{1}{4}(1 + 3^2 + 1 + 3^2) = 5$. Alternatively, a 4-PAM constellation of power P can be synthesized by *superposition* of two BPSK constellations of power P in the following way:

$$z_4 = \sqrt{\alpha} z_{B1} + (1 - \sqrt{\alpha}) z_{B2} \tag{5.23}$$

where z_4 is the equivalent 4-PAM symbol, z_{B1} and z_{B2} are two independent BPSK symbols. The coefficient α is a *power coefficient* with $0 \leq \alpha \leq 1$. Since z_{B1} and z_{B2} are independent, the power P_4 of the 4-PAM constellation is calculated as:

$$P_4 = \alpha P + (1 - \alpha)P = P. \tag{5.24}$$

The process of constructing 4-PAM constellation by superposition is depicted in Figure 5.5(c). Note that both BPSK constellation points z_{B1} and z_{B2} need to lie on the

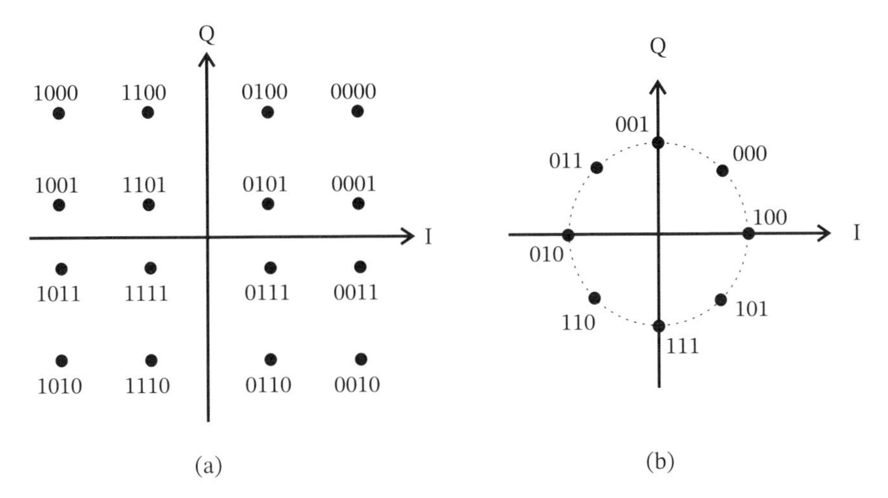

Figure 5.6 Higher-order constellations. Axis labels: I (in-phase) and Q (quadradure). (a) 16-QAM with Gray mapping; (b) 8-PSK with Gray mapping.

same line of the complex plane. In other words, they need to have the same *phase*, which is the angle of the complex constellation point when expressed in polar coordinates. In order to obtain the same constellation from Figure 5.5(b), the power coefficient should be chosen as $\alpha = \frac{8}{10}$ (or, equivalently, $\alpha = \frac{2}{10}$). It is interesting to note the following: no matter how the bits 0 and 1 are mapped on the constituent BPSK constellations, the bit-to-symbol mapping of the resulting 4-PAM constellation is not a Gray mapping. In other words, the Hamming distance between the bit vectors mapped to two neighboring constellation points is not necessarily equal to 1.

The PAM constellation can be applied in each of the complex dimensions and, similar to the way in which QPSK is synthesized from two BPSK signals, the two constellations can be superposed, leading to *quadrature amplitude modulation (QAM)*. Figure 5.6(a) depicts a 16-QAM constellation. In contrast to the superposition in (5.23), the two PAM signals that are superposed to obtain QAM do not have the same phase, but they are rotated by 90° with respect to each other. As a special case, one lies on a real and the other on the imaginary axis. This is identical to the superposition of two BPSK signals onto a QPSK signal. Note that in Figure 5.6 we have labeled the axes *I (in-phase component)* and *Q (quadrature component)*, respectively, which is the standard terminology in digital communication.

Considering Figure 5.6(a), if the bit-to-symbol mapping is made after the constellation is created, then it is possible to obtain a Gray mapping, as the one used in Figure 5.6(a). Nevertheless, it should be noted that the desirable property of Gray mapping (Hamming distance proportional to the Euclidean distance) is preserved only *locally*. For example, the bit vectors 1000 and 1010 have a Hamming distance of 1 and can be considered to be neighbors in the digital domain, while their corresponding symbols are not neighbors in terms of Euclidean distance, observed within the baseband domain. On the other hand, if the 16-QAM symbol is obtained by superposition of two constituent 4-PAM constellations, one as the *I*-component and one as the *Q*-component, then it is not possible to obtain Gray mapping; this is similar to the situation in which 4-PAM constellations are created by superposition of two BPSK signals.

There are other possible constellations, out of which we mention the *phase shift keying (PSK)* constellation. Note that not all of the 16-QAM symbols have the same amplitude, while in certain cases the underlying hardware requires that each constellation symbol has the same amplitude. PSK achieves exactly that; see Figure 5.6(b) that depicts the 8-PSK constellation. The name is due to the fact that information is carried solely in the phase of the symbol, not its amplitude.

5.4.4 Generalized Mapping to Many Symbols

A common characteristic of the modulation methods discussed so far is that each complex symbols is modulated in an identical way. For example, in 16-QAM the set of possible 16 constellation points for each symbol is identical and the 4-bit array of a particular value, e.g. 0000, is always mapped to the same constellation point.

A way to generalize this approach is to redefine the way in which the power constraint is taken into account. Namely, the constellations discussed until now satisfy the constraint of the average power calculated per symbol. Let us, instead, consider two complex symbols at a time (u, v) to represent a *hypersymbol* of size two, such that the constellation point is a *vector* of two complex numbers. The power of a specific constellation point of the hypersymbol is $|u|^2 + |v|^2$, while the *average* power per complex symbol per one transmitted constellation point is:

$$P(u, v) = \frac{1}{2}(|u|^2 + |v|^2). \tag{5.25}$$

Since each complex number has two degrees of freedom, the hypersymbol has in total four degrees of freedom. Hence, the design of a constellation that satisfies (5.25) is design of a constellation in a four-dimensional space, under a given power constraint.

For example, one can specify a set $\mathcal{S}$ of 256 possible constellation points, such that for each possible transmitted pair it holds that $(u, v) \in \mathcal{S}$. Each group of 8 bits is mapped to one constellation point and, as the 8-bit groups are equiprobable, each constellation point is transmitted with probability $\frac{1}{256}$. The constellation points are selected in such a way to satisfy the average power constraint P_{T} per single symbol or $2P_{\mathrm{T}}$ per hypersymbol of size two:

$$\frac{1}{256} \sum_{(u,v)\in\mathcal{S}} P(u, v) = 2P_{\mathrm{T}}. \tag{5.26}$$

We should think about 16-QAM as being *only a special way* to satisfy this constraint by choosing $u \in \mathcal{S}_Q$ and $v \in \mathcal{S}_Q$, where $\mathcal{S}_Q$ is the 16-QAM constellation and the set $\mathcal{S}$ is constructed as $\mathcal{S} = \mathcal{S}_Q \times \mathcal{S}_Q$. In other words, when designing the set $\mathcal{S}$ under the constraint (5.26), the values that u can take are not necessarily identical with the values that v can take. If one finds the best modulation set $\mathcal{S}$, where each constellation points is a vector of two complex symbols and the average power per symbol satisfies (5.26), then this modulation can never be worse than the choice of 16-QAM.

In fact, the best way to pack a given number of points in a four-dimensional space is to use a four-dimensional lattice, rather than a composition of two-dimensional lattices. A more general formulation of this approach can be stated as follows. The hypersymbol can consist of K symbols, such that we need to deal with a $2K$-dimensional lattice. Given that there is a power constraint P_T per symbol, then the power constraint per hypersymbol is KP_T. If one hypersymbol should contain $\log_2 M$ bits, then the size of the constellation set $\mathcal{S}$ is $|\mathcal{S}| = M$. The problem can be formulated as follows: Find M constellation points on a $2K$-dimensional lattice, such that the average power across all those points is less or equal to KP_T. The actual design is quite involved, but it illustrates an important principle: expanding the number of dimensions by spreading over multiple symbols and then designing a constellation in the space with extended dimensions. We will see in Chapters 6, 7, and 8 that this is the main principle underlying the techniques for reliable communication over unreliable channels.

5.5 Symbol-Level Interference Models

The representation of data packets through symbols significantly enriches the models discussed in Chapter 3, and that in itself offers new possibilities for design of communication schemes.

We start by considering how interference or collision is modeled at a symbol level. We refer again to Figure 3.1 from Chapter 3, where we introduced the concepts of capture effect and successive interference cancellation (SIC). For simplicity, we assume that Zoya and Xia are synchronized, such as in slotted ALOHA, and they both start transmitting towards Basil simultaneously. Each transmitted packet requires L complex symbols, which corresponds to a duration of a slot T. If the symbols are sent in a time-division manner, then each symbol has a duration of $\frac{T}{L}$. We can represent the ith symbol received by Basil as:

$$y_{B,i} = h_{ZB} \cdot z_i + h_{XB} \cdot x_i + n_{B,i} \tag{5.27}$$

- $y_{B,i}$ is the ith received symbol at Basil
- z_i and x_i are the ith transmitted symbols from Zoya and Xia, respectively
- n, i is the noise received by Basil during the ith symbol reception
- h_{ZB} and h_{XB} are the complex channel coefficients from Zoya to Basil and from Xia to Basil, respectively.

When there is no danger of confusion, we will drop the index i and write z to denote a symbol sent by Zoya. Note that the situation in which Xia is not transmitting, corresponds to having $x_i = 0$ for all i. In that case, the system model from (5.27) falls back to the single transmitter–single receiver system, described as

$$y'_B = h_{ZB} \cdot z + n_B. \tag{5.28}$$

Another assumption is that h_{ZB} and h_{XB} are constant during all L symbols and, furthermore, these coefficients are known by Basil. Let Basil receive

$$y_B = h_{ZB} \cdot z + h_{XB} \cdot x + n_B \tag{5.29}$$

and let us assume for a moment that he somehow knows the data bits of Xia (we will see in the following two subsections how that could happen). Then Basil can recreate the baseband signal x locally and, since he knows h_{XB}, he can create $h_{XB} \cdot x$, cancel it from y_B and obtain:

$$y'_B = y_B - h_{XB} \cdot x = h_{ZB} \cdot z + n_B. \tag{5.30}$$

In other words, if Basil knows the signal of Xia in advance, then the situation the Basil faces in decoding Zoya's signal is equivalent to the case in which Xia does not transmit/interfere at all. This leads to a seemingly trivial, but often overlooked fact: *the situation in which the information content of an interfering signal is known is equivalent to the situation in which there is no interfering signal at all.*

In the following two subsections we will represent the four scenarios depicted in Figure 3.1 and provide new insights brought by the symbol-level interference models.

5.5.1 Advanced Treatment of Collisions based on a Baseband Model

We first treat the cases depicted in Figures 3.1(c) and (d), where Zoya is closer to Basil compared to Xia, such that Basil receives a stronger signal from Zoya. In contrast to the packet-level model, having the symbol-level interference model (5.27) enables us to represent the stronger signal of Zoya by putting the following condition on the channel coefficients:

$$|h_{\mathrm{ZB}}| > |h_{\mathrm{XB}}|. \tag{5.31}$$

Let us for a moment ignore the received noise and, then given $h_{\mathrm{ZB}}, h_{\mathrm{XB}}$, as well as the constellations used to send z and x, we can speak about a *received constellation*, produced by the superposition:

$$h_{\mathrm{ZB}} \cdot z + h_{\mathrm{XB}} \cdot x. \tag{5.32}$$

Figure 5.7(a) shows the QPSK constellation of power P_{T}, used by both transmitters. Figure 5.7(b) depicts the channel coefficient h_{ZB} and the scaled version of the constellation $h_{\mathrm{ZB}} \cdot z$, while Figure 5.7(c) depicts the scaled version of the constellation $h_{\mathrm{XB}} \cdot x$. Note that the latter constellation is rotated as h_{XB} is a complex number with a phase that is different from 0.

The received constellation has 16 different points, as depicted in Figure 5.7(d). Each of the 16 points in the received constellation can be obtained from a *unique pair* of transmitted symbols $(s_{\mathrm{Z}}, s_{\mathrm{X}})$, where s_{Z} is sent by Zoya and s_{X} is sent by Xia. This means that, if Basil is able to correctly decode the point of the received constellation (5.32), then Basil can correctly decode s_{Z} and s_{X}, both of them simultaneously. This is termed *joint decoding* and it represents a new insight provided by the symbol-level interference model. Recall that the packet-level models treated in Chapter 3 are not capable of representing joint decoding in a similar way. Figure 5.7(d) shows the situation in which the SNR is very high, such that the noise clouds are not overlapping. Hence, all 16 constellation points can be reliably distinguished at the receiver. From each received symbol, Basil can jointly decode the symbols of Zoya and Xia, and these symbols will be correct with high probability[3].

A similar outcome can be achieved by intra-collision SIC. In contrast to joint decoding, which arises as a possibility when we consider a symbol-level interference model, this type of decoding is also possible within the packet-level model, described in Figure 3.1(d), where it is referred to simply as SIC. At first Basil tries to decode the packet of Zoya, since her signal is stronger. For our example, this means that, at first, Basil tries to determine in which quadrant each of the received symbols belong. Determining the quadrant corresponds to

3 We will sometimes use the abbreviation w.h.p. to denote "with high probability".

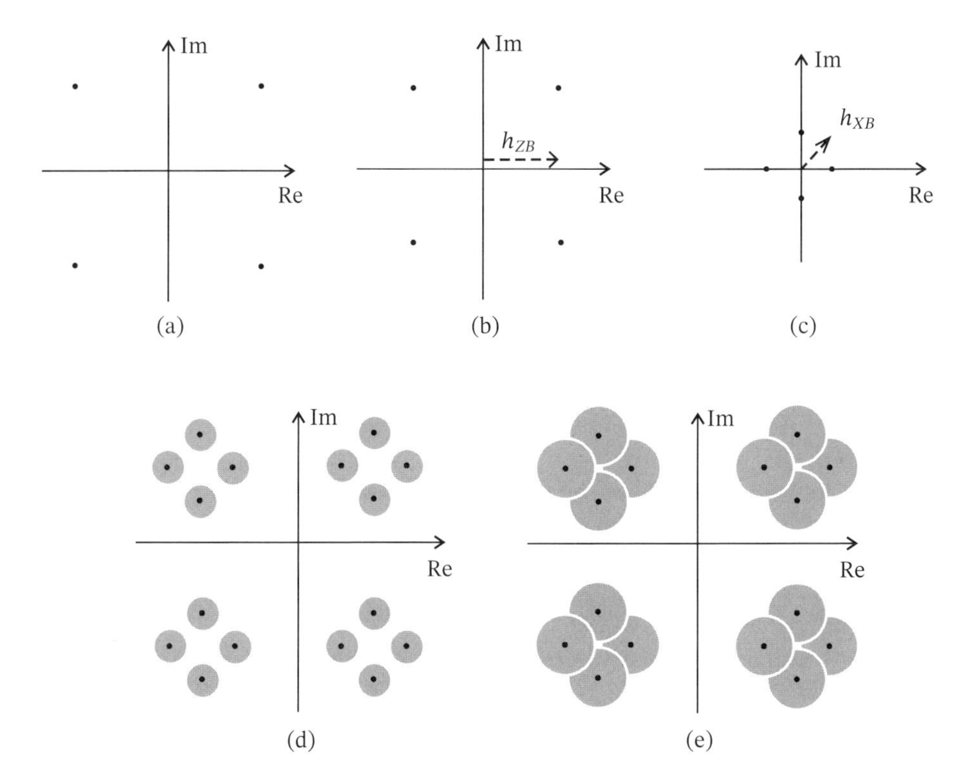

Figure 5.7 Received constellation from two interfering transmitters, each using QPSK signaling. (a) The original QPSK constellation; the power of this constellation is not plotted in proportion to the received constellations and in practice it is much larger than them. (b) The channel coefficient h_{ZB} (dashed) and the scaled constellation $h_{\mathrm{ZB}} \cdot z$. (c) The channel coefficient h_{XB} (dashed) and the scaled constellation $h_{\mathrm{XB}} \cdot x$. (d) The received constellation under high SNR. (e) The received constellation under low SNR, where it can be seen that $h_{\mathrm{XB}} \cdot x$ contributes to an expanded noise cloud.

determining the symbol s_{Z} sent by Zoya. After all L symbols s_{Z} are decoded, then Basil verifies if the packet is correct by using error detection (CRC check). If the CRC check is negative, then the process stops and neither the packet of Zoya nor the packet of Xia is decoded correctly. On the other hand, if the CRC check is positive and no error is detected in Zoya's packet, then Basil can recreate all symbols z and, with the knowledge of h_{ZB}, Basil can perfectly cancel the contribution of Zoya in the received signal (5.27). With that, Basil obtains a signal whose ith symbol is given by:

$$y'_{\mathrm{B},i} = y_{\mathrm{B},i} - h_{\mathrm{ZB}} \cdot z = h_{\mathrm{XB}} \cdot x + n_{\mathrm{B},i}. \tag{5.33}$$

Since this equation also describes a single-user, point-to-point channel, for which we have already described how demodulation/decoding is done, it follows that Basil can decode the packet sent by Xia.

It is important to note that, when SIC[4] is used, the receiver observes that the signal $h_{\mathrm{ZB}} \cdot z$ is received with a cloud that consists of noise plus interference and is thus an expanded

4 Unless explicitly stated otherwise, when we use the term SIC, we will refer to intra-collision SIC.

version of the original noise cloud. Hence, we can treat it as an expanded noise cloud, where the interfering signal $h_{XB} \cdot x$ is treated as part of the noise, such that the total noise observed when decoding z_i is $h_{XB} \cdot x + n_B$. In this example, x_i can take four different values, while $n_{B,i}$ is Gaussian, which means that $h_{XB} \cdot x + z_B$ is *not* Gaussian. However, in practice, the mix of noise and interference is often approximated to be a Gaussian random variable. This assumption enables us to approximate the error probability by using the known expressions from the Gaussian case. In this case, the variance of the resulting, approximately Gaussian, noise is:

$$|h_{XB}|^2 P_T + P_N \tag{5.34}$$

where P_T is the power of x_X and P_N is the power of the original noise. Now we can define the *signal-to-interference-and-noise ratio (SINR)*:

$$\frac{|h_{ZB}|^2 P_T}{|h_{XB}|^2 P_T + P_N} = \frac{\gamma_{ZB}}{1 + \gamma_{XB}} \tag{5.35}$$

where the last equation has been obtained by dividing the numerator and the denominator by P_N, while the remaining SNR variables are:

- $\gamma_{ZB} = \frac{|h_{ZB}|^2 P_T}{P_N}$ is the SNR of the received signal from Zoya when Xia is silent $x = 0$
- $\gamma_{XB} = \frac{|h_{XB}|^2 P_T}{P_N}$ is the SNR of the received signal from Xia when Zoya is silent $z = 0$.

Figure 5.7(e) shows a situation of a lower SNR: noise clouds are selected such that the quadrant of the received symbol can be decoded reliably; however, the received point within the quadrant cannot be reliably determined. If the receiver applies SIC, then this means that the data of Zoya will be correctly decoded, and Zoya's signal is subtracted from the received signal in order to obtain (5.33). However, the SNR of the obtained single-user channel is not sufficient to decode the signal of Xia correctly. This corresponds to the *capture effect*, discussed in Chapter 3, where only the packet of Zoya is decoded correctly.

The reader should note that the results of joint decoding and intra-collision SIC are not completely identical. With intra-collision SIC it can happen that the packet of Zoya is not decoded correctly in the first step, such that if the CRC check for the decoded packet of Zoya does not work, Basil does not proceed to decode the signal of Xia. In joint decoding, both packets are simultaneously decoded, while CRC check is run for each of the packets separately; therefore, it may happen that Xia's packet is received correctly even if Zoya's is not, which is never the case with intra-collision SIC.

As a final remark, we have presented the ideas of SIC by assuming perfect interference cancellation. In practice, it can happen that the signal of Zoya cannot be completely removed from the received signal, even if Zoya's data is decoded correctly and her signal is perfectly recreated locally by Basil. One reason for non-ideal interference cancellation can be the non-ideal knowledge of the channel coefficient h_{ZB}.

5.6 Weak and Strong Signals: New Protocol Possibilities

In this section we address the two remaining cases from Figure 3.1. The case in Figure 3.1(b) corresponds to *weak received signals*, where the SNRs γ_{ZB} and γ_{XB} are low, such that even in the absence of interference, the individual signal of Zoya or Xia is very unlikely to be decoded correctly. Using the example with QPSK from Figure 5.7, it follows that the constellation for each of the signals will be received with very weak power. In this case, the corresponding diagram of the joint received constellation, depicted in Figures 5.7(d) and (e), will be completely drowned in noise, such that neither joint decoding nor SIC is likely to produce at least one correct packet.

Figure 3.1(a) depicts the case in which each of the signals is received by Basil as a strong signal, such that the SNRs γ_{ZB} and γ_{XB} of Zoya and Xia, respectively, are high. In other words, the SNR of each signal is sufficient to perform correct decoding when the signal of the other user is absent. However, in the presence of the interfering signal, the SINR for both Zoya and Xia is low, such that intra-collision SIC cannot be initiated. Indeed, the SINR of Zoya and Xia, respectively, is:

$$\mathrm{SINR_Z} = \frac{\gamma_{ZB}}{1+\gamma_{XB}} \tag{5.36}$$

$$\mathrm{SINR_X} = \frac{\gamma_{XB}}{1+\gamma_{ZB}}. \tag{5.37}$$

As an example, if $\gamma_{ZB} = \gamma_{XB}$ is very large, then

$$\mathrm{SINR_Z} = \mathrm{SINR_X} \to 1 = 0 \quad \text{(dB)}. \tag{5.38}$$

In this case the reception of the signals from Zoya and Xia is said to be *interference limited* rather than *noise limited.*

This point is further illustrated by an extreme example in which Basil can have ambiguity about the signals sent by Zoya and Xia. Assume that $h_{ZB} = h_{XB} = 1$ and there is no noise $n_B = 0$, while both Zoya and Xia use QPSK symbols from the constellation $s_Z, s_X \in \{1+j, 1-j, -1+j, -1-j\}$. It can be easily checked that in such a setting, the received constellation has 9 points, rather than 16. If Basil receives $y_B = 0$, then there are four possible symbol pairs (s_Z, s_X) that could have produced that outcome:

$$(s_Z, s_X) \in \{(1+j, -1-j), (-1+j, 1-j), (-1-j, 1+j), (1-j, -1+j)\}.$$

In this case Basil has an ambiguity about the exact pair (s_Z, s_X) that produces $y_B = 0$, such that error-free decoding is not possible. The reader can note that there are other four received constellation points $y_B \in \{2, 2j, -2, -2j\}$ for which Basil has an ambiguity about the symbols transmitted by Zoya and Xia. For completeness, we note that there are four received constellation points for which Basil can recover the transmitted symbols of Zoya and Xia without ambiguity: $y_B \in \{2+2j, 2-2j, -2+2j, -2-2j\}$. However, since both Zoya and Xia are selecting their transmitted symbol in an independent and uniform random manner, there is a high probability of obtaining y_B, which leads to ambiguity and thus a decoding error.

5.6.1 Randomization of Power

Here we look at the implications that our observations about SNR and SINR can have on the design of random access protocols.

In the simple collision model, when two users collide, none of them can increase their chances for correct transmission by deterministically and persistently retransmitting their packet, as identical action from two or more users will continue to cause collisions. For this type of model, the reception is all-or-nothing, such that the only option is that the accessing devices use some ALOHA-like or probing mechanism: apply randomization in selecting the time for retransmission and hope that next time there will be no collision.

Having a symbol-level perspective on the interference, one can think of ALOHA as a protocol that applies randomization in the power domain rather than in the time domain. To see this, consider the example in which Zoya sends her packet to Basil in slot 1 using power P, experiences collision, randomly selected to wait during slots 2 and 3, and retransmit the packet in slot 4. This protocol is originally specified as a randomization in time.

Nevertheless, one can think about it as follows: Zoya has two possible levels for the transmitted power, P and 0. She sends the packet using power P, and gets feedback that the packet has not been received correctly. In slot 2 she flips a coin in order to decide whether to use power P or 0. She selects power 0, which trivially implies that her packet will not be received correctly. As a side note, when Zoya uses power 0 and does not transmit, Basil does not need to send feedback to her and tell her about the unsuccessful reception. She flips a coin in slot 3 and selects power 0 again and finally the coin flip in slot 4 results in a transmission decision, which means that she applies a transmit power of P.

The described power randomization seems to be an unnecessary complication of the ALOHA protocol. However, this impression can be improved by considering that the power can be controlled in a more fine-tuned manner, using multiple rather than only two levels, which leads to new opportunities for protocol design that rely on the symbol-level model. An equivalent of collision in the collision model is a wasted, undecodable interference in the complex baseband model, where each of the two users has a high SNR, but low SINR. Instead of randomizing the time instant of retransmission, both Zoya and Xia can decide to transmit in the next slot, but make a randomized choice of the power level that is used for transmission. This will again result in a collision in the next slot, but the new randomized instances of the transmit power of Zoya and Xia may result in favorable conditions for initiating an intra-collision SIC, such that the collision will eventually not be wasted.

More concretely, assume that for the retransmission Zoya keeps the power at the same level and γ_{ZB} stays the same, while Xia decreases her power and thus attains $\gamma_{XB,1} < \gamma_{XB}$ during the retransmission. Let us assume that $\gamma_{XB,1}$ is chosen in a way that leads to the following properties. $\gamma_{XB,1}$ should be sufficiently low, such that SINR_Z is sufficiently high so that Zoya's packet can be decoded under the interference from Xia. After Zoya's packet is decoded and canceled, then Xia's packet is ready for decoding without any interference. This can be carried out successfully if the choice of Xia's transmit power is such that $\gamma_{XB,1}$ is still sufficiently high.

The previous example is clearly illustrating the fact that, by enriching the collision model into a baseband interference model, the design space for random access protocols expands by allowing fine-tuned randomization in the power domain. This strategy is, in principle, also applicable to the case of weak signals. One can think of a scenario where Zoya and Xia starts sending with minimal power. If no feedback is received from Basil, then one reason is that the signals are too weak for Basil to detect anything. As a next action, Zoya can increase her power to a randomized higher layer, and Xia can do the same, such that the received SNR and SINR can come to a level that can result in capture and, possibly, intra-collision SIC.

5.6.2 Other Goodies from the Baseband Model

The representation at the level of baseband symbols has also implications on the techniques based on inter-collision SIC, described in relation to coded random access in Section 3.4.2. From our discussion in Section 3.4.2 related to the example in Figure 3.4, the reader can get some intuition into why it is possible to have inter-collision SIC. However, the idea of inter-collision SIC becomes much clearer when it is described in the context of an interference model with complex symbols. For the example in Figure 3.4(a), the collision received in slot 1 is represented as an interference between two uplink signals. The received signal in slot 1 is given by:

$$y_{\mathrm{B},i,1} = h_{\mathrm{ZB}} \cdot z_i + h_{\mathrm{XB}} \cdot x_i + n_{\mathrm{B},i,1}. \tag{5.39}$$

The ith received symbol in the collision-free slot 2 is described by

$$y_{\mathrm{B},i,2} = h_{\mathrm{ZB}} \cdot z_i + n_{\mathrm{B},i,2}. \tag{5.40}$$

It is important to note the noise instance $n_{\mathrm{B},i,2}$ in slot 2 is independent and, generally, different from $n_{\mathrm{B},i,1}$. From this slot, Basil decodes packet **Z**, then reconstructs Zoya's baseband symbol z_i and subtracts $h_{\mathrm{ZB}} \cdot z_i$ from the signal buffered from slot 1. Now Basil obtains:

$$y'_{\mathrm{B},i,1} = h_{\mathrm{XB}} \cdot x_i + n_{\mathrm{B},i,1} \tag{5.41}$$

and based on this attempts to decode Xia's packet **X**.

An important advantage of the symbol-level interference model is that it can easily scale to more than two interfering users. If there are K transmitters sending simultaneously to Basil, the ith received symbol is:

$$y_{\mathrm{B},i} = \sum_{k=1}^{K} h_{k\mathrm{B}} x_{k,i} + n_{\mathrm{B},i} \tag{5.42}$$

where $x_{k,i}$ is the ith transmitted symbol from transmitter k, while $h_{k\mathrm{B}}$ is the channel coefficient from transmitter k to Basil. There is still a single noise sample per symbol, $n_{\mathrm{B},i}$.

The concepts of joint decoding, SIC, capture can be generalized to the case of K transmitters and the associated probabilities can be systematically derived starting from (5.42). Without going into detail, we note that when the interfering signals contains contribution from multiple independent transmitters, then one can invoke the central limit theorem and thus justify the treatment of the interference as Gaussian noise. For example, when user 1 is about to be decoded from (5.42), the interference from $K-1$ users where K is large looks like a Gaussian noise, unlike the case in which there is a single interferer.

Finally, we briefly discuss how the model is changed when the packets sent by different transmitters do not necessarily fully overlap. This is another aspect that is very difficult to account for in a packet-level model, while it can naturally be represented in the symbol-level baseband model. Recall that when the model (5.27) was introduced, we assumed that the packets of Zoya and Xia have an identical number of L symbols and both packets start at the same time. Instead of presenting a general model, we provide an example to show what is changed when the previous assumptions do not hold. Let us assume that Zoya's packet consists of $L_{\mathrm{Z}} = 100$ symbols, while Xia's packet consists of $L_{\mathrm{X}} = 40$ symbols. Let us also assume that Xia starts her transmission when Zoya transmits the 11th symbol. Using the Gaussian approximation from above we can state the following:

- The first 10 and the last 50 symbols of Zoya are received with SNR of γ_{ZB}.
- The symbols 11–50 sent by Zoya are received with SINR of $\frac{\gamma_{ZB}}{1+\gamma_{XB}}$.
- All the symbols sent by Xia are received with SINR of $\frac{\gamma_{XB}}{1+\gamma_{ZB}}$.

If the SINR of Xia is high, then Xia's signal is decoded first, subtracted, and then Zoya's signal is decoded free of interference. With our current assumptions about the symbol generation process, all the symbols sent by Zoya are independent, i.e. knowing something about symbol 1 of Zoya tells us nothing about symbol 31 of Zoya. This is changed when *error control coding* is introduced, which is described in the upcoming chapters, where the symbols sent within the same packet do have certain dependency. In this case, a viable approach would be to decode at first the interference-free symbols of Zoya and then use that information, along with the received symbols 11–50, to decode the whole packet of Zoya. If Zoya's packet is decoded correctly, then SIC is started and Xia's packet is decoded in the absence of interference.

5.7 How to Select the Data Rate

The definition of a wireless link given under the collision model, illustrated in Figure 1.1, is very much idealized and the value of the data rate R is not selected based on the physical properties of the model. Furthermore, once R is fixed, it is received perfectly when the receiver Yoshi is within a given distance d from the transmitter Zoya, provided that there is no collision. On the other hand, no data is received correctly when Zoya and Yoshi are separated for a distance larger than d. In Chapter 3 we have softened the link model by introducing probabilistic elements: the data rate R is still fixed, but the probability of receiving the packet decreases as the distance to the receiver increases, which is a better approximation of the physical reality. The symbol-level model allows for a more sophisticated modeling of the distance effects by relating it to the SNR and the probability of error.

5.7.1 A Simple Relation between Packet Errors and Distance

Before examining how the error probability depends on the distance, we need to look closer into the definition of a data rate. Let us consider a packet of L symbols. If we choose to use BPSK, then $D = L$ bits are sent during time T; however, if 16-QAM is chosen, then $D = 4L$ bits are transmitted during the same time. The respective data rates are given by:

$$R_{\mathrm{BPSK}} = \frac{L}{T} \qquad R_{16-\mathrm{QAM}} = \frac{4L}{T} = 4R_{\mathrm{BPSK}}.$$

It should be noted that the data rates exemplified above are *nominal rates*, which reflect what the receiver gets when we assume that there are no errors. The actual data rate depends on the *packet error probability (PEP)*, which is determined according to the modulation applied by the transmitter and the SNR at the receiver. In our current model for packet transmission, the data bits mapped to one symbol are independent of the data bits mapped to the other symbols. Hence, errors occur independently for each symbol reception and in order to receive the packet correctly, the receiver needs to receive correctly *all* symbols belonging to that packet. Note that, if only part of the symbols belonging to the packet are received correctly, then they are of no use, since the packet has a single integrity check that tells whether it has been received correctly or not. In other words, for a packet that is not received correctly there is no some "correctness localization" property that would indicate which bits are received correctly with certainty.

Let P_s denote the probability of symbol error. For a packet of L symbols, PEP is given by the probability that there is at least one symbol error, which is given by:

$$\text{PEP} = 1 - (1 - P_s)^L. \tag{5.43}$$

It is important to note that, for fixed modulation used by the transmitter Basil, P_s increases as the SNR of Zoya decreases.

The baseband model can bring novel insights about how the data rate is affected by the received power of the useful signal. In order to do that, at first we need to identify how the received power depends on the distance. The basic physical law implies that electromagnetic power decreases as a square of the distance from the power emitting source. Nevertheless, as Chapter 10 shows, this simplified law is valid under idealized assumptions: empty space around the emitter, no obstacles or scatterers, etc.

Yet, we can use this propagation law, as well as the speech analogy, to make the following simple approximation: if Zoya is closer to Basil compared to Xia, then a signal sent by Basil will be received by Zoya with a higher power compared to the signal of Basil that is received by Xia. In contrast to the received power of the useful signal, the noise power is a phenomenon related to the receiver and it is independent of the transmitted signal. Consider the situation in which Basil transmits to Zoya and Xia. Assuming that Zoya and Xia have identical receivers, then they have identical noise power, but since Zoya receives a stronger signal from Basil, it follows that the SNR that Zoya has at her disposal to decode the signal from Basil is higher compared to the SNR available to Xia.

Let us assume that Basil fixes the modulation and uses a certain modulation constellation to transmit. This also results in a fixed nominal data rate R. As the distance between Basil and Zoya increases, the power at which Zoya receives Basil's signal decreases. This means that the constellation that is observed by the receiver Zoya, without considering the noise, shrinks towards $(0, 0)$ in the complex plane. In other words, the distances among the constellation points decrease, while the noise power and therefore the noise cloud size stays the same. This results in a decreased SNR. Note that the described, distance-dependent decrease in SNR, is different from the mechanism illustrated in Figure 5.7(d) and Figure 5.7(e), where the received constellation remains constant, while the noise cloud increases. The decreased SNR explains the higher error rate experience by Zoya as she moves to a larger distance from the sender Basil, which worsens the overall data reception.

At this point the reader may raise an objection to the consistency of the uncoded transmission and the packet integrity check, used to obtain equation (5.43). In order to have an integrity check, the data (information) bits should be appended with check bits that are used for error detection. The check bits are calculated based on a certain code, say CRC, and each check bit is a deterministic function of the information-carrying bits. Therefore, a symbol carrying check bits is dependent on the symbols that carry information bits, which could potentially be used to increase the overall probability of successful packet reception.

In this chapter we ignore this possibility and we assume that each symbol is decoded individually at the decoder and mapped onto bits. These bits are not altered further based on the side information that can be received from the dependent symbols, but instead only an error detection check is made and, if it fails, the packet is deemed incorrect. Clearly, this type of operation offers a sub-optimal performance, but the expression for PEP in (5.43) is correct. Otherwise, the dependency among the bits provides the basis for error correction, which is discussed in Chapter 7.

5.7.2 Adaptive Modulation

Having introduced the PEP, the next question is how to select the nominal data rate. Let Zoya transmit to Yoshi using a fixed transmission power and let Yoshi receive the signal of Zoya at an SNR denoted by γ. Zoya has an option to use BPSK, QPSK or 16-QAM and the respective symbol error probabilities are denoted by $P_{\mathrm{B}}(\gamma), P_{\mathrm{Q}}(\gamma)$, and $P_{16}(\gamma)$. Recall that the probability of symbol error is inversely proportional to the Euclidean distance between the constellation points. Therefore, the following relation for the symbol error probabilities holds *for any SNR* γ:

$$P_{\mathrm{B}}(\gamma) < P_{\mathrm{Q}}(\gamma) < P_{16}(\gamma). \tag{5.44}$$

If the number of symbols in a packet is fixed to L, then the previous relation immediately implies that the following holds for any SNR:

$$\mathrm{PEP}_{\mathrm{B}}(\gamma) < \mathrm{PEP}_{\mathrm{Q}}(\gamma) < \mathrm{PEP}_{16}(\gamma) \tag{5.45}$$

for the respective PEPs that can be calculated using (5.43). Thus, at any SNR and for a fixed number of symbols L, the packet sent by using BPSK is the most reliable one and the reliability decreases as the modulation order increases.

Nevertheless, this does not imply that Zoya should always choose BPSK. If Zoya has, for example, a large file of B_{F} bits to send to Yoshi, then Zoya is interested in minimizing the total time required to send the file. The transmission of the large file takes long time, such that the minimization of the total transmission time T_{F} corresponds to the maximization of the average goodput $\mathrm{G} = \frac{B_{\mathrm{F}}}{T_{\mathrm{F}}}$. Observed over many slots, this becomes statistically equivalent to the *expected number of bits* that Yoshi receives correctly in a single time slot of duration T, written as follows:

$$\mathrm{G}_M(\gamma) = \frac{B_M}{T}(1 - \mathrm{PEP}_M(\gamma)) \tag{5.46}$$

where M denotes the modulation used and can have value B, Q or 16, such that $B_B = L$, $B_Q = 2L$, and $B_{16} = 4L$. If Zoya knows γ and can calculate $\mathrm{PEP}_M(\gamma)$, then Zoya compares $\mathrm{G}_B, \mathrm{G}_Q$, and G_{16}, and selects the modulation that provides the highest goodput.

The properties of the PEP are such that there is no single modulation that offers the highest throughput at all SNRs. Figure 5.8 shows a typical relationship of the goodputs offered by different modulation schemes at different SNRs, assuming that a packet has the same length in terms of number of baseband symbols. The main idea behind *adaptive modulation* is that Zoya should adapt her selection of modulation to the current SNR. For the example depicted in Figure 5.8, at low SNR and until SNR $= \gamma_1$ BPSK should be used. For SNR between γ_1 and γ_2, QPSK should be used. Finally, when the SNR is above SNR $= \gamma_2$, 16-QAM should be used. A condition to apply the described adaptation mechanism is that

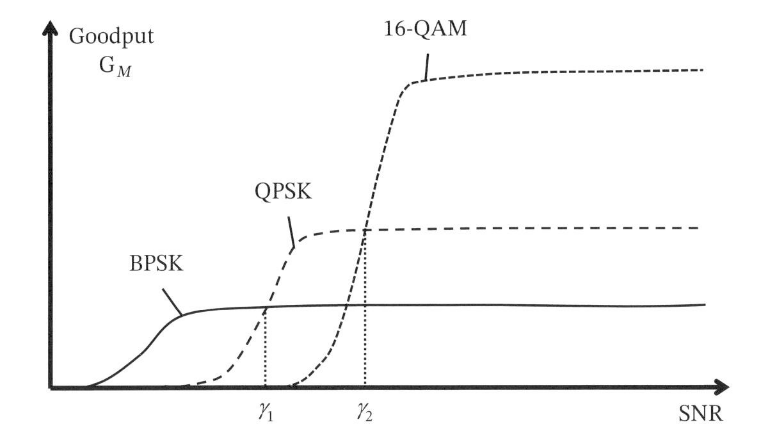

Figure 5.8 Curves of the goodput G_M for three different modulations BPSK, QPSK, and 16-QAM versus the SNR. The SNRs at which the modulation should be changed are denoted by γ_1 and γ_2.

Zoya knows the SNR at which her signal is received by Yoshi. Since, for fixed transmission and noise power, the SNR changes only due to the channel coefficient, we can say that Zoya can apply adaptive modulation only if she has *channel state information (CSI)*, which is in fact the knowledge of γ.

We note that the baseband model works with the assumption that the time period T_s between two symbols is fixed, such that the number of symbols sent over a time interval of duration T is also fixed to $L = \frac{T}{T_s}$. The choice of the symbol time period T_s cannot be described within the baseband model and is based on the physical constraints of the communication signals. This reveals the limitations of the baseband model, which hides the underlying physical phenomena and the associated analog signals that are used for communication.

Another interesting factor related to adaptive modulation is the packet length. When defining the rate in (5.43), we have assumed that the packet length is fixed in terms of number of symbols L, rather than number of data bits. If the data packet is fixed in terms of number of data bits, then different modulations will result in different number of baseband symbols. For example, if the packet contains D bits, then this corresponds to $L_B = D$ symbols with BPSK, $L_Q = \frac{D}{2} = \frac{L_B}{2}$ symbols with QPSK and $L_{16} = \frac{L_B}{4}$ symbols with 16-QAM. Using again (5.43) to calculate the PEP, it is seen that now not only the probability of symbol error P_s changes, but also L changes: the higher the modulation order, the higher the P_s, but the lower the L. While a higher P_s increases PEP, a lower L decreases it. This will lead to different thresholds for adaptive modulation, $\gamma_1, \gamma_2, \gamma_3$ from Figure 5.8, compared to the case when the thresholds are determined by assuming a fixed number of baseband symbols per packet.

These observations can be used to optimize the packet length along with the modulation in a particular setting. The advantage of using a shorter packet when the modulation order is higher should be applied cautiously, since there is a lower limit to the packet length and that is determined by the amount of *overhead* in the packet. As discussed in the previous chapters, not all the bits in a packet are actual information carrying bits, but some of them are used to carry signaling information, such as source/destination address, CRC bits, etc.

As the packet length decreases, the percentage of the overhead increases, thus contributing to the decrease in the goodput.

5.8 Superposition of Baseband Symbols

In the way we have used modulation constellation until now, a group of bits is mapped to a constellation point and the total power of the baseband symbol is used to carry this constellation point. However, the baseband model, unlike the packet-level models, offers the possibility to split the total power and allocate it to multiple independent streams of data bits. This leads to the idea of *superposition coding*, a baseband procedure in which the total available power is allocated to multiple independent data packets and these packets are transmitted simultaneously.

In order to illustrate the basic idea, we can start from equation (5.23), which involves superposition of two simplest modulated signals (BPSK) and generalize to the superposition of S independent data packets. Each of those packets consists of L baseband symbols. The ith symbol transmitted by Zoya can be represented as:

$$z_i = \sum_{s=1}^{S} \sqrt{\alpha_s} z_{s,i} \tag{5.47}$$

where $\sqrt{\alpha_s} z_{s,i}$ is the contribution to the transmitted symbol from the ith symbol of the packet belonging to the sth data stream. The coefficient α_s stands for the fraction of total power that is allocated to the sth packet. Since the transmission power of Zoya is limited, the following must be satisfied:

$$\sum_{s=1}^{S} \alpha_s = 1. \tag{5.48}$$

More generally, the sum of all alphas in the previous equation needs to be less or equal to 1. However, it is intuitively clear that we should be using all the available power in order to get the highest throughput. It should be noted that the selection of the modulation for the sth packet z_s can be independent from the modulation used for the other packets.

Figure 5.9 illustrates a simultaneous transmission of three packets from Zoya to Yoshi by using superposition coding. Assume, for simplicity, that the channel coefficient between

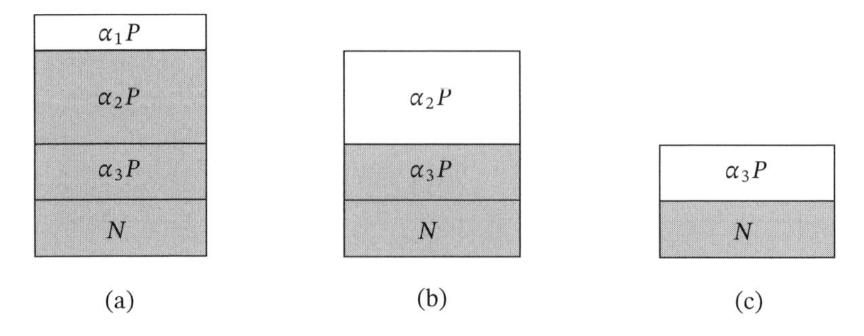

Figure 5.9 Three superposed packets with powers $\alpha_1 P$, $\alpha_2 P$, and $\alpha_3 P$, while N is the noise power. For each decoding step in SIC the shaded boxes represent the power of the noise. (a) Decoding of packet 1. (b) Decoding of packet 2. (c) Decoding of packet (3).

Zoya and Yoshi is 1, such that the symbols received by Yoshi have the same power as the symbols sent by Zoya. Decoding is done by SIC. Yoshi decodes the first signal z_1, with power $\alpha_1 P$, by treating z_2 and z_3 (powers $\alpha_2 P$ and $\alpha_3 P$, respectively) as additional contributions to the noise. The total noise and interference that is perceived during the decoding of z_1 is a sum of the powers of z_2 and z_3, as well as the power of the true noise n. The SINR available for decoding z_1 is:

$$\text{SINR}_1 = \frac{\alpha_1 P}{\alpha_2 P + \alpha_3 P + N}. \tag{5.49}$$

After z_1 is decoded, it is cancelled and the receiver proceeds to decode z_2, but now the SINR is:

$$\text{SINR}_2 = \frac{\alpha_2 P}{\alpha_3 P + N}. \tag{5.50}$$

Finally, after z_1 and z_2 are decoded and canceled, z_3 is decoded in the presence of noise only, such that the SINR is in fact the SNR and is equal to

$$\text{SINR}_3 = \text{SNR}_3 = \frac{\alpha_3 P}{N}. \tag{5.51}$$

As the modulation of each superposed packet can be chosen independently, it follows that the data rate of each of those packet can be different. Recalling the principles of adaptive modulation explained in the previous section, the data rate R_s of the sth data packet is selected based on the power allocated to that packet as well as the power of the noise/interference that affects its decoding. In this case, the additional noise is the sum of the powers of $z_{s+1}, z_{s+2}, \ldots z_S$, and n. If all the data packets are decoded correctly, the achieved data rate is

$$R = \sum_{l=1}^{S} R_s. \tag{5.52}$$

However, if during the decoding procedure the first $K-1$ packets are decoded correctly, but the Kth is not, then the decoding is stopped and the achieved data rate in that particular transmission is:

$$R' = \sum_{s=1}^{K-1} R_s. \tag{5.53}$$

Superposition coding is creating *self-interference* among multiple packets, such that different packets are disturbing each other's reception. Although it seems that such an operation is sub-optimal, we present two problems for which the idea of superposition coding offers elegant solutions.

5.8.1 Broadcast and Non-Orthogonal Access

The term "broadcast" is used in communication engineering in different ways, not always consistent with each other[5]. For example "broadcast traffic" refers to, for example, a transmission of a TV tower where the same data is intended to be received by all terminals that are in range of the transmitter. In information theory a "broadcast channel" is a

5 This is also addressed in Chapter 1 when multicast is introduced.

communication channel in which Basil's transmission is received by two or more receivers (Zoya, Xia, etc.), but the transmitted signal may contain separate data for each of the receivers, unlike the usual use of a "TV broadcast".

An example of a broadcast channel is a base station (Basil) whose signal is received by more than one terminal in its range. A single transmission made by Basil, consisting of L baseband symbols, may simultaneously carry *common data*, which corresponds to the "broadcast traffic" as in TV broadcast, as well as *individual data* for Zoya and/or Xia. Common data is intended to be received by both Zoya and Xia, and in that sense it represents broadcast traffic. As already indicated in Chapter 1, we will also use the term *multicast traffic* to explicitly denote the case in which there are more than one intended recipients of the message. The individual data is part of the *unicast traffic*. The individual data for Zoya *should* be decoded by Zoya, while it *may* be decoded by Xia, provided that it helps Xia to decode her individual data or the common data. Clearly, the identical statement is valid when the roles of Zoya and Xia are reversed.

Let us consider the case of a broadcast channel in which there is no common data (multicast traffic), but only individual data intended for Zoya and Xia. A straightforward way to do it would be to use time sharing, as in the downlink transmissions described in Chapter 1, such that in one time slot only one user is served. We can present such time sharing, although in a convoluted way, through superposition coding: in the slot in which Zoya is served, there is superposition coding that allocated all the power to the packet of Zoya and zero power to the packet of Xia. In the next slot the situation is the opposite one. This representation implies that the time-shared downlink transmission is a degenerate form of superposition coding, where the superposition coefficients can be varied from one transmission to another.

The previous interpretation motivates us to try to improve the throughput by exploring the design space of superposition coding. In each transmission there are $S = 2$ data packets, for Zoya and Xia, with power fractions α_Z and α_X, respectively. In contrast to the degenerate case, we can set both α_Z and α_X to have non-zero values. Furthermore, note that the values of α_Z and α_X can be changed in each slot and this should be either known by the receivers in advance or communicated through the transmission. For example, the transmitted superposed packet can be preceded by a common preamble, modulated without superposition coding, which carries information about the coefficients α_Z and α_X and is a common data, intended to be decoded by both Zoya and Xia. The transmission methods in which both α_Z and α_X are not zero represent a class of methods, sometimes referred to as *non-orthogonal multiple access (NOMA)*.

Now let us assume that Zoya receives a signal that is stronger compared to the signal received by Xia. Furthermore, let us assume that both data packets are modulated with QPSK and the coefficients are chosen, for example, $\alpha_Z = 0.1$ and $\alpha_X = 0.9$. The overall baseband signal transmitted by Basil, denoted by b, is given by:

$$b = \sqrt{\alpha_Z} z + \sqrt{\alpha_X} x \tag{5.54}$$

where z is the part of the symbol sent by Basil to Zoya, while x is the part of the symbol sent by Basil to Xia. The nominal data rate of each signal z and x is set to $R_Z = R_X = 2$ bits/symbol. The noise power is assumed to be identical for both receivers. In this setup, the first signal to be decoded is x and, after canceling it, z is decoded.

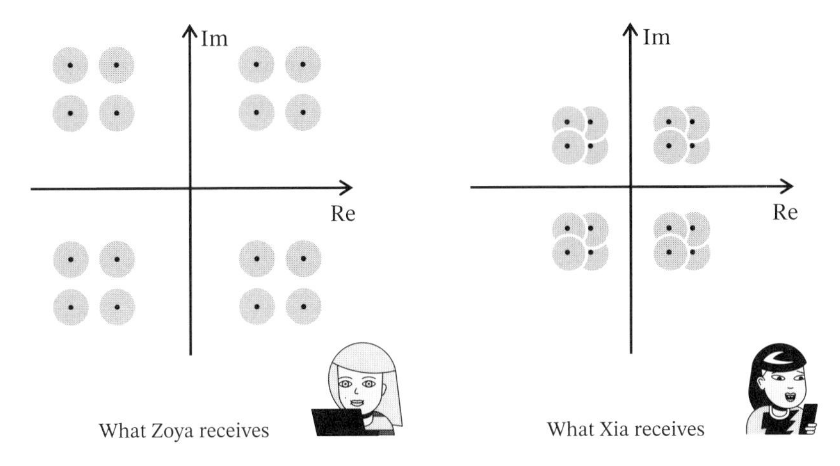

Figure 5.10 Received constellations of Zoya and Xia when Basil broadcasts two superposed QPSK signals. The signal received by Xia is weaker than the signal received by Zoya.

Figure 5.10 depicts the received signals for Zoya and Xia, respectively. By choosing $\alpha_X > \alpha_Z$, as in our example, we are ensuring the following type of operation. Both Zoya and Xia can decode x by determining in which quadrant the received signal lies; however, only Zoya can reliably decode z. If x contains the data for Xia and z contains the data for Zoya, then the goal of the broadcast is achieved, as each receiver manages to receive the desired data. In addition, Zoya is also decoding the data of Xia, since that is part of the process for getting the desired data.

The performance of non-orthogonal access, implemented as a broadcast with superposition coding, can be tuned by choosing several parameters: $\alpha_Z, \alpha_X, R_Z, R_X$. On the other hand, the definition of a suitable performance objective can be more involved. If we want to optimize the overall throughput, then the solution is to set $R_X = 0$ and $\alpha_X = 0$, such that the weaker receiver is ignored. Another performance objective could be to maximize the overall throughput by guaranteeing minimal throughput for both Zoya and Xia; in this case we must allocate some power to Xia by setting $\alpha_C > 0$ as her data rate cannot be $R_X = 0$. In any case, the performance that can be obtained through the optimization process will be always at least as good as the reference scheme that uses time sharing and serves one user at a time.

The decoding of someone else's data may raise the issue of data security. Indeed, Zoya is able to decode the data of Xia, but she may not be able to *interpret* the data, since the data bits obtained by decoding the baseband signals can be encrypted with a secret key known only by the transmitter Basil and the receiver Xia. Yet, once Zoya has the data for Xia, she may try to break the cipher and decrypt Xia's data. The situation is completely opposite with the data of Zoya: to start with, Xia cannot decode it, such that one can argue that Zoya's data is protected even more than what is offered by the encryption key.

5.8.2 Unequal Error Protection (UEP)

When the transmitted data represents multimedia content, such as an image or video, then not all the parts of the data have the same importance. As an example, consider a video encoded in the following way: basic data D_B that enables the video to be reconstructed at a lower quality and enhancement data D_E that improves the resolution. It is essential to receive the basic data D_B, which makes it a must-have. The reception of the enhancement data improves the quality, such that it is nice-to-have. This is different from the common way of treating all bits equally at the reception and can be reflected in a suitable utility function $u(D_B, D_E)$, which puts higher value on the bits belonging to the basic data compared to the enhancement data. As a simple example, the utility function u gets the following values for different outcomes:

- $u = 0$ if no basic data D_B is received, regardless of whether D_E is received
- $u = 1$ if D_B is received, but D_E not
- $u = 2$ if both D_B and D_E are received.

Let q_B be the probability that D_B is received, but not D_E; while let q_{BE} be the probability that both D_B and D_E are received. Then the expected utility at the receiver is:

$$E[u] = q_B \cdot 1 + q_{BE} \cdot 2. \tag{5.55}$$

Zoya sends a video to Yoshi over a wireless link and the objective is to maximize the expected utility. Zoya uses superposition coding, such that the basic data is modulated on z_B with α_B, and the enhancement data z_E is superposed to it with α_E. By choosing α_B and α_E, Zoya can tune the probabilities q_B and q_{BE} in order to maximize the expected utility, which supposedly reflect the overall video experience for Yoshi. Increasing α_B increases q_B, but at the same time α decreases, thus decreasing q_{BE}. Note that decoding with SIC is particularly suitable in this case, as Yoshi never attempts to decode D_E without having decoded D_B.

The idea of tuning the superposition coding to optimize the expected utility is particularly suitable for multicasting multimedia traffic to multiple receivers. This can be related to our broadcast discussion from the previous section as follows. The common message that is multicast contains the low-quality video that should be received by the worst receiver (Xia). The message unicast only to the stronger receiver (Zoya) contains the quality enhancement data. This represents an example of cross-layer optimization, as it both involves determination of the type of data (common or individual) jointly with the optimization of its transmission. The example can be generalized by applying superposition of multiple packets, which would correspond to multiple levels of enhancement. Instead of adjusting the transmission to the worst receiver, superposition coding enables graceful degradation of the multimedia quality across the receivers, such that better receivers get better quality, but worse receivers get at least some low-quality content.

5.9 Communication with Unknown Channel Coefficients

The central assumption in using the AWGN channel model $y = hx + n$ between the transmitter Xia and the receiver Yoshi is that Yoshi knows the channel coefficient h. This means that the channel h should be constant for a reasonably long time in order to allow both Yoshi and Xia to learn the value of h and then use this knowledge during the actual data communication.

The time during which the channel can be treated as constant is referred to as *coherence time* and let us assume that this corresponds to the duration of L_C symbols. Let L_P be the required number of symbols to be used as pilots, such that Yoshi can get a satisfactory estimate of h. Then the AWGN communication model is feasible to be used if $L_P \ll L_C$ and thus the channel estimation overhead is negligible. However, in some cases this assumption is not applicable. For example, this is the case when Yoshi moves relatively quickly with respect to Xia, such that the channel changes rapidly and L_C is low. In that case the value of L_P may be close to L_C, leading an inefficient operation: most of the resources (symbols) are spent on learning h, which is only an auxiliary step and not a goal per se, while few resources are left for the real goal, which is the communication of data from Xia to Yoshi.

In order to cope with this situation, Xia and Yoshi should resort to *non-coherent* communication, where the communication takes place under the assumption that h is unknown to Yoshi. Working with unknown h means that the receiver Yoshi treats h as a part of the generalized noise, which is a term that denotes the set of all factors that introduce random disturbance to Yoshi's reception. Note that h plays the role of a multiplicative noise instead of an additive one.

Let us start from the extreme case $L_C = 1$, where h changes independently from one symbol to another. The ith received symbol is given by

$$y_i = h_i x_i + n_i \tag{5.56}$$

but here both h_i and n_i are random unknown disturbances. The simplest communication scheme that Xia can use is *ON-OFF* keying, also known as binary *amplitude shift leying (ASK)*: Xia sends the bit value 0 by setting $x_i = 0$ and 1 with $x_i = \sqrt{P}$, where P is the maximal allowed power. Yoshi calculates the amplitude of the received signal:

$$r = |y_i| = |h_i x_i + n_i|. \tag{5.57}$$

The received signal needs to be tested with respect to a fixed threshold ρ, chosen in a way to minimize the probability of error. Recall that in the BPSK over AWGN channel, error occurs only if the noise n_i has an excessive value, regardless of whether the bit value 0 or 1 is sent. In the case of ASK, when 0 is sent, again the additive noise is the only factor that contributes to error, as $r = |n_i|$ and an excessive value of the noise, either positive or negative, can bring the value of r over the threshold ρ, leading to an erroneous decision by Yoshi. However, when the point 1 of the ASK constellation is sent, then $r = |h_i\sqrt{P} + n_i|$. In this case, even if there is no noise $n_i = 0$, an error can occur if the value of $|h_i|$ is very low, resulting in $r < \rho$. We note that binary ASK is the basic option; there can be Mary ASK schemes that use M different power levels, thus transmitting $\log_2 M$ bits per symbol.

If the coherence time $L_C > 1$, then other non-coherent communication schemes are possible, taking advantage of the fact that h is unknown, but constant. One idea is to use *differential modulation*, which works as follows. Let us assume that Xia sends the same symbols as in BPSK, that is $x = 1$ or $x = -1$, where for simplicity we have assumed that the average (and in this case the maximal as well) transmit power of Xia is also one. Let the first symbol x_1 be arbitrary and it does not contain any information for Yoshi; but let us fix $x_1 = 1$ and assume that this is known both to Xia and Yoshi. Then Xia modulates the first bit when sending x_2. The information is not in x_2 itself, but in the *difference* between x_2 and x_1: Xia sends the bit value $b_1 = 0$ by setting $x_2 = x_1$ and $b_1 = 1$ by setting $x_2 = -x_1$. Yoshi creates the decision variable:

$$r = |y_2 - y_1| = |hx_2 + n_2 - hx_1 - n_1| = |h(x_2 - x_1) + n_2 - n_1|. \tag{5.58}$$

When $b_1 = 0$ then $|x_2 - x_1| = 0$, while when $b_1 = 1$, then $|x_2 - x_1| = 1$ and, similar to the case with ASK, Yoshi should compare the decision variable with a threshold ρ. The second bit b_2 is modulated in identical way, by choosing x_3 relative to x_2. This type of modulation is called *differential BPSK (DBPSK)*. Continuing in the same fashion, Xia can send L differentially modulated bit values by sending $L + 1$ symbols.

Differential modulation requires h to be constant throughout the whole packet transmission, since it should be $h_2 = h_1 = h$, then $h_3 = h_2 = h_1 = h$, etc. In order to break this chain of equality with the previous symbol, let us consider the case with coherence time $L_C = 2$. We show how to construct a scheme that takes advantage of the fact that h is constant over two symbols only. Let us use the term *vector symbol* to denote a vector of two symbols $\mathbf{x} = (x_1, x_2)$. Using Figure 5.1 from the beginning of this chapter and the layering terminology, the use of vector symbol can be seen as creating an extra layer that resides above the layer TXbaseband–RXbaseband and DataSender–DataCollector.

In the simplest case, let the vector $\mathbf{x}$ sent by Xia take only two possible values, $\mathbf{x} = (1, 0) = \mathbf{s}_1$ and $\mathbf{x} = (0, 1) = \mathbf{s}_2$. Furthermore, in order to facilitate two-dimensional illustrations, let us simplify and assume for the following discussion that the values of x, h and n are real. We can now represent the two possible symbols in a two-dimensional coordinate system, as in Figure 5.11, where the horizontal axis represents the value of hx_1 and the vertical axis the value of hx_2. The gray circle represents the cloud of the additive noise n; note that here the cloud is two-dimensional, not because n is complex, but because the figure represents the two real noise samples of two consecutive symbols. Figure 5.11(a) illustrates the case $h = 1$ and the dashed line is used to mark the decision region, which is half-plane. Clearly, if $h = 1$ and Yoshi knows it, then the choice of the two constellation points in Figure 5.11(a) is not optimal, as they could be separated by a larger distance. This is not valid in the non-coherent regime, as Yoshi does not know h and the value of h can change to a negative one, as in Figure 5.11(b). However, regardless of the actual (real) value of h, what remains invariant is that $h\mathbf{s}_1$ lies always on the horizontal axis, while $h\mathbf{s}_2$ is always on the vertical axis. In other words, the one-dimensional linear subspace in which $h\mathbf{s}_i$ resides remains invariant. This means that the decision criterion, determined by the dashed line, stays identical in both Figure 5.11(a) and Figure 5.11(b). Once the

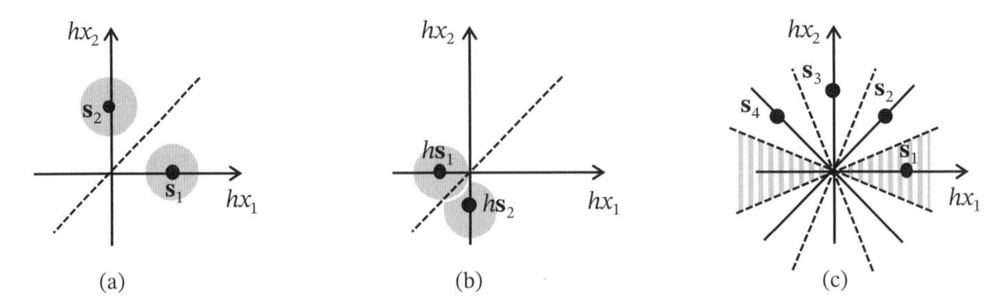

Figure 5.11 Illustration of non-coherent communication with linear sub-spaces. The values of the symbols, channel coefficients and noise are assumed to be real. (a) Case $h = 1$ and two constellation points. (b) When $h \neq 1$ each of the constellation points remains on the same axis, unless disturbed by noise. (c) Quaternary non-coherent modulation; $h = 1$ is assumed.

decision criterion is determined, then the probability of error can be determined as the joint probability that h and n lead to the event in which the noise cloud crosses the decision region.

In the previous scheme Xia sends one bit by using two symbols, such that the nominal data rate is halved compared to, say, BPSK in which one bit is modulated in each symbol. This can be compensated by using a quaternary non-coherent modulation, as illustrated in Figure 5.11(c), where it is assumed $h = 1$, since two bits are sent over two symbols. The decision region for each possible symbol spans 90°, as is illustrated for the decision region related to the symbol $\mathbf{s}_1$. The fact that the linear subspace does not change can be used to design modulation schemes in multi-dimensional spaces, with complex h, x, z and using coherence time $L_C > 2$.

We end this section by discussing briefly the possibilities for designing hybrid schemes. When the coherence time is very long, $L_C \gg 1$, then it is acceptable to waste some symbols as pilots that carry only dummy data in order to apply coherent communication afterwards. On the other hand, if L_C is very close to 1, then it is not feasible to explicitly invest resources in pilots. However, the following opportunity should be noted: if Xia sends several symbols to Yoshi to be received non-coherently and Yoshi buffers those symbols, then after he decodes the data correctly, he can reuse the known data in order to estimate the channel. Hence, when L_C is in the intermediate region, not too large and not too close to one, then one can think of a communication protocol in which the part of the packet that should contain pilots also contains data that is detected non-coherently. The non-coherent data can have its own integrity (CRC) check, such that Yoshi can verify its correctness and use it to estimate the channel h. The non-coherent part of the packet uses $L_P < L_C$ symbols, where L_P is known in advance by Xia and Yoshi. After that, for the remaining $L_C - L_P$ symbols Xia uses coherent modulation. This is a simple example of a scheme with *blind channel estimation*, where the estimation is not done by dummy data, previously agreed between Xia and Yoshi.

5.10 Chapter Summary

This chapter has established the connection between, on the one hand, *networking*, where the basic building blocks are packets, and, on the other hand, the *digital communication theory*, where the main themes are bits, symbols, modulation, and noise. We have worked with the baseband model of the system, where the basic units are complex symbols sent by the transmitter and their noisy versions at the receiver. The baseband model qualitatively expands the models from the previous sections in multiple ways: *(i)* it captures the central role that the SNR has in digital communications; *(ii)* it allows the error performance to be related to the received power and thus introduce effects that a physical distance has on the signal reception; *(iii)* it reveals new possibilities for design of protocols and transmission schemes based on control of the transmit power or creating multiple streams through superposition coding. Finally, the chapter has introduced the important idea of non-coherent communication that occurs when the receiver cannot learn the channel before decoding the useful data.

5.11 Further Reading

The material discussed in this chapter can be further explored in Gallager [2008], which discusses the principles of digital communication as well as Proakis and Salehi [2008], which provides a comprehensive discussion on signals and modulation. The ideas of superposition coding for non-orthogonal access in a broadcast channel have been introduced in Cover [1972], while it has been popularized under the term NOMA in Saito et al. [2013]. Unequal error protection in broadcast channels has become a part of the Digital Video Broadcasting (DVB) standard, see [2004–11].

5.12 Problems and Reflections

1. *Collision outcomes with actual modulation*. The introduction of a symbol level model allows us to assume that the packets contain symbols with a particular modulation type and then derive the probabilities of successful decoding and capture. Assume that Zoya and Xia are transmitting simultaneously in the uplink to Basil. The signal received by Basil is:

$$y_{\mathrm{B}} = h_{\mathrm{ZB}}z + h_{\mathrm{XB}}x + n \tag{5.59}$$

   where n is the noise with variance P_{N}. Both Zoya and Xia transmit with a fixed power P_{T} and send packets consisting of L symbols. The communication channel is slotted and one slot contains L symbols. A packet transmission can start only at the start of a slot, such that the transmissions of Zoya and Xia are packet synchronous. Each symbol contains QPSK modulated data and it is assumed that a packet is decoded successfully if all its L symbols are decoded successfully.
   The SNRs of Zoya and Xia are denoted by γ_{ZB} and γ_{XB}, respectively, and are given by:

$$\gamma_{\mathrm{ZB}} = \frac{|h_{\mathrm{ZB}}|^2 P_{\mathrm{T}}}{P_{\mathrm{N}}} \qquad \gamma_{\mathrm{XB}} = \frac{|h_{\mathrm{XB}}|^2 P_{\mathrm{T}}}{P_{\mathrm{N}}}. \tag{5.60}$$

   There are three possible outcomes of this collision between Zoya and Xia, as Basil will be able to decode 0 packets correctly, 1 packet correctly (through capture), and 2 packets correctly (capture and successive interference cancellation). Investigate how these probabilities are affected by changing $\gamma_{\mathrm{ZB}}, \gamma_{\mathrm{XB}}$, and L.
   Hint: Fix at first L. Then find a value of γ_{ZB} that offers a high probability of successful decoding of Zoya's packet for that L under the assumption that the interference from Xia is absent. Then consider the interference from Xia and vary γ_{XB}. In deriving the probabilities make the simplifying assumption that any interference can be treated as a Gaussian noise.
2. *Flexible transmit power*. Assume that the same model as in the previous assignment is used, but now Zoya and Xia are allowed to change the transmit power at each transmission. The transmit power can be at most P_{T}. Investigate the strategies that Zoya and Xia can use, while assuming that there is no feedback from Basil.

3. *Flexible transmit power and feedback.* Repeat the previous assignment, but now assume that there is a feedback from Basil after each slot. The feedback tells us whether Basil decoded no packet, one packet (also telling us whether this decoded packet was from Zoya or Xia), or both packets.
4. *Adaptive modulation over a broadcast channel.* Basil transmits downlink packets to Zoya and Yoshi. Assume that the available modulation constellations are BPSK, QPSK, and 16-QAM. A packet consists of 8 control bits and D bits of data, where D can be 32, 64 or 128 bits. The control bits contain, besides the other information, an information about the:
   - Packet destination, which can be Zoya, Yoshi, or both of them (broadcast packet)
   - The actual amount of D data bits following the control bits
   - The type of modulation (BPSK, QPSK, and 16-QAM) used to send the data bits.

   As in the previous assignments, the SNRs of Zoya and Yoshi are denoted by γ_{ZB} and γ_{XB}, but now they refer to the downlink transmission.
   (a) Which modulation should be chosen for the control bits? Discuss the pros and cons.
   (b) Following the choice in (a), find the SNR thresholds for using adaptive modulation for this downlink transmission. Keep in mind that some of the packets are broadcast, intended for both Zoya and Yoshi, such that for those packets the adaptive modulation should take into account both γ_{ZB} and γ_{XB}.
5. *Scheduling of NOMA downlink.* The objective of this assignment is to investigate the scheduling strategies for downlink transmission by using NOMA, discussed in Section 5.8.1. Consider a setup in which Basil needs to transmit downlink data to K users. Some of this data is intended for all K users, while some of the data is intended for each of the individual K users. Hence, there are $K + 1$ data packets that Basil needs to send. Assume that Basil knows the SNR of each user. You are free to make assumptions about the packet sizes as well as the available modulation constellations.

   Discuss the trade-offs that need to be considered when Basil decides how to schedule the users. Should he, for example, pair two users with strong SNRs and transmit to them using superposition coding? How should the broadcast packet be sent? What about the fairness among the users?

a uses her communica-
n channel to Yoshi
ce per day.
consists of a single
ece of paper,
ngle pencil…

…and a single white pigeon.
Y

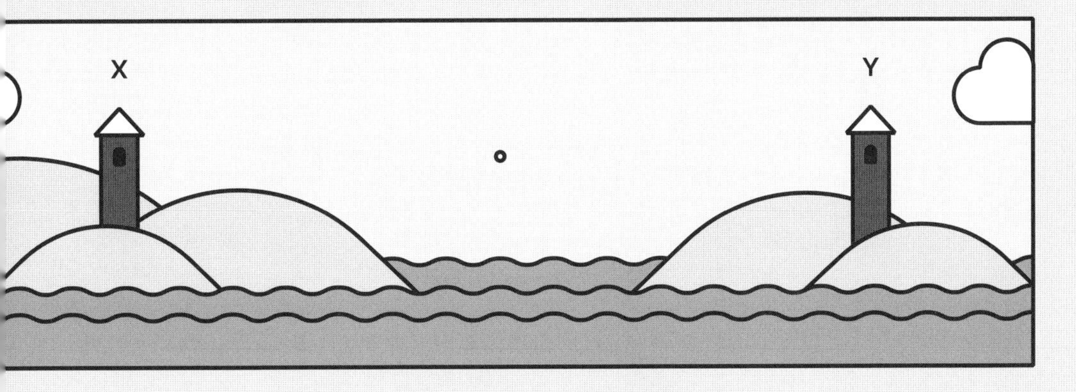
X
Y

er communication channel
n be made different by
owing her to use more than
e piece of paper per day…

…or more pencils per day…

…or more pigeons per day
or even pigeons with different
colors.

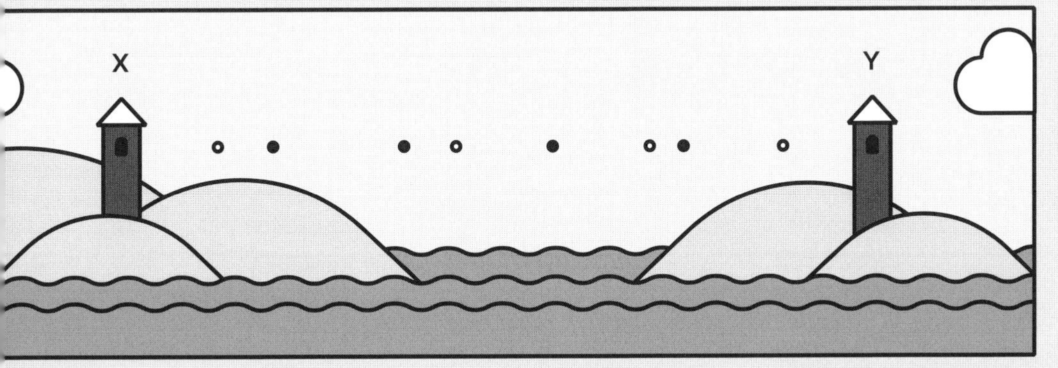
X
Y

Story by Petar Popovski / Art by Peter Gregson

6

A Mathematical View on a Communication Channel

The communication subsystem involving the TXbaseband and the RXbaseband defines a communication channel in a sense of a *mathematical model for communication*, as formulated by Shannon in his groundbreaking paper from 1948. The schematic diagram of the general communication system considered by Shannon is shown in Figure 6.1. One of the most succinct descriptions of the models and methods used in communication and information theory is stated as follows[1]:

> *A channel is that part of the communication system that one is "unwilling or unable to change".*

This simple definition can also be extended to describe the layering principle, presented in Chapter 4, since a layer is a black box with well defined interfaces, which we are "unwilling or unable to change". In Shannon's model in Figure 6.1 it can be noticed that the system model features two layers, precisely as the one we have used to introduce layering in Figure 4.1. Specifically, at the higher layer the information source communicates with the destination; at the lower layer the transmitter communicates with the receiver. The noise source is part of the channel between the transmitter and the receiver and it represents all random disturbances that can affect the reception of the transmitted signal. On the other hand, once the transmitter and the receiver are fixed, then the transmitter, noise source, and the receiver can be grouped together and considered as a channel that is defined between the information source and the destination.

In the following discussion, we will see the importance of the above definition in order to identify what constitutes a communication channel at a given layer/part of the communication system.

6.1 A Toy Example: The Pigeon Communication Channel

In order to introduce the basic idea behind channel models as well as the terminology, we take an example of a channel that is rather unorthodox in communication engineering. This channel is illustrated in the cartoon at the beginning of this chapter. Xia and Yoshi are

1 This lucid statement has often been used by the great information theorist James L. Massey, but he himself attributed the definition to J. L. Kelly.

Wireless Connectivity: An Intuitive and Fundamental Guide, First Edition. Petar Popovski.

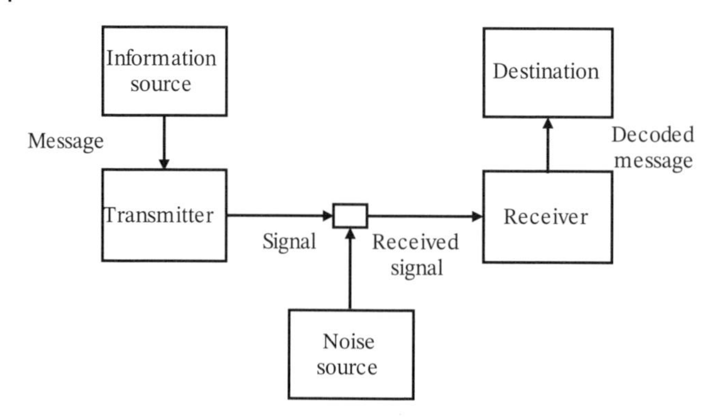

Figure 6.1 The general model of communication system considered by Shannon.

imprisoned in two towers that are several kilometers apart. Every morning at Xia's window there is a pigeon that can carry a single sheet of paper from Xia to Yoshi. After the pigeon delivers the paper to Yoshi, it returns to Xia, without getting any paper or message from Yoshi. Xia uses a pencil and handwriting in order to compose the messages to Yoshi. Xia does have a radio tuner in her tower, but Yoshi does not and the only way to get news updates is through the messages sent by Xia. The question is: *how much news per day can Yoshi learn?* Or, considering that the news is eventually encoded into bits and bytes, the question becomes: *what is the data rate, expressed in bits per day or bytes per day, at which Xia can supply news to Yoshi?*

A basic term used in this context is *channel use*: the smallest, atomic unit of communication that can be sent from the transmitter to the receiver. In the example, the channel use is a single piece of paper and we are interested in the maximal amount of information that can be sent by having the pigeon carry the paper from Xia to Yoshi. Note that the fact that the paper is sent once per day *is not part of the channel.* Even if the pigeon can fly between Xia and Yoshi several times during the day, the description of the channel use stays the same: flying more times per day may increase the data rate per day, but not the data rate per single channel use. We are thus separating the question "how frequently can you send data from Xia to Yoshi?" from the question "how many data bits can be sent in a single channel use?". Therefore, it is meaningful to speak about *bits per channel use*, in short bits/c.u.

We note that the communication channel between Xia and Yoshi is still not completely defined, as we have not put any constraint on how the messages are written and read. Here is an example how to define such *channel constraints and costs.* Assume that Xia can use up to one whole pencil per day: every day Xia returns the pencil remaining from the previous day and gets a new one. The pencil has a finite stroke width and the total amount of writing that can be done with a pencil, such as the total length of the lines, is limited. For this specific example of a channel, a higher cost does not necessarily result in a higher number of bits per channel use. To see this, we can think of the paper and the pencil being dimensioned in such a way that, if the whole pencil is used, then the paper becomes completely gray, not carrying any information.

On the way from Xia to Yoshi, the pigeon is subject to external disturbances that can, for example, damage the piece of paper or make the text less readable. These disturbances fall under the general notion of *noise*. Let us denote by x the original text that Xia sends. This is an *input* to the pigeon channel. What the pigeon delivers to Yoshi is the *output* of the channel. The noise impacts the paper in a random way, such that if two pieces of paper with identical writing are sent at two different times, then Yoshi may observe two different outputs or, in other words, read two different texts from the two received sheets. An essential property of a communication channel is that, for any possible input there is a set of possible output values, such that a particular value y occurs with a conditional probability $p(y|x)$. That is, for a given input text on the sheet sent by Xia, there are multiple options about what Yoshi can receive and the actual output observed by Yoshi is a random outcome.

6.1.1 Specification of a Communication Channel

The communication channel between Xia and Yoshi is specified by the following:

- The set $\mathcal{X}$ of possible input symbols x that Xia can transmit. One possible input symbol is represented by a text that is possible to write on a single piece of paper, such that the number of input symbols is equal to the number of possible texts. Since the number of readable letters that can be written on a piece of paper is finite, we can assume that the number of possible input symbols is finite.
- The set $\mathcal{Y}$ of possible output symbols y that can be received by Yoshi. It can happen that one, more or even all of the input symbols from $\mathcal{X}$ can appear unaltered as output symbols. This happens, for example, if the sheet is not affected in any way while being transported from Xia to Yoshi. However, there can be other possible output symbols. For example, part of the paper may be torn apart during the pigeon flight and consequently missing when the paper arrives at Yoshi. A damaged paper is an output symbol y that is different from any of the input symbols in $\mathcal{X}$. Another type of noise could be manifested in that the pigeon loses the paper along the way and nothing is delivered to Yoshi, which results in a channel output that is an empty symbol.
- The probabilities $p(y|x)$ for all possible input/output pairs. This is the most essential part of the communication channel, as it describes the external and uncontrollable factors that affect the transmission, sublimed under the term *noise*. Clearly, we would like to have a perfect channel where, for each input symbol $x \in \mathcal{X}$ we have $p(y = x|x) = 1$ and $p(y \neq x|x) = 0$, but this is rarely the case. In the mathematical theory of communication we always assume that $p(y|x)$ is given and we are unwilling or unable to change it.
- The cost and constraints imposed on how to use the channel. It should be noted that in this particular example the single-pencil-per-day constraint is already reflected in the possible input symbol $\mathcal{X}$.

A question that can be asked is: can we improve the capacity of the channel by using, for example, higher quality paper? This is similar to asking if we can improve the capacity of a wireless channel by changing the antenna. While the capacity to communicate in the system, understood in broader terms, is indeed increased, the question does not make sense if we consider the capacity per channel use that can be achieved once the communication channel is given as above. Namely, the communication channel is defined once the paper quality (or the antenna) is fixed and represents the part of the channel that we are unwilling or unable to change. The specific choice of paper or antenna will affect the specification of the probability $p(y|x)$.

Another question could be: can we improve the capacity by finding a faster pigeon or let the pigeon carry two/more pieces of paper? Again, the overall data rate of the system will be increased, but neither the pigeon speed nor the number of paper pieces is a part of the communication channel that we have defined. The speed of the pigeon corresponds to how often there is a transmission of a channel use. Recalling our baseband model, the capacity of a channel use describes the maximal number of bits that can be sent through a single baseband symbol. A different question is how often can we send that symbol and, as stated previously in this chapter, in our discussion these two questions are separated.

As illustrated in the cartoon, there are different ways to create new channels, for example, by allowing Xia to use more items per day, such as more than one piece of paper or more than one pencil. In a more extended example, Xia may be able to use multiple pigeons per day. Assume that Xia can freely decide to send a white or a black pigeon; then the choice of the pigeon color also carries information to Yoshi. For example, white pigeon stands for "0" and black pigeon stands for "1". Of all these channel extensions, in the remainder of this section we look how the constraint on the number of pencils per day affects the ability to communicate through the channel.

6.1.2 Comparison of the Information Carrying Capability of Mathematical Channels

The constraint of a single-pencil-per-day is rather simple and let us refer to that channel as 1pen-channel for brevity. Let us define two new channels by changing the constraint on the used pencils, while still assuming that the pigeon flies once per day:

- $\frac{1}{3}$pen-channel: one pencil is given every third day.
- 3pen-channel: three pencils are given every third day.

Which of these three channels is able to carry more information per channel use?

Let us at first compare the 1pen-channel and the $\frac{1}{3}$pen-channel. If one pencil is sufficient to have a sheet completely colored gray, then with the 1pen-channel we can create more different input symbols as compared to the $\frac{1}{3}$pen-channel. In order to see this, observe three consecutive days. If one pencil is available per day, then on any day Xia can create any symbol she wishes on the piece of paper. However, if one pencil is given per three days, then the symbols that can be created on the third day depend on how much of the pencil has been spent in the first two days. Consequently, anything that could be sent, in three consecutive days, over the $\frac{1}{3}$pen-channel, can be also sent over the 1pen-channel, but not vice versa. Thus, we expect that the 1pen-channel is more capable of carrying information than the $\frac{1}{3}$pen-channel.

Now let us compare the 1pen-channel with the 3pen-channel. Both channels use, on average, one pencil per day. Any input that can be created by the 1pen-channel can also be made by the 3pen-channel by deciding to use one pencil per day for the latter. However, there are other inputs that can be created in the 3pen-channel and not in the 1pen-channel. To see this, assume that a single pencil is not sufficient to make the paper fully gray, such that there can be advantage of using more than one pencil per day. In that case, the 3pen-channel can create inputs that are based on, for example, using more than one pencil the first day, but less than a single pencil in the second day. These inputs cannot be created by the 1pen-channel.

It can be concluded that the 3pen-channel is more capable of carrying information than the 1pen-channel.

In order to relate this discussion to a wireless example, consider two different baseband communication channels. For the first channel, each transmitted symbol has to have a power of at most P. The following is an example of a set of symbols powers that can be used in a sequence of symbols sent over that channel:

$$P \qquad 0.8P \qquad P \qquad 0.7P \qquad 0.5P \qquad P. \tag{6.1}$$

Let us now look at another, second channel, where the *average power* used over a group of three symbols can be at most P. An example of allowable symbol sequence through that channel is:

$$2P \qquad P \qquad 0 \qquad 1.8P \qquad 0.5P \qquad 0.7P. \tag{6.2}$$

The reader should note that the first channel is a *special case* of the second channel, as fixing the power of each symbol to P is *only one way* in which we can ensure that the average power used over three symbols is at most P. Thus, any combination of symbols that can be sent over the first channel, can also be sent over the second channel, but not vice versa. In other words, the sequence of input symbols (6.1) can appear in both channels, but the sequence (6.2) only in the second channel. We conclude that the amount of information that can be carried by the second channel can always be made at least as high as the one carried by the first channel. Observations of this type are essential in order to understand how the channel is defined in a particular situation and whether we are losing/gaining something from the ways in which the channel constraints are specified.

6.1.3 Assumptions and Notations

In this chapter we will always refer to point-to-point channels and links, where Xia transmits to Yoshi, unless explicitly stated otherwise. Recalling the terminology from Chapter 4, we will observe the communication channel in a connection oriented mode, after Xia and Yoshi have already agreed to communicate. Specifically, we assume that all the signaling has been exchanged and each channel use carries only pure data. The channel uses are numbered $1, 2, \ldots$. For example, one packet transmission can consist of L channel uses and we can also refer to the ith channel use, where $i = 1, 2, \ldots$.

6.2 Analog Channels with Gaussian Noise

We use the phrase *analog channels* for communication channels in which both the input and the output symbols are continuous, that is, complex or real numbers. A single channel use for the baseband channel, described in Chapter 5, is represented as:

$$y = hx + n. \tag{6.3}$$

The input symbol is x, the output symbol is y. The core parameters of this channels are h and n. We need to specify what Xia and Yoshi know about these parameters, as well as what constraints/costs are imposed on the usage of x. The channel coefficient h can be given as an exact value, learned before the communication channel starts to get used, or specified through a statistical characterization. In the latter case it can also be seen as a multiplicative noise, as it contributes to the random disturbances that define the channel.

6.2.1 Gaussian Channel

Let us now define a standard *AWGN (additive white Gaussian noise) channel*, often called simply a *Gaussian channel*. Xia and Yoshi have agreed that one packet transmission consumes L channel uses. During all the L channel uses, the value of h is constant $h = h_0$. For the ith channel use, n_i is an independent Gaussian random variable with a mean value zero and a variance of σ^2. The average power of the transmitted signal x is P, such that the following must be satisfied:

$$\frac{1}{L}\sum_{i=1}^{L} |x_i|^2 \leq P \tag{6.4}$$

where x_i is the symbol sent by Xia in the ith channel use. Both Xia and Yoshi know h_0, σ^2, P and L through some prior communication and/or measurement.

An important property of the Gaussian channel (6.3) is that it is *memoryless*: for a given ith channel use, the output y_i depends only on the input x_i and the noise n_i, but not on the previous inputs $x_{i-1}, x_{i-2}, \ldots$. Since the noise n_i is independent of the noise that occurs in the other channel uses, it follows that y_i does not depend on noise instances that affect the jth channel use, where $j \neq i$. Unless stated otherwise, it will be always assumed that we speak about memoryless channels.

The Gaussian channel is illustrated in Figure 6.2(a). The channel specification given above may seem exaggerated, as the researchers and engineers often refer to it as "Gaussian channel with an SNR equal to γ", which in our case would be $\gamma = \frac{|h_0|^2 P}{\sigma^2}$. The reason is that such a shortened specification makes a number of implicit assumptions, such as the knowledge of the channel coefficient, variance, etc. In order to see the impact of these assumptions, note that for the Gaussian channel Xia knows the exact mean value of the received signal y, as it is dependent on the symbol x she is transmitting. In other words, for a fixed x, Xia observes y as a Gaussian random variable with known mean value, equal to hx, and known variance σ^2. Based on that, Xia can use appropriate coding and modulation methods in order to achieve, for example, the highest goodput (recall the adaptive modulation from the previous chapter). However, if Xia does not know h, then Xia cannot treat y as a Gaussian random variable with known mean, but a random variable with a completely different statistics, which also depends on the statistics of h. Hence, when Xia does not know h, one cannot guarantee that she should use the same coding/modulation methods as for the case where the channel h is known. We will make a more detailed example of this later in this chapter.

6.2.2 Other Analog Channels Based on the Gaussian Channel

For completeness, we remark here that in the usual model of a Gaussian channel, Yoshi observes y perfectly, or more formally, taken from Yoshi's perspective, there is no uncertainty about y, while there is an uncertainty about x and, of course, n. We can also model

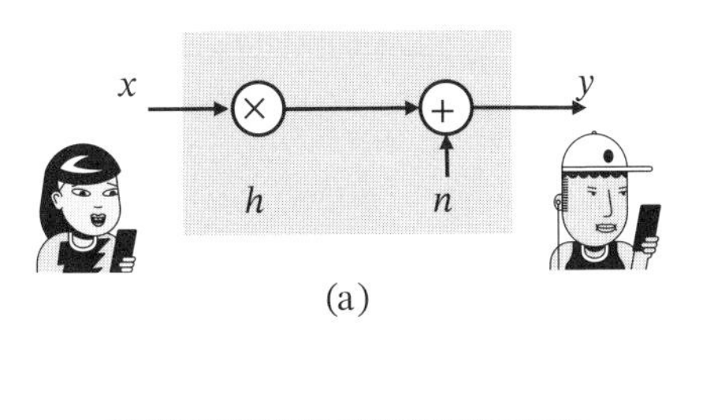

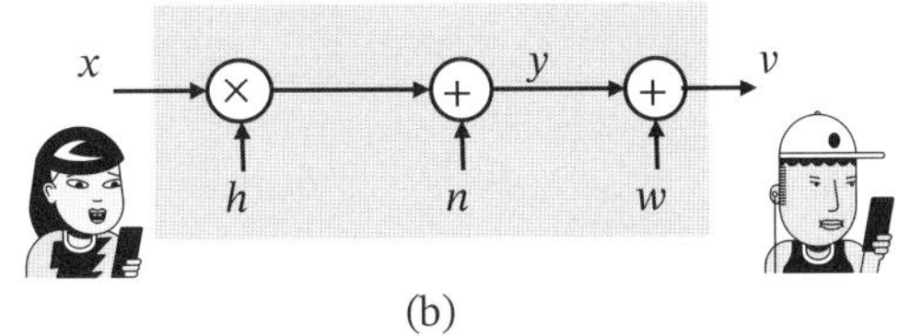

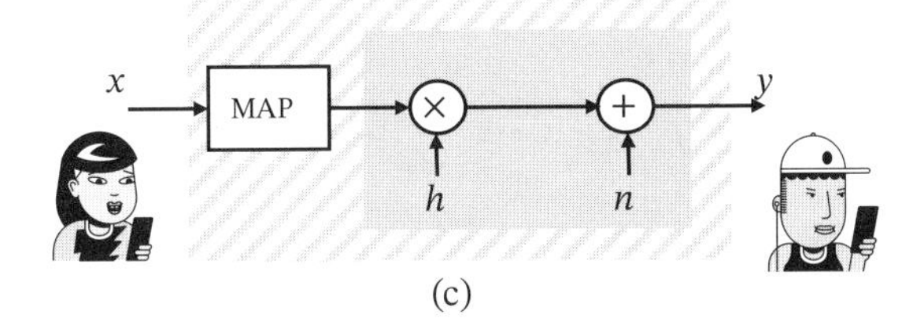

Figure 6.2 Three baseband communication channels between Xia and Yoshi with additive Gaussian noise. (a) Gaussian channel with known channel coefficient h at the transmitter and the receiver. (b) Channel with an imperfect observation at the receiver. The imperfect observation is modeled as additional Gaussian noise w. (c) Channel with additive Gaussian noise in which the inputs are constrained to belong to a discrete constellation, e. g. 8-PSK.

the situation in which Yoshi cannot perfectly observe y, but this would constitute a different channel. In order to illustrate this point, assume that Yoshi can perfectly observe v, which is a randomly altered version of y and specified through a probability density function $p(v|y)$. The reader can now think as if y is sent through a different communication channel, specified by $p(v|y)$; however, Xia does not have direct control over the input y of that channel. As an example, the channel y–v may be again a channel with Gaussian noise:

$$v = y + w \tag{6.5}$$

where w is a Gaussian noise, independent of the noise n and with variance σ_w^2 that is, in general, different from σ^2. As Figure 6.2(b) illustrates, now the channel between Xia and Yoshi is a *cascade* of two channels x–y and y–v. Ultimately, the channel between Xia and Yoshi is x–v, as we can hide y in the model and take into account only its statistical impact. For example, if the imperfect observation is modeled as an added noise, see (6.5), then the channel between Xia and Yoshi is represented as

$$v = hx + n + w. \tag{6.6}$$

If both w, n are Gaussian, then the channel $x - v$ is again a Gaussian channel, but with higher noise power $\sigma^2 + \sigma_w^2$. The equivalent channel is represented in Figure 6.2(b).

For the model of Gaussian channel that we have defined, we are allowed to design any possible modulation method that satisfies the power constraint (6.4) within the prescribed value of L symbols. A *specific* way to satisfy the constraint would be, for example, to use 8-PSK modulation, such that each of the eight constellation points has a power of P (recall the discussion with the pencils for the pigeon channel). By constraining that each transmitted symbol must belong to a 8-PSK modulation, we create a new channel that *overlays* the original Gaussian channel, as shown in Figure 6.2(c) and this channel is referred to as 8PSK_P. We can think of that channel as being constructed by adding a gadget that generates 8-PSK symbols over the original Gaussian channel and we are "unable or unwilling to change it". In a similar way, we can define a BSPK_P channel by using the Gaussian channel to send only two possible symbols, $-\sqrt{P}$ and $\sqrt{P}$.

It should be noted that whenever we fix the power of the modulation constellation to P and use the same constellation for each transmitted symbol, we are using the Gaussian channel in a sub-optimal way. Recall the discussion from Section 6.1.1 on the use of the constraints/cost in specifying a communication channel. Let us call the set of symbols sent in the L channel uses a *super-symbol.* For example, if the constellation contains S points, then there are S^L different super-symbols that can be transmitted within L channel uses. On the other hand, if Xia only respects the constraint (6.4), then she is allowed to send many more different super-symbols. To see why this is the case, let the power of that ith symbol that is part of the lth super-symbol be P_{li}. Then P_{li} can be any non-negative real number as long as:

$$\frac{1}{L}\sum_{i=1}^{L} P_{li} \leq P. \tag{6.7}$$

Since we have the freedom to select L real numbers to constitute a symbol, then the number of possible super-symbols that can be sent is uncountably infinite. Let us now assume that we want to be able to send S^L different super-symbols; then fixing the constellation of S points that needs to be the same for all L symbols is *only one possible way* to make that selection of S^L super-symbols. However, if we are allowed to select S^L from all super-symbols that satisfy (6.7), then the best set, selected in such a way, will be at least as good as the one that uses a fixed Sary constellation for all the symbols in the super-symbol. Here "goodness" is measured according to a communication criterion, such as probability of error in sending the entire super-symbol.

This argument shows the general principle that restricting the freedom at the channel input can only decrease the capability of a communication channel to carry information.

6.3 The Channel Definition Depends on Who Knows What

Changes in the assumptions about h can lead to significant deviations of the channel model from the model of a Gaussian channel. In the context of wireless communications, knowing what each variation brings is essential, as the channel coefficient is subject to random change due to *fading*, which summarizes the effects of the propagation environment, antennas, etc. More on the physics behind those changes is given in Chapter 10.

Here we consider a rather artificial, but sufficiently illustrative model of random *binary fading*, where the channel can only be in two states. Assume that, in each channel use, the coefficient h is selected randomly to be $h = 0.5$, with probability $p = 0.625$, and $h = -1.5$, with probability 0.375. The standard terminology used to describe this situation is to say that the channel changes its *state*, represented by h, randomly from one channel use to another. In this section we will be concerned with the knowledge of h that is available to

Table 6.1 A random sample of five consecutive channel uses.

Channel use number	**1**	**2**	**3**	**4**	**5**
Channel state, h	0.5	0.5	−1.5	0.5	−1.5

Xia and Yoshi before the actual channel use occurs. If Xia has that knowledge, we say that the channel state information (CSI) is available to the transmitter; if Yoshi knows h, we say that the CSI is available at the receiver.

In order to make the example more concrete, let us assume that that the transmission power is $P = 1$. If not explicitly stated otherwise, Xia uses BPSK modulation, where -1 stands for bit value 0 and 1 for bit value 1. The average received power at Yoshi is:

$$0.625 \cdot 0.5^2 \cdot 1 + 0.375 \cdot 1.5^2 \cdot 1 = 1 \tag{6.8}$$

that is, identical on average to the case in which the coefficient is constant $h = 1$ for all channel uses. Finally, we observe a random sample of five channel uses, whose channel states are depicted in Table 6.1.

We have the following four cases of CSI knowledge under which the system can operate:

- *Both Xia and Yoshi have CSI*. This falls back to the Gaussian channel as we have described it above. Figure 6.3(a) shows the probability density of the output signal when $h = 0.5$ and Xia sends 1, while Figure 6.3(b) shows the density when Xia sends -1. The decision rule at Yoshi's side is 1 when for positive signals, -1 for negative signals. According to the illustration, when the channel is in state $h = 0.5$, the noise power is sufficiently high to cause a significant probability of error, as Yoshi gets a negative signal when Xia sends 1. This probability is much lower when $h = -1.5$, as can be seen from Figure 6.3(c) and Figure 6.3(d). However, note that now, receiving a positive signal indicates that -1 has been sent. Since both Xia and Yoshi know the CSI in that channel use, Yoshi can apply the correct decision rule and invert the sign.

  If we remove the assumption that BPSK is used for transmission, then Xia can apply adaptive modulation in order to take advantage of the stronger channel. Specifically, Xia has two communication strategies, denoted by strategy 1 and 2, respectively. Strategy 1 uses a lower-order modulation, for example QPSK, and is applied when the channel is weaker with $h = 0.5$. Strategy 2 relies on a higher-order modulation, such as 8-PSK, and is used when when the channel has $h = -1.5$.

  Referring to Table 6.1, let us at first optimistically assume that Xia and Yoshi know the sequence of five channel states in advance, before the first channel use. Then Xia uses strategy 1 in the channel uses $1, 2, 4$ and strategy 2 in uses $3, 5$. Since Yoshi knows which strategy is used when, this is equivalent to the following case: the channel in state 0.5 for three consecutive uses and Xia and Yoshi communicate with strategy 1; then the channel is in state -1.5 for two consecutive uses and strategy 2 is used. Thus, the overall channel consists of two time-multiplexed Gaussian channels.

  Let us now look at the case in which Xia and Yoshi learn the channel state just before the channel use in which it is applicable. We still need one strong assumption, namely that Xia and Yoshi know before the first channel use that the state $h = 0.5$ will occur three

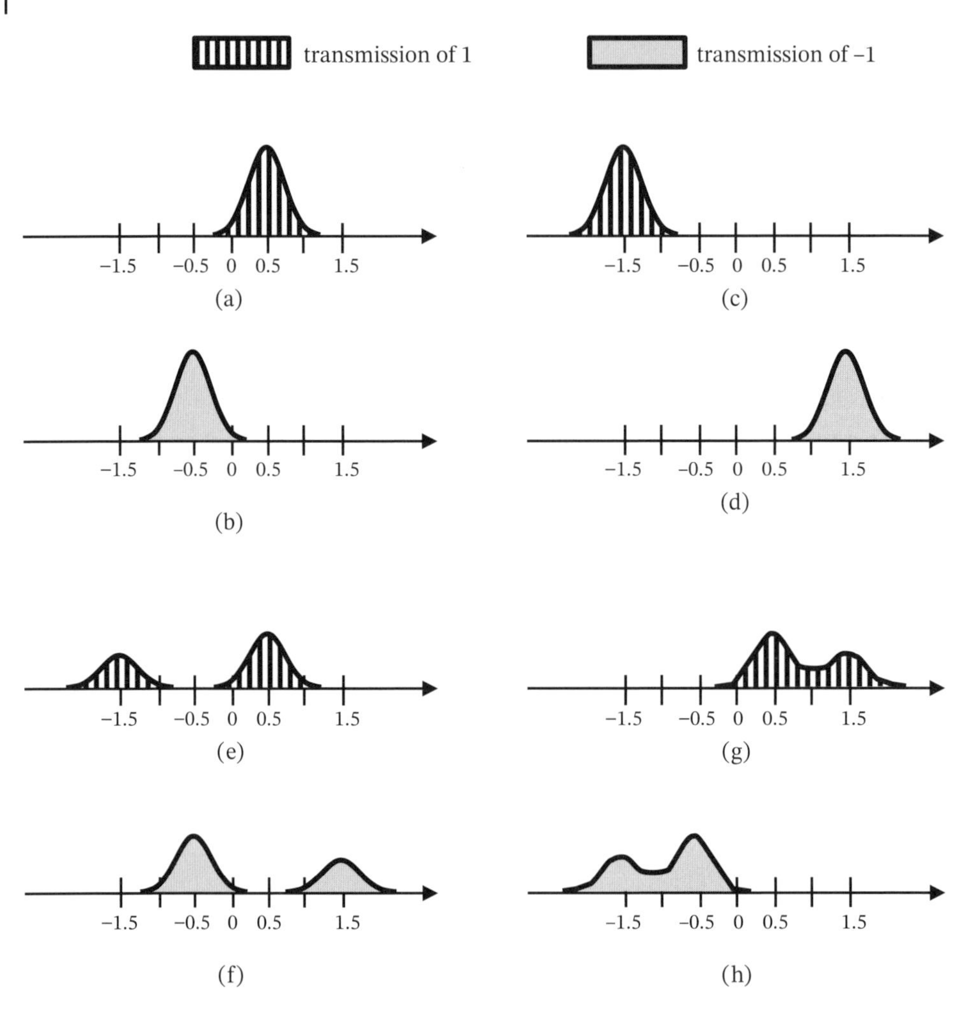

Figure 6.3 Distribution of the received/output signal at Yoshi's side. (a) Known $h = 0.5$ and 1 is sent. (b) Known $h = 0.5$ and -1 is sent. (c) Known $h = -1.5$ and 1 is sent. (d) Known $h = -1.5$ and -1 is sent. (e) Unknown h and 1 is sent. (f) Unknown h and -1 is sent. In (g) and (h), Xia knows h, Yoshi does not, and Xia represents the bit value 1 with 1 when $h = 0.5$ and -1 when $h = -1.5$. (g)/(h) shows the output distribution as seen by Yoshi when bit value 1/0 is sent, respectively.

times and $h = -1.5$ two times; however, Xia and Yoshi do not know in advance which state will be applicable in a particular channel use. This is a strong assumption, but is used here only for illustrative purposes. If, before the first channel use, both Xia and Yoshi know how many times will each channel state occur, but not when it will occur, then learning the channel state just before the channel use is as good as knowing the complete channel state sequence before the first channel use. The reason is that Xia again prepares for three channel uses with $h = 0.5$ and two with -1.5, and, more specifically, prepares three symbols created according to strategy 1 and two symbols according to strategy 2. When Xia learns that the next channel state will be, for example, $h = 0.5$, she picks the next symbol to be transmitted with strategy 1 and sends it. Since Yoshi has also learned the channel state, he knows that what is arriving is the next transmission symbol created according to strategy 1.

In summary, the common knowledge that Xia and Yoshi have about the states of the channel with binary fading allows them to use time-multiplexing between two communication strategies.

- *Both Xia and Yoshi do not have the CSI*. Figure 6.3(e) shows the distribution of the output signal when Xia sends 1. The channel h is a random factor, unknown to both Xia and Yoshi, such that it has essentially the same effect as a random noise, except that it appears as a multiplicative rather than an additive factor in (6.3). Figure 6.3(f) shows the distribution of the output signal when Xia sends -1. The distributions in Figures 6.3(e) and (f) are valid for all five channel uses from Table 6.1. Note that the height of the bell curve at 1.5 is lower than at 0.5, thus reflecting the fact that $h = 0.5$ occurs with a higher probability. With the distributions in Figures 6.3(e) and (f), we cannot use the same decision rule, as when 1 is sent it is likely to get positive as well negative value at Yoshi's side.

  One can think of what a good decision rule would be, but what is more important is to rethink the transmission strategy. Namely, having an average symbol power of 1 and having two input symbol, there is a better strategy than selecting $\{-1, 1\}$ as transmitted symbols. One possibility could be to use symbols with different power, e.g. power 0 when 0 is sent and power 2 when 1 is sent, such that the average power stays equal to 1. This is similar to the strategies for non-coherent communication discussed in the previous chapter; the reader is encouraged to use the examples from Figures 6.3(e) and (f) and try to devise a suitable decision criterion.
- *Xia has the CSI, but Yoshi does not*. If Xia sends 1, then, according to Xia, the distribution of the output signal is either given by Figure 6.3(a) or by Figure 6.3(c), since the distribution is conditioned on the fact that Xia knows h. But this is not true for Yoshi and thus the distribution of the output signal for Yoshi is given in Figure 6.3(e).

  However, this is not the same situation as when neither Xia nor Yoshi has the CSI and, in fact, Yoshi can still use the same simple decision rule. Since Xia knows h, then she can control her transmission and can thereby control the distribution of the output signal. A possible communication strategy could be as follows. If Xia would like to send the bit value 1, then she looks at the channel state and if $h = 0.5$, she transmits 1, otherwise she transmits -1. Using such a transmission strategy, whenever Xia wants to send the bit value 1, the distribution of the output signal for Yoshi is given by Figure 6.3(g). The output distribution that corresponds to bit value 0 is shown in Figure 6.3(h). Hence, now Yoshi can decide that Xia wants to send a bit value 1 whenever he observes a positive signal. Note that here Yoshi does not decide on the actual symbol sent by Xia, but the bit value behind it, and Yoshi is ignorant of whether the actual analog value sent by Xia is -1 or 1. In our example from Table 6.1, the bit value 0 is represented by -1 in the channel uses 1, 2, 4 and by 1 in the channel uses 3, 5.
- *Xia does not have the CSI, but Yoshi does*. In this case, if Xia transmits 1, then the distribution of the output signal to Xia looks as in Figure 6.3(e). On the other hand, Yoshi knows what the channel state is, such that to him, the output distribution is either according to Figure 6.3(a) or according to Figure 6.3(c). Yoshi can again apply the simple decision rule for positive/negative signals, since he knows h and can act exactly the same as in the case when both Xia and Yoshi know the CSI. What is different, though, is that Xia cannot

have two different strategies, 1 and 2, each to be used with a different channel state. In other words, by not having the CSI, Xia cannot adapt to the channel in the sense of using adaptive modulation and she has to use a fixed transmission scheme.

6.4 Using Analog to Create Digital Communication Channels

In introducing modulation and baseband symbols in the previous chapter, we did not think of the data as being directly represented by analog baseband symbols. The data is originally represented through bits and a group of bits is mapped onto a specific constellation point, thereby resulting in a modulated baseband symbol. This naturally puts forward the *binary channel* as one of the most fundamental communication channels, with a single bit as an input symbol and a single bit as an output symbol.

We can create a discrete binary channel by using a baseband module that sends a BPSK channel over a channel with AWGN. This is illustrated in Figure 6.4(a). The module "BPSK map" maps 0/1 to $-\sqrt{P}/\sqrt{P}$, respectively. At the output of the channel polluted with Gaussian noise, there is another block that makes a decision whether the transmitted signal has been positive or negative. We assume that the receiver Yoshi knows h, applies a matched filter, and obtains the decision variable (recall the discussion on BPSK from the previous chapter):

$$|h|^2 x + h^* n. \tag{6.9}$$

The next block outputs 1 if the decision variable is positive and 0 otherwise. As discussed in the previous chapter, by fixing the decision threshold, we also determine the error probabilities $P(1|0)$ and $P(0|1)$, which completely specifies the digital binary channel, as we can find the remaining probabilities as follows:

$$P(0|0) = 1 - P(1|0) \qquad P(1|1) = 1 - P(0|1).$$

Since the noise variance in the underlying Gaussian channel does not depend on the symbol that has been transmitted, the probability of error is identical for the two possible transmitted symbols $P(1|0) = P(0|1) = p$. This is one of the most frequently used channel models; it is known as the *binary symmetric channel (BSC)* and is commonly represented as shown in Figure 6.4(b). Note that the impact of the noise from the underlying Gaussian channel is indirect, as it determines the probability of error. In general, any communication channel with a discrete set of inputs and a discrete set of outputs can be described by specifying the probability $P(y|x)$ for all possible pairs (x, y), as depicted in Figure 6.4(a). We remark that the binary symmetric channel is memoryless.

6.4.1 Creating Digital Channels through Gray Mapping

Let us now consider the digital channel that is obtained by putting the message bits through a baseband module that uses QPSK symbols and sends them over a channel with AWGN. Let us fix the power of the QPSK constellation and select the constellation points from the set $S_{\text{QPSK}} = \{s_1, s_2, s_3, s_4\}$, as depicted in Figure 6.4(c). At each channel use, Xia provides two bits to the TXbaseband module, while the RXbaseband module outputs two bits to Yoshi.

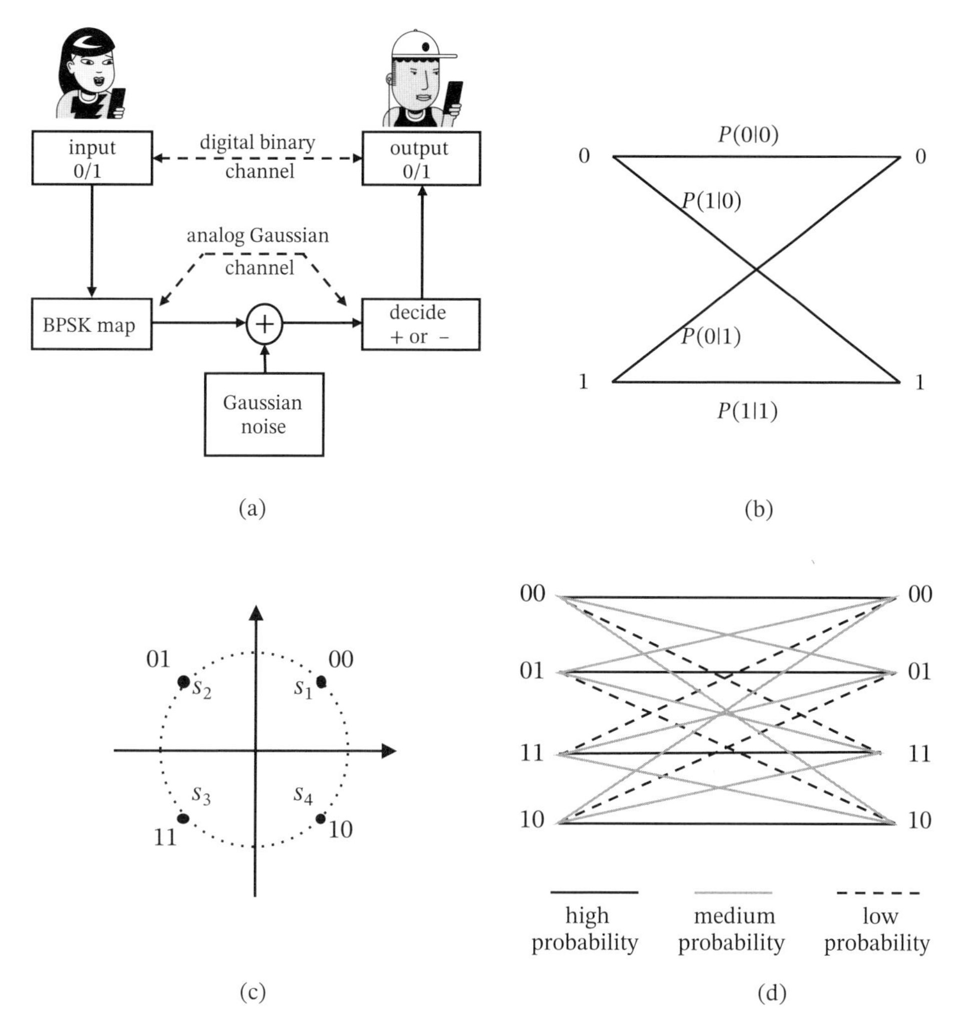

Figure 6.4 Creating digital channels from Xia to Yoshi by using BPSK or QPSK modulation. (a) System blocks are used to create a digital binary symmetric channel (BSC) that uses an underlying channel with Gaussian noise. (b) Common representation of a BSC. (c) QPSK modulation with Gray mapping of the bit pairs. (d) Equivalent digital channel obtained with the QPSK mapping from (c) and a symbol-by-symbol decision.

According to Figure 5.1, the user of the baseband channel is the DataSender that has to map the message/packet bits into baseband symbols x.

In a single channel use, two bits can be mapped to one input symbol. One common way to map the bits is shown in Figure 6.4(c) and is referred to as Gray mapping, introduced in the previous chapter. By assuming that the QPSK symbols are sent over a Gaussian channel and fixing the Gray mapping, we define a new digital channel that has four inputs and four outputs. We can represent the input and the output as two-bit vectors:

$$\mathbf{x}_i, \mathbf{y}_i \in \mathcal{B} = \{00, 01, 10, 11\} \tag{6.10}$$

and the digital channel is specified through the conditional probabilities:

$$P(\mathbf{y}_i|\mathbf{x}_i) \tag{6.11}$$

for all $\mathbf{x}_i, \mathbf{y}_i \in \mathcal{B}$. Note that these probabilities will change if the bits-to-symbol mapping is not Gray mapping, thus leading to a definition of a *different* discrete channel with four inputs and four outputs.

The probabilities in (6.11) can be computed using the distribution of the Gaussian noise and referring to Figure 6.4(c). Let us assume that $\mathbf{x}_i = (00)$ has been sent, which corresponds to the baseband transmission of the symbol $x_i = s_1$, then:

$$P(\mathbf{y}_i = 00|\mathbf{x}_i = 00) = P(00|00) = P(s_1|s_1)$$
$$P(01|00) = P(s_2|s_1) \qquad P(10|00) = P(s_4|s_1) \qquad P(11|00) = P(s_3|s_1). \tag{6.12}$$

Following the properties of the Gaussian noise, such as the symmetry and the fact that it is more likely that the noise will keep the received signal closer to the original point rather than to a more distant constellation point, it follows that the probabilities from (6.12) can be ordered as:

$$P(00|00) > P(01|00) = P(10|00) > P(11|00). \tag{6.13}$$

This is indicated in Figure 6.4(d), which shows the equivalent digital communication channel with four inputs and four outputs. To create this channel we can refer again to Figure 6.4(a), with the following changes: (1) “input 0/1” is replaced with “input 00/01/10/11”; (2) “BPSK map” is replaced with “QPSK map”; (3) “decide + or −” is replaced with “decide s_1, s_2, s_3 or s_4”, and (4) “output 0/1” is replaced with “output 00/01/10/11”. Note that when $\mathbf{x}_i = 00$ is sent, then if $\mathbf{y}_i = 01$ there is only one bit error, but receiving $\mathbf{y}_i = 11$ leads to two bit errors. From this the main idea behind Gray mapping becomes evident: it minimizes the average number of bit errors that occur in a single symbol transmission. It is desirable to have a lower number of bit errors as those are more likely to be repairable by using error correction codes; this will be discussed in the next chapter.

Another property of QPSK is that each of the two bits mapped to the same symbol experiences errors that are independent of the value of the other bit. If we denote the input/output of the ith bit by x_i/y_i, where $i = 1, 2$, then we can write:

$$P(y_1y_2|x_1x_2) = P(y_1|x_1) \cdot P(y_2|x_2). \tag{6.14}$$

Therefore, a single use of a digital channel that is defined over a QPSK baseband transmission is equivalent to a simultaneous use of two independent binary symmetric channels that have identical bit error probabilities.

The latter property is not transferred to higher order complex constellations that use Gray mapping. For example, a 16-QAM with Gray mapping is depicted in Figure 6.5. It should be noted that there are other maps that have the properties of Gray mapping and this is only one of them. Each use of the digital channel defined over the 16-QAM baseband module has 4-bit input/output, denoted as follows:

$$\mathbf{z} = x_1x_2x_3x_4 \qquad \mathbf{y} = y_1y_2y_3y_4 \tag{6.15}$$

where each x_i and y_j corresponds to a bit value 0 or 1. Regardless of what kind of bit-to-symbol mapping is used, the probability of error does depend on the actual

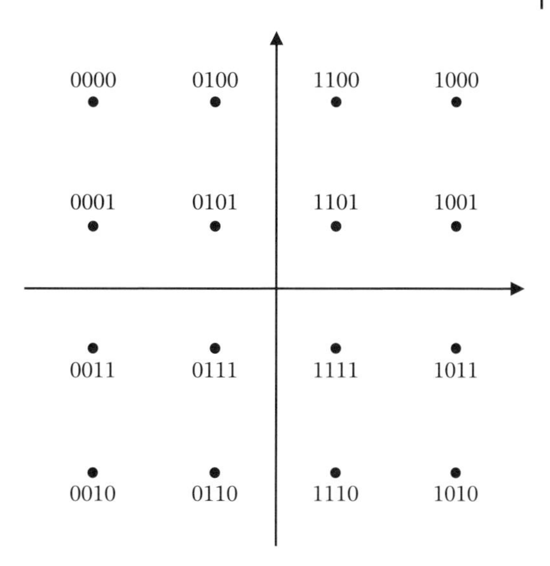

Figure 6.5 16-QAM constellation with Gray mapping.

transmitted symbol. For example, when the transmitted symbol corresponds to one of the corner constellation points, labeled 0000, 1000, 0010, 1010 in Figure 6.5, the probability of symbol error is lower compared to any other from the remaining twelve symbols. This is because the corner points have the least number of neighboring constellation points and thus a lower chance of the noise bringing the received signal into a decision region that belongs to another symbol. Another important difference with QPSK is that the four bits that constitute a single channel use are not equally reliable and do not have the same error probabilities. For example, for the mapping used in Figure 6.5, it can be checked that:

$$P(y_2 = 1|x_2 = 0) > P(y_3 = 1|x_3 = 0). \tag{6.16}$$

Finally, the transmission of four bits through the Gray-mapped constellation in Figure 6.5 *does not* correspond to a simultaneous transmission of four bits through four independent binary symmetric channels. This means that, in general:

$$P(y_1y_2y_3y_4|x_1x_2x_3x_4) \neq P(y_1|x_1)P(y_2|x_2)P(y_3|x_3)P(y_4|x_4). \tag{6.17}$$

The implication of this is that, for example, the probability of error when 0001 is sent is not the same with the probability of error when 0010 is sent, which can be checked by the reader. On the other hand, 0010 is a permutation of 0001 and if those four bits are sent through four identical binary symmetric channels, then, for example, the following should hold: $P(0110|0010) = P(0101|0001)$ and, overall, the error probability when 0001 is sent should be identical to the one when 0010 is sent.

6.4.2 Creating Digital Channels through Superposition

A digital channel can also be obtained when the high-order constellations, such as 16-QAM are created by superposing symbols of lower-order constellations. A simple example of a constellation created by superposition is 4-PAM (pulse amplitude modulation).

Figure 6.6(a) shows a 4-PAM constellation with power P. Once this constellation is fixed, then, strictly speaking, the channel is no longer Gaussian, but a 4-PAM input constrained analog channel with Gaussian noise. The type of digital channel that one can get by using this 4-PAM analog channel depends on the type of bit-to-symbol mapping. Figure 6.6(b) shows a possible Gray mapping for this constellation.

However, the same 4-PAM constellation can be obtained by superposing two binary 2-PAM signals, as shown in Figure 6.6(c). Each of the constituent 2-PAM signals is modulated independently with a single bit. It can be seen that the resulting bit-to-symbol mapping is not a Gray mapping, which is a consequence of the independent modulation of the constituent binary streams. As discussed in the previous chapter, no mapping of the constituent binary signals can result in a Gray mapping.

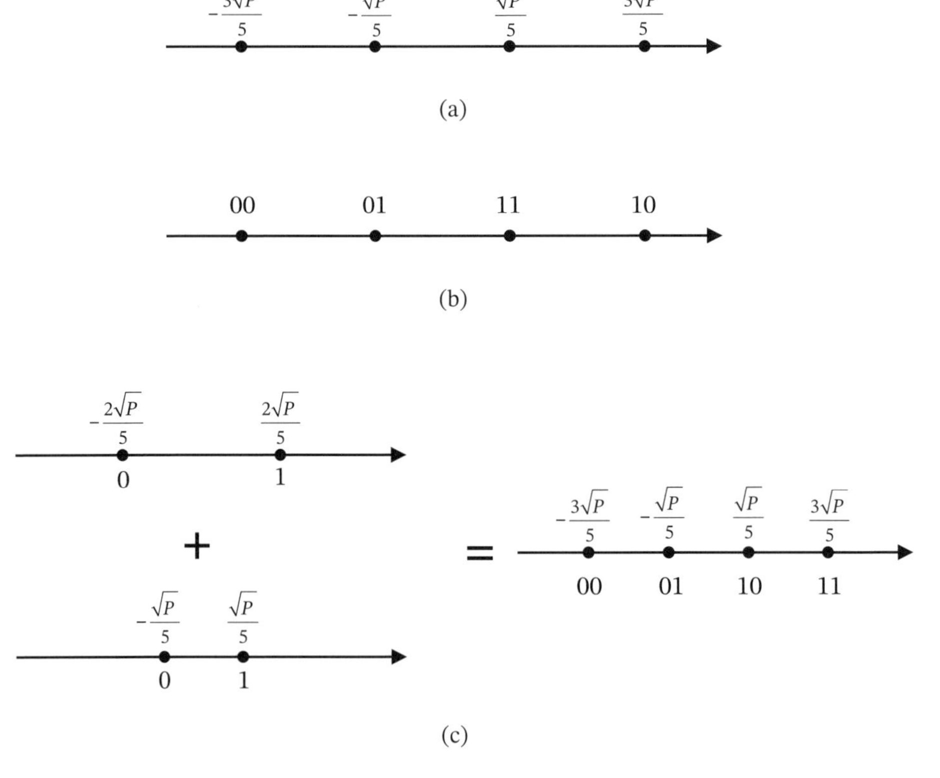

Figure 6.6 Difference between Gray mapping and the bit-to-symbol mapping obtained when a higher-order constellation is obtained by superposition of lower-order modulations. (a) 4-PAM modulation with power P. (b) A possible Gray mapping for the 4-PAM modulation. (c) Synthesis of a 4-PAM modulation by superposition of two 2-PAM modulated signals; the result is not a Gray mapping.

6.5 Transmission of Packets over Communication Channels

6.5.1 Layering Perspective of the Communication Channels

In the previous section we saw that it is possible to create a new channel by taking an existing communication channel and putting certain gadgets at the input and/or the output of that channel. Figure 6.4(a) can be interpreted as follows: the BSC operates by relying on the service provided by the BPSK/Gaussian channel. This irresistibly reminds us of the layering paradigm, described in Chapter 4, and we can thus extend the paradigm and built more channels in a hierarchical manner. For the example in Figure 6.7, channel 2 uses the black-box service provided by channel 1 and provides a black-box service to channel 3. In terms of layering, as discussed in Chapter 4, we can treat channel 1 as representing the lowest layer and channel 3 a representing the highest layer.

An important property that can be used in creating a new channel through the layered approach can be described as follows. For Figure 6.7, a single channel use of channel 2 can consist of one or more channel uses of channel 1. As an example, assume that the channels have the following inputs/outputs:

- Channel 1 is a BSC with inputs/outputs $\{0, 1\}$.
- Channel 2 has three inputs/outputs $\{A, B, C\}$.
- Channel 3 has eight inputs/outputs $Z_1, Z_2, \dots Z_8$.

Since the number of inputs for channel 2 is larger than two, the input block of channel two should map A, B, C to more than one binary symbol. A possible mapping can be the following:

$$A \mapsto 0, B \mapsto 10, C \mapsto 11.$$

In a similar way, each input symbol of channel 3 should be mapped to more than one symbol of channel two, for example:

$$Z_1 \mapsto AA, Z_2 \mapsto AB, Z_3 \mapsto AC, Z_4 \mapsto BA$$

$$Z_5 \mapsto BB, Z_6 \mapsto BC, Z_7 \mapsto CA, Z_8 \mapsto CB.$$

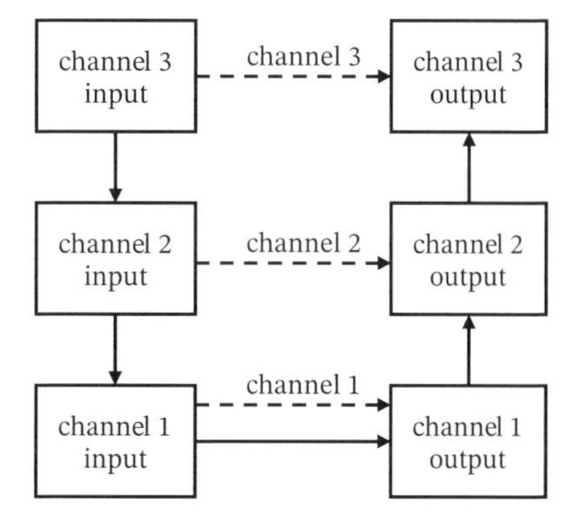

Figure 6.7 Layered representation of communication channels.

If each use of channel 1 is independently affected by noise, then we can compute the transition probabilities for channel 2 and channel 3 using the probabilities specified for channel 1.

It should be noted that the sequence $Z_3Z_5Z_8$ sent over channel 3 will result in a binary sequence sent over channel 1 that looks as follows: 01110101110, taking 11 bits. We have on purpose introduced three inputs in channel 2 in order to introduce inefficiency in the way the symbols from one channel are mapped to another. Note that, had the channel 3 been directly connected to channel 1, we could have represent each symbol Z_i with three bits and thus represent $Z_3Z_5Z_8$ with 9 bits. This is an example of *cross-layer optimization*, already introduced in the previous chapters.

In general, let us consider a communication channel that resides at a lower layer (LL), termed LLChannel, and its duty is to transfer each of its input symbols to the output, independently of all the other symbols before or after that symbol. Let LLChannel have S input/output symbols. We can create a channel, termed HLChannel, that resides at a layer that is higher than the LLChannel and each channel use of the HLChannel is mapped to L channel uses of the LLChannel. HLChannel can be seen as a black box that has S^L possible inputs/outputs and a single symbol for the HLChannel is in fact a *packet* that consists of L Sary symbols of the LLChannel. If the time between two channel uses for the LLChannel is T, then a single channel use of the HLChannel has a duration of $L \cdot T$. However, in order to devise a communication strategy over the HLChannel, we do not need to consider the timing (recall that timing is not part of the definition of a channel use) and we can thus think of HLChannel as a regular communication channel, with a remark that it has a large number of inputs and outputs.

6.5.2 How to Obtain Throughput that is not Zero

Bundling multiple channel uses of the LLChannel to represent a single use of the HLChannel is the essential operation towards enabling reliable communication, which will be elaborated in details in the next chapter. In order to see why this operation is essential, let us look at a LLChannel defined as a BSC that has a probability of error $p > 0$. Let Xia try to send data to Yoshi by making a single use of the LLChannel: she picks the next data bit, which can have value 0 or 1 with probability 0.5, and sends it. The receiver will get either 0 or 1, but has no way of knowing whether the bit is correct or not. Therefore, the throughput that can be achieved by sending one-by-one bit through this channel is *strictly zero*.

Let us now enhance the communication channel and allow the following type of operation: after receiving a single bit from Xia, then Yoshi immediately feeds back to Xia the bit value that he has received. The feedback from Yoshi is ideal and without errors. For the moment we are not concerned how this feedback is implemented, but we just introduce it to see how it can be used to get a non-zero throughput. The main observation is that Xia can still not achieve a positive throughput by making a single use of the LLChannel. This is because the feedback can only help Xia to understand that error has occurred, but Yoshi still lacks the knowledge on which bits are correct.

The problem with the BSC is that there is nothing certain about the data transmission that takes place in a single channel use. In order to see how this can be circumvented, let us assume that the LLChannel is a *binary erasure channel*, which completely changes the

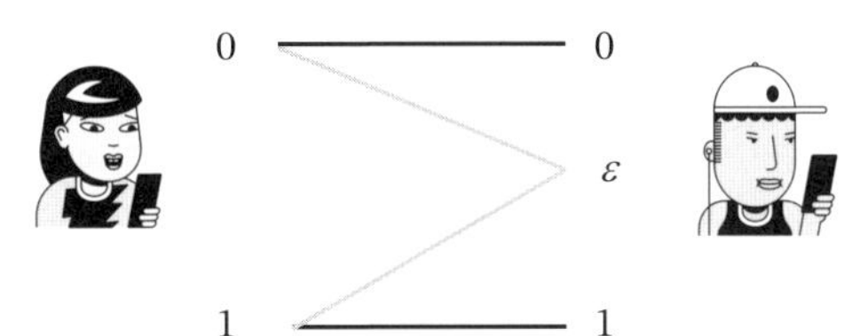

Figure 6.8 The binary erasure channel where Yoshi knows for sure if an error has occurred.

situation. The erasure channel is depicted in Figure 6.8 and the probability that a bit will get correctly to the destination is $(1 - p)$, while it is erased with probability p. If Yoshi receives 0 or 1, he knows that this is a correct value; however, Xia cannot know whether the bit she sent was erased or not. One may argue that in this case the throughput of a single channel use is $(1 - p)$, however, we somehow want to be sure that the information arrives at Yoshi and Xia knows that it has happened. If we now enhance the channel model and allow ideal feedback from Yoshi to Xia, then we can certainly say that the throughput is $(1 - p)$ bits per channel use: if an error occurs, both Xia and Yoshi know it and Xia can retransmit the same bit value in the next channel use.

The main conclusion from the previous discussion is that one can get a positive throughput by having a reliable detection of erasure on the receiver side. This brings us back to the importance of creating a new communication channel by grouping multiple channel uses of another, LLChannel, and ensuring that there is detectable erasure in the newly created channel. Let us assume that Xia again uses BSC to communicate to Yoshi, but now takes $b > 1$ bits to send them in a packet. Xia appends c bits for error detection to those b bits and thus obtains a packet of $(b + c)$ bits. These c check bits are uniquely determined by the other b bits. We have thus created a new channel that has 2^b possible inputs, denoted by $\omega_1, \omega_2, \dots \omega_{2^b}$, which correspond to the possible patterns of the b data bits. The packet is sent by making $(b + c)$ channel uses of the BSC, such that there are 2^{b+c} possible outputs. Among them, 2^b outputs are equal to the inputs $\omega_1, \omega_2, \dots \omega_{2^b}$. The remaining $(2^{b+c} - 2^b)$ outputs can only appear if at least one bit error has occurred in the packet.

Ideally, we would like the error check to operate as follows. If Xia sends ω_i, then Yoshi either receives the same output and declares a correctly received message or he receives another packet from the set of $(2^{b+c} - 2^b)$ and declares an erasure (packet error). However, this ideal operation assumes that it is impossible that an input ω_i produces an output equal to another valid input ω_j, where $\omega_j \neq \omega_i$. This is denoted as an undetected error. Nevertheless there is always a non-zero probability for undetected error as long as the LLChannel is a channel in which an input sequence can produce any output sequence with non-zero probability. For example, the BSC has this property: the combination of bit errors can produce any output bit sequence, regardless of which input bit sequence has been sent, though not all with equal probability (this latter property will turn out to be the key point in attaining reliable communication). In other words, when $(b + c)$ bits are sent, then any of the possible 2^{b+c} bit sequences can be received with non-zero probability. It is thus impossible to construct an ideal error check when the LLChannel is a BSC and $(b + c)$ is a finite number, since there may always be a bit error pattern that tricks Yoshi into falsely assuming that he correctly decodes ω_j when Xia actually sends ω_i.

More formally, let P_{d} be the probability that a packet is erased, that is, an error is successfully detected. Let P_{u} denote the probability of undetected error, described as an event in which Xia sends ω_i, Yoshi receives ω_j with $j \neq i$. Strictly speaking, and in line with the discussion above, no positive throughput can be achieved over the BSC. However, in practice, P_u can be made to be several orders of magnitude lower than P_d, such that we can approximate the error detection to operate as an ideal one. Given that, Yoshi can map all the erased packet to a single symbol, thus creating a channel that has a set of $2^b + 1$ outputs, out of which 2^b outputs correspond to the valid packets and a single output that stands for an erased packet. We have thus obtained a generalization of the binary erasure channel, where a packet will be erased if there is at least one error in the $(b + c)$ bits, such that the probability of erasure is:

$$P_e = 1 - (1 - p)^{b+c} = P_d + P_u \approx P_d \tag{6.18}$$

since $P_u \approx 0$. The *nominal data rate* of this channel, expressed in terms of the channel uses of the BSC, corresponds to the data rate that Xia obtains if there are no errors in the channel. This allows us to find the ideal goodput, calculated only by using the contribution from the information carrying bits:

$$R = \frac{b}{b+c} \quad \text{(bits/c.u.).} \tag{6.19}$$

If we assume that Xia gets an ideal feedback from Yoshi and therefore learns if the packet has been erased or not, then a fraction of $(1 - P_e)$ packets goes through, such that the goodput is:

$$G = R(1 - P_e) = \frac{b}{b+c}(1 - P_e) \quad \text{(bits/c.u.).} \tag{6.20}$$

Coming back to the layering discussion, we can think of the packet erasure channel created in this way as being an HLChannel that is based on $(b + c)$ uses of the BSC LLChannel.

In reality, P_u is not zero and it can happen that an error goes undetected. The approach to tackling this is to have extra error checks at the higher layers; this implies that the b data bits are carrying less than b bits of information, as a fraction of them are error detection bits applied by the higher layer to check the integrity of the data. As a general rule, whenever error detection is put towards the lower layers, an overhead is introduced for every packet sent at the lower layer, which increases the overall overhead. Nevertheless, the advantage is that an error is detected with a lower delay. Conversely, when error detection is placed at the higher layers, the receiver needs to wait to receive all the data parts and, if an error is detected and there is an ideal feedback to the sender, then all of the data needs to be retransmitted.

The LLChannel can be generalized to the case in which it has Sary input and Sary output; for example, $S = 4$ for QPSK. Since a single channel use can carry $\log_2 S$ bits, the number of channel uses required to send a packet with b bits and c check bits is $L = \frac{b+c}{\log_2 S}$. The nominal data rate is:

$$R = \frac{b}{L} = \frac{b}{b+c}\log_2 S \quad \text{(bits/c.u.).} \tag{6.21}$$

The HLChannel constructed in this way has $S^L = 2^{b+c}$ different outputs. By using an ideal error check, only 2^b of those outputs are accepted as correct data packets, while all the remaining received symbols are mapped into the erased symbol.

6.5.3 Asynchronous Packets and Transmission of "Nothing"

The communication with packets, as described in the previous section, makes an implicit assumption that Xia and Yoshi are synchronized with respect to the packet transmission. In other words, Xia and Yoshi have a prior agreement at which channel use the new packet transmission starts and how many channel uses belong to the same packet. We will not go through all the details of link establishment and how Xia and Yoshi agree on the communication parameters; the reader can use Chapter 1 in order to get an idea about the control packets that are necessary for this process. Here we will address a more fundamental issue, which sets the basis for relating the definition of a communication channel to the problem of sending a particular type of control messages and packets.

A common view on wireless packet transmission is that if the sender is not transmitting, the receiver gets "nothing" and observes an idle channel. This mode of operation is not supported by the model of synchronous packet transmission, which underlies the discussion from the previous section, since each channel use over the binary symmetric channel is a valid transmitted symbol and it is not possible to transmit "nothing". This follows from Shannon's model of communication from Figure 6.1, which involves the tacit assumption that Xia and Yoshi have already agreed to communicate and every symbol that is sent over the channel is intentionally sent by Xia, there is no unintentional transmission of an idle symbol. Then, how can we model the asynchronous packet transmission, in which the start instant of a packet is not predefined? How to define a transmission of "nothing", which can be interrupted by a packet transmission? In that case, the receiver first needs to detect existence of a packet and then decode the information bits carried in the packet.

We use the layered framework depicted in Figure 6.9. Xia has the DataSender as an entity that is the user of the packet channel and, in a given channel use, DataSender passes b data bits to the PacketSender. The role of the PacketSender is to create a packet by supplementing the bits received from DataSender with r bits to denote the packet start/stop and c bits for error checking. After that, PacketSender passes on $r + c + b$ bits for transmission over the

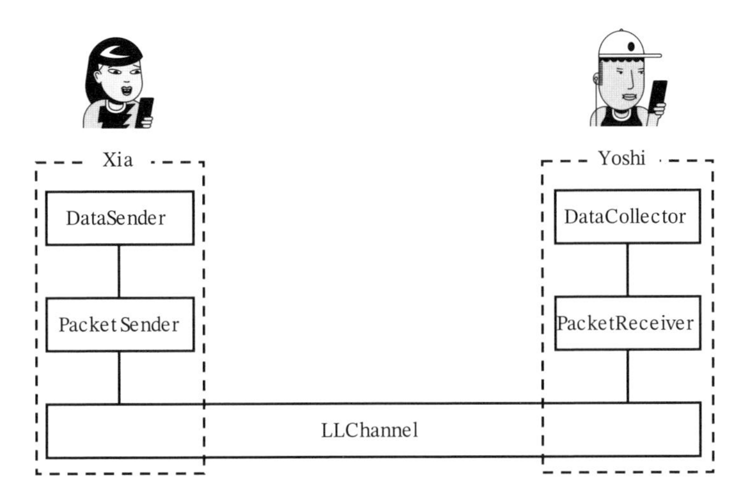

Figure 6.9 Layered model for asynchronous packet transmission.

LLChannel. Yoshi has a PacketReceiver that needs to detect if there is a packet and, if yes, collects the $c + b$ bits, runs an error check and, if the packet is correctly received, it passes the data bits to the DataCollector. If the packet is not correct, then the PacketReceiver passes on an erasure signal to the DataCollector. In the following we will investigate how to enable the PacketSender to send packets to the PacketReceiver asynchronously, when requested by the DataSender. This boils down to the choice of the r bits that need to mark the packet existence and duration. Note that, if the packet duration is fixed and known in advance, then the r bits are only required to denote the start of the packet; those bits are often termed *preamble*.

Let us at first assume that the low layer channel (LLChannel) is a binary channel, not necessarily symmetric. We can think of a channel based on non-coherent transmission: the symbol 0 is sent during a given channel use by staying silent, while power transmission in that channel use corresponds to sending 1. The symbol value 0 has a dual role: it can represents an idle state (no ongoing packet transmission) as well as the bit value 0. In other words, when there is no packet transmission and there are no errors in the channel, the receiver gets 000 …. The next transmitted packet interrupts this zero sequence and this should be detected by Yoshi. If there are no errors in the LLChannel, a preamble consisting of a single 1 is sufficient to reliably interrupt the idleness (transmission of "nothing") and mark the packet start. However, if there are errors in the channel, then the receiver may experience false positives and incorrectly start to receive a packet. Therefore, in practice, the preamble consists of multiple symbols in order to reliably delimit the packet start, but it is clear that it is always possible to have transmission errors and thereby an imperfect preamble detection.

The problem of detecting existence/start of a packet is known as *frame synchronization*. A simple variant of the problem is illustrated in Figure 6.10. It is assumed that a packet

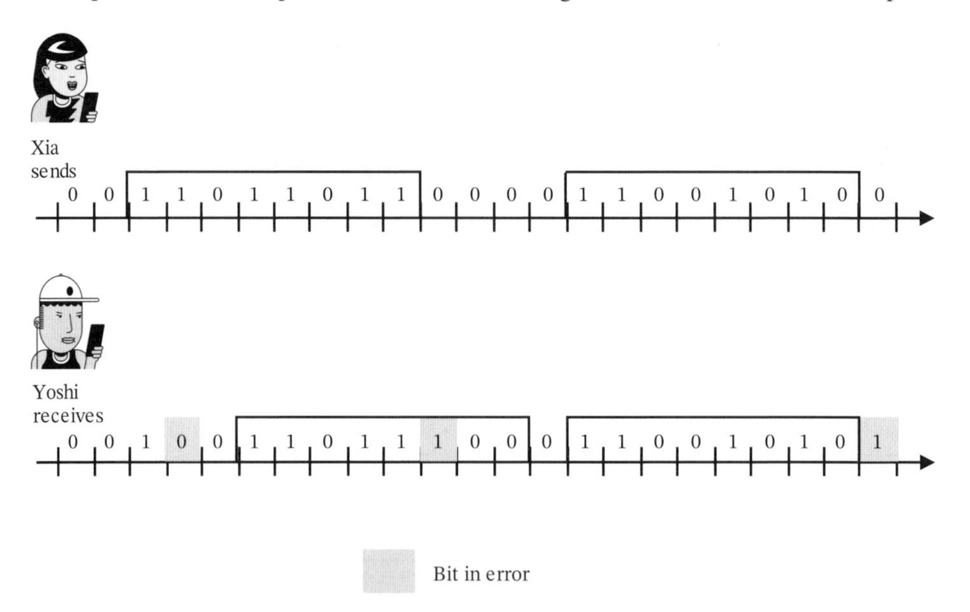

Figure 6.10 An illustration of the frame synchronization problem with packets of length 8 that include a preamble 11, five data bits, and a parity check bit.

sent by Xia has a predefined length of 8 bits, such that there is no need for a group of bits within the packet that describes the packet length. At this point it is useful to recall the discussion on making packets in the discussion in Section 3.5. The 8 bits include: $r = 2$ bits for the preamble 11, $b = 5$ data bits passed on by the DataSender and a $c = 1$ parity check bit for error detection. For the example in Figure 6.10, the first packet 11011011 of Xia is not detected correctly by Yoshi; instead Yoshi detects a "phantom" packet with content 11011100; however, the group of bits 011100 will not pass the parity check and will be discarded or Yoshi will request a retransmission through feedback to Xia. Nevertheless, if the probability to receive 1 if 0 is sent is $P(1|0) > 0$, then there is always a chance that an error goes undetected and an erroneous packet is delivered to the DataCollector. Reliability can be improved if additional error check is used at the higher layer at which the DataSender communicates with the DataCollector. In that case, the total data that needs to be sent by the DataSender is supplemented by an error check, such that the erroneous data packet delivered from the PacketReceiver to the DataCollector will be detected when the DataCollector runs an error check over the total received data.

6.5.4 Packet Transmission over a Ternary Channel

Another model that can be used to treat asynchronous data transmission is to have TXbaseband for which the signals used for 0 and idle are different. For example, when TXbaseband sends data, it uses uses non-zero power and BPSK modulation with input symbols $\{-1, 1\}$. When the TXbaseband is idle, it sends zero power, which corresponds to a baseband symbol 0. Now the analog channel that resides in the lowest layer has a *ternary* input $x \in \{-1, 0, 1\}$, while the output is obtained after adding Gaussian noise to the input. Following the discussion in Section 6.4 about making a digital channel by using an underlying Gaussian channel, the receiver of Yoshi uses a detecting device that operates as follows. For each received symbol, the detecting device produces a discrete output y_d with the following possible values:

- $y_\mathrm{d} = 0$, if the detecting device believes that -1 has been sent
- $y_\mathrm{d} = 1$, if the detecting device believes that 1 has been sent
- $y_\mathrm{d} = \epsilon$, if the detecting device believes that the channel is idle.

The question now is how to carry out frame synchronization for a channel with ternary input and output. One may conjecture that this case is somewhat easier compare to the case in which 0 is interpreted as idle ϵ, as now there is a dedicated input symbol to denote an idle channel. A straightforward way to use this channel for sending packets asynchronously would be to set the input to ϵ whenever there is no packet and send either 0 or 1 when there is a packet. However, note that the detecting device is looking only at a single received symbol y at a time and it is agnostic to the fact whether that symbol belongs to a packet or to an idle period. For example, noise can cause y to be detected as ϵ; this results in a channel use that is detected as idle in the middle of a packet. Conversely, the receiver may detect 0/1 when the channel is idle.

The latter observation leads to another idea to create a different type of a discrete communication channel on top of the Gaussian channel. For convenience, let us also put a mapping device at the input of the baseband channel in Xia's transmitter. The discrete input to this

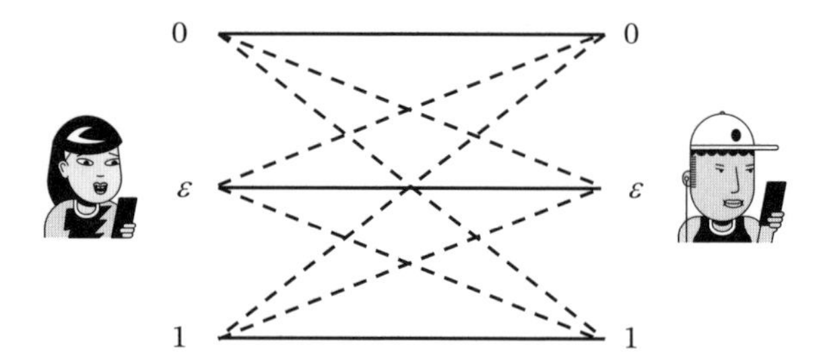

Figure 6.11 Description of the channel with ternary inputs and outputs, where the idle channel symbol is also a valid, information carrying input.

Table 6.2 Possible mapping of three data bits into two ternary symbols.

Bit combination	Transmitted symbols
000	00
001	01
010	0ϵ
011	10
100	11
101	1ϵ
110	$\epsilon 0$
111	$\epsilon 1$

device is denoted by x_d and it can get three possible values $x_\text{d} \in \{0, 1, \epsilon\}$, where ϵ is a valid discrete input symbol that also has the role of an idle symbol. The mapping between x_d and the baseband symbol x is given as follows:

$$x_\text{d} \to x: \quad 0 \to -1; \quad 1 \to 1; \quad \epsilon \to 0. \tag{6.22}$$

The new insight offered by this channel is that Xia can use the idle symbol ϵ as a legitimate symbol to send data. A single use of the channel is described in Figure 6.11; the transition probabilities can be determined in a way that is similar to the one we have used to create digital channels from analog channels with Gaussian noise. Since ϵ is now also an information carrying symbol, we need to use a preamble for frame synchronization. For simplicity, let us assume that the packet preamble is 11 and it is followed by five information bits and a parity check bit, as for the example in Figure 6.10. If only BPSK is used for packet transmission, then Xia needs to send 6 inputs 0/1. However, if ϵ may also be used to send information, then each group of three bits can be encoded into two ternary symbols $0, 1, \epsilon$ by following Table 6.2. Therefore, a single packet, including the preamble, now consumes a total of 6 channel uses instead of 8, provided that ϵ is also a legitimate data carrying symbol.

The latter example illustrates the conceptual separation between the data networking community and the information–theoretic view on the communication channel. While in data networking the approach is always to use only the "legal" symbols to transmit data packets, the approach in information theory is to use all the possibilities provided by a given communication model. Keeping the erasure symbol only to denote an idle channel could be more practical from the viewpoint of implementation, energy consumption, or communication architecture. However, if our goal is to send the maximal amount information over the communication channel, one needs to treat the idle symbol as another modulation/input symbol for sending data. More on the information–theoretic notion of channel capacity in Chapter 8.

6.6 Chapter Summary

The notion of a communication channel can be made mathematically precise by specifying the channel inputs, outputs, the probabilistic relation between them, as well as any constraint imposed on the way the channel is used. This chapter has emphasized the importance of understanding a communication channel as a mathematical object that can be defined at any layer of the communication system, as long as there is an agreement what we, as system designers, are unwilling or unable to change. This chapter has also illustrated the following property of a communication channel: the way the channel is defined is tightly related to the assumptions about which knowledge is available at the transmitter and the receiver. This aspect is an essential bridge between the proper engineering assumptions about the system in question and the mathematical model used to design and optimize communication schemes. A communication channel at a higher layer can be constructed by using one or more channel uses of a channel defined at a lower layer. An example of this is the creation of a BSC with binary inputs and outputs by using an underlying Gaussian channel. The chapter has illustrated other possible mappings between the data bits and various baseband modulation schemes. The last part of the chapter makes a connection between the communication channel, as a mathematical object, and the data packets, as main entities in the communication protocols.

6.7 Further Reading

A must-read for understanding the fundamentals and the idea behind the mathematical models used in communication engineering is the first paper by Shannon [1948]. For the readers that are interested to explore further the information–theoretic treatment of communication channels, two standard references are Gallager [1968] and Cover and Thomas [2012]. The latter also covers mathematical models for multi-user channels, which have not been discussed here. A very insightful reading on communication channels and information theory can be found in the Lecture Notes by Massey in Massey [1980] (Chapter 4 discusses the channel as an entity that one is "unwilling or unable to change"). A primer on a definition of a channel based on the underlying packets (which already carry data from other communication channels) is given in Anantharam and Verdu [1996], while a wider discussion on the relation between mathematical models for communication channels and communication networks is given in Ephremides and Hajek [1998]. Communication channels are universal in nature and technology; the reader is referred to Nakano et al. [2012] for the more recent interest in molecular communication and nano-communication channels.

6.8 Problems and Reflections

1. *Using all channel outputs versus combining the outputs.* Zoya transmits through a wireless communication channel to Walt. She transmits one complex symbol z every T_s s. There are two different receiver configurations for Walt:

   Configuration 1. Every T_s s Walt gets a noisy version of the symbol sent by Zoya, such that for the ith transmitted symbol of Zoya, Walt gets the ith received symbol $w_i = z_i + n_i$, where n_i is a Gaussian noise.

   Configuration 2. At Walt's receiver there is an additional module that combines two noisy received symbols into one, such that Walt gets one new received symbol every $2T_s$ s. Thus, if Zoya transmits $z_1, z_2, z_3, \ldots$, Walt gets:

$$
\begin{aligned}
w_1 &= z_1 + z_2 + n_1 + n_2 \\
w_2 &= z_3 + z_4 + n_3 + n_4 \\
\vdots &\quad \vdots \quad \vdots \\
w_{i/2} &= z_{i-1} + z_i + n_{i-1} + n_i.
\end{aligned}
$$

   Compare the two channels and argue which one has a larger capability to carry information.

2. *Communication channel based on packets.* Zoya transmits data to Yoshi through data packets. Each packet consists of D bits of data and a header. Each transmitted packet is either received without errors by Yoshi or is lost and never arrives at Yoshi. The packets that arrive at Yoshi arrive in order, such that if Zoya sends packet l before the packet $l+1$, then it cannot happen that Yoshi receives first the packet $l+1$ and then the packet l. Discuss the differences between the two communication channels that can be defined for the following two cases:
   (a) The header of a packet contains "Yoshi" as a destination as well as the ID of the transmitted packet.
   (b) The header of a packet only contains "Yoshi" as a destination, but there is no packet ID transmitted.

   Hint: The number of inputs for the communication channel is 2^D.

3. *Communication with timed packets.* This is similar to the previous problem, but here Zoya and Yoshi agree that she sends one packet with D bits each T_p s. Each packet either arrives immediately to Yoshi or is lost. Discuss the differences between the two communication channels that can be defined for the following two cases, as well as the difference with the two channels from the previous problem:
   (a) The header of a packet contains "Yoshi" as a destination as well as the ID of the transmitted packet.
   (b) The header of a packet only contains "Yoshi" as a destination, but there is no packet ID transmitted.

   Hint: Zoya can get an additional channel input by deciding not to send a packet. If the packet ID is available, think of how Zoya can use that by intentionally not sending the packets in the proper order.

4. *Signaling with antenna activation.* Xia has two transmit antennas, denoted by TXA_1 and TXA_2, while Yoshi has two receive antennas denoted by RXA_1 and RXA_2, respectively.

At a given time, Xia can only transmit through one of the antennas. When Xia transmits through TXA_1, only RXA_1 receives a signal, while there is no signal at RXA_2. When Xia transmits through TXA_2, both antennas of Yoshi receive signals. Establish a mathematical model for this communication channel and devise communication strategies for the following cases:

(a) Each of the receiving antennas can only detect if there is power or not. Assume that there is no noise.

(b) Each transmission of Xia is a QPSK modulated signal, while the received signal has an added Gaussian noise. When Xia sends x through TXA_1, then the received signal at RXA_1 is $x + n_1$, while RXA_2 stays idle. When Xia sends x through TXA_2, then the received signal at RXA_1 is $x + n_1$ and the received signal at RXA_2 is $x + n_2$. Here n_i is a Gaussian noise.

5. *Bit reliabilities with 16-QAM*. Consider the Gray mapped 16-QAM depicted in Figure 6.5. For a fixed SNR, study the reliability of each of the 4 bits transmitted with this modulation. Verify the statement given in the text that the four binary channels created with a 16-QAM transmission are not four identical and independent binary symmetric channels.

ya wants to shop online. Her father wrote the number of his credit card on a piece of paper.

One of the digits is unreadable.
54 72 5 21

ya tries to type "7" for the unreadable digit,
t this is incorrect.
7

She then tries "4", but this is again incorrect.
4

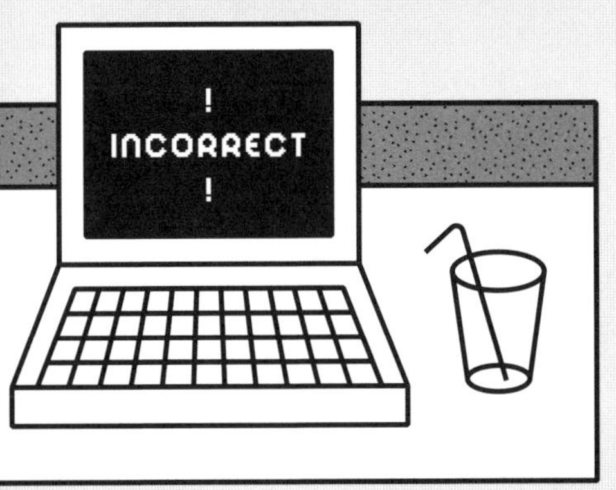

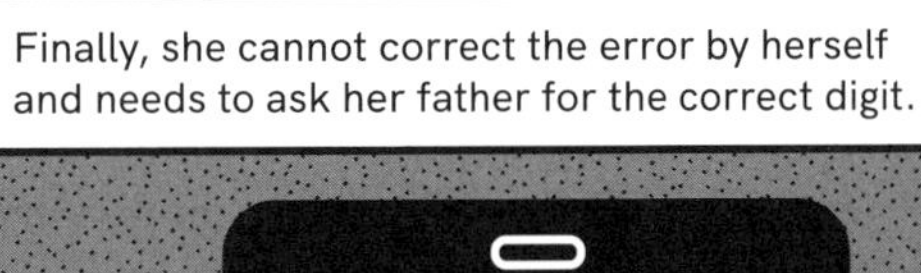

Finally, she cannot correct the error by herself
and needs to ask her father for the correct digit.

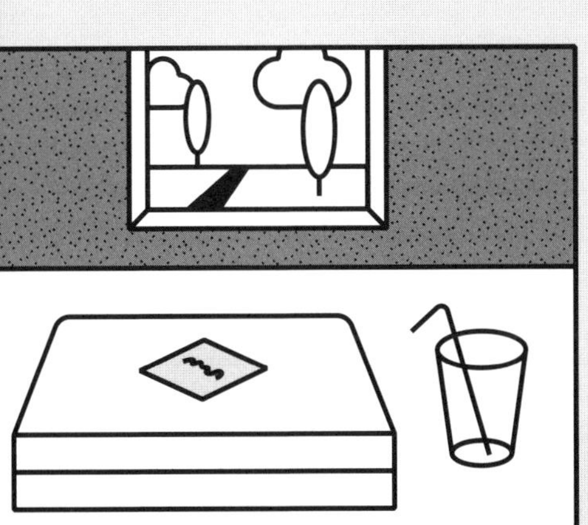

Story by Petar Popovski / Art by Peter Gregson

7 Coding for Reliable Communication

The fundamental problem in information and communication theory is how to achieve reliable communication over channels that are originally unreliable in a sense that each individual transmission over the channel has the chance to be received incorrectly. We have seen in the previous chapter that, if a single channel use is unreliable, then non-zero throughput can only be achieved by grouping the data bits into packets, where each packet spans multiple channel uses and contains redundant bits for checking the integrity of the packet.

It is important to note that checking for errors does not directly improve the reliability of the data bits. For example, consider a binary symmetric channel (BSC) where the probability of error is $p = P(1|0) = P(0|1)$. Xia transmits a packet to Yoshi, containing b data bits and c check bits. The error check is assumed ideal: if there is at least one erroneous bit in the packet, Yoshi will detect a packet error and the packet will be considered to be erased. The probability of packet error is given by (6.18) and the average goodput is given by (6.20), and here it is written in a more convenient form:

$$G = \frac{b}{b+c}(1-p)^{b+c}. \tag{7.1}$$

This mode of transmission, where the probability of error for a single data bit remains unchanged, is termed *uncoded transmission*. We remark here that this term is not completely correct, as the addition of the c bits for error detection is a form of coding[1]. The way to improve the reliability of the individual data bits is to use redundancy in a form of a *forward error correction (FEC)* code. We start to introduce the methods for error correction through some simple examples involving the BSC.

7.1 Some Coding Ideas for the Binary Symmetric Channel

7.1.1 A Channel Based on Repetition Coding

A trivial way to improve the reliability is to repeat the same data bit/symbol multiple times. Let us take the BSC with probability of error p, depicted in Figure 7.1(a) and referred to as a p-channel in this discussion. For easier explanation, let us assume that one channel use

1 Strictly speaking, an error detection code can also be used for error correction, but so far we have considered only its error detecting functionality.

Wireless Connectivity: An Intuitive and Fundamental Guide, First Edition. Petar Popovski.

takes 1 s. Xia increases the transmission reliability by using the following strategy: if she needs to send the bit value 1, she uses the channel three times and transmits 111. Similarly, she transmits 000 for the data bit with value 0. Since the BSC can flip any of the transmitted bits, Yoshi receives the triplet $y_1y_2y_3$ that can take any of the 8 possible values:

$$y_1y_2y_3 \in \{000, 001, 010, 011, 100, 101, 110, 111\}.$$

Yoshi applies majority voting in order to decide whether 000 or 111 has been transmitted. If there is at most single bit error, the decision of Yoshi is correct; otherwise, it is incorrect. With this type of decoding, the probability that an error will occur is:

$$p_E = 3p^2(1-p) + p^3. \tag{7.2}$$

Once the transmission/decision rules are fixed in the way described above, then it can be said that we have created a *new binary symmetric channel* that has a probability of error equal to p_E. This channel is depicted in Figure 7.1(b) and is referred to as the r-channel. It can be easily checked that $p > p_E$ whenever $p \leq 0.5$ [2].

Nevertheless, there is a price to pay for decreasing the probability of error. A single use of the new channel takes three seconds instead of one. Let us now try to find the goodput, in bits per second, that can be achieved by using each of the channels above. As discussed

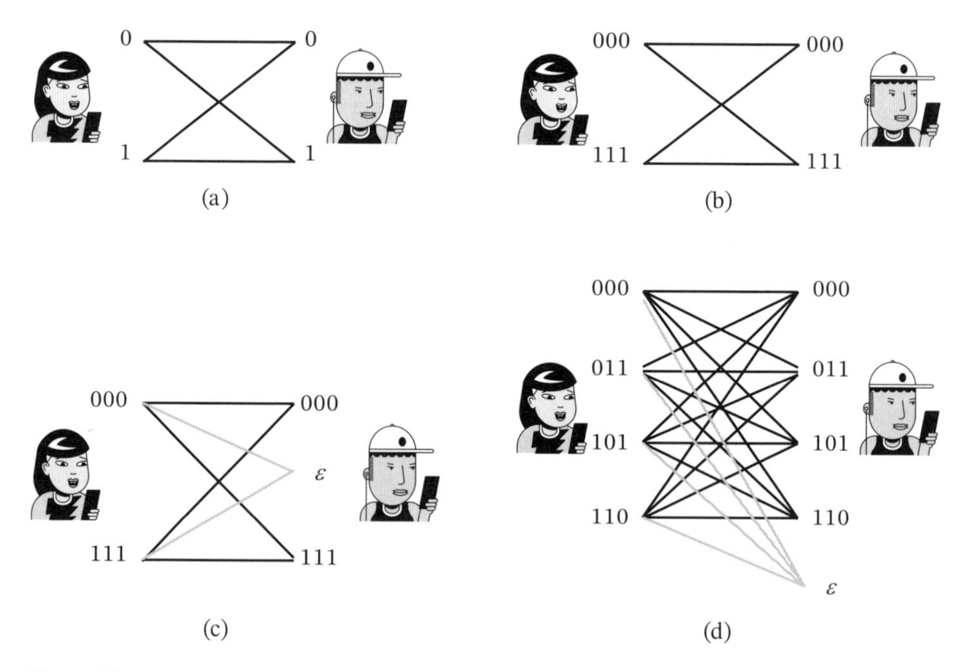

Figure 7.1 Binary symmetric channel (BSC) between Xia and Yoshi and three channels derived from the BSC through repetition/coding. (a) p-channel: BSC with probability of error (p); (b) r-channel: repetition coding with majority voting; (c) ϵ-channel: repetition coding with erasures; (d) c-channel: coding with four inputs.

2 Note that it is sufficient to consider the BSC with $p \leq 0.5$, since if the opposite is true, then the receiver should always interpret the received symbol 0 as 1 and vice versa, which brings us back to the case of a BSC with $p \leq 0.5$.

in the previous section, obtaining non-zero goodput requires grouping the data bits, adding error check bits in order to obtain a packet with checkable integrity, and then sending it by making use of multiple channel uses. Xia takes b data bits and adds c check bits (the error check is assumed perfect), and the packet length is $l = b + c$. If the p-channel is used, then the achieved goodput is:

$$G_{\mathrm{p}} = \frac{b}{l}(1-p)^l \quad \text{(bit/s)}. \tag{7.3}$$

When the r-channel is used, we still need to make l uses of that channel, which corresponds to $3l$ uses of the p-channel, such that the transmission is three times longer. This leads to the following goodput:

$$G_r = \frac{b}{3l}(1-p_{\mathrm{E}})^l \quad \text{(bit/s)}. \tag{7.4}$$

We can write G_r in a different way:

$$G_r = G\frac{1}{3}\left(\frac{1-p_{\mathrm{E}}}{1-p}\right)^l \tag{7.5}$$

from which it can be seen that the repetition coding will lead to benefit in terms of goodput only if

$$\frac{1}{3}\left(\frac{1-p_{\mathrm{E}}}{1-p}\right)^l > 1. \tag{7.6}$$

As stated above, we only need to consider the case $p \leq 0.5$. It turns out that G_r is not always higher than G, such that the reliability gained in repetition coding does not necessarily translate into a better goodput. For example, for $l = 100$ and $p = 0.495$ it can be calculated that $G_r = 0.546G$. However, if we increase the packet size to $l = 300$ and keep the other parameters unchanged, it follows that $G_r = 1.15G$ and the channel obtained by repetition coding leads to a higher goodput. When the original channel is very reliable and $p = 10^{-6}$, then repetition reveals its inefficiency. The reader can check that even with $l = 1000000$, it is still $G_r < G$, as the improved reliability cannot repair the damage made by decreasing the nominal transmission rate to $\frac{1}{3}$.

7.1.2 Channel Based on Repetition Coding with Erasures

Now we create a different channel by using again three repetitions over the BSC, but change the decision rule at Yoshi's side. If Yoshi receives 000 or 111, then he decides 000 or 111, respectively. Otherwise, if Yoshi receives any of the symbols 001,010, 011,100, 101,110, he announces an erasure. The channel created in such a way is shown in Figure 7.1(c) and is referred to as the ϵ-channel. The probabilities that define the ϵ-channel can be determined as follows:

- Probability of successful transmission:

$$p_{\mathrm{S}} = (1-p)^3. \tag{7.7}$$

- Probability that error occurs (all three bits in error):

$$p_{\mathrm{E}} = p^3. \tag{7.8}$$

- Probability of erasure is:

$$p_{\mathrm{ERS}} = 3p(1-p)^2 + 3p^2(1-p). \tag{7.9}$$

Using the erasures, Yoshi can now apply a different strategy for decoding the packets. Let Yoshi receive a packet over l uses of the ϵ-channel, corresponding to $3l$ BSC channel uses, which has a duration of $3l$ s. Let us assume that p of the BSC channel is small, such that the probabilities of error p_{E} and erasure p_{ERS} are also small. More specifically, assume that l is large and $p_{\mathrm{ERS}} = \frac{1}{l}$, such that only a single erasure is expected in the received packet. Yoshi tries to reconstruct the packet: he puts 0 at the bit position that has been erased and runs the error check again. Note that we are still assuming that the error check is ideal. If no error is indicated, then Yoshi accepts 0 at the erased position and the packet is correctly received. If not, Yoshi puts 1 at the bit position, runs the error check again and, if correct, Yoshi puts 1 and the erased position, thereby recovering the packet correctly. If neither 0 or 1 work, then Yoshi has to conclude that some of the other bits are erroneous and he discards the packet as incorrect.

The strategy based on flipping bit combinations until the error check flags that the packet is correct does not need to be used only with the erasure channel. The same strategy can be used for the p-channel, which is without repetition and erasures. If the received packet is flagged by the error checks as erased, then Yoshi starts to flip the bits one by one and, after each flip, he runs an error check. If no correct packet is obtained, then Yoshi starts to try all possible 2 bit combinations at all possible pairs of positions. Eventually, for some m, Yoshi will try all of the 2^m combinations for all possible sets of m out of l positions and the error check will indicate a correct packet.

The reasoning in the previous paragraphs should imply that Yoshi will eventually find "the correct" bit error pattern and decode the packet correctly, although the packet was originally erased. However, this reasoning is flawed. The problem is that we have assumed that the error check, created by adding c bits to the b information bits, is ideal. However, that assumption may only be approximately correct if the number of errors is very small and becomes completely incorrect as the number of bit errors increases. To see why this is the case, note that when Yoshi receives the packet of $l = b + c$ bits, there are 2^l possible bit sequences that Yoshi can receive and there are 2^b among them that represent the correct codewords. If an erroneous codeword is received and m bits are flipped to look for the right combination, then as m grows it becomes increasingly more likely that by flipping bits Yoshi will obtain one of the 2^b codewords that is different from the one that has originally been sent.

We therefore limit only to flipping the bits that correspond to the erasures and, if the number of erasures is small and the error check is ideal, then Yoshi will obtain the correct packet. Thus the probability that a packet of l bits is received correctly is $(1-p^3)^l$ and the goodput is:

$$G_\epsilon = \frac{b}{3l}(1-p^3)^l = G\frac{1}{3}\left(\frac{1-p^3}{1-p}\right)^l \quad \text{(bit/s)}. \tag{7.10}$$

Taking a packet of length $l = 200$ and $p = 0.006$, one gets $p_{\mathrm{S}} = 0.9821$, $p_{\mathrm{E}} = 2 \cdot 10^{-7} \approx 0$ and $p_{\mathrm{ERS}} = 0.018$, such that the expected number of erasures in a packet is less than 4, i.e. $l \cdot p_{\text{ERS}} = 3.6$. For the chosen values one can calculate:

$$G_\epsilon = 1.11G > G_r > G.$$

One can conclude that there is a region of values for p and l for which repetition with erasures offers the highest goodput. However, can we do even better while still using three BSC channel uses per one data bit?

7.1.3 Coding Beyond Repetition

At first glance, repeating the bit and applying majority voting at the receiver seems to be the only possible way to improve channel reliability when three channel uses are bundled together. Assuming that one symbol is defined through three BSC uses, one can possibly choose from 8 different input symbols 000,001,010,011,100,101,110,111. Let us call them 3-BSC symbols, for brevity. Differently from this, when repetition coding is used, we are restricted to choosing either 000 or 111.

With this insight, we can now try to choose four possible input symbols, for example 000,001,011,111. For each use of this new channel, Xia picks two new data bits and uniquely assigns one of the input symbols. This is very similar to the transmission over a baseband channel that uses QPSK, except that now the four constellation points are chosen among the eight possible patterns that can be created by three bits. If the packet contains l bits, then it can be sent by using $\frac{l}{2}$ 3-BSC symbols, which corresponds to a total of $\frac{3l}{2}$-BSC channel uses and therefore a duration of $\frac{3l}{2}$ s. If we assume that the probability of error of the original channel is $p = 0$, then the nominal goodput of this channel is:

$$\frac{b}{\frac{3l}{2}} = \frac{2}{3}\frac{b}{l} \quad \text{(s)}.$$

The four input 3-BSC symbols can be selected in $\binom{8}{4} = 70$ different ways. Each of the different selections of the set of four 3-BSC symbols defines a different communication channel with four inputs. The number of outputs from this channel depends on how the erasures are treated, but in general, there can be 8 possible output symbols, that is, all possible 3-BSC symbols. The four selected 3-BSC symbols that can act as valid inputs to the newly defined channel can be seen as length 3 *codewords* for the original BSC channels.

Following our discussion in Chapter 5 about selecting the modulation constellations and the assumption that $p \leq 0.5$, we should select the four points that have the highest separation among each other. With that criterion, the best choice of the input symbols is:

$$000 \quad 011 \quad 101 \quad 110 \tag{7.11}$$

where the Hamming distance between each pair of codewords (valid 3-BSC symbols) is two. Similar to the second channel created with repetition coding, we define the channel outputs as follows. Yoshi uses a decision rule that is based on erasures, such that whenever he receives a 3-BSC symbol that is not part of the input symbols, 001,010, 100,111, then Yoshi announces an erasure. The channel created in that way is illustrated in Figure 7.1(d) and will be referred to as the *c*-channel, defined with the probabilities:

- Probability of successful transmission:

$$p_{\text{S}} = (1 - p)^3. \tag{7.12}$$

- The probability of error is more involved, as there are four input symbols. For example, when the symbol 101 is sent, the probability to receive a specific incorrect symbol, say 011, corresponds to the probability that exactly two errors have occurred:

$$p^2(1-p) \tag{7.13}$$

and this is the probability associated with the edge, in this case 011–101. However, the probability that an error has occurred when a symbol, say 101, is sent is three times higher, as error occurs if either of the three remaining triplets 000,011, 110 is received:

$$p_{\mathrm{E}} = 3p^2(1-p). \tag{7.14}$$

- Probability of erasure is:

$$p_{\mathrm{ERS}} = 3p(1-p)^2 + p^3. \tag{7.15}$$

If we assume again that p is small, the number of erasures is small, the error check is ideal, and Yoshi attempts different bit patterns at the erased positions until getting a correct packet, then the goodput can be calculated as:

$$G_{\mathrm{c}} = \frac{2}{3}\frac{b}{l}(1-p_{\mathrm{E}})^{\frac{l}{2}} \quad (\mathrm{s}). \tag{7.16}$$

It can be noted that G_{c} increases the goodput with respect to repetition coding in two ways. First, the factor $\frac{1}{3}$ is increased to $\frac{2}{3}$, as now two data bits per 3-BSC symbol (a triplet) are sent. Second, the number of required triplets is $\frac{l}{2}$, which decreases the packet length measured in number of uses of the c-channel. However, the total duration of the packet in terms of BSC uses is $\frac{3l}{2} > l$, which indicates redundancy. The objective in coding is to reduce this redundancy while still meeting the target error performance.

7.1.4 An Illustrative Comparison of the BSC Based Channels

Figure 7.2 provides illustrative results on the goodput provided by each of the channels defined above: p-, r-, ϵ-, and c-channels. The figure shows the value of the goodput normalized by $\frac{b}{l}$, such that, for example, what is plotted for G_{p} is $(1-p)^l$. The packet length is fixed to $l = 128$. It can be noticed that when the error probability of the original BSC is very low, then using *uncoded transmission* and sending the bits one by one through the BSC leads to the highest goodput G_{p}. However, as p increases, G_{p} rapidly decreases, such that the goodput G_{c}, that is not based on repetition, leads to the highest goodput. The reason is that the use of code balances the rate loss that occurs due to coding (which is $\frac{1}{3}$ for the repetition code), while still improving the reliability of each individual transmission; yet, less compared to the repetition code.

Figure 7.3 is supplementary and should be used for sanity assessment of the assumption about an ideal error check. It shows the expected number of erased symbols for the channels that feature erasures, which are the ϵ-channel and the c-channel. Note that each symbol is represented by a triplet of transmitted values 0, 1. However, the ϵ-channel uses $l = 128$ triplets, while the c-channels uses 64 triplets. Therefore, the expected number of erased symbols for the c-channel is approximately half that for the ϵ-channel, but otherwise the fraction of erased symbols is almost the same for both in the considered range of p. The reader should also notice that a single erased symbol for the c-channel corresponds

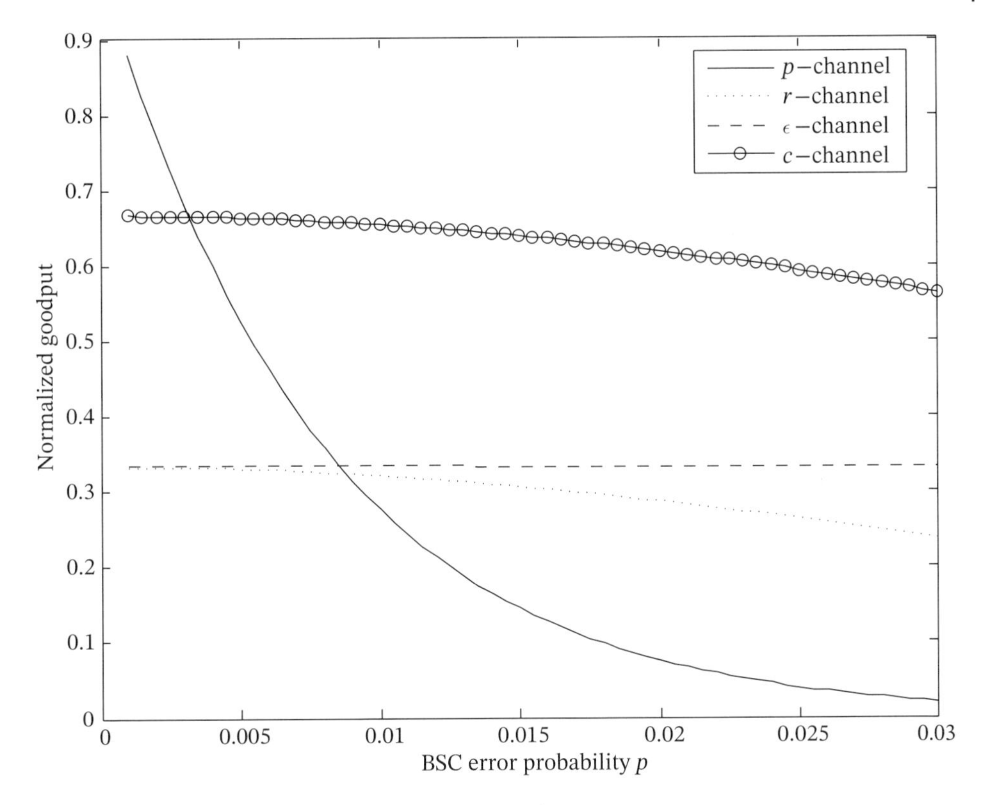

Figure 7.2 Comparison of the normalized goodput $G\frac{l}{b}$ for each of the four communication channels. The packet length is $l = 128$ bits.

to two erased bits in the packet. In any case, the average number of erasures stays low for $p < 0.015$ and we can use the ideal error check assumption to interpret the relation among the goodputs. A realistic comparison would require that, in addition to l, we specify b and the actual used c bit error checking code.

7.2 Generalization of the Coding Idea

Using the simple case of the BSC, we have illustrated the main idea behind FEC or *channel coding*, which can be summarized as follows. Given a channel with an unreliable single channel use, group multiple channel uses to create an expanded version of the channel. Then select a subset of the possible inputs for the expanded channel to represent valid input symbols. Those valid input symbols for the expanded channel are codewords for the original channel. This leads to a channel that is more reliable than the original one. The main tension in the design is explained as follows. On the one hand, how to select as many as possible inputs for the expanded channel and in this way prevent the decrease of the nominal data rate. On the other hand, how to select as few inputs as possible in order to guarantee high reliability for the transmission of a single input in the expanded channel.

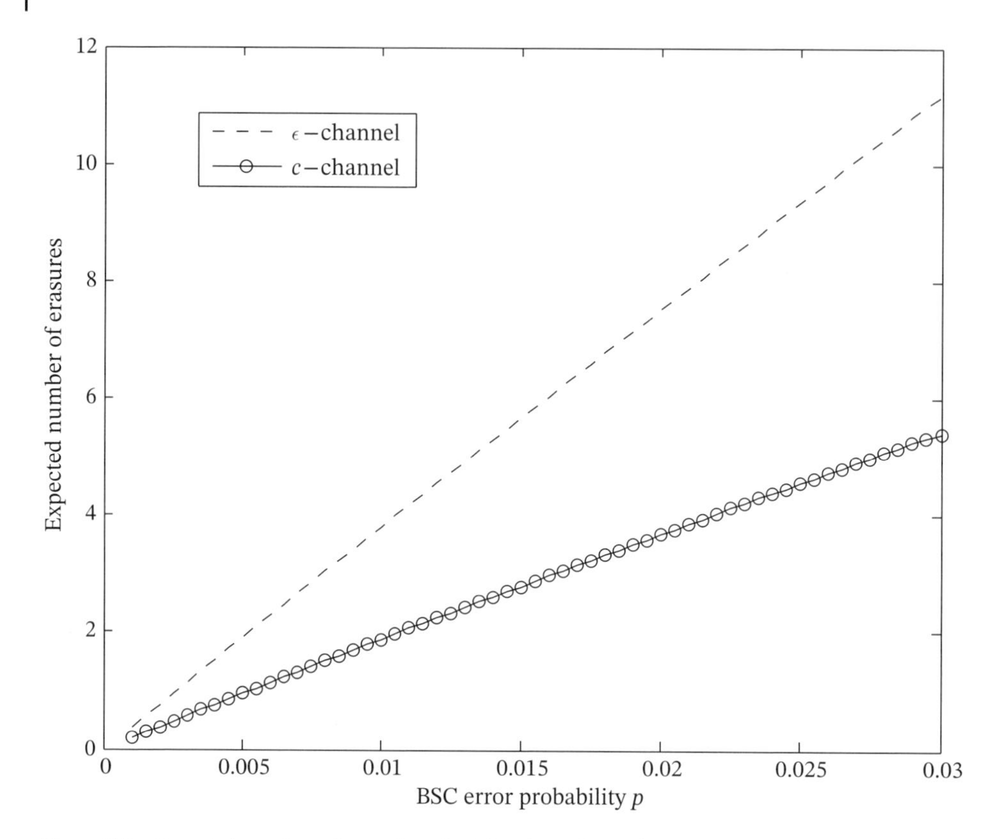

Figure 7.3 Expected number of erased symbols for the ϵ-channel and the c-channel. The number of bits is $l = 128$, which corresponds to 128 uses of the ϵ-channel and 64 uses of the c-channel.

We take now the idea to a discrete (digital) channel that is not the most general one, but it is sufficiently more general than a BSC to show how coding ideas lead to reliable communication. Let the channel between Xia and Yoshi have S input symbols and S output symbols, identical to the input ones. An *uncoded* transmission over that channel is done by taking $\log_2 S$ data bits at a time, mapping them to one of the S symbols and transmitting them. In this way, during the l uses of the channel, the total number of transmitted bits is $l\log_2 S$. The input symbols that Xia sends over l channel uses with a l-dimensional vector are

$$\mathbf{x} = (x_1, x_2, \dots x_l) \tag{7.17}$$

where each x_i is one of the S possible input symbols. In uncoded transmission, any x_i can take any of the S values, such that the total number of possible vectors that can be sent over l channel uses is S^l. In the absence of errors, this is the best possible communication strategy, as in each channel use the transmitter is *multiplexing* the maximal possible number of $\log_2 S$ bits.

However, if there are errors, then the reliability of each of the S^l possible transmitted vectors decreases, in a sense that it can be confused with another vector at the receiver. Figure 7.4(a) depicts the situation in which a symbol transmitted in a single channel use is received correctly with probability $p_1 < 1$ and incorrectly with probability $(1 - p_1)$. Specifically, when the symbol is not correctly received, the receiver Yoshi gets instead any

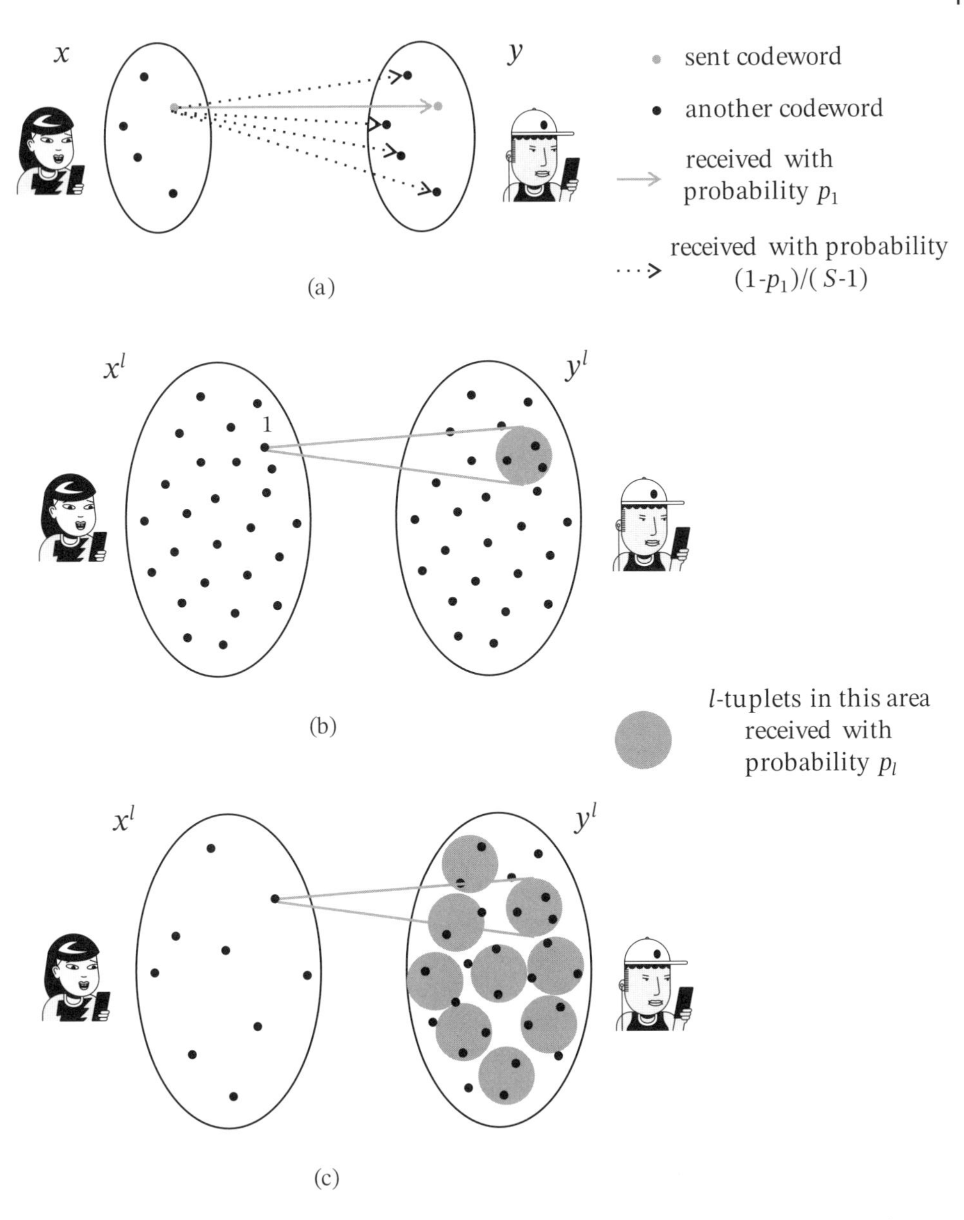

Figure 7.4 The main plot in error correction coding. (a) The original communication channel between Xia and Yoshi with S inputs and identical set of S outputs. (b) Probabilistic outcome upon transmission of a given l-tuplet over l uses of the original channel. (c) New channel between Xia and Yoshi created by selection of l-tuplets that are codewords and have non-overlapping shaded areas (where the received signal will lie with high probability).

of the remaining $(S-1)$ symbols. The easiest channel model to think of is the one where the probability that a particular incorrect symbol is received is $\frac{1-p_1}{S-1}$. Hence, considering only a single channel use does not leave us with many options to seek more reliable transmission.

Now let us consider Figure 7.4(b), which is a new communication channel, created by using the original channel l times. Let us call the original channel a 1-channel and the one obtained by l channel uses the l-channel. The l-channel has S^l possible inputs and S^l possible outputs. However, one important thing is different. For many types of 1-channels, the derived l-channel does not have the property that the correct symbol is received with probability q, while *any* of the incorrect symbols is received with equal probability $\frac{1-q}{S^l-1}$. Instead of that, when a symbol over the l-channel is sent, there is a subset of outputs (l-tuplets) that are more likely to be received compared to the rest of the possible outputs. Referring to Figure 7.4(b), if Xia sends the l-tuplet x^l labeled as "1", then Yoshi receives an l-tuple that belongs to the shaded area with probability p_l, where $p_l > p_1$. The shaded area plays the role of a noise cloud, discussed in the previous chapters, and contains the correct l-tuple as well as a subset of l-tuples that are likely to be received. The exact shaded area changes depending on which l-tuple has been sent by Xia. For simplicity, in this discussion we assume that the shaded area has the same shape regardless of which l-tuple has been sent.

Let us assume that Xia can select M different l-tuples whose shaded areas do not overlap, as in Figure 7.4(c). Each l-tuple is called a codeword and can be understood as a *modulation symbol* that can carry $\log_2 M$ bits. Let Yoshi make the following decision rule. If he receives an l-tuple, then he looks for a codeword $\hat{x}^l$, such that the shaded area associated with $\hat{x}^l$ contains the received l-tuple. If such a codeword is not found, then Yoshi announces an erasure. The decision rule will be made more precise in the next section.

The newly created channel, after deciding the codewords and Yoshi's decision rule, has a probability of error at most $(1-p_l)$. On the other hand, in each use of this new channel Xia can send $\log_2 M$ bits and that corresponds to

$$R = \frac{\log_2 M}{l} \tag{7.18}$$

bits per single use of the original channel. How do you select M? What is the probability of error in relation to l? These and similar questions lie at the heart of information theory and will be discussed in Chapter 8. However, what is important is that for many meaningful channels, as l increases, the probability of error $(1-p_l)$ can be made arbitrarily small, while multiplexing $R > 0$ bits per single use of the original channel.

The reader might have noticed the following terminological subtlety. Namely, once we create the new channel that takes l uses of the original channel and has a defined set of inputs/outputs (that we are unwilling or unable to change), then a transmission of $\log_2 M$ bits is an *uncoded transmission* over the new channel, while it is a coded transmission over the original channel.

Finally, in our example channel with S inputs and S outputs, redundancy is reflected in the fact that for sending M different messages we are using l Sary symbols. In the absence of errors only $\log_S M$ Sary symbols are sufficient to represent all M messages and therefore the redundancy is:

$$l - \log_S M$$

channel uses.

7.2.1 Maximum Likelihood (ML) Decoding

In this section we look closely at the decision rule, loosely defined above as the "valid codeword belonging to the shaded area". There are S^l possible channel inputs to the channel expanded over l channel uses, but Xia and Yoshi have agreed that only M of them are valid codewords and they are denoted by $\mathbf{x}_1, \mathbf{x}_2, \dots \mathbf{x}_M$. There are S^l possible outputs $\{\mathbf{y}\}$ that can be observed by Yoshi, each consisting of l outputs of the original channel. Given that a specific vector $\mathbf{y}$ has been received, Yoshi makes a decision that $\hat{\mathbf{x}}$ has been sent. Similar to the discussion in the previous chapter, related to the QPSK modulation and Gray mapping, here we would also like to decide $\hat{\mathbf{x}}$ such that it is the most likely candidate that can produce the observed $\mathbf{y}$. A decoding rule is established by specifying which codeword $\hat{\mathbf{x}}$ should be decided by Yoshi for every possible received vector $\mathbf{y}$. If each codeword is sent with equal probability of $\frac{1}{M}$, then for given $\mathbf{y}$ Yoshi needs to find $\mathbf{x}_i$ such that the following probability is maximized:

$$P(\mathbf{x}_i|\mathbf{y}) = \frac{P(\mathbf{y}|\mathbf{x}_i)P(\mathbf{x}_i)}{P(\mathbf{y})} = P(\mathbf{y}|\mathbf{x}_i)\frac{1}{MP(\mathbf{y})}. \tag{7.19}$$

Since $P(\mathbf{y})$ does not depend on the decision we associate with $\mathbf{y}$, the maximization of $P(\mathbf{x}_i|\mathbf{y})$ is equivalent to the maximization of $P(\mathbf{y}|\mathbf{x}_i)$. The latter corresponds to the likelihood of observing $\mathbf{y}$ when $\mathbf{x}_i$ is sent; thereby the term *maximum likelihood (ML) decoding*. Considering a memoryless channel, we get

$$P(\mathbf{y}|\mathbf{x}_i) = \prod_{j=1}^{l} P(y_j|x_{ij}) \tag{7.20}$$

where x_{ij} is the input in the jth channel use of the ith codeword and the probabilities $P(y_j|x_{ij})$ are specified in a single use of the original channel. Instead of maximizing directly (7.20) it is equivalent and more convenient to work with the log-likelihood function:

$$\log P(\mathbf{y}|\mathbf{x}_i) = \sum_{j=1}^{l} \log P(y_j|x_{ij}). \tag{7.21}$$

For the communication channels that are commonly used, such as a BSC, the smaller the Hamming distance between $\mathbf{y}$ and $\mathbf{x}$, the higher the value of the log-likelihood $\log P(\mathbf{y}|\mathbf{x}_i)$. Therefore we can represent the outputs $\mathbf{y}$ that are decoded into a particular $\mathbf{x}_i$ as a shaded area around $\mathbf{x}_i$ in Figure 7.4. This can be directly related to the concept of noise clouds, used in the previous chapters, where upon transmission of a symbol, it is most likely to receive a symbol that lies within a noise cloud that surrounds the transmitted symbol. In other words, the noise cloud around $\mathbf{x}_i$ contains[3] $\mathbf{y} = \mathbf{x}_i$ and all other outputs $\{\mathbf{y}\}$ for which the ML decoding rule outputs $\mathbf{x}_i$.

The major obstacle in building practical codes with error probability that goes to zero is the complexity of decoding. In order to completely specify the ML decoding rule, for each of the S^l possible received symbols we need to compute M different likelihoods and pick $\mathbf{x}$ that leads to the maximal one. Hence, the complexity grows exponentially with l. Indeed, if the output y belongs to a discrete alphabet, then the number of possible outputs is countable and finite, such that the receiver can store the decoding rule as a lookup table for each

3 There can be communication channels in which a given input does not appear as an output; however, for the meaningful channels considered here, the input symbol always appears as a valid output symbol in the absence of errors.

possible **y**. This somewhat lessens the real-time computation burden, but it does not essentially decrease the need for an exponentially large search. Therefore, the main trade-off in practical coding is to find sufficiently separated codewords, while keeping the decoding complexity at a reasonable level. Coding history has witnessed many outstanding attempts in this direction, including convolutional codes, turbo codes, low-density parity check codes (LDPCs), polar codes, etc. In the next section we illustrate the coding mechanism for different examples of channels.

7.3 Linear Block Codes for the Binary Symmetric Channel

We continue to introduce ideas in error control coding in the context of a BSC. There are many different code constructions and in this section we keep the discussion on the important class of linear block codes. Another important and widely used class of codes is the one of *convolutional codes*, briefly treated in the next section in relation to the *trellis codes*.

A *block code* for a BSC gets b data bits, which constitute a *message*:

$$\mathbf{d} = (d_1, d_2 \dots d_b)$$

and for each message there is a codeword, which is an output l-tuple of binary values:

$$\mathbf{x} = (x_1, x_2, \dots x_l)$$

where $l > b$, and both d_i and x_j are binary values. We thus obtain a (l, b) code with *code rate* equal to:

$$R = \frac{b}{l} \quad \text{(bit/c.u.)} \tag{7.22}$$

where "c.u." stands for "channel use", which corresponds to the nominal data rate that is achieved in the absence of errors. Following the terminology from the previous sections, we can say that creating a (l, b) block code corresponds to creating a c-channel by gathering l channel uses to represent a single use of the c-channel. The obtained c-channel has 2^b channel inputs and 2^l channel outputs.

During the transmission of a particular codeword **x**, there can be some channel uses in which the bit values x_j are flipped from 0 to 1 or vice versa, thereby introducing error in the received codeword. Referring to Figure 7.4(c), Xia hopes that the channel errors will not move the received codeword outside of the shaded area around **x**, such that Yoshi can correctly decide the codeword that has originally been sent. Refocusing from a codeword to a bit level, the idea behind a good code is that a particular message bit d_i affects multiple coded bits in **x**. In that way the same bit of information d_i experiences diversity in the transmission and arrives through multiple independent channel uses. The simplest form of diversity is repetition coding, which is fundamentally inefficient, as each transmitted symbol x_j carries information for only one information bit d_i. The art of error-correcting codes is to find smart ways to introduce diversity: each d_i should affect multiple transmitted symbols $\{x_j\}$, while each transmitted symbol x_j should be affected by multiple information bits $\{d_i\}$.

Consider again the example of a c-channel in Figure 7.1(d). The c-channel has four possible inputs and five possible outputs, since whenever an error is detected in the received

Table 7.1 Encoding rule for the simple code of rate 2/3.

Message d	Codeword x
00	000
01	011
10	101
11	110

codeword, the channel output is an erasure. Once the c-channel is specified, and assuming that the input message is specified in quaternary and equally probable symbols, in principle, we do not need to go back to the BSC and deal with bits. However, a bit is the main currency of digital communication, the original messages are specified in bits, and channel coding is about finding systematic ways to map a sequence of bits into a sequence of transmitted symbols. This will lead us to methods for creating a c-channel with 2^b inputs, where b and the code rate are arbitrary. Therefore, we can see the c-channel as a channel that applied an error-correction code that maps 2-bit messages into three transmitted binary symbols and thus has a nominal rate of $R = \frac{2}{3}$ (bit/c.u.). The *encoding* is specified through the mapping of each possible 2 bit sequence into a codeword of three binary transmitted symbols and is depicted in Table 7.1.

If there is a single bit error, Yoshi is capable of detecting it and considers a received triplet from the set $\{001,010,100,111\}$ as being erased. However, Yoshi has no way of correcting a single bit error: for example, when 001 is received and ML decoding is applied, then it is equally likely that one of the following codewords has been sent $\{000,011,101\}$. In this simple case each noise cloud contains a single received bit triplet, which is the originally transmitted codeword, while the four received triplets that are erased do not belong to any cloud. Furthermore, if two bit errors occur in the codeword, then the error is *undetectable*. In order to be capable to correct and detect more errors, the code construction needs to have larger b and l, as will be shown later in this section.

The code described in Table 7.1 belongs to the class of *linear block codes*, where linearity is defined with respect to addition and multiplication of binary numbers, that is, within the Galois Field GF(2). These codes have a convenient representation using vectors and matrices. For this particular example, a message is a 1×2 binary vector $\mathbf{d}$, a codeword is a 1×3 binary vector $\mathbf{x}$ and the encoding process is described as:

$$\mathbf{x} = \mathbf{d} \cdot \mathbf{G} \tag{7.23}$$

where $\mathbf{G}$ is a *generator matrix* specified as follows:

$$\mathbf{G} = \begin{bmatrix} 1 & 0 & 1 \\ 0 & 1 & 1 \end{bmatrix}. \tag{7.24}$$

Linear codes have multiple interesting properties. For example, due to linearity, the sum of two or more codewords is a valid codeword. This means that a sum of a codeword with itself is always a valid codeword. In other words, $\mathbf{x} = [000 \cdots 0]$ is always a valid codeword that is associated with the message $\mathbf{d} = [000 \cdots 0]$.

Perhaps one of the most important property of linear codes is related to the Hamming distance. The *Hamming distance spectrum* for a particular codeword $\mathbf{x}_0$ is represented by the histogram of Hamming distances between $\mathbf{x}_0$ and all the other codewords. The minimal Hamming distance tells us how far away the closest codeword is and the multiplicity of the minimal Hamming distance tells us how many of these closest codewords there are. For linear codes, the Hamming distance spectrum for each codeword is identical, and, specifically, equal to the distance spectrum of the codeword $[000 \cdots 0]$. Therefore, the noise clouds of all codewords are identical, although in the binary case and discrete space in which the codewords are placed it is hard to argue that their geometric shape is a circle cloud.

Linear codes represent a subclass of all possible codes and there are nonlinear codes that are at least as good as the linear ones. For example, by changing the mapping in Table 7.1 into

$$00 \rightarrow 011 \qquad 01 \rightarrow 000 \qquad 10 \rightarrow 101 \qquad 11 \rightarrow 110$$

we create a nonlinear code that has identical performance in terms of communication and errors, but cannot be generated using relation (7.23).

An well known example of a code that can correct at most one bit error is the Hamming code with $b = 4$ and $l = 7$, denoted as the $(7, 4)$ Hamming code. The generator matrix of this code is given by:

$$\mathbf{G} = \begin{bmatrix} 1 & 0 & 0 & 0 & 1 & 1 & 0 \\ 0 & 1 & 0 & 0 & 1 & 0 & 1 \\ 0 & 0 & 1 & 0 & 0 & 1 & 1 \\ 0 & 0 & 0 & 1 & 1 & 1 & 1 \end{bmatrix}. \tag{7.25}$$

The Hamming code is a systematic code since the original message bits appear in the associated codeword. For the matrix $\mathbf{G}$ in (7.25) the reader can check that the first four bits of the codewords are identical to the 4 bit message. Each of the 16 possible codewords is generated by using (7.23). Conversely, given a 7-tuple of binary values, we can check whether it is a valid codeword by using the *parity-check matrix* $\mathbf{H}$, which is uniquely determined by the generator matrix $\mathbf{G}$. Namely, each codeword $\mathbf{x}$ should satisfy the following:

$$\mathbf{x} \cdot \mathbf{H}^T = \mathbf{0} \tag{7.26}$$

where H^T is a transposed parity-check matrix and the result $\mathbf{0}$ is a row vector of size 3. The parity-check matrix that corresponds to the generator matrix (7.25) is:

$$\mathbf{H} = \begin{bmatrix} 1 & 1 & 0 & 1 & 1 & 0 & 0 \\ 1 & 0 & 1 & 1 & 0 & 1 & 0 \\ 0 & 1 & 1 & 1 & 0 & 0 & 1 \end{bmatrix}. \tag{7.27}$$

The parity-check property (7.26) provides a tool for the receiver for checking if errors have occurred. A convenient way to represent the received 7-tuple of binary values $\mathbf{y}$ is:

$$\mathbf{y} = \mathbf{x} \oplus \mathbf{e} \tag{7.28}$$

where $\oplus$ is a binary addition, for each vector component, and the binary 7-tuple is termed *error vector*. If the channel is BSC with probability of error p, then each component of $\mathbf{e}$ takes the value 0 with probability $(1-p)$ and 1 with probability p. Upon receiving $\mathbf{y}$, Yoshi applies (7.26) and gets:

$$\mathbf{s} = \mathbf{y} \cdot \mathbf{H}^T = (\mathbf{x} \oplus \mathbf{e}) \cdot \mathbf{H}^T = \mathbf{e} \cdot \mathbf{H}^T \tag{7.29}$$

where $\mathbf{s}$ is a *syndrome vector*. In (7.29) we have used the linearity and the party-check property (7.26) in order to arrive at the fact that the value of the syndrome vector depends only on the error vector. If $\mathbf{s} \neq \mathbf{0}$, then this is an indication that an error has occurred, that is, at least one of the components of $\mathbf{e}$ is not 0.

The objective of Yoshi is to find the correct transmitted vector $\mathbf{x}_c$, or more precisely, the message $\mathbf{d}_c$ that produced $\mathbf{d}_c$. However, $\mathbf{s}$ does not carry any information about the transmitted vector, but only about the error vector. This means that any of the 2^b possible messages could have been sent, since for a given vector $\mathbf{x}$, we can always find an error vector that will produce a given syndrome vector $\mathbf{s}$. Hence, there are 2^b possible error vectors that result in $\mathbf{s}$. Among all the possible error vectors, Yoshi needs to select the one that has the highest likelihood of occurrence, which in the case that the error probability in the BSC is $p < 0.5$, corresponds to the error vector that has the lowest Hamming weight. In other words, Yoshi selects the error vector that is closest to *some codeword* $\hat{\mathbf{x}}$ in terms of Hamming distance and then decides that Xia has transmitted $\hat{\mathbf{x}}$.

The $(7, 4)$ Hamming code is capable of correctly decoding the transmitted message when the received binary vector has up to one bit error. That means that the noise cloud around a valid codeword contains in total 8 binary vectors: the codeword itself (Hamming distance 0) and 7 binary vectors that correspond to the 7 possible error vectors of Hamming weight 1. Since there are 16 codewords and each of them has a noise cloud of size 8, the total number of binary vectors of size 7 that are covered by the noise clouds is $16 \cdot 8 = 128$. On the other hand, this corresponds to the number of all possible binary vectors of size 7. It can be concluded that any possible received vector is within a Hamming distance of one from exactly one codeword and there are no received vectors that can, with equal likelihood, be associated with two or more codewords, as it was the case for the simple $(3, 2)$ code from Table 7.1. This is a very interesting feature of the Hamming code, which brings it into the class of *perfect codes*; the only other code that belongs to this class, besides the Hamming codes, is the $(23, 12)$ Golay code.

The decoding problem in linear block codes is equivalent to a search for an error vector with minimal Hamming weight and there is a long track of research works aimed at reducing the complexity of that type of search. Hamming codes represent only one possible class of linear bock codes. A detailed discussion on other possible constructions is out of scope for this book; we only remark here that another important class of codes are the *cyclic block codes*, whose members are the cyclic redundancy check (CRC) codes, extensively used for error detection. We have indicated in the previous chapter, Section 6.5.2, that no error detection can be perfect and this can be further clarified by the discussion on the linear block codes and the error vectors. Since the error vector of size l can get any of the 2^l possible values with non-zero probability, it follows that there is a non-zero probability that the error vector shifts a transmitted codeword into the noise cloud of another codeword, thus resulting in an undetected error.

7.4 Coded Modulation as a Layered Subsystem

We shift our focus from BSC to Gaussian channels. A codeword $\mathbf{x}$ of size l in a Gaussian channel is represented by (7.22); however, now each x_i is a complex number rather than a binary value. We recall from Section 6.2.2 that in Gaussian channels there is a restriction imposed by the average power P of the transmitted symbols such that a single transmitted symbol x_i can have an arbitrary amplitude, as long as the average power of all transmitted symbols in a given codeword satisfies the power constraint. Let us call such an approach *analog coding*, to emphasize the fact that each message $\mathbf{d}$ is directly mapped to a codeword with analog values $\mathbf{x}$. In other words, analog coding combines error control coding and modulation in a single operation. In the next chapter we show that analog coding, focused on the full codewords rather than the individual symbols in the codeword, is essential for achieving the highest possible rate for a given Gaussian channel.

In Chapter 5 we introduced the idea of adaptive modulation. This can be generalized to the concept of *adaptive modulation and coding (AMC)* with the same objective: use a higher rate, measured in bits per symbol, when the SNR increases. In order to illustrate this, recall the uncoded adaptive modulation, in which a BPSK transmission sends 1 data bit per symbol and a QPSK transmission sends 2 data bits per symbol. Assume that a rate-1/2 FEC code is available, such that each data bit produces two coded bits. The combination of this code and QPSK modulation results in transmission of 1 data bit per symbol, exactly the same nominal goodput as in uncoded BPSK. However, the reliability of coded QPSK transmission is usually superior to uncoded BPSK transmission. The same nominal rate can be achieved by using a rate-1/4 FEC code and 16-QAM modulation. The important point is that the combination of modulation and coding schemes lead to a larger number of transmission modes to select from and thus a better granularity for adaptation, which would result in smaller "stairs" in Figure 5.8.

In order to design a transmission mode with FEC coding and a specific modulation, we can apply a layered approach for the Gaussian channel, which means that the coding and modulation are designed separately. Refer to Section 6.5.1 for the concept of layering applied to communication channels. In order to avoid confusion, we introduce the following notation. The original message $\mathbf{d}$ of b bits is mapped to a binary codeword

$$\mathbf{x}_d = (x_{d_1}, x_{d_2}, \cdots x_{d_l}) \tag{7.30}$$

which is further mapped onto another codeword of u complex symbols:

$$\mathbf{s} = (s_1, s_2, \cdots s_u). \tag{7.31}$$

In the most common approach, the constellation used for modulation is fixed, such that each s_i gets a value from a set of M predefined complex values and the average power of each s_i is P. In the usual case, M is of a form $M = 2^m$, such that l and u satisfy the following relation:

$$l = m \cdot u. \tag{7.32}$$

The mapping of the original message $\mathbf{d}$ to the transmitted codeword $\mathbf{s}$ can be seen as a *concatenation* of two codes. This is illustrated in Figure 7.5, where the output of code 1 $\mathbf{x}_d$ acts as an input to code 2. Code 1 is a binary code with a rate $R_d = \frac{b}{l}$, as it gets b message

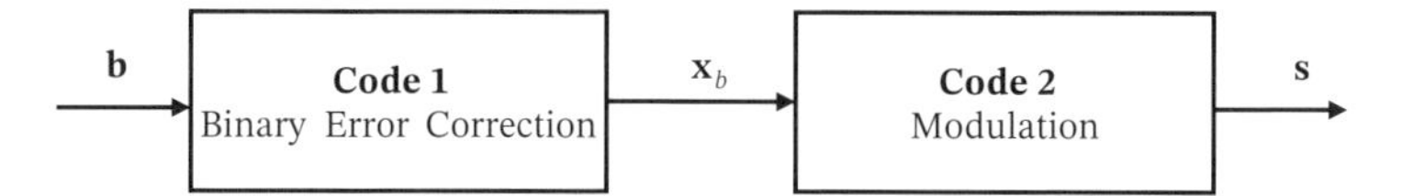

Figure 7.5 Representation of coding and modulation as a concatenation of two codes.

bits and outputs l binary values. Since both the input and the output are binary values, this code clearly introduces redundancy. The output of code 2 is in a form of analog values and one cannot tell easily whether redundancy has been introduced or not. If (7.32) is satisfied, then a possible implementation of code 2 is to take a group of m bits from $\mathbf{x}_d$ and map them to an Mary output symbol using, for example, Gray code. Hence, looking only at code 2, its nominal transmission rate is $m = \frac{l}{u}$ bits per symbol. The overall transmission rate of the concatenated code is

$$R = R_d \cdot m = \frac{b}{l} \cdot \frac{l}{u} = \frac{b}{u} \quad \text{(bit/c.u.).} \tag{7.33}$$

Figure 7.5 indicates a clear separation between code 1 and code 2, representing coding and modulation, respectively. This is due to the fact that, for example, the Gray coding of code 2 is done independently from the error correction method used in code 1. Having this perspective, it comes natural to ask: what can be attained by not separating code 1 and code 2 and designing them jointly through some form of cross-layer optimization?

This question has motivated *trellis coded modulation*, a shining example of technology breakthroughs in communication engineering. In order to understand the basic idea of it, let us at first look at the concept of a trellis code. Figure 7.6 shows an example of a simple trellis code that encodes 2 bit messages into codewords of length $l = 6$, thus leading to a code rate of $R = \frac{1}{3}$. The codewords of a trellis code are created by a finite-state machine and in this example there are two states, S0 and S1. At the start of the codeword generation, the encoder is in state S0. Given the 2 bit input message $\mathbf{d} = (d_1, d_2)$, the codeword $\mathbf{x} = (x_1, x_2, x_3, x_4, x_5, x_6)$ is generated in the following way. In the first step, Xia takes the first message bit d_1; if $d_1 = 0$ then she sends $(x_1, x_2) = (0, 0)$ and transits to state S0, otherwise, if $d_1 = 1$ she sends $(x_1, x_2) = (1, 1)$ and transits to state S1. The next steps proceed analogously, following the structure of the *trellis graph* in Figure 7.6. Each branch of the graph is labeled by $d_i, x_j x_{j+1}$ and specifies the output binary values that are generated by the jth message bit. If the message bit is set to $*$, then the transition is made with a dummy message bit, which in this example is the “third” message bit.

This is a very simple example of a trellis code and it can be generalized in many ways. For example, the number of branches going out of a state is not necessarily two, as the encoder can take b message bits in a single step and produce l output binary values of the codeword. In that case, each branch is labeled $(d_1 d_2 \cdots d_b, x_1 x_2 \cdots x_l)$. The concept of trellis

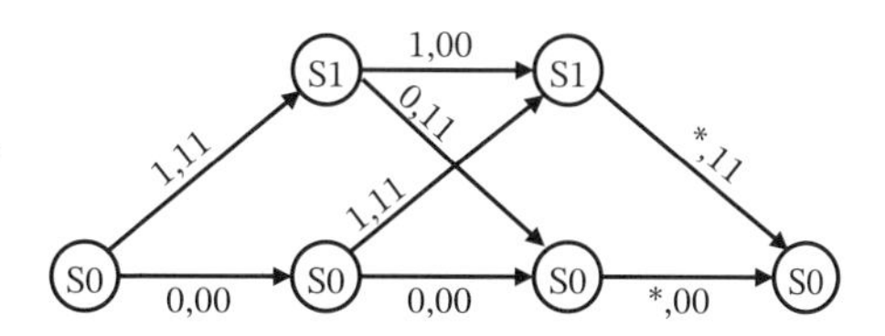

Figure 7.6 Example of a trellis code with two states. On the left is the trellis diagram and on the right the code of rate $R = \frac{1}{3}$ produced by the trellis.

coding was originally introduced to describe *convolutional codes*, where the state transitions are implemented by shift registers; however, the trellis can also be used to represent other codes, such as linear block codes. The trellis representation has had widespread usage due to the fact that the decoder can use the seminal Viterbi algorithm for maximum likelihood decoding.

The strength of a trellis code is tightly related to the number of states used. The idea is that, the more states there are, the larger number of output bits become affected by a specific input bit, which is in line with the principles of good code design stated earlier in this chapter. Furthermore, multiple states represent the key to the joint design of coding and modulation in trellis-coded modulation. In order to see this, assume that 8-PSK modulation is used and the symbols are $s_1, s_2, \dots s_8$. Let us assume that there are two data bit sequences $\mathbf{d}_1$ and $\mathbf{d}_2$ that act as inputs to code 1 from Figure 7.5. Let us denote the respective outputs of code 1 as follows:

$$\mathbf{d}_1 \mapsto \mathbf{c}_{11}, 000001, \mathbf{c}_{12}$$
$$\mathbf{d}_2 \mapsto \mathbf{c}_{21}, 000001, \mathbf{c}_{22}$$

where $\mathbf{c}_{ij}$ represent sequences of coded bits associated with the data bits $\mathbf{d}_i$. For the present discussion the actual content of the coded bit sequences $\mathbf{c}_{ij}$ is irrelevant; it is only important that the coded bits of both $\mathbf{d}_1$ and $\mathbf{d}_2$ contain the subsequence 000001.

Let us at first look at the case in which coding and modulation are designed separately. In that case, the mapping of the bit triplets from the code output onto the symbols is fixed: for example, 000 is always mapped to s_1, 001 is always mapped to s_2, etc. Therefore, the mapping of the data bits into modulated symbols looks as follows:

$$\mathbf{d}_1 \mapsto \mathbf{s}_{11}, s_1, s_2, \mathbf{s}_{12}$$
$$\mathbf{d}_2 \mapsto \mathbf{s}_{21}, s_1, s_2, \mathbf{s}_{22}$$

where $\mathbf{s}_{ij}$ represent the other modulation symbols associated with $\mathbf{d}_i$.

In contrast, in trellis coding the system has different states and the mapping of a coded bit sequence onto a modulation symbol depends on the state the system. For example, there can be a trellis-coded modulation where the mappings are made as follows:

$$\mathbf{d}_1 \mapsto \mathbf{t}_{11}, s_1, s_2, \mathbf{t}_{12}$$
$$\mathbf{d}_2 \mapsto \mathbf{t}_{21}, s_3, s_4, \mathbf{t}_{22}.$$

This means that the state the trellis-coded system is in at the step when the modulation sequence $\mathbf{t}_{11}$ has been produced is different from the state the system is in at the step at which the modulation sequence $\mathbf{t}_{21}$ has been produced. Hence, 000001 results in two different pairs of transmitted symbols, s_1, s_2 and s_3, s_4, respectively.

There are certainly other coding methods in which 000001 can result in different modulation symbols, but the important advantage of trellis-coded modulation is that its structure preserves the possibility of using the efficient Viterbi algorithm for decoding.

7.5 Retransmission as a Supplement to Coding

The methods for channel coding or FEC that emerge from the basic model of a communication system by Shannon, depicted in Figure 6.1 do not consider the possibility of feedback from Yoshi to Xia. By contrast, in practice, Xia can believe that Yoshi has received the data packet correctly if she gets an acknowledgement from him. In Chapter 4, when discussing the reliable transmission service provided by a lower layer to the layer above, we have assumed that the lower layer ensures successful transmission of an acknowledgement. In this section we discuss concepts and ideas that Xia can use for retransmission upon receiving feedback from Yoshi.

At this point we are completing the story of the cartoon from the beginning of this chapter. At first Zoya notices that one of the digits is erroneous and her capability for pattern recognition, combined with previous experience, plays the role of FEC decoding. The credit card number has an embedded error checking capability, which helps to detect that the digit that Zoya "decodes" from the piece of paper is incorrect. Finally, after not succeeding to decode the correct credit number by only using FEC, Zoya asks her dad to tell her the correct digit. The call placed to her dad can be understood as a feedback that asks for retransmission. Here it should be noted that Zoya can ask her father to read again all the digits of the credit card (inefficient) or only the digit that was unreadable (efficient).

Nevertheless, receiving an ACK (acknowledgement) from Yoshi is not a guarantee to Xia that he has received the packet correctly. The non-ideality of the error checking code can lead to error patterns that can trick Yoshi into believing that he has received a packet correctly, while in reality he is acknowledging the reception of a wrong packet. However, if receiving an ACK is not sufficient, is there a way for Yoshi to send richer feedback to Xia and guarantee the reception of the correct packet?

To answer this question, assume that Xia sends the data packet $\mathbf{d}_1$. Yoshi decodes the packet

$$\mathbf{d}_2 = \mathbf{d}_1 \oplus \mathbf{e} \tag{7.34}$$

where $\mathbf{e}$ is the bit error pattern and $\oplus$ is a binary addition, as in (7.28). Since $\mathbf{e}$ is a product of a random process, assume that it happens to be such that $\mathbf{d}_2$ satisfies the error check and Yoshi believes that this is the correct packet. Now, instead of sending ACK, Yoshi decides to use (very) rich feedback and transmit the whole packet $\mathbf{d}_2$ to Xia. To do that, Yoshi may also use some form of FEC, which does not need to be the same FEC used by Xia to send $\mathbf{d}_1$. Now the random error process happens to produce the same error pattern $\mathbf{e}$. Note that this event can occur, although with very low probability. Then Xia receives:

$$\mathbf{d}_2 \oplus \mathbf{e} = \mathbf{d}_1 \oplus \mathbf{e} \oplus \mathbf{e} = \mathbf{d}_1 \tag{7.35}$$

since $\mathbf{e} \oplus \mathbf{e} = \mathbf{0}$, which is an all-zero vector. Clearly, $\mathbf{d}_1$ satisfies the error check and Xia believes that this is the packet received and decoded by Yoshi. However, the actual packet decoded and accepted as correct by Yoshi is an incorrect one.

The previous result shows the theoretical impossibility of perfectly reliable communication within finite time; the next chapter will present the conditions under which communication can become perfectly reliable in an asymptotic regime. To facilitate the discussion, for the rest of this section we will neglect the imperfectness of the error check and assume that ACK always represents the correct packet.

7.5.1 Full Packet Retransmission

The most elementary way for Xia to react upon not receiving an ACK from Yoshi is to retransmit the whole packet again. Even when Xia does not use any form of FEC, retransmission will eventually lead to reliable transmission, provided that the error check is perfect. If Xia does use FEC for sending the original packet, then the retransmission protocol is called *hybrid ARQ (HARQ)*. In the simplest HARQ version, Yoshi discards all versions of the packet that were not decoded correctly and waits until he gets a packets that is decoded correctly. An improved version is the one in which Yoshi does not waste the erroneously received copies of the packets and instead combines them with the present version of the packet in order to improve the decoding reliability.

To see how this works, let $\mathbf{s} = (s_1, s_2, \dots s_u)$ represent the packet sent by Xia through the respective baseband symbols. Yoshi does not receive the first packet transmission correctly, Xia does not receive an ACK and she retransmits the same $\mathbf{s}$. Let us look at the same received symbol from both packet transmissions:

$$y_{i,1} = hs_i + n_{i,1}$$
$$y_{i,2} = hs_i + n_{i,2} \tag{7.36}$$

where the index j stands for the jth packet transmission; $y_{i,j}$ is the ith symbol received by Yoshi and $n_{i,j}$ is the ith noise sample. It is assumed that the channel h stays constant during the transmission and the retransmission. Using Chase combining or maximum ratio combining (MRC), Yoshi creates:

$$y_i = y_{i,1} + y_{i,2} = 2hs_i + n_{i,1} + n_{i,2}. \tag{7.37}$$

If the noise samples are independent, then MRC makes the SNR of y_i be double the original SNR under which the data is attempted to be decoded from the individual $y_{i,j}$. Assuming that the feedback from Yoshi is ideal and instantaneous, then L retransmissions will increase the overall transmission time of the packet L times, which decreases the nominal throughput L times. On the other hand, using MRC will increase the decoding SNR L times. These are two opposing mechanisms that are well-sublimed in Shannon's expression for the capacity of a Gaussian channel (see the next chapter).

This is illustrated in Figure 7.7(a), where Xia sends the same packet $\mathbf{s}_1$. The jth transmitted version of the packet is denoted as $\mathbf{S}_{1,j}$. After the first transmission, Yoshi sends NACK. However, Yoshi is not able to detect the packet $\mathbf{S}_{1,2}$, therefore the short reception time at the beginning of the packet. Hence, he is not in a position to send NACK and Xia sends the version $\mathbf{S}_{1,3}$ due to the absence of NACK. Nevertheless, the fact that Yoshi has not detected $\mathbf{S}_{1,2}$ has the consequence that he is not able to combine all three transmitted versions of the packet, that is, the transmission of $\mathbf{S}_{1,2}$ is wasted. This figure also illustrates the fact that retransmissions effectively decrease the throughput by consuming more time to deliver the same portion of data.

As a final remark, there is one statistical subtlety related to the Chase combining in (7.37). Namely, the statistical nature of $n_{i,1}$ and $n_{i,2}$ is different. This is because, after retransmission is requested, then Yoshi knows something about $n_{i,1}$, since the set of all noise instances in the first transmission have been such that they have resulted in an erroneous reception. This creates a dependence among $n_{1,1}, n_{2,1}, n_{3,1}, \ldots$, while $\{n_{i,2}\}$ remain independent from each other, as well as from the noise instances from the first transmission. As a result, strictly speaking, the SNR of the combined signal is not doubled, but has a more involved statistical relationship with the first and the second transmissions.

7.5.2 Partial Retransmission and Incremental Redundancy

Instead of retransmitting the same $\mathbf{S}_{1,1} = \mathbf{S}_{1,2} = \mathbf{S}_{1,3} = \mathbf{s}$, Xia can retransmit another set of symbols $\mathbf{R}_{1,2}$ for the first retransmission, $\mathbf{R}_{1,3}$ for the second retransmission, etc. This is depicted in Figure 7.7(b), where it is seen that the retransmission of smaller set of symbols results in a throughput improvement, provided, of course, that the smaller packets $\mathbf{R}_{1,2}, \mathbf{R}_{1,3}$ do not deteriorate the decoding reliability. If one looks at the entirety of $\mathbf{S}_{1,1}, \mathbf{R}_{1,2}, \mathbf{R}_{1,3}, \ldots$, then $\mathbf{R}_{1,2}, \mathbf{R}_{1,3}, \ldots$ can be seen as elements contributing to an additional error correction capability. However, the redundancy of $\mathbf{R}_{1,2}, \mathbf{R}_{1,3}, \ldots$ is not introduced in the first transmission, but upon feedback from the receiver. This is why this method of retransmission is called *incremental redundancy*.

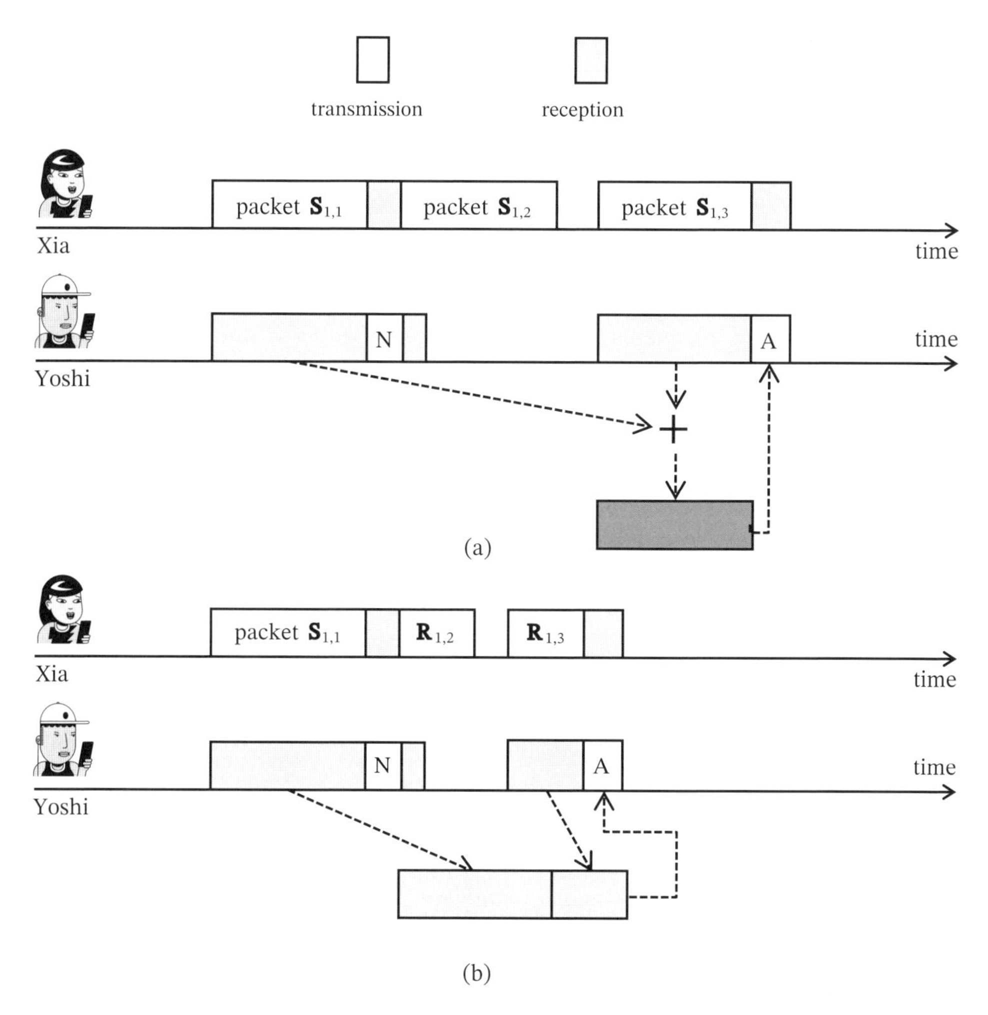

Figure 7.7 Comparison of full and partial retransmission. In both cases the packet of the first retransmission is not detected by Yoshi and therefore not used in the coding process. (a) Full retransmission. (b) Partial retransmission and incremental redundancy.

Incremental redundancy utilizes the fact that a full retransmission may introduce excessive redundancy, in a sense that Yoshi could have been able to decode the data with less information received through the retransmission. This is why it can be possible to improve the throughput as in Figure 7.7(b) without decreasing the reliability. The ideal redundancy should be just sufficient to supplement the previously received version of the packet and thus enable Yoshi to decode the data packet correctly.

In principle, this could be achieved by rich feedback instead of a simple NACK: after failing to decode the packet, Yoshi sends feedback to Xia and this feedback contains information "I am missing b' bits of information to be able to decode". The key phrase is "in principle", as there is a major difficulty in finding coding/decoding methods that enable Yoshi to measure how much information he is missing to decode the packet correctly. Despite the potential of the rich feedback and the suitability of some of the modern codes to quantify the amount of information required to decode the packet, the norm in wireless communication is the use of a simple binary ACK/NACK feedback.

The simplest way to use NACK for incremental redundancy is to retransmit only a subset of the symbols $\mathbf{S}_{1,1} = \mathbf{s}$ sent in the original transmission. For example, Xia retransmits s_1, s_2, s_3 only and during the retransmission, Yoshi receives $y_{i,2}$ for $i = 1, 2, 3$, as given by (7.36). Then one option is that Yoshi replaces $y_{i,1}$ from the previously received packet with the respective $y_{i,2}$ for $i = 1, 2, 3$, while he reuses the remaining $y_{i,1}$ for $i > 3$ and attempts to decode the packet again. Another option is that Yoshi attempts partial MRC: he replaces $y_{i,1}$ with $y_{i,1} + y_{i,2}$ for $i = 1, 2, 3$, retains $y_{i,1}$ for $i > 3$ and attempts to decode the packet. The problem with this type of retransmission is that all symbols of $\mathbf{S}_{1,1} = \mathbf{s}$ are statistically identical and there is no special reason to select s_1, s_2, s_3 upon reception of NACK. In other words, the retransmission of s_1, s_2, s_3 is a guess; in a similar way, Xia could have tried a guess by retransmitting, for example, s_2, s_5, s_7, s_9. More generally, Xia could decide to select $l_1 < l$ symbols for retransmission, but Yoshi should either know the positions of those l_1 symbols in advance or Xia should inform him about them in the packet. The latter would represent an additional overhead.

A more general approach would be the one following the very definition of incremental redundancy; Xia applies a FEC code to an input message $\mathbf{d}$ and obtains the codeword $\mathbf{c}$, but does not transmit the whole codeword. Let us assume that the codeword consists of three parts $\mathbf{c} = (\mathbf{c}_1, \mathbf{c}_2, \mathbf{c}_3)$. In the first transmission, Xia modulates $\mathbf{c}_1$ and transmits it. After getting NACK, she modulates $\mathbf{c}_2$ and transmits it, and, after the second NACK, she transmits $\mathbf{c}_3$. A common way for doing this when a linear code is used is based on *puncturing*, by which some of the parity bits generated by the code are omitted in the first transmission.

Another generalization step can be seen in the adaptation of the transmit power. For example, Xia can start optimistically, by transmitting with a lower power and, upon receiving NACK, retransmit with a higher power. This is of interest in applications with a constraint on latency. By starting with a low power and gradually increasing it in the retransmissions, Xia keeps the average power acceptable, while investing a significant power as the deadline is approaching; this is what most of us do with project deadlines as well.

Incremental redundancy plays also a role in AMC. Given a set of transmission modes and their associated "stairs" in Figure 5.8, incremental redundancy can act as a bridge between different stairs. In fact, the use of feedback and retransmissions can be interpreted as a method that helps the transmitter to find out the proper combination of modulation and coding that is suitable for the current SNR of the channel.

7.6 Chapter Summary

This chapter has introduced the concepts of error control coding or FEC. The main underlying idea is to use the channel multiple times and thus define a new channel, call it a super-channel, with larger possible number of inputs and outputs. The coding consists of restricting the subset of inputs that are allowed to be used over the super-channel. Those allowed inputs are called codewords. The fact that not all possible inputs can be used over the super-channel can be interpreted as a source of redundancy. The good coding methods are selecting the inputs of the super-channel in such a way as to enable the receiver to be able to correctly guess, with high probability, the actually transmitted channel input. We have also introduced the relation between the coding and modulation and exemplified the cross-layer design principle through trellis-coded modulation. Finally, we discussed the role of feedback and retransmission as a way to improve the communication reliability.

7.7 Further Reading

Error control coding has been one of the areas of major intellectual and technological advances in communication engineering. There are a number of textbooks dedicated to it, such as Shu and Costello [2004] and Richardson and Urbanke [2008], but also the more general book on digital communications Proakis and Salehi [2008]. Two interesting reads by the original authors who made giant intellectual leaps in coding theory are Ungerboeck [1987] in trellis-coded modulation and, more recently, Arıkan [2009] in polar codes. Regarding feedback and retransmissions, a revisionist view on HARQ with a richer feedback is presented in Trillingsgaard and Popovski [2017].

7.8 Problems and Reflections

1. *Machine learning for the coding problem.* The problem of an exhaustive search for good codes of a given length is quickly becoming infeasible due to the vast space of codebooks that should be searched. Try to devise a procedure for searching for good codebooks of a given rate and given (short) length for a BSC. You can use some of the advanced tools from machine learning and algorithms for searching large data sets.
2. *Multiple error checks.* The common way of carrying block coding is the one in which the sender Xia takes a message of b bits, adds c check bits for error detection and then produces an FEC codeword. After FEC decoding, the receiver Yoshi has only one possibility to check, that is, he will either accept all $b + c$ bits as correct or none. As an intermediate step, Xia can add multiple error checks to the original b bit messages. Analyze the costs and the benefits from introducing multiple error checks and make examples for coding/decoding.

3. *Rich feedback with error checks.* The approach from the previous problem can be changed as follows. Xia sends a message $\mathbf{d}$ that consists of b bits and c check bits. Assume for the moment that Xia does not use any FEC coding. Yoshi receives the $b + c$ bits and runs an error check. If it is correct, he sends an ACK. If it is not correct, Yoshi sends a NACK plus an additional feedback created as follows. Yoshi computes error check bits for the first half of the received packet $\mathbf{d}'$, that is, for the first $\frac{b+c}{2}$ bits, and obtains the checksum $\mathbf{c}_1'$. Then he computes the checksum for the second half of the bits and obtains $\mathbf{c}_2'$. Yoshi sends $\mathbf{c}_1', \mathbf{c}_2'$ as feedback to Xia. On her side, Xia knows $\mathbf{d}$ and she can compute $\mathbf{c}_1, \mathbf{c}_2$ for the correct data as well as the checksum. If the error in the packet to Yoshi were only the first half of the packet, then Xia would find out that $\mathbf{c}_1' \neq \mathbf{c}_1$ and would retransmit only the first half of $\mathbf{d}$.
   (a) Discuss the pros and cons of this scheme.
   (b) Generalize the scheme to more than two checksums sent from Yoshi.
   (c) Generalize the scheme to work with a FEC code and discuss the trade-offs.
4. *Interleaving.* An interleaver is a system element that often appears in coding schemes and its role is to permute (shuffle) the message bits at the transmitter side, while the suitable de-interleaver brings them back to the original order. Investigate and compare the use of interleaver in error control coding in three different contexts:
   (a) In fading channels.
   (b) In turbo codes.
   (c) In bit-interleaved coded modulation (BICM).
5. *Superposition coding and retransmission.* In Chapter 5 we introduced the idea of superposition coding. Assume that Zoya transmits data to Yoshi by splitting her power into multiple superposed packets. Design retransmission schemes that utilize the structure of the transmission. *Hint*: Recall that the decoding and canceling of one of the packets makes it easier to decode another of the superposed packets.

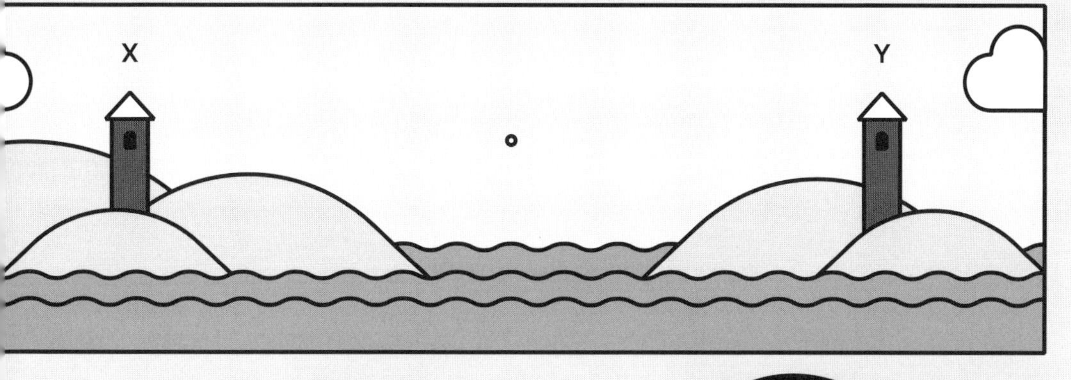

he probability that a pigeon
rops the piece of paper Is 0.1.
Y

Xia cannot use this fact in any way if she sends the pigeon only once.

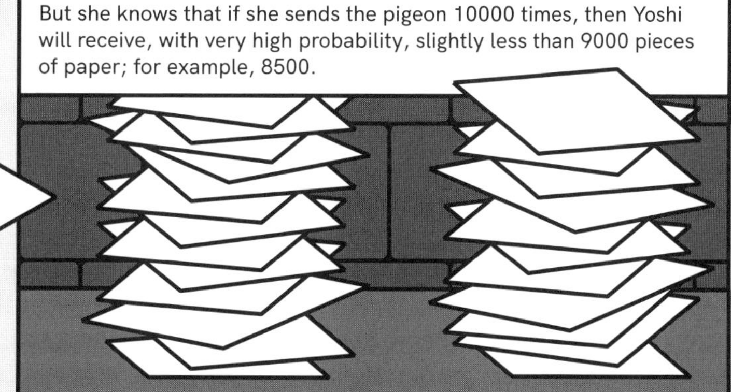

But she knows that if she sends the pigeon 10000 times, then Yoshi will receive, with very high probability, slightly less than 9000 pieces of paper; for example, 8500.

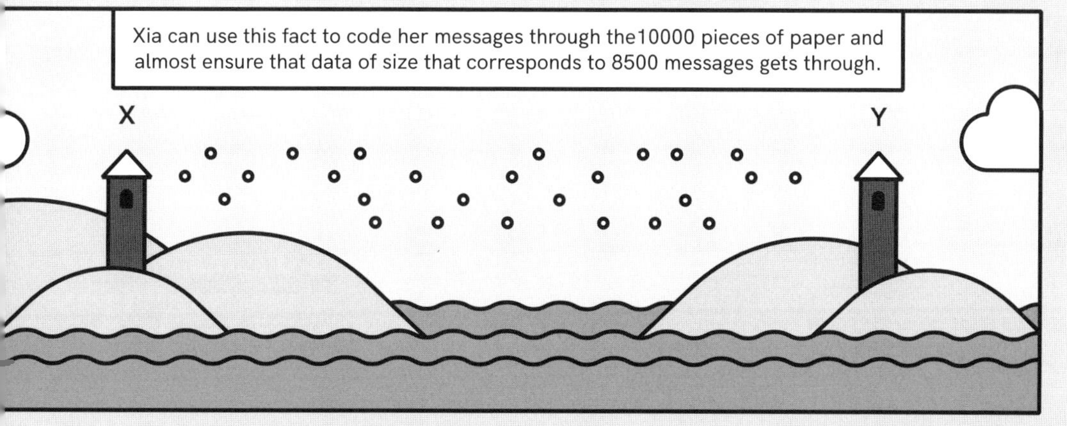

Xia can use this fact to code her messages through the10000 pieces of paper and almost ensure that data of size that corresponds to 8500 messages gets through.
X
Y

Story by Petar Popovski / Art by Peter Gregson

8

Information-Theoretic View on Wireless Channel Capacity

Using various error correction codes, we can decrease the probability of decoding error and thus improve the reliability of reception, at the expense of additional redundancy and a decrease in the transmission rate. However, in order to achieve almost perfectly reliable transmission, do we need to increase the redundancy ad infinitum and decrease the data rate to almost zero? The answer is no. The reason is that, as we create codewords over increasingly more channel uses l, then, referring to Figure 7.4, it is possible to choose M codewords whose shaded areas do not overlap. The important thing is that M can grow exponentially with l as 2^{lC}, with $C > 0$, such that the data rate in terms of bits per channel use becomes $\frac{\log_2 M}{l} = C$. The highest possible value C for a given channel is called the *channel capacity* and it is a quantity that enables to assess the communication capability of a given channel without explicitly constructing a code; which is rather surprising. The notion of channel capacity is often used and misused in assessing system performance. In this chapter we introduce the information-theoretic ideas behind the concept of channel capacity and provide interpretations that facilitate its correct use. The basic information-theoretic notions are first introduced for a binary symmetric channel (BSC), extended to Gaussian, and finally to fading channels.

8.1 It Starts with the Law of Large Numbers

The basic ideas of information theory are rooted in the law of large numbers (LLN), which is also illustrated in the cartoon for this chapter. For the purposes of our discussion, the LLN can be introduced as follows. Assume there is a random bit generator that outputs the binary value 1 with probability p and 0 with probability $(1 - p)$. Let this generator generate l binary values, each of them independent of the others. We say that the generator creates a sequence of l independent identically distributed (i.i.d.) random variables. The LLN tells us what to expect to see when l becomes very large. Specifically, we expect to see around lp values 1 and around $l(1 - p)$ values 0 in the generated sequence and if the generated sequence indeed has those properties, we call it a *typical sequence*. In that sense, if $p = 0.1, l = 10000$, then it is non-typical and highly unlikely to observe a sequence with 9000 values 1, but it is very likely that the number of zeros will be larger than 8500.

However, the reader should not be mislead into thinking that if a sequence is typical, then it is very probable to observe that particular sequence. For example, if $p = 0.1, l = 100$,

Wireless Connectivity: An Intuitive and Fundamental Guide, First Edition. Petar Popovski.

a typical sequence would consist of 10 ones and 90 zeros, but the probability to observe that particular sequence is

$$0.1^{10} \cdot 0.9^{90} \approx 8 \cdot 10^{-15}$$

which is very small. However, what we should expect to see is not a particular sequence, but *any* sequence that is typical and has ≈ 10 ones and ≈ 90 zeros. The probability of observing a particular typical sequence is

$$P_{\mathrm{T}} = p^{lp}(1-p)^{l(1-p)}. \tag{8.1}$$

However, this should be taken with caution and only as approximation, since a typical sequence is likely to have approximately, but not strictly lp ones. With a simple arithmetic manipulation we get:

$$P_{\mathrm{T}} = 2^{\log_2(p^{lp}(1-p)^{l(1-p)})} = 2^{-l(p\log_2\frac{1}{p}+(1-p)\log_2\frac{1}{1-p})} = 2^{-lH(p)} \tag{8.2}$$

where the last equality follows from the definition of the entropy[1]. Since all typical sequences have approximately equal probability, given by (8.2), and since we will almost certainly observe a typical sequence when the number of random trials l is very large, then the number L_{T} of typical sequences must satisfy the basic probability law:

$$L_{\mathrm{T}} \cdot P_{\mathrm{T}} \approx 1 \tag{8.3}$$

from which it follows that

$$L_{\mathrm{T}} \approx 2^{lH(p)}. \tag{8.4}$$

This reveals a very interesting feature of the entropy $H(p)$. On the one hand, the entropy is a function that is calculated based on the *local property* of the i.i.d. random process and this property is the probability to observe 1 in a single generation of a binary value. On the other hand, thanks to the LLN, this local property is capable of describing a *global property* of the sequences generated according to the described i.i.d. random process. This is manifested as a high probability of observing a typical sequence, since the total number of typical sequences is high.

8.2 A Useful Digression into Source Coding

In this section we make a digression in the relationship between entropy and source coding, which appears as a large deviation form the main topic of this book. However, the insights obtained from the meaning of entropy and its connection to the LLN are essential to understand and to correctly use the notion of channel capacity, discussed in the following sections. The notions of entropy, source coding and compression have already been introduced in Chapter 5; here we discuss it from the perspective of the LLN.

---

1 This is, of course, more complicated, and we need to consider a typical sequence to be not only the one that contains exactly lp ones, but also the sequences that contain approximately lp ones. We are not aiming here for a rigorous exposition and the reader is referred to standard textbooks on information theory for a precise definition of typicality.

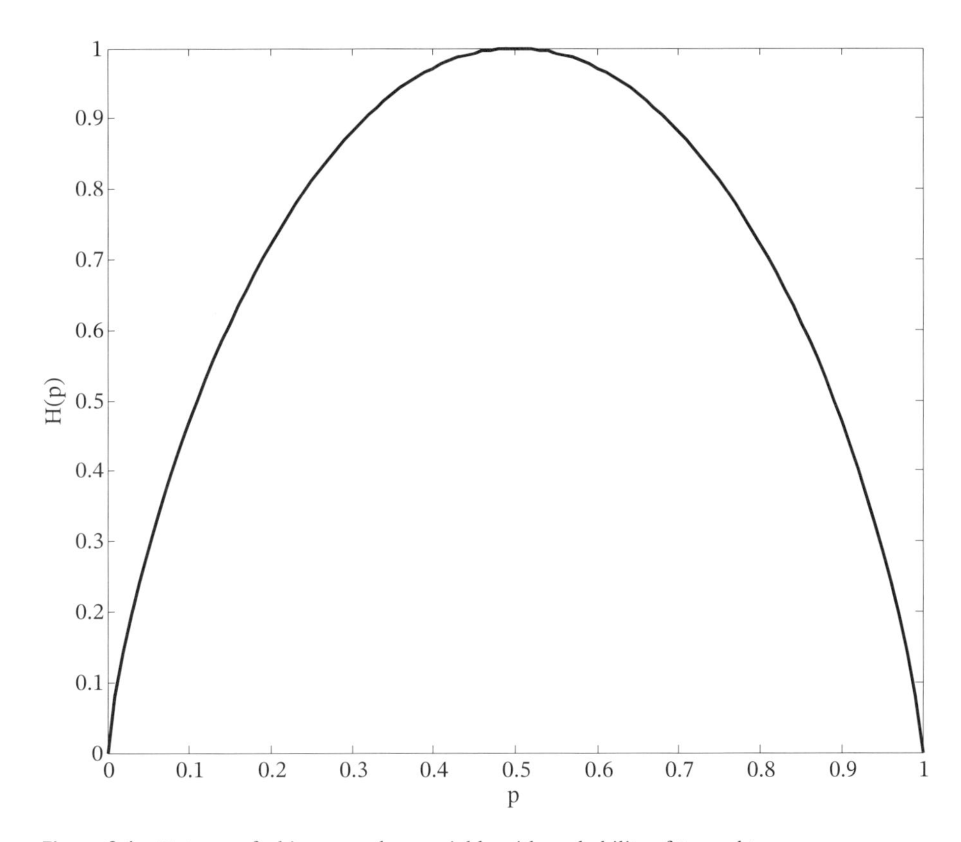

Figure 8.1 Entropy of a binary random variable with probability of 1 equal to p.

In describing the random process we have insisted on the term "binary value" rather than "bit", in order to be consistent with the definition of bit given in the previous chapter and use "bit" to denote the binary values only if $p = 0.5$. Figure 8.1 shows how the entropy function $H(p)$ depends on the probability p, attaining its maximal value 1 when $p = 0.5$. This leads us to conclude that a binary value carries less than one bit of information when $p \neq 0.5$. However, we still need a single binary value to represent it as for the one with $p = 0.5$, so where is the saving of the information resources?

The answer is that we cannot save an information resource locally, for a single binary value, but we can save if we observe long sequences of binary values generated according to the random process (recall the cartoon and the pigeons). We can ask the following question: *How many bits of information are required, on average, to represent the binary sequence produced by a random generator for which the probability of getting 1 is p?* To give a meaningful answer to this, we first need to recognize that a sequence will be typical with a probability that is less than 1:

$$L_T P_T = 1 - P_e \tag{8.5}$$

where L_T is the number of typical sequences. $P_e > 0$ is the probability that the observed sequence is not typical and, according to the LLN, becomes vanishingly small as l goes to infinity. With this, we revise the approximation (8.4) and state that L_T can be at most

$$L_T \leq 2^{l(H(p)+\epsilon)} \tag{8.6}$$

where ϵ is a small positive number, which also goes to 0 as $l \to \infty$. In order to represent a typical sequence, we need

$$D_T = \lceil \log_2 L_T \rceil < \log_2 L_T + 1 = l(H(p) + \epsilon) + 1 \qquad \text{(bits)}. \tag{8.7}$$

On the other hand, any of the non-typical sequences can be represented by simply reproducing its l bits. With these observations, we can encode the output of the random generator into a data packet as follows:

- If a typical sequence occurs, we put a header consisting of a single bit with value 1 at the beginning of the packet to denote that what follows is an D_T bit description of a typical sequence.
- If a non-typical sequence occurs, we put a header with bit value 0, followed by the l bits of the originally generated sequence.

Note that it is assumed that the receiver of that packet knows l and D_T in advance; otherwise information about them should be put as an additional header in the data packet. The average number of bits contained in a data packet will be:

$$\overline{D} = (1 - P_e)(1 + D_T) + P_e(l + 1) = 1 + (1 - P_e)(l(H(p) + \epsilon) + 1) + P_e l. \tag{8.8}$$

What we are interested in is the source encoding rate R, which is the average length of the data bits that represent the random sequence with respect to the total length of the sequence l. Recalling our assumptions that P_e and ϵ become zero as l goes to infinity, we get from (8.8):

$$R = \frac{\overline{D}}{l} \approx H(p). \tag{8.9}$$

In fact, R is always slightly higher than $H(p)$. The conclusion implied by (8.9) is very profound: the local property $H(p)$, which tells us how many bits the random process produces per single sample, which is approximately equal to the data rate that should be used to encode long sequences produced by that random process. Thus, we can only take advantage of the fact $H(p) < 1$ in terms of saved data resources if we create data packets by *compressing* long sequences of the random process that has $H(p)$ in each sample.

It is interesting to see the interpretation of the compressing operation when $p = \frac{1}{2}$ and all l length binary sequences occur with equal probability 2^{-l}. In that case each binary value of the source sequence carries exactly one bit of information $H(0.5) = 1$ such that the source sequence cannot be compressed. However, following the LLN, a typical sequence would have $\approx \frac{l}{2}$ ones and zeros. Hence, for example, the all-zero sequence is not typical, although it occurs with probability 2^{-l}, equal to the probability of any typical sequence.

In order to explain this "paradox", we have to look again at what happens in the asymptotic case $l \to \infty$. Namely, the number of typical sequences will be $\approx \binom{l}{\frac{l}{2}}$ and the probability that a typical sequence will occur is $\approx \binom{l}{\frac{l}{2}} 2^{-l}$, which goes to 1 as l goes to infinity. On the other hand, the probability of observing a sequence that has k ones where k differs

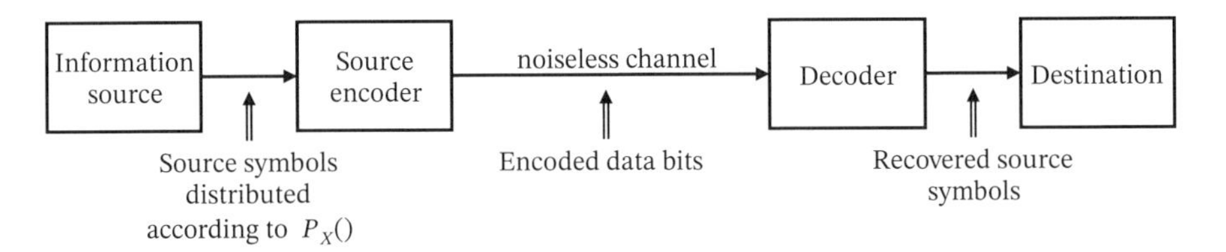

Figure 8.2 Diagram of the source coding problem.

from $\frac{l}{2}$ is $\approx \binom{l}{k} 2^{-l}$ and tends to zero as l grows. Hence, although all sequences are equally probable, most likely a sequence with $\approx \frac{l}{2}$ ones will be observed, but the total number of those sequences tends to 2^l as l goes to infinity, such that l bits are required to "compress" those sequences, which resolves the seeming paradox.

In summary, what we have presented is a solution to the source coding problem, whose block diagram is shown in Figure 8.2. The random generator has the role of information source and generates symbols, not necessarily binary, distributed according to some probability distribution $P_X(\cdot)$. The source encoder maps the source to data packets that are transferred over an error-free channel to a decoder. Through this, we have introduced the concept of *lossless source coding*, in which the decoder is capable to completely recreate the source message. Another option would be to have *lossy source coding*, in which the decoder recreates the source message, but only approximately. In the case of lossy source coding we need to have a certain criterion of how good the approximation recovered by the decoder is. A simple example would be the following: if in total l binary values are produced by the source, then the recovered version at the decoder should have at most F binary values that are different from the original ones. The value F represents the allowed *distortion* in the lossy compression. There are many ways to define distortion, but the objective is always to capture the level of dissimilarity between the recovered and the original source.

8.3 Perfectly Reliable Communication and Channel Capacity

We take the BSC with crossover probability $p > 0$, such that in any packet of finite length, an error occurs with non-zero probability. We use the XOR representation of the binary symmetric channel, introduced in the previous chapter, where the x, y, υ are binary variables that represent the input sent by Xia, the output received by Yoshi and the binary digital noise, respectively:

$$y = x \oplus \upsilon \tag{8.10}$$

such that an error occurs in a channel use for which $\upsilon = 1$. Assuming that the transmitted packet (codeword) consists of l binary values, the l noise instances are binary i.i.d. variables and we can put them in an l-dimensional vector, such that we can write (8.10) in vector form:

$$y^l = x^l \oplus \upsilon^l. \tag{8.11}$$

Following the LLN, we expect that the noise vector υ^l contains approximately lp values 1, and each value 1 represents a transmission error. For simplicity, we assume that lp is an integer[2]. In other words, we observe a *typical* noise sequence that contains $\approx lp$ errors and there are approximately $\binom{l}{lp}$ such typical sequences. Following the discussion in the previous section, a fixed value of x^l is likely to produce one of the $\binom{l}{lp}$ sequences y^l at the output, each corresponding to a particular noise pattern with lp errors.

We refer again to Figure 7.4 and the associated discussion. We can see the input/output relation $x^l \rightarrow y^l$ to represent a single use of a newly constructed super-channel, while at the same time it represents a particular codeword for the original BSC. The fact that the noise sequence is typical implies that the shaded area in y^l in Figure 7.4(c) contains approximately $\binom{l}{lp}$ different outputs y^l. In other words, for a given fixed x^l there are approximately $\binom{l}{lp}$ outputs y^l that can occur with non-zero probability when we take into account the impact from the typical noise sequences.

Using the approximation from the previous section for the number L_{T} of typical sequences, we can say that each transmitted sequence x^l is associated with a shaded area of size $\approx 2^{lH(p)}$ outputs. The key question is: *how many different sequences x^l are we allowed to transmit, such that no two sequences produce shaded regions in y^l that are overlapping?* Since there are 2^l possible sequences y^l, the maximal number of possible codewords that can be sent is:

$$M = \frac{2^l}{2^{lH(p)}} = 2^{l(1-H(p))}. \tag{8.12}$$

In that case, each codeword can carry $\log_2 M$ bits, such that the data rate, expressed in terms of information bits carried per channel use of the BSC is:

$$R = \frac{\log_2 M}{l} = 1 - H(p) \qquad \text{(bit/c.u.)}. \tag{8.13}$$

This is the maximal data rate that one can possibly get over the BSC for which the probability of error will go to zero as l goes to infinity. Note that for finite l the probability of error is not zero, as there is a small probability that codewords x^l will produce an output y^l that is outside the shaded region.

The only remaining question is whether it is possible to select M different codewords x^l such that their shaded areas are not overlapping. This is perhaps the climax of the idea that Shannon put in his landmark work from 1948. He did not provide an explicit way of selecting M codewords; instead, he showed that such a code exists without constructing the code. The existence has been shown by using the *probabilistic method*, whose main idea can be explained as follows. Let there be m different objects of different length. The lengths are unknown, but what we know is that, if we pick an object at random, then the expected length is $\leq \overline{L}$. We can conclude that there must be at least one object that has length $L_i \leq \overline{L}$.

Shannon defined the notion of a *random codebook*. Such a codebook contains M different codewords $\{x^l\}$. In order to create the first codeword, we toss a *fair coin* l times and generate l values 0/1, where 0 or 1 occurs with equal probability. Such obtained sequence of random binary values is the first codeword x_1^l. We repeat the same process for $x_2^l, x_3^l, \ldots x_M^l$, thereby obtaining the complete codebook. Note that, in the process of random codeword generation, it can happen that we generate the same codeword twice; however, the probability for getting such an outcome is very low. What is important to keep in mind is that we are trying

2 For given p, we can always choose a sufficiently large l to make lp arbitrarily close to an integer value.

to characterize the *average codebook*, which is representative of all possible codebooks, and see what the expected behavior of that codebook is. The probability of error associated with this codebook is determined by the level of overlap between the shaded areas of $x_1^l, x_2^l, \ldots x_M^l$. Shannon showed that the probability of error for such a randomly chosen codebook goes to zero as l goes to infinity, provided that the number of allowed codewords is below a certain value related to the channel capacity. It turns out that the way in which we have created the average codebook is equivalent to generating all possible codebooks and selecting randomly one of them. Then, following the probabilistic method, there must be at least one actual codebook for which the probability of error goes to zero as l goes to infinity.

Although the proof of Shannon does not aim to recommend a specific construction, it has inspired the principle "random codes are good". Many actual code constructions, such as turbo codes, follow this principle and create codewords that look random.

We have thus provided non-rigorous arguments to show that no higher rate than (8.13) can be used if we want to ensure that the communication becomes perfectly reliable as the number of channel uses l per codeword goes to infinity. Furthermore, we have also shown that one can achieve the rate R given by (8.13). We therefore say that (8.13) is the *capacity of the BSC*:

$$C = 1 - H(p) \qquad \text{(bit/c.u.)}. \tag{8.14}$$

We emphasize again the meaning of capacity C: it is the maximal rate at which Xia can send information to Yoshi in a way that is perfectly reliable (zero error probability) as the packet size goes to infinity. Therefore, one should be very careful in comparing capacity to throughput, as throughput can also be measured when the probability of packet error is not zero.

8.4 Mutual Information and Its Interpretations

8.4.1 From a Local to a Global Property

It turns out that, similar to the role of entropy in compression, the channel capacity is a local property of the communication channel, as it can be determined through the statistical description of a single channel use. Specifically, it is determined by the conditional probability distribution $P(y|x)$, along with the set of K possible inputs $x \in \{a_1, a_2, \ldots a_K\}$ and the set of L possible outputs $y \in \{b_1, b_2, \ldots b_L\}$. If we fix the input to $x = a_1$, then the output of the channel is a random variable Y with a probability distribution $P(y|a_1)$ and we can calculate the entropy based on that distribution as:

$$H(Y|x = a_1) = \sum_{j=1}^{L} P(b_j|a_1) \log_2 \frac{1}{P(b_j|a_1)}. \tag{8.15}$$

Now, instead of assuming a fixed value of x, let us assume that there is a random variable X that has a certain probability distribution $P(x)$, defined over the set of possible inputs $\{a_1, a_2, \ldots a_K\}$. The entropy $H(X)$ of a random variable X depends on its probability distribution and is a measure of uncertainty that we have about X before making any observation related to X. Using that interpretation, $H(Y|x = a_1)$ is the uncertainty that we have about the channel output that can be observed after we apply $x = a_1$ at the channel input. However, if, in single channel use the value of the input is selected randomly according to $P(x)$, then it makes sense to calculate the average uncertainty that we have for the output given the input value:

$$H(Y|X) = \sum_{i=1}^{K} P(a_i)H(Y|X = a_i). \tag{8.16}$$

This is called *conditional entropy*. Another observation is that, once we fix $P(x)$ and given $P(y|x)$, we obtain the distribution of the output:

$$P(Y = b_j) = \sum_{i=1}^{K} P(a_i)P(Y = b_j|X = a_i) \tag{8.17}$$

which enables us to compute $H(Y)$. Thus, by choosing a particular distribution of the input $P(x)$ we are able to control the distribution of the output. On the other hand, having $P(y)$ we can also compute $P(x|y)$ and thereby the conditional entropy $H(X|Y)$. In the context of a communication channel, the conditional entropy $H(X|Y)$ has a convenient interpretation: it denotes the average remaining uncertainty that the receiver has about X after observing Y.

This finally leads us to the definition of *mutual information*, which is the information that a random variable Y can provide about another random variable X:

$$I(X;Y) = H(X) - H(X|Y). \tag{8.18}$$

For a given channel $P(y|x)$ we can control the value of $I(X;Y)$ through the selection of the probability $P(x)$, as it is used to determine $P(y)$ and thereby $P(x|y)$. If we choose $P(x)$ such as to maximize the mutual information, then $I(X;Y)$ is equal to the *capacity of the communication channel.*

Similar to the problem of compression, a property that is local and pertaining to a single channel use determines the capability of the channel to carry information over many channel uses. However, the connection to the "global property" of the codewords is more intricate compared to the compression, since now $P(y|x)$ is given and we use another distribution $P(x)$ to control the mutual information. For a BSC it can be shown that $I(X;Y)$ is maximized when $P(x) = 1/2$. Recall that a capacity achieving codeword, described in the previous section, consists of l input symbols $x_1, x_2, x_3, \ldots x_l$. The components of this codeword have been selected by tossing a fair coin. However, after the codebook is generated in a randomized way, the codeword that represents certain message is deterministic and the only uncertainty (randomness) that remains is about which message will be selected to be sent by the sender Xia. Let us observe the following process: Xia selects randomly one of the M messages, finds the codeword that corresponds to that message, and starts to send the values $x_1, \ldots x_l$ one by one to Yoshi. Let us assume that Zoya is an observer that does not know the codebook used by Xia and Yoshi; then to her the outputs $x_1, \ldots x_l$ look as if they are generated by the toss of a fair coin, one by one. Let us assume that Zoya is an observer that does not know the codebook used by Xia and Yoshi; then to her the outputs $x_1, \ldots x_l$ look as if they are generated by the toss of a fair coin, one by one. In other words, the observer sees a typical sequence generated by fair coin tossing. Therefore, following the LLN, the total mutual information carried over the l channel uses becomes approximately $lI(X;Y)$, or $I(X;Y)$ per channel use, which makes the required connection. Therefore, following the LLN, the total mutual information carried over the l channel uses becomes approximately $lI(X;Y)$, or $I(X;Y)$ per channel use, which makes the required connection.

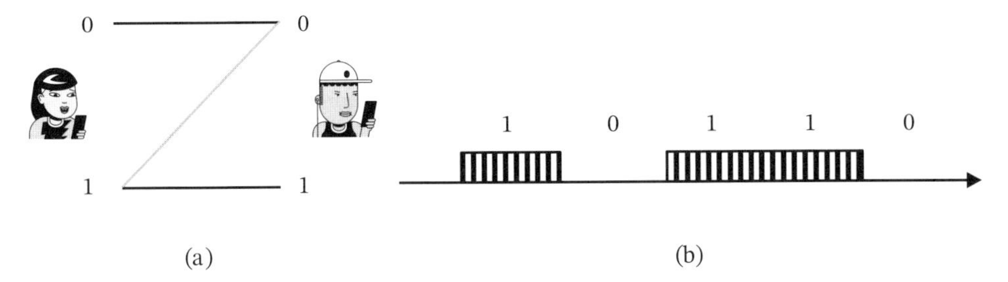

Figure 8.3 Communication through a Z-channel. (a) Z-channel with perfectly reliable zero. (b) An example of creating a Z-channel by the presence/absence of packets.

8.4.2 Mutual Information in Some Actual Communication Setups

In creating the random codebook that achieves the capacity of a BSC, we have chosen to toss a *fair* coin. However, this is not always the way in which we should create the random codebooks. To see this, let us consider the following digital communication channel, termed the *Z-channel*, and shown in Figure 8.3(a). Xia and Yoshi are connected through a communication system in which Xia is able to send packets to Yoshi. Then a new communication channel, the Z-channel, can be created as follows: by sending a packet of duration T, Xia send 1, while by staying silent for time T, she sends zero. This is depicted in Figure 8.3(b). It can be noticed that the probability of having an error when 1 is sent is higher compared to the case when 0 is sent, because it is very difficult for the noise to create something that will look like a packet from the original system. Taking this observation to the extreme, we arrive at the Z-channel model depicted in Figure 8.3(a): no error occurs when 0 is sent and the probability of error when 1 is sent is p. Clearly, it is desirable to use 0 more often, as it is more reliable than 1, but 1 must be used if we want to send any information. Yet, it is not immediately clear what is the exact ratio between the frequency of using 0 and 1.

In order to maximize the mutual information, the components of the codeword should simulate the probability distribution $P(x)$ that is matched to the channel in question. it happens that this probability is $\frac{1}{2}$ when the channel is BSC. The distribution that achieves the capacity of a Z-channel is determined by

$$P(X = 0) = 1 - \frac{1}{(1-p)(1 + 2^{\frac{H(p)}{1-p}})} \tag{8.19}$$

and therefore the capacity achieving strategy should use codewords that simulate this probability distribution, in which $P(X = 0) > P(X = 1)$, rather than fair coin tossing.

There are other communication setups for which mutual information is appealing for intuitive interpretations. Let Basil broadcast information to Zoya and Walt, as illustrated in Figure 8.4(a). Let us assume that Zoya and Walt are connected by a very high-speed link, for example a wire, through which they can cooperate towards receiving successfully the signal from Basil. In this setting, the communication channel is specified by the probability distribution $P(z, w|b)$ where b is the symbol (channel input) sent by Basil, while z/w is the output received by Zoya/Walt, respectively. Let B stand for the random variable that denotes the input symbol sent by Basil, and Z/W is the output random variable observed by Zoya/Walt. Since Zoya and Walt can perfectly cooperate, Zoya is able to provide the value

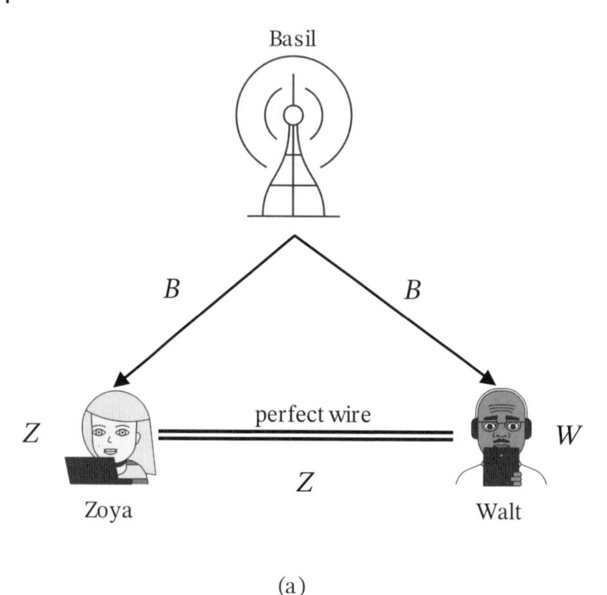

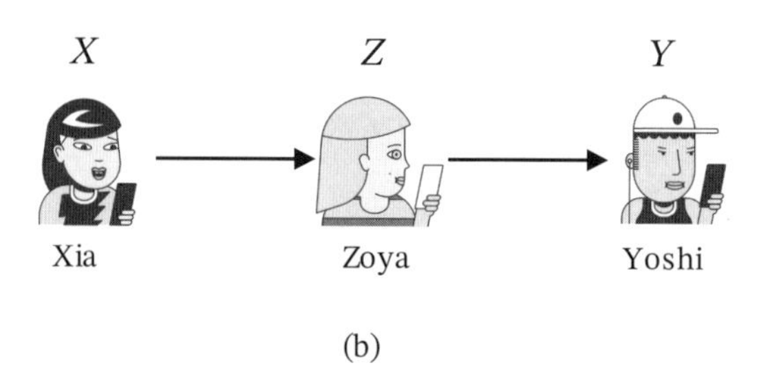

Figure 8.4 Two wireless communication setups for interpretation of mutual information. (a) Broadcast from Basil to Zoya and Walt. Zoya uses a perfect wire to send her observation Z to Walt. (b) Multi-hop communication between Xia and Yoshi through Zoya.

of Z to Walt. This enables Walt to observe both random variables Z and W, such that the mutual information transferred from Basil to Walt is written $I(B; ZW)$. One can think of Z and W as being single, composite random variables. The following holds:

$$I(B; ZW) = I(B; W) + I(B; Z|W) \tag{8.20}$$

where $I(B; Z|W)$ is the mutual information between Basil and Zoya, assuming that they both know what Walt has received. The interpretation is that the total information is what Basil can send to Walt plus what Zoya can see about Basil in addition to what Walt has seen.

Another important intuitive interpretation is associated with the *data processing inequality*, which can be directly related to layering and creation of communication channels, discussed in the previous chapters. To put forward the main idea, consider the multi-hop setup in Figure 8.4(b), Xia is able to communicate with Yoshi only through Zoya as an intermediate node. In other words, think of a multi-hop communication in which Xia provides a message to Zoya and Zoya provides the message to Yoshi. The data processing inequality states that Yoshi cannot learn more about Xia's message than what Zoya can learn about

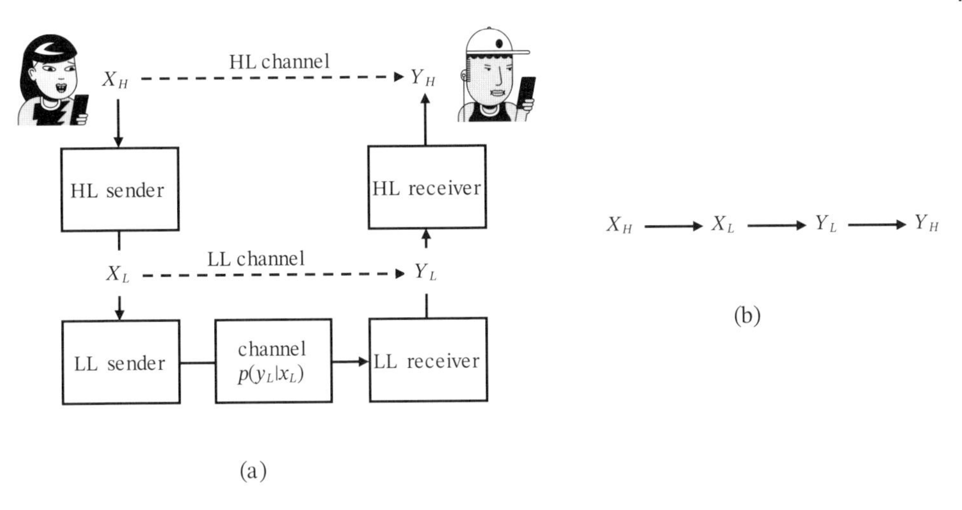

Figure 8.5 Illustration of data processing inequality for a layered communication system between Xia and Yoshi. (a) Two-layered protocol with a high layer (HL) and a low layer (LL). (b) Markov chain that represents the inputs/outputs of the layered channels.

Xia's message. This is because everything that Yoshi can know about Xia's message has to come through Zoya. To state this more formally, use X as the random variable that denotes the input symbol sent by Xia, Z as the output symbol in the channel Xia–Zoya and simultaneously an input symbol for the channel Zoya–Yoshi and, finally, Y is the random variable of the output symbol observed by Yoshi. Then the mutual information satisfies:

$$I(X;Y) \leq I(X;Z) \tag{8.21}$$

and similarly $I(X;Y) \leq I(Z;Y)$. The mathematical reason for this is that the random variables X and Y are *conditionally independent* given Z, which is expressed as

$$P(Y = y|Z = z, X = x) = P(Y = y|Z = z) \tag{8.22}$$

and X–Z–Y form a Markov chain.

Figure 8.5 can be used to establish the relationship between the data processing inequality and layered communication channels. Figure 8.5(a) shows a two-layer communication system between Xia and Yoshi with a higher layer (HL) and a lower layer (LL). The HL channel X_H–Y_H is created by using the LL channel X_L–Y_L. For example, X_L–Y_L can be an analog channel (we will discuss mutual information for analog channels in the next section), while X_H–Y_H is a digital channel. Here the data processing inequality can be expressed as:

$$I(X_H;Y_H) \leq I(X_L;Y_L) \tag{8.23}$$

that is, the information transferred at a higher layer between Xia and Yoshi cannot be higher than the information transferred throughout the lower layer. This is because the HL sender has no other way of reaching the HL receiver except through the LL channel. To extend the analogy with layered protocols, we can say that the ultimate objective of cross-layer optimization is to have equality in (8.23).

8.5 The Gaussian Channel and the Popular Capacity Formula

8.5.1 The Concept of Entropy in Analog Channels

The definition of entropy for discrete random variables, associated with digital communication channels, is not directly applicable when one or more of the involved random variables are continuous. For example, the input variable X that has a power constraint, as in the case of a Gaussian channel, can take any real (or complex) value, as long as the power constraint is satisfied. A continuous variable has a probability density function $f(x)$ and, since the number of states in which X can be is uncountable, the entropy is infinity and thereby not directly applicable to assess the uncertainty in X in terms of number of bits.

Yet, the mutual information for a channel with an analog input X and analog output Y can be meaningfully measured in terms of number of bits. Let us assume that X should have an average power 1, let Z be a random noise, uniformly distributed in $[-0.5, 0.5]$, and let the output be $Y = X + Z$. For a fixed value of $X = x_0$, the received signal is in the noise cloud of diameter 1, centered in x_0. Since we are able to control the probability density of X, as long as the power constraint is satisfied, we choose the following: with probability 0.5, X takes values uniformly in the interval $[0.6, 1.35]$ and with probability 0.5 uniformly in the interval $[-1.35, -0.6]$ (the reader can check that the average power of X is 1). It can be observed that X has a continuous probability density, but from the received signal Y, one can always conclude with certainty whether the sent value of X was positive or negative. Thus, the mutual information $I(X;Y)$ is at least equal to a single bit. On the other hand, we can intuitively see that there must be an upper bound on $I(X;Y)$, since the noise clouds of different values of X overlap.

The important point of the previous example is that the mutual information $I(X;Y)$ can still be meaningfully measured in bits, although X and Y are continuous. In order to quantitatively arrive at the same conclusion, we define differential entropy of a continuous random variable X with probability density function $f(x)$ as:

$$h(X) = -\int f(x)\log_2 f(x)\mathrm{d}x. \tag{8.24}$$

Analogous to ordinary entropy, we can define a conditional differential entropy $h(Y|X)$ that can be calculated based on the conditional probability density $f(y|x)$ and $f(x)$. And, from $f(y|x)$ and $f(x)$ one can determine $f(y), f(x|y)$ and therefore the differential entropy $h(X|Y)$, which can be used to calculate the mutual information:

$$I(X;Y) = h(X) - h(X|Y). \tag{8.25}$$

The differential entropy $h(X)$ can also be interpreted as a measure of uncertainty, but this interpretation should be used with caution, as $h(X)$ can be negative and we cannot directly say that it represents the information contained in a continuous random variable. Geometrically, $2^{lh(X)}$ is approximately the volume of the l-dimensional space that contains most of the probability of the random variable X. Hence, negative $h(X)$ reflects a lower region of spread of the probability mass of X. Another property that cannot be directly transferred is the entropy maximization. Namely, for a discrete random variable X, the entropy $H(X)$ is always maximized if X is uniformly distributed. However, making a similar statement for a continuous random variable X requires an additional constraint. For example, out of all random variables that have a fixed variance $E[|X|^2] = \sigma^2$, the variable X_G with Gaussian distribution attains the maximal differential entropy, given by:

$$h(X_G) = \frac{1}{2}\log_2 2\pi e\sigma^2. \tag{8.26}$$

Therefore the use of Gaussian distribution for additive noise can be interpreted as the worst assumption that we can make about the noise with a predefined power, as it maximizes the uncertainty in the received signal. On the other hand, if instead of $E[|X|^2]$ we observe all the random variables that have a given mean value $E[X] = \mu$, then the exponential distribution has the highest differential entropy.

8.5.2 The Meaning of "Shannon's Capacity Formula"

In order to find the maximal mutual information that Xia can transmit over a real Gaussian channel, she needs to properly choose the probability distribution of X. Additionally, we assume that the power of X is limited $E[|X|^2] \leq P$ and let us denote the power of the Gaussian noise[3] by $N = \sigma^2$. It can be shown that, if X is chosen to be Gaussian with zero mean and power P, then the receiver Yoshi observes a random variable Y that is also Gaussian with zero mean and power $P + N$. The maximal mutual information for the real Gaussian channel is:

$$C = \frac{1}{2}\log_2\left(1 + \frac{P}{N}\right) = \frac{1}{2}\log_2(1 + \gamma)$$

where $\gamma = \frac{P}{N}$ is the SNR of the channel. If we consider the complex Gaussian channel, we get the usual form used in wireless systems:

$$C = \log_2(1 + \gamma) \quad \text{(bit/c.u.)}. \tag{8.27}$$

We recall from our discussion on QPSK/BPSK modulation that a complex symbol can be seen to be composed of two orthogonal real symbols, such that a capacity of a complex Gaussian channel is double the capacity of the corresponding real channel. Therefore, without danger of causing confusion, we can continue the discussion by using the real channel, but if necessary apply the result straightforwardly to the complex channel.

We note that (8.27) is the celebrated analytical expression for capacity of a Gaussian channel, perhaps the most widely used and misused formula in wireless communication engineering. To be precise, (8.27) only refers to the number of bits that can be transmitted in a single channel use, while the full expression should also include the used frequency bandwidth and be measured in bits per second. We will make that connection in the next chapter.

Similar to the discussion on the BSC, we need to provide a way of achieving the capacity by creating codewords over multiple channel uses. This is achieved again by using random codebook generation, now with a Gaussian distribution with zero mean and variance P, denoted by $\mathcal{N}(0, P)$. Let the length of the codeword be fixed to l channel uses. To create the first codeword, Xia takes l samples of $\mathcal{N}(0, P)$ and obtains x_1^l. She repeat the process to obtain the codewords $x_2^l, x_3^l, \ldots x_M^l$. In contrast to the BSC channel, here we first need to take

3 We have previously used P_N to denote the noise power, but here we use N instead to reflect the common notation used in Shannon's formula.

care that the constraint for average power per codeword is not violated. If l is sufficiently large, then we can use LLN to guarantee that the average power of each codeword will not exceed P with very high probability. In order to determine the maximal number M of codewords that Xia should select while not causing error at the decoder, we need to look at the l-dimensional volumes of the relevant signals. Recalling the interpretation of the differential entropy, we can say that the l-dimensional volume likely occupied by the noise n^l given the transmitted codeword x_i^l, is approximately

$$V_n \approx 2^{lh(n)} = (2\pi el\sigma^2)^{\frac{l}{2}} = (2\pi elN)^{\frac{l}{2}}. \tag{8.28}$$

Since each component x_i^l of each codeword is Gaussian, then the distribution of each received symbol y is Gaussian with zero mean and variance $P + N$. Therefore, the size of the l-dimensional volume that is likely occupied by the received signal y^l is

$$V_y \approx 2^{lh(y)} = (2\pi el(P+N))^{\frac{l}{2}}. \tag{8.29}$$

The maximal number of noise volumes that can be packed within the total volume of the received signal, without intersecting with each other, is

$$M = \frac{V_y}{V_n} \tag{8.30}$$

which corresponds to the data rate of:

$$\frac{\log_2 M}{l} = \frac{1}{2}\log_2\left(1 + \frac{P}{N}\right) \quad \text{(bits/c.u.)} \tag{8.31}$$

and is thus equal to the capacity of a Gaussian channel. At this stage, the discussion that we had in relation to the capacity of a BSC carries to the Gaussian channel. For example, as l increases to infinity, the choice of codewords with random Gaussian samples will, with probability tending to 1, produce a codebook that has a zero error probability. Furthermore, the local property, which is the mutual information over a single channel use, can be only operationalized by long codewords that spread over many channel uses.

However, why did we say above that the capacity expression (8.27) is often used and misused? On the one hand, it is a very handy formula for indicating the data performance based on a purely physical parameter, such as the SNR, without having to assume anything specific about the coding/modulation method that is used. However, the conditions under which this expression is derived are often neglected. Let us look again at the definition of a Gaussian channel:

$$y = hx + n \tag{8.32}$$

with SNR given by $\gamma = \frac{|h|^2 P}{N}$. Then to achieve the capacity, we need to send codewords over many, practically infinite, channel uses. Let us consider a transmission period of $l = 10\,000$, which can be considered to be practically infinity. If h is constant over 10 000 channel uses, the receiver Yoshi knows h and the transmitter Xia knows γ, then indeed we can say that, with probability almost 1, Xia can successfully transmit $10\,000 \cdot C(\gamma)$ bits to Yoshi. However, if h changes after every 50 channel uses, then there will be 200 different values of SNR $\gamma_1, \gamma_2, \dots \gamma_{200}$; then we cannot use exactly the same strategy as when h is constant for 10 000 channel uses, and claim that the number of bits sent successfully over the transmission period is $50\sum_{i=1}^{200} C(\gamma_i)$. The problem is that having only 50 channel uses is not sufficient to put the LLN to work. Another practical issue is that providing γ to Xia and h to Yoshi incurs overhead costs in terms of resources, which are not visible in the simple communication model that we have described. If h is constant for $l = 10\ 000$ channel uses, then the overhead is billed only once, while if h changes every 50 channel uses, it is billed 200 times.

A final remark is on the use of the Shannon capacity formula with practical constellations, such as 16-QAM. Recall our discussion in Chapter 6 where it was explained that using a specific constellation for each transmission symbol is only one possible, and sub-optimal, way of meeting the power constraint of the Gaussian channel. Strictly speaking, in order to achieve the capacity $C(\gamma)$, for each value of γ we need to have a different codebook that is designed for that γ. This is the ultimate adaptive modulation and coding (AMC), for which we need to have an uncountably infinite set of codebooks corresponding to the different values of γ in order to be able to ideally adapt to the channel transmit at the channel capacity. Fixing the constellation and changing only the code rate can only approach the capacity at specific values of SNR, but not all SNR values.

8.5.3 Simultaneous Usage of Multiple Gaussian Channels

Let us assume that Xia has two different and independent Gaussian channels to communicate with Yoshi. This can occur, for example, if there are two radio interfaces that operate at different frequency channels, such that their channel uses occur in parallel. A single use of the two channel is described as:

$$\begin{aligned} y_1 &= h_1 x_1 + n_1 \\ y_2 &= h_2 x_2 + n_2 \end{aligned} \tag{8.33}$$

where the noise has equal power N in each channel. If each channel has an independent power constraint, say $E[|X_i|^2] \leq P_i$ for $i = 1, 2$, then the capacity of these parallel Gaussian channels is easily found to be

$$C\left(\frac{|h_1|^2 P_1}{N}\right) + C\left(\frac{|h_2|^2 P_2}{N}\right).$$

However, it is also reasonable to assume that Xia has a limit of the total power applied in both channels:

$$E[|X_1|^2] + E[|X_2|^2] \leq P. \tag{8.34}$$

This changes the problem, as Xia now has an additional degree of freedom, namely where to invest the power. Intuitively, if the second channel is very weak $|h_2| \approx 0$ and $|h_1| \gg |h_2|$, then Xia should put all the power in the first channel and not use the second one.

In order to proceed with this problem, we will reformulate the channel in a rather standard form. Once the power in a channel is fixed, then the capacity is dependent only on the SNR, but not on the actual power of the noise. We therefore assume that $h_1 = h_2 = 1$, while the noise power for the first and the second channels, N_1 and N_2, are different. Without loss of generality, we assume that the second channel is more noisy $N_2 > N_1$. Let us start by assuming that power of 0 is allocated to each of the channels. Then Xia takes a small quantum of power $\Delta P = \upsilon$ and allocates it to the channel that will lead to the highest increase in rate. In order to find the increase of the rate, we look at the achievable rate for the ith channel when power x is allocated

$$C_i(v) = \log_2\left(1 + \frac{v}{N_i}\right) \tag{8.35}$$

and evaluate its derivative at $v = 0$, which is:

$$\left.\frac{dC_i}{dv}\right|_{v=0} = \frac{\log_2 e}{N_i}. \tag{8.36}$$

Clearly, the first quantum of power should be allocated to the better channel, channel 1, as $\frac{1}{N_1} > \frac{1}{N_2}$. Let us continue to allocate power to to channel 1 and assume that we are at a stage in which the power allocated to the first channel is $v = v_1$, while the second channel has a zero power. How large does v_1 need to be before we start to allocate power to the second channel? We find that the speed of increase of the data rate when the next quantum of power is allocated to channel 1 is:

$$\left.\frac{dC_1}{dv}\right|_{v=v_1} = \frac{\log_2 e}{v_1 + N_1}. \tag{8.37}$$

If v_1 has increased to a level such that $v_1 + N_1 = N_2$, then the speed of rate increase equalizes for both the channels. From that point on, power should be allocated to both channels equally, until all available power P is consumed.

The described process can be visualized as a *water filling*, depicted in Figure 8.6. We can think of the noise in the channel as being represented by a brick, where the height of the brick is proportional to the noise power. Allocating power is analogous to pouring water in a container that contains the two "bricks" with heights N_1 and N_2, respectively. The total available power corresponds to the total available water; it then becomes clear that only if the total volume of the available water is sufficiently high does the water start filling on top of the higher brick, thus determining the final level of the water.

The water filling principle can be extended to M channels, where $M > 2$. It can be shown that it is optimal to allocate the power by putting bricks with heights that are proportional to the noise levels in the channels, $N_1, N_2, \ldots N_M$; the reader can easily construct an example

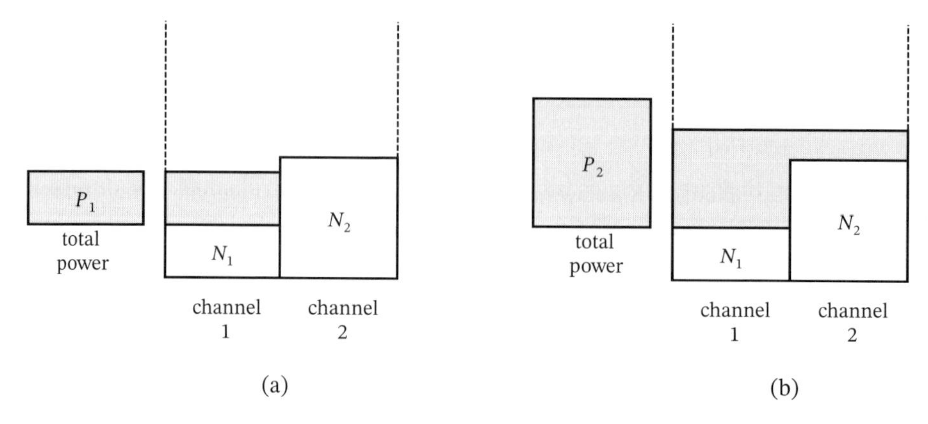

Figure 8.6 Illustration of water filling for two channels, where channel 2 is more noisy than channel 1, for two different values of the total power. (a) The total power P_1 is small and all of it is allocated to channel 1. (b) If the total available power is larger $P_2 > P_1$, then both channels get allocated power.

using Figure 8.6. It should be noted that the final water level does depend on the noise levels and can only be determined if $N_1, N_2, \dots N_M$ are known in advance, before starting to pour the water.

8.6 Capacity of Fading Channels

For a model of a channel with fading, a single channel use is described by:

$$y = hx + n. \tag{8.38}$$

The dynamics of the fading its expressed through the rate of change of the channel coefficient h. For given h, the SNR is equal to $\gamma = \frac{|h|^2 P}{N}$. If h, and therefore the SNR γ, is constant for many channel uses, such that the channel can be considered quasi-static, then we refer to it as a *slow fading channel*. To be more precise, the number of channel uses L for which the channel stays constant should be so large that Xia can send a sufficiently long codeword to Yoshi at a rate equal to the capacity $\log_2(1+\gamma)$ and Yoshi is able to decode it with almost probability 1. We know from the discussion in the previous sections that for this to be completely true the channel cannot change at all and $L \to \infty$. However, in order to get insight in the communication strategies that take place over a fading channel we can think of two "different infinities": the "shorter infinity" L that denotes the period during which the channel is static and the "longer infinity", where we consider a large (infinite) number of periods V in which the total number of channel uses is VL.

If the channel changes faster, then L cannot be assumed to be asymptotically large and we cannot assume that a codeword can be reliably sent at the channel capacity. In that case we are speaking about *fast fading*; in the most extreme case, $L = 1$ and the channel coefficient changes in each channel use. In the text that follows we will discuss the communication strategies and the suitable performance measures for fast and slow fading under different assumptions about knowledge at the transmitter Xia. As a rule, we will assume that the receiver Yoshi knows the channel coefficient h perfectly, unless explicitly stated otherwise. Furthermore, we will ignore the use of resources (channel uses, power, etc.) that are required to obtain channel knowledge such that when it is available, it comes at zero cost.

8.6.1 Channel State Information Available at the Transmitter

In this case the transmitter Xia knows the channel coefficient before the transmission starts. Let us start by assuming slow fading. Since both Xia and Yoshi know h, then during a block of L channel uses, they are in fact communicating over a Gaussian channel. The duration of the block L during which the channel h is constant is referred to as *coherence time*. Let us assume that Xia uses a constant power P for each codeword; then knowing the channel coefficient, automatically determines the SNR to be $\gamma = \frac{|h|^2 P}{N}$. Thus Xia can transmit reliably at a rate $C(\gamma) = \log_2(1+\gamma)$ and use the usual Gaussian codebook that has the appropriate rate. If we observe multiple blocks V, then the average rate achieved over those blocks is:

$$R_P = \frac{1}{V}\sum_{\upsilon=1}^{V} C(\gamma_\upsilon) \tag{8.39}$$

where the index P denotes a constant power P and $\gamma_\upsilon = \frac{|h_\upsilon|^2 P}{N}$ is the SNR in the υth block that has a channel coefficient h_υ.

In a common theoretical setting, the fading coefficient h_v is assumed to be a random value that is independently sampled from a certain probability density function (pdf) at the beginning of each new transmission block. Since the rate is related directly to the SNR, it is more convenient to speak about the pdf of the SNR γ_v instead of the pdf of h_v. For example, let us consider a Rayleigh fading: the coefficient h_v consists of a real and imaginary component, each of them with Gaussian distribution, while the envelope $|h_v|$ has a Rayleigh distribution. However, the squared envelope $|h_v|^2$ has an exponential distribution and therefore the random value of the SNR γ_v has also an exponential distribution. In order to emphasize the dependence of the SNR on the transmitted power P used by Xia, we write the SNR as:

$$\gamma_v = P\frac{|h_v|^2}{N} = Pg_v \tag{8.40}$$

where g_v is the normalized SNR when the transmitter power is $P = 1$ and its pdf is:

$$p(g) = \frac{1}{G}e^{-\frac{g}{G}} \tag{8.41}$$

where G is the mean SNR when the transmitter power is $P = 1$. When power P is applied, the average SNR is PG and the distribution of the SNR γ is given by:

$$p(\gamma) = \frac{1}{PG}e^{-\frac{\gamma}{PG}}. \tag{8.42}$$

If we observe a large number of blocks V, then the average of the sum in (8.39) becomes an integral that takes the expectancy with respect to the pdf of the SNR:

$$R_P = \int_0^\infty C(\gamma)p(\gamma)d\gamma = \int_0^\infty C(Pg)p(g)dg. \tag{8.43}$$

We continue with the slow fading, but now we allow the power to differ from block to block, as long as the *average power constraint* over many blocks V is met. If we denote by P_v the power applied in block v, then

$$\frac{1}{V}\sum_{v=1}^{V} P_v \leq P. \tag{8.44}$$

It is intuitively clear that the power P_v that is applied in a given block should be a function only of the channel quality g_v. In that case two channel instances with identical g_v will use identical transmission rates (at a capacity level) equal to $C(P_v\gamma_v)$.

In order to see how P_v should depend on g_v, let us take a degenerate example of binary fading, in which g_v can have only two fading values, G_1 and G_2, where $G_1 \geq G_2$ and they can occur with probability (not pdf!) $P(G_1)$ and $P(G_2) = 1 - P(G_1)$, respectively. Assume that we observe only two blocks $V = 2$, one of them has G_1, the other has G_2. Let us enforce the average power constraint over the two blocks, which can be written as $P_1 + P_2 \leq 2P$ and is thus equivalent to (8.44). We can represent $G_1 = \frac{1}{N_1}$ and $G_2 = \frac{1}{N_2}$ and we can think of the two channel instances as having different noise level, rather than same noise level and different channel coefficient. Then it seems that we can straightforwardly apply water filling with total power $2P$.

Nevertheless, there is a catch. Recall that water filling was defined for parallel channels and the noise levels of all channels should be known before the power is allocated to any of them. For the case of slow block fading with $V = 2$, this would mean that the transmitter already knows at the start of the first block what will be the channel coefficient of the next block. This is not possible if the channel coefficients are random and independent of each other.

Here the LLN again comes to the rescue. Namely, if we observe a large number of blocks V, then with high probability the total number of channel instances with $g = G_1$ and $g = G_2$ will be $VP(G_1)$ and $VP(G_2)$, respectively. Hence, the LLN allows us to know in advance, almost precisely, how many channel instances with G_1 and G_2 will occur in the next block of size V. This setting becomes similar to the parallel Gaussian channels, as we know the noise levels of all V channels over which power needs to be allocated and we need to allocate a total power of VP to those channels. However, in contrast to the parallel channels, we do not allocate power to all the fading instances in advance, but *we can use the knowledge of how many channel instances will occur in order to determine the water level!* Once the water level is known, we can determine how much power should be allocated when the channel instance is G_1 or G_2. We illustrate this by example in the following subsection.

8.6.2 Example: Water Filling for Binary Fading

Let the channel have binary fading, with two possible values of the normalized SNRs that correspond to two possible states $G_1 = 5$ and $G_2 = 1$, which occur with probability q_1 and q_2, respectively. Once the values of G_1 and G_2 are fixed, the water level in power allocation not only depends on the total available power P, but also on the values of q_1 and $q_2 = 1 - q_1$. Recalling our discussion on water filling for two parallel Gaussian channels, we first need to allocate power only to the first channel. When the power allocated to the first channel is equal to

$$N_2 - N_1 = \frac{1}{G_2} - \frac{1}{G_1} \tag{8.45}$$

then the rate increase that each channel offers with any additional power becomes identical. Let us observe a large number of blocks V such that the total available power for allocation is VP. In the set of V blocks, there are Vq_1 channel instances with normalized SNR of G_1 and Vq_2 blocks with G_2. We start to pour power into the Vq_1 instances of G_1 and assume that the power in each channel is

$$P_1 < \frac{1}{G_2} - \frac{1}{G_1} \tag{8.46}$$

such that the power allocated to the second channel is zero or, in other words, the second channel is not activated. Then it must be valid that the total power allocated to the blocks with G_1 is equal to the total available power VP:

$$Vq_1P_1 = VP \quad \Rightarrow P_1 = \frac{P}{q_1}. \tag{8.47}$$

Using the assumption (8.46), the allocated power to the first channel is $P_1 = \frac{P}{q_1}$ only if the second channel is note activated, and then the following must be satisfied:

$$\frac{P}{q_1} < \frac{1}{G_2} - \frac{1}{G_1}. \tag{8.48}$$

From (8.48) we can conclude that whether the second channel will be activated or not depends on the total power P, and also on the statistics of the fading states, represented by q_1. Hence, if (8.48) is satisfied, then Xia transmits according to the following strategy:

- If the channel state is G_1, then Xia uses power $P_1 = \frac{P}{q_1}$ and transmits at a rate $C(P_1G_1) = \log_2(1 + P_1G_1)$.
- If the channel state is G_2, then Xia stays silent.

If condition (8.48) is not satisfied, then Xia should also allocate power and transmit during the channel states G_2. Following the water filling principle, the power allocated to the state G_1 and G_2 is P_1 and P_2, respectively, given by:

$$P_1 = \frac{1}{G_2} - \frac{1}{G_1} + \upsilon \qquad P_2 = \upsilon \tag{8.49}$$

where υ is determined considering that $P_1(P_2)$ will be allocated to $Vq_1(Vq_2)$ blocks, respectively:

$$\begin{aligned} VP &= Vq_1P_1 + Vq_2P_2 \\ P &= q_1\left(\frac{1}{G_2} - \frac{1}{G_1} + \upsilon\right) + q_2\upsilon \\ \upsilon &= = P - q_1\left(\frac{1}{G_2} - \frac{1}{G_1}\right) \end{aligned} \tag{8.50}$$

and, using $q_2 = 1 - q_1$, the total allocated powers can be found to be

$$P_1 = P + q_2\left(\frac{1}{G_2} - \frac{1}{G_1}\right) \qquad P_2 = P - q_1\left(\frac{1}{G_2} - \frac{1}{G_1}\right).$$

Thus, if (8.48) is not satisfied, the transmission strategy of Xia is to use power P_i and rate $C(P_iG_i)$ when the channel state is G_i, where $i = 1, 2$.

It is interesting to note that, for given average power P and normalized SNRs G_1 and G_2, the activation of the worst channel G_2 depends on the channel statistics, given by q_1. If q_1 is small then the good state appears not so often, such that Xia must also use power in the bad channel state. Another important message is that optimal water filling in time requires not only knowing channel state information (CSI) at the transmitter, but also knowing the channel statistics.

8.6.3 Water Filling for Continuously Distributed Fading

We can follow the intuition gained through the binary fading and formulate the water filling solution for a channel in which the fading follows certain pdf $p(g)$. Since now the channel state g can take any non-negative value, there will be a *threshold value* G_0, such that no transmission is made if $g < G_0$, i.e. those channel states have zero allocated power. Conversely, the larger g is from G_0, the larger the power allocated to that state. Putting this in more precise terms, the power $P(g)$ that is applied when the normalized SNR is equal to g is:

- If $g < G_0$, then $P(g) = 0$ and no transmission occurs.
- If $g \geq G_0$, then transmit with power $P(g) = \frac{1}{G_0} - \frac{1}{g}$ and select the rate equal to the capacity of the equivalent Gaussian channel:

$$C(P(g)) = \log_2\left(1 + \frac{g}{G_0} - \frac{g}{g}\right) = \log_2\left(\frac{g}{G_0}\right). \tag{8.51}$$

The value of the threshold is determined based on the average power constraint P and the statistics of the channel, given by $p(g)$:

$$\int_{G_0}^{\infty} \left(\frac{1}{G_0} - \frac{1}{g} \right) p(g)dg = P. \tag{8.52}$$

Figure 8.7(a) illustrates the water filling in time. It should be noted that the slow fading assumption allows us to treat the number of channel used in a block as sufficiently high to support a reliable transmission of a full packet (codeword).

The average rate achieved over many blocks is:

$$R_{\mathrm{WF}} = \int_{G_0}^{\infty} \log_2 \left(\frac{g}{G_0} \right) p(g)dg. \tag{8.53}$$

It is interesting to see when water filling leads to a significant advantage in the average rate R_{WF} over the average rate R_P, given by (8.43), achieved when the power in each block is constant and equal to P. It turns out that when the average value $\overline{g}$ of the normalized SNR is high and the high values of g are likely to occur more frequently, then R_{WF} is not significantly higher than R_P. This can be explained by the diminishing return of the log function for high values, such that small adjustments in the overall SNR do not really matter when the SNR is high. On the other hand, when $\overline{g}$ is is low and bad values occur frequently, then the transmitter should save power and use the good channel states, such that the existence of a channel quality threshold, below which no transmission is made, is crucial. In fact, for low $\overline{g}$ it is not that important to follow strictly the water filling principle. The average rate is not very much decreased if constant power is applied when g is above the threshold and 0 when it is not. What is important is to have a threshold below which power is saved.

8.6.4 Fast Fading and Further Remarks on Channel Knowledge

Finally, let us depart from the assumption that the fading is slow and let us assume that the block length L during which the channel is constant is small, going down to changing randomly even at each channel use. Nevertheless, we retain the assumption that both Xia and Yoshi know the channel coefficient h for each channel use. Under this assumption, Xia can use exactly the same water filling strategies and codebook adaptation as in the case of slow fading. The difference is that now a full packet (codeword) cannot be sent during the L channel uses.

This is illustrated in Figure 8.7. For Figure 8.7(a) the fading is slow and L is sufficient to send the entire packet. The fast fading case is depicted in Figure 8.7(b) and we focus on the channel instance in which the normalized SNR is g_1. Let us assume that the water filling principle dictates applying power P_1, such that the transmission rate used in blocks with g_1 is $C(P_1 g_1)$ and the total number of bits that can be sent during L channel uses is $LC(P_1 g_1)$ bits, where L is relatively small. On the other hand, the size of the packet that is intended to be sent through the channel instances with normalized SNR of g_1 is $VLC(P_1 g_1)$, where $V \gg 1$ and thus we can assume that the packet is, in fact, a capacity achieving codeword. The first group of $LC(P_1 g_1)$ bits of the packet are sent in the first block that has g_1, the second group of $LC(P_1 g_1)$ bits in the second block that has g_1, etc. Therefore, the packet is fully transmitted after there have been V instances of normalized SNR with value g_1. Since both

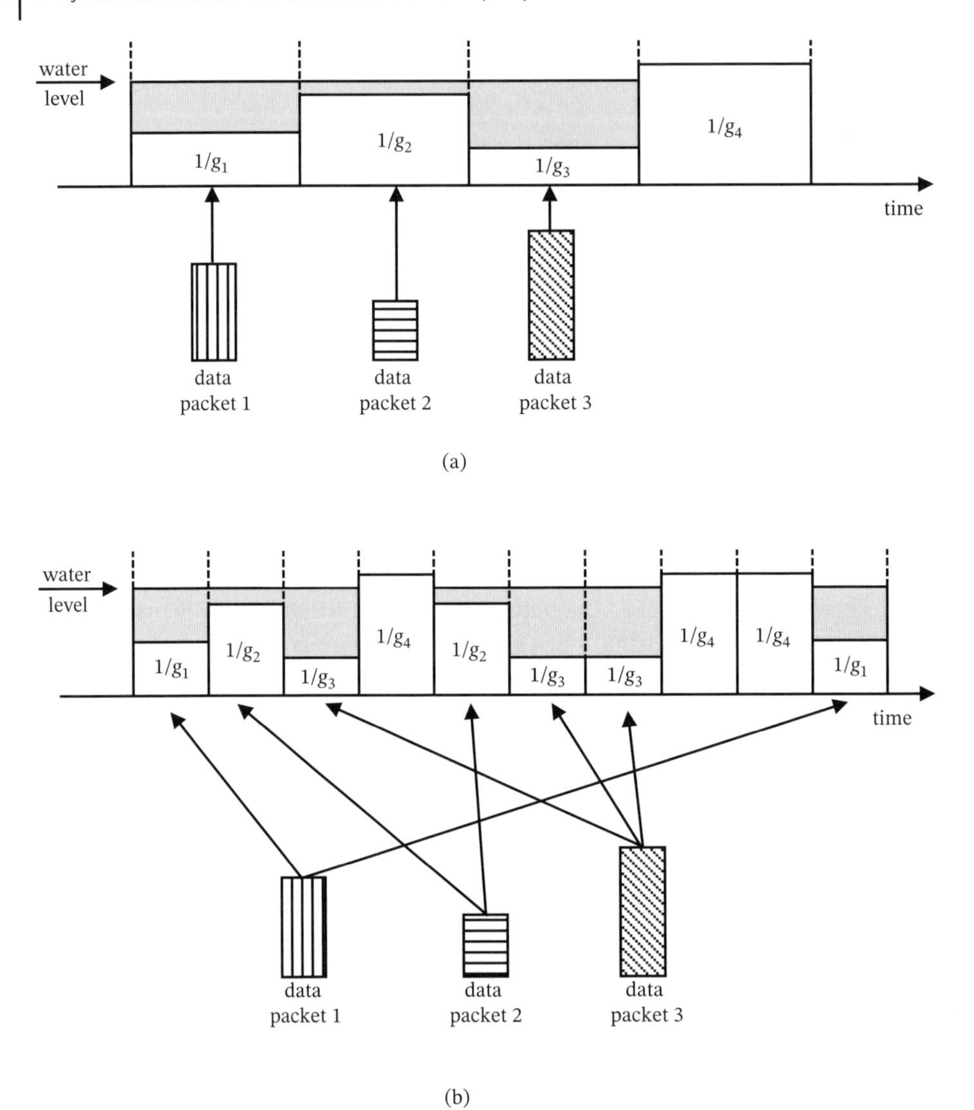

Figure 8.7 Water filling in fading channels with known CSI at the transmitter. In each block, g_i is the normalized SNR and $1/g_i$ is a measure of how noisy the channel is in that block. (a) Slow fading, where a full packet (codeword) fits in a single block. (b) Fast fading, where only a part of the packet fits within a block, packet transmission is continued in the next block that has the same normalized SNR g_i.

Xia and Yoshi are assumed to track perfectly the CSI, then both of them perfectly know which part of which packet is sent during a given fading instance.

A parameter that can be of interest in practice is the average packet delay, expressed as the average time it takes to send the full packet from Xia to Yoshi. For illustration, let us fix L to a small value and denote the coherence time of the fast fading channel with L, while the one of the slow fading channel with VL, where $V \gg 1$. Let the packets be created such that they require VL channel uses. Then the delay incurred by the slow fading is lower compared to the fast fading, since in the slow fading case the V required transmission blocks are adjacent, while in the case of fast fading they are separated in time.

From the previous description it follows that, if CSI is assumed to be a common knowledge for both Xia and Yoshi, it can act as a scheduling protocol or as a pointer that describes what is currently being transmitted. However, this should be regarded cautiously, as we have ignored the protocol resources that are required to ensure that both sides have the correct value of CSI. The protocol signaling ensures that both Xia and Yoshi know the same CSI that is used to schedule the data packet that Xia sends. This signaling must be taken into account when the coherence time L is small. For example, let us assume that Yoshi perfectly estimates the channel coefficient h and he needs to send F bits to inform Xia about the value of h. Then, if fading is too fast and the typical values of $LC(\gamma)$ become comparable to F, we cannot ignore the cost of providing CSI to the transmitter. On the other hand, if $LC(\gamma)$ is almost always significantly larger than F, then it is justifiable to ignore the cost of F in the system.

It is legitimate to ask how to send reliably a few F bits from Yoshi to Xia, which is a usual assumption that is adopted for transmission of the bits that describe the CSI. That is, if we knew how to send those F bits reliably, why are we then not using the same method to fit short packet in the coherence time when the fading is fast? This seeming paradox is a result of the underlying assumptions. It can be explained by the fact that the F bits are sent at a very low rate, far below the channel capacity and thus ensuring that the probability of error is almost zero. Recall that also our assumption of transmission with capacity achieving codewords is idealized; in practice there will be always a certain probability of error P_e. We can thus apply a low-rate code for the feedback bits, whose error probability p_f is much lower than P_e and we can conveniently assume that the F bits are transmitted free of errors. More detailed analysis of these issues is outside the scope of the book.

8.6.5 Capacity When the Transmitter Does Not Know the Channel

We start again with slow fading, where the length L of the coherence time is sufficiently long in a sense that if Xia knew the CSI h, she could have sent reliably a codeword at the channel capacity. The problem is that now Xia does not know it and cannot apply the proper codebook. One can still ask the question: what is the maximal rate at which Xia can send her data, using one codeword per coherence time, although she does not know the CSI?

8.6.5.1 Channel with Binary Inputs and Binary Fading

Let us again consider binary fading, but this time with a channel with binary inputs/outputs 0/1 rather than complex ones. Specifically, let the channel have two states, good and bad, respectively. In the good state the channel is a (BSC) with error probability p. In the bad state, the channel output is always 0, regardless of the channel input. These two states are depicted in Figure 8.8. One can think of this as a channel defined on top of a channel with fading, such that in the good state the receiver can differentiate reception of "something" (1) from "nothing" (0), while in the bad state no signal reaches the receiver and it outputs always "nothing". During each fading block the state of the channel is randomly and independently chosen to be good, with probability p_G, or bad, with probability $1 - p_G$.

Figure 8.8 Representation of the good and the bad states for a binary channel with binary fading.

We first look at the case of slow fading, where the channel state lasts for L channel uses and L is sufficiently long to transmit a whole codeword. For this channel we cannot find a positive data rate that can be supported with probability of error that is arbitrarily close to zero. This is easy to see, as in the bad fading blocks, which occur during a fraction of $(1 - p_G)$ of the time, the channel cannot transfer any information. On the other hand, during the good channel blocks, the codeword can transmit codewords at a data rate

$$R_G = 1 - H(p) \qquad \text{(bits/c.u.)]}. \tag{8.54}$$

Xia can use the strategy where in each block she sends a codeword at the data rate R_G, but she knows in advance that the probability that Yoshi will receive it correctly is p_G, while he will lose it in the remaining blocks. The standard terminology for this is that the *probability of outage* in a block is $(1 - p_G)$. Although the capacity in each block, in a sense defined by Shannon, is zero, it is still possible that Xia achieves a non-zero average throughput. In fact, this setting fits naturally for application of the retransmission mechanisms, discussed at the end of the previous chapter: after each block, Yoshi sends feedback to Xia, stating whether the packet was received correctly (ACK) or not (NACK). The average throughput achieved over many blocks is:

$$\overline{R} = R_G p_G = (1 - H(p))p_G. \tag{8.55}$$

We note here the alternative interpretation of the ACK/NACK feedback, by which it corresponds to provision of the CSI to the transmitter, is only *after* the transmission has been done.

Let us now look at the other extreme, fast fading with block duration $L = 1$, such that in each channel use, the channel is in the good or bad state with probability p_G or $(1 - p_G)$, respectively. We denote by p_{ij} the probability that Yoshi receives $j \in \{0, 1\}$ when Xia sends $i \in \{0, 1\}$ and these probabilities are determined as follows:

$$\begin{aligned} p_{00} &= p_B \cdot 1 + p_G \cdot (1 - p) \\ p_{01} &= p_G \cdot p \\ p_{10} &= p_B \cdot 1 + p_G \cdot p \\ p_{10} &= p_G \cdot p. \end{aligned} \tag{8.56}$$

Let us at first assume that the receiver Yoshi does not know the channel state as well. In that case the channel to both Xia and Yoshi looks like a binary (non-symmetric) channel without fading, specified with the transition probabilities (8.56). In other words, it does not matter for Xia and Yoshi whether error occurs because the channel is bad or due to an error in a good channel state. If we are allowed to use codewords that span over many more blocks rather than a single block (which is a single channel use in this case), then we know the capacity achieving strategy: Xia should use the codewords designed for a binary channel specified by (8.56).

We take a numerical example with $p_G = 0.9$ and $p = 0.01$. We can use (8.56) to calculate the transition probabilities of the new binary channel, obtained when there is fast fading. One can calculate that the capacity of the fast fading channel, when the channel state is unknown both to Xia and Yoshi, is 0.7104 (bits/c.u.)], achieved with a codebook in which the probability of sending 0 is 0.5366. Note that the 0 symbol is more reliable than the 1 symbol in (8.56) and it is expected that the capacity achieving strategy uses 0 more frequently. We can also say that the average throughput achieved over the fast fading channel, over a long period, is also 0.7104 (bits/c.u.).

Let us now look at how things change when Yoshi knows the channel state. In order to see which strategy needs to be used for communication, we represent the system through an equivalent cascade of channels, see Figure 8.9. Since Yoshi knows the channel state, he can differentiate which 0 is coming from a bad channel state and which one due to error in a good channel state, such that the 0 from a bad state can be treated as an erasure and denoted by ϵ. The capacity of this channel is given by (8.55), its value can be calculated to be 0.8273 (bits/c.u), and the communication strategy that achieves it can be described as follows. Xia creates codewords of rate $(1 - H(p))$ that would achieve the capacity of the BSC. It can be checked that the capacity achieving distribution of the BSC outputs is uniform. This corresponds to uniformly distributed outputs that appear as uniformly distributed inputs for the erasure channel and are thus capacity achieving. During l uses of the channel, the erasure channel is able to send 2^{lp_G} codewords reliably. Thus, the codeword size for the BSC is lp_G channel uses and therefore the number of source messages that can be sent through the cascade is $2^{lp_G(1-H(p))}$.

We can conclude that the average throughput achieved when only Yoshi knows the channel state is the same with the average throughput achieved for slow fading and ACK/NACK feedback available to Xia after the transmission. If we assume that in the fast fading case Yoshi also provides perfect feedback after each channel use (which corresponds to a block) on whether the channel was bad or good, then the average throughput that is achieved is again 0.8273 (bits/c.u) and is achieved in a simpler way compared to the double coding strategy used for the cascade in Figure 8.9. Xia takes a codeword with code rate $(1 - H(p))$, sends the first bit and waits for feedback. If the channel is bad, Xia resends the first bit, otherwise she sends the second bit. In this case a fraction $(1 - p_G)$ of the channel uses will be ignored by both Xia and Yoshi and the communication will effectively take place only over the good channel states, leading to long-term average throughput of $Rp_G = 0.8273$ (bits/c.u).

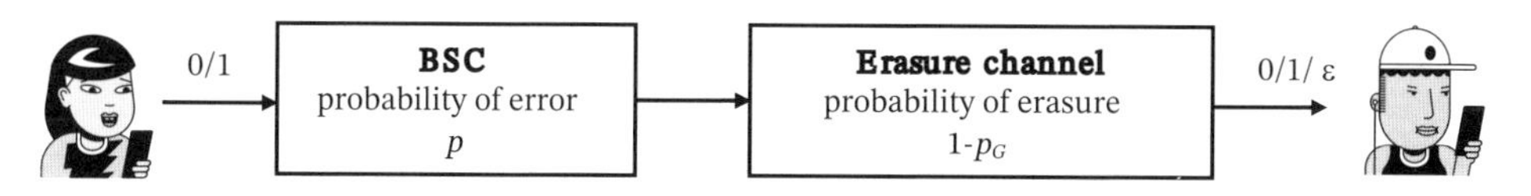

Figure 8.9 Equivalent representation of the channel with two states, when the receiver knows the channel state, with a cascade of a BSC and an erasure channel.

The reason why Xia could achieve a positive rate of reliable transmission in the case of fast fading is that a codeword spans many channel states, averaging out the effect of bad channel states. According to the law of large numbers, the channel visits all the states for a typical number of times during the l channel uses of the codeword. In other words, if the codeword is sufficiently long, then the statistical behavior of the channel during the block (e.g. average in time) represents the "true" statistical behavior of the channel. We refer to this as *ergodic* behavior. In fact, we can achieve something similar in the case of slow fading, provided that we take sufficiently large codewords. Consider again slow fading with $L \gg 1$, but let the codeword length l be such that $\frac{l}{L} \gg 1$, such that the channel behavior is ergodic. This implies that each state is visited for a typical number of times and there are approximately $p_G \frac{l}{L}$ good blocks and $(1 - p_G)\frac{l}{L}$ bad blocks. Xia picks a codeword of length l of rate equal to the capacity of the channel specified by (8.56). This codewords is sent through an interleaver, which randomly permutes the positions of the codeword. Yoshi has a de-interleaver that brings the received symbols in the same order as in the original codeword. Then Yoshi decodes the codeword and the probability of error is vanishingly small.

Figure 8.10 illustrates this process. Note that once the interleaver is fixed, then we can define a new communication channel between Xia and Yoshi, which includes the interleaver and the de-interleaver. A codeword that Yoshi receives through such a channel, after de-interleaving, looks statistically equivalent to the case in which the codeword is sent through a fast fading channel. The role of the interleaver in this setting is to break the correlated channel states within a block of size L and average out the effects of the bad channel

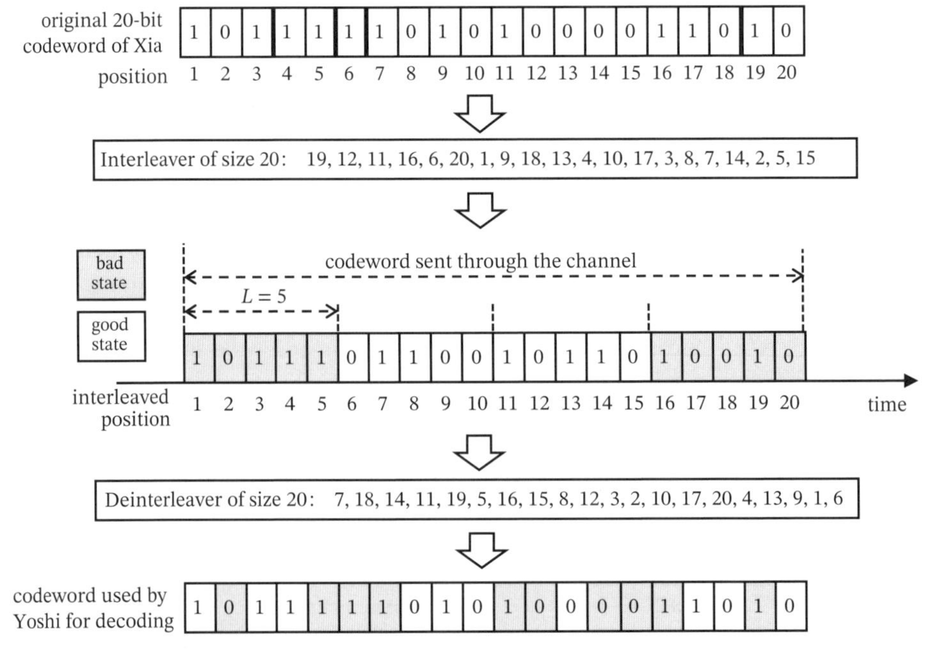

Figure 8.10 Illustration of the equivalent fast fading channel between Xia and Yoshi created by an interleaver. The coherence time is $L = 5$ and the interleaver of size 20 spans four coherence periods. The codeword that Yoshi uses as an input to the decoder looks as if it has passed through a channel with fast fading, such that there is a random channel state change in each channel use.

state. We can thus use exactly the same communication strategy as in the fast fading case, both for the cases when the channel state is known or unknown to Yoshi.

8.6.5.2 Channels with Gaussian Noise and Fading

We can transfer the insights obtained for the simple binary fading channel to channels with continuous input/output, fading and Gaussian noise, given by (8.38).

Let us assume that Yoshi knows the channel state h. In the case of slow fading, h is constant during the whole codeword transmission and unknown to Xia. First of all, since she does not know h, she cannot apply any channel dependent strategy for power allocation across blocks and uses constant power P for each codeword. Xia also knows that, during each block, the channel is Gaussian and the capacity achieving distribution of the input is complex Gaussian, with mean value zero and variance P, denoted by $\mathcal{CN}(0, P)$. However, since Xia does not know h, and thereby the SNR $\gamma = \frac{|h|^2 P}{\sigma^2}$ that is applicable to that block, it follows that she does not know the instantaneous capacity of that block:

$$C(\gamma) = \log_2(1 + \gamma). \tag{8.57}$$

In other words, to Xia the value of γ and therefore the capacity (8.57) is a random variable. Xia can select the random codebook using the capacity achieving distribution $\mathcal{CN}(0, P)$. However, the problem is that Xia must also select the rate R without any knowledge of γ and generate 2^{lR} codewords using the Gaussian distribution. If the channel is bad and γ is low, then the channel does not allow Yoshi to differentiate among 2^{lR} codewords. This leads almost certainly to a decoding error, which in this case is referred to as *outage*, in order to emphasize that the error is caused by a deep fade. On the other hand, if the capacity of the channel is at least $C(\gamma) = R$, then Yoshi is always able to reliably differentiate between the codewords. This is generalization of the concept of outage described in the case of binary fading. The probability that outage occurs is:

$$\Pr(\text{outage for given rate } R) = \Pr(C(\gamma) < R) = \Pr(\gamma < 2^R - 1). \tag{8.58}$$

In the standard example of Rayleigh fading, the distribution of the SNR is given by $p(\gamma) = \frac{1}{PG} \mathrm{e}^{-\frac{\gamma}{PG}}$ and the outage probability is:

$$P_{\text{outage}}(R) = \int_0^{2^R - 1} p(\gamma)\mathrm{d}\gamma = 1 - \mathrm{e}^{-\frac{2^R - 1}{PG}}. \tag{8.59}$$

Reliable communication over this channel can be supported by assuming that Yoshi can feed back ACK/NACK after each codeword. Since we assume that whenever $R < C(\gamma)$ the transmission is perfect, the transmission of ACK/NACK corresponds to 1 bit quantization of the channel state information: ACK= 1 denotes the values $|h| > \frac{\sigma^2}{P}(2^R - 1)$ and NACK= 0 the remaining values.

One may object that the use of Gaussian distributed codebooks is only optimal if Xia knows the instantaneous channel capacity and in order to get a positive rate when that is not satisfied, Xia should perhaps use a different distribution to generate the codewords. Unfortunately, it can be shown that there is no such distribution and the fact that there can be badly faded blocks, keeps the probability of error always positive.

In the case of fast fading, assuming that Yoshi knows the instantaneous channel state h, it is always possible to achieve positive rate when we allow the codeword to spread over many blocks and instances of channel fading. We take the extreme case with $L = 1$, such that h changes independently in each channel use. Xia knows that, regardless of the value of h, the distribution that maximizes the mutual information over the channel is the Gaussian $\mathcal{CN}(0, P)$, and therefore uses it to generate the codewords. However, the codewords should be much longer that the codewords used for a Gaussian channel with constant and known h, since the law of large numbers needs to be put to work both for the fading instances and the noise. This situation is more involved compared to the situation with the binary fading, in which Yoshi knew exactly when erasures were occurring. Here we cannot use the same trick with the erasure channel and outer erasure code, since the channel state is continuous. However, we can put the following interpretation. Since Yoshi knows h, he also knows the channel capacity $C(\gamma)$ in that block and he knows that this block contributes $C(\gamma)$ bits to the decoding process that operates over a large number of blocks. If the codeword is of a length $l \gg 1$, then the fading state with SNR equal to γ will occur approximately $lp(\gamma)\mathrm{d}\gamma$ times[4]. Thus, there is no outage and the capacity of the fast fading channel is:

$$E[\log_2(1+\gamma)] = \int_0^{\infty} p(\gamma)\log_2(1+\gamma)\mathrm{d}\gamma. \tag{8.60}$$

When the fading blocks are longer $L > 1$, we can use the same trick with the interleaver, described for the binary fading channel, and achieve capacity (8.60) as long as the codeword length l satisfies $l \gg L$.

8.6.6 Channel Estimation and Knowledge

Knowing the channel is not a cost-free operation and this fact has to, somehow, be included in the communication model. We first note that for the fading channel model (8.38), Yoshi needs to know the exact, complex value of h in order to be able to apply reception with matched filtering h^*y. On the other hand, in the Gaussian model, the transmitter Xia needs only to know $|h|$ in order to determine γ, but not the phase of the complex value h. The reason is that the capacity achieving codewords are always generated according to the distribution $\mathcal{CN}(0, P)$, which only depends on the power constraint at Xia. The value of γ is necessary for Xia to be able to select a proper number of $M = 2^{lC(\gamma)}$ codewords that constitute the codebook applicable in that block.

A possible protocol for communication between Xia and Yoshi can include three phases:

1. Channel estimation, in which Yoshi estimates h
2. Channel reporting, where Yoshi reports γ to Xia
3. Actual data communication, in which Xia transmits to Yoshi, adapting the rate and the codebook to the channel.

In the first phase Xia transmits to Yoshi a known signal x, called a *pilot signal*, using υ channel uses. In each channel use Yoshi receives $y = hx + n$, but now the unknown parameter is h, not x. For simplicity, assume that the pilot signal is a constant signal of power P, such that $x = \sqrt{P}$. Then based on υ noisy observations, where the ith observation is affected by the ith independent noise value:

$$y_i = h\sqrt{P} + n_i \tag{8.61}$$

4 A mathematically inclined reader would have a problem with this statement, since we work with a continuous distribution and $p(\gamma)\mathrm{d}\gamma \to 0$, but it is sufficient to bring the idea forward.

Based on the υ observations, Yoshi creates an estimate of the channel coefficient, denoted by $\hat{h}$. Abstracting from the concrete estimation procedure, it is intuitively clear that the larger the number υ of noisy observations, the closer $\hat{h}$ gets to the true value h. Note that $\hat{h}$ is a random value, since each y_i is a random value, such that the closeness between $\hat{h}$ and h can be modeled statistically. For example, we can use $\hat{h}_\upsilon$ the estimate obtained after υ noisy observations and write:

$$\hat{h}_\upsilon = h + w_\upsilon \tag{8.62}$$

where w_υ is a random variable that represents the *estimation noise* and models the estimation error. Restating the idea from above, we can say that as υ increases, the randomness of the noise w_υ decreases. For example, if w_υ is modeled as a Gaussian noise, it means that the power (variance) of w_υ decreases as υ increases.

After Yoshi has estimated the channel, he needs to use r channel uses to report the value of γ to Xia. Now γ plays the role of data, which can be encoded, packetized and sent as a codeword. However, we are entering in a loop here, since the channel from Yoshi to Xia can also be represented as a fading channel

$$x = gy + n \tag{8.63}$$

where g is now the channel coefficient applicable when Yoshi sends to Xia, y is the signal transmitted by Yoshi and x is the signal received by Xia. So, now Yoshi needs to send a pilot signal to Xia, such that she can estimate g, then she reports g to Yoshi such that Yoshi can send the value of γ in the most efficient way. But wait; if Xia needs to report g, or in fact $|g|$, then we are back where we started.

To get out of this loop, we can relax the requirements put on Xia about her adapting the codebook to the channel state. Instead of having uncountable infinite set of codebooks, each for different value of γ, Xia can use a discrete set of K different codebooks with rates $R_1 < R_2 < \cdots < R_K$. The rate R_k codebook is not applied only for a specific value of γ, but for a range of values $[\gamma_k, \gamma_{k+1})$. This closely parallels our discussion on adaptive modulation and coding in Chapter 5. This choice does not decrease the estimation burden on Yoshi, since he still needs to learn the value h as precisely as possible, but now Yoshi needs only to report $\log_2 K$ bits to Xia to tell her which codebook to apply, instead of sending her the precise value of γ. The value of $\log_2 K$ is usually small and can be reliably sent over r channel uses at a very low rate $\frac{\log_2 K}{r}$. The value of r can be commonly assumed to be much smaller than the codeword length l (recall that we use capacity achieving codewords of large length).

In the previous discussion it was assumed that, in general, the channel coefficient h from Xia to Yoshi is not equal to the channel coefficient g from Yoshi to Xia. In many practical scenarios the channel has the property of reciprocity and $h = g$; see Section 10.9 for further explanation of this feature.

How large should υ be chosen? Let us assume at first that the channel changes in a way that is sufficiently slow and is constant during the three phases of the protocol listed above. For fixed υ, Yoshi estimates $\hat{h}_\upsilon$ and applies matched filtering, such that the signal that is used for decoding can be represented as:

$$y_\upsilon = \hat{h}_\upsilon^* y = (h^* + w_\upsilon^*)y = (h^* + w_\upsilon^*)(hx + z) = |h|^2 x + n_{\upsilon x} \tag{8.64}$$

where $n_{\upsilon x}$ is an equivalent noise that contains the uncertainty due to Gaussian noise, channel estimation, but it is also dependent on the transmitted signal, which is different from a standard Gaussian channel. Hence, Gaussian codebooks are not necessarily capacity achieving for the equivalent channel, but we do not need that: we only need to ensure that the codebook with a suitable chosen rate R_k will get through reliably, with probability of almost one. We can thus approximate n_{vx} as a Gaussian noise with noise power σ_{vx}^2, assume that the SNR of the equivalent channel is $\gamma_{vx} = \frac{|h|^2 P}{\sigma_{vx}^2}$, and report the kth codebook of rate R_k that needs to be used by Xia. In designing the thresholds for using different modulation levels, one should take into account that n_{vx} is only a Gaussian approximation and be conservative in the rate selection. This means, for example, to report R_{k-1} for some γ that would normally be associated with R_k under perfect Gaussian noise.

The two auxiliary procedures from phase 1 and phase 2 consume in total $v + r$ channel uses, such that the effective data rate (goodput) achieved when Xia transmits at rate R_k is:

$$G_k = \frac{l}{v + r + l} R_k. \tag{8.65}$$

By choosing to use a discrete set of codebooks, we are effectively creating another communication channel on top of the original fading channel; yet another use of the layering principle. From Xia's perspective, the new channel has K different states, where the kth state occurs with probability P_k and corresponds to the applicability of the codebook of rate R_k. The probability P_k can be calculated from the fading probability distribution of the original channel, the channel estimation procedure and the selected SNR thresholds. The average long-term goodput is then:

$$\overline{G} = \frac{l}{v + r + l} \sum_{k=1}^{K} P_k R_k. \tag{8.66}$$

One immediate drawback of creating a new channel by constraining the number of states for the original channel is that there is an upper bound on the achievable rate, represented by R_K, which means that Xia cannot take advantage of very good channel states.

If the channel changes sufficiently slowly, then $l \gg v, r$ and the scheme works almost perfectly. Note that here we implicitly assume that the data amount to be sent is very large, as required by the information-theoretic codebooks that use the LLN; these results are not usable for short data packets. On the other hand, if the channel changes quickly, then the channel estimation made during v uses is invalid, as the channel will likely change by the time the data transmission takes place.

8.7 Chapter Summary

This chapter dealt with the information-theoretic notion of capacity of a communication channel. Information theory is normally associated with a very rigorous mathematical treatment and that is the only way to properly establish all the fundamental claims. Nevertheless, many of the information-theoretic results, most notably the expression for a capacity of a Gaussian channel, are widely used and not always in a correct way. This chapter was an attempt to highlight the conditions under which the information-theoretic conclusions are derived by removing the mathematical rigor and focusing on the ideas and the intuition behind them. In that sense, central concept of mutual information was discussed, with a special emphasis on the local (single channel use) statistical property and the operational interpretation that is global (many channel uses). In the case of fading channels, we used the simple model of a binary fading, which is far from practical but has a pedagogical value. We have touched upon the problem of channel knowledge and its impact on the channel capacity. Wireless information theory is a vast area that has grown significantly in the last two decades and it is hard to capture all important ideas and results; for example, we have not discussed the capacity for the case of non-coherent transmissions or for multi-user channels. Nevertheless, the material presented in this chapter, as well as the previous two chapters, equips the reader with the proper perspective to study wireless information theory for the (many) cases not covered here.

8.8 Further Reading

The concepts of channel capacity for a single link and some treatment of the multi-user channels, also known as a network information theory, is given in Cover and Thomas [2012]. A book dedicated to network information theory is El Gamal and Kim [2011]. A broader discussion on wireless communications with information-theoretic flavor is available in Tse and Viswanath [2005].

8.9 Problems and Reflections

1. *Using all channel outputs versus combining the outputs.* We introduced this problem as Problem 1 in Chapter 6. Redo the problem and find the maximal rate at which Zoya can transmit to Walt for the cases (a) and (b).
2. *Multi-user diversity.* Basil communicates in the downlink with K terminals. Each downlink transmission consists of a large number l of channel uses, such that we can assume that Basil can transmit at a capacity. Let the SNR of the kth terminal be γ_k. For each downlink transmission, Basil wants to find the user $k*$ that has the best SNR γ_{k*} and transmit to this user at a rate $\log_2(1+\gamma_{k*})$.
   Propose a system design based on a TDMA frame that can support this operation. Discuss how it is related to the coherence time of the channel, expressed in terms of the number of channel uses for which the channel is constant. Also, take into account the fact that Basil needs to learn the channels before transmitting.
3. *Multiple access.* The best understood multi-user channel is a multiple access channel. Let Zoya and Yoshi transmit to Basil such that the signal received by Basil is:

   $$b = z + y + n \tag{8.67}$$

   where z is transmitted by Zoya, y by Yoshi and n is the Gaussian noise. When only one user transmits, the other user sets the transmitted symbol to 0 and we get the single-user Gaussian channel. Zoya and Yoshi transmit at rates R_Z and R_Y, while their SNRs to Basil are γ_Z and γ_Y, respectively. Assume that the coherence time of the channel is sufficiently larg, such that, for example, if only Zoya transmits and Yoshi is silent, she can transmit to Basil at the rate equal to a single-link capacity $R_Z = \log_2(1+\gamma_Z)$. In order to characterize the achievable rates in a multiple access channel, we need to characterize R_Z and R_Y simultaneously; see, for example, chapter 6 from Tse and Viswanath [2005] for a detailed explanation. If Zoya and Yoshi transmit to Basil simultaneously, then their transmissions are decoded free of errors (in asymptotic sense) if the rates satisfy the following:

$$\begin{aligned} R_Z &\leq \log_2(1+\gamma_Z) \\ R_Y &\leq \log_2(1+\gamma_Y) \\ R_Y + R_Z &\leq \log_2(1+\gamma_Y+\gamma_Z). \end{aligned} \tag{8.68}$$

In multiuser channels, instead of capacity, we speak about a capacity region, such as the one described using (8.68).

Describe the possible strategies to for communication based on zuccessive interference cancellation (SIC). For this, assume that the transmission of the other user can be treated as an additional Gaussian noise.

4. *Multiple access with changed assumptions.* Here we consider again the multiple access channel from the previous problem and see how the model is affected by changing some of the assumptions.
   (a) The rate region (8.68) is valid if both Zoya and Yoshi know both γ_Z and γ_Y. Discuss how the communication strategies change if this assumption is not valid.
   (b) Assume a setup similar to random access, where the activation of a user is uncertain: for a given transmission, Zoya does not know if Yoshi is active and vice versa. Discuss the communication strategies for this setup.
5. *Outdated channel knowledge and retransmission.* Xia transmits to Yoshi at a fixed rate R determined as follows:

$$R = \log_2(1+\beta). \tag{8.69}$$

Xia does not know the channel before the transmission. The SNR of the channel during the transmission is γ. After the transmission is finished, then

- If $\gamma \geq \beta$, Yoshi receives the transmission correctly and sends ACK.
- If $\gamma < \beta$, Yoshi sends a NACK, along with the value of γ.

Upon receiving NACK, Xia gets outdated CSI, as we assume that at the time of her next (re-)transmission, the channel has changed. On the other hand, by learning γ, Xia learns that the amount of missing information for Yoshi to decode the packet is

$$\log_2(1+\beta) - \log_2(1+\gamma). \tag{8.70}$$

How can this be knowledge used to improve the overall efficiency of retransmission? For inspiration, the reader can look into the concept of BRQ from Trillingsgaard and Popovski [2017]. How can this be generalized to a multiple access scenario, where both Xia and Zoya transmit to Yoshi?

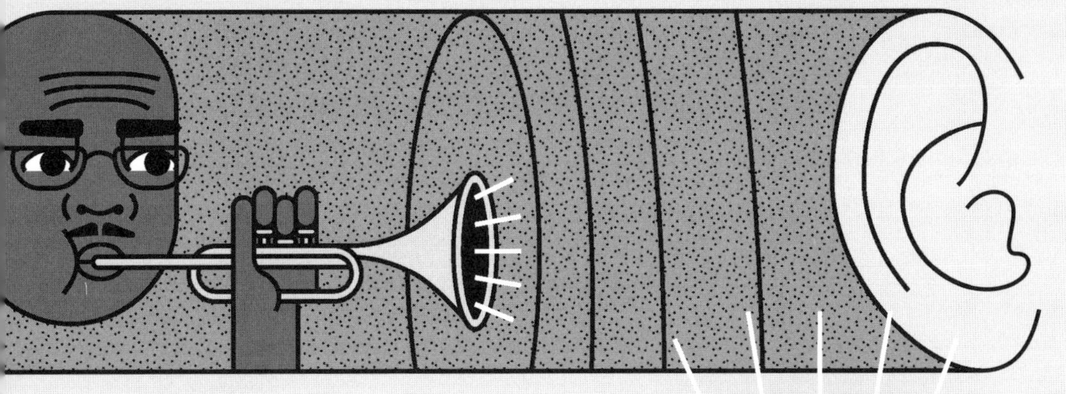

t still plays the trumpet.
hi puts away the trumpet, gets a tuba and plays along with Walt.

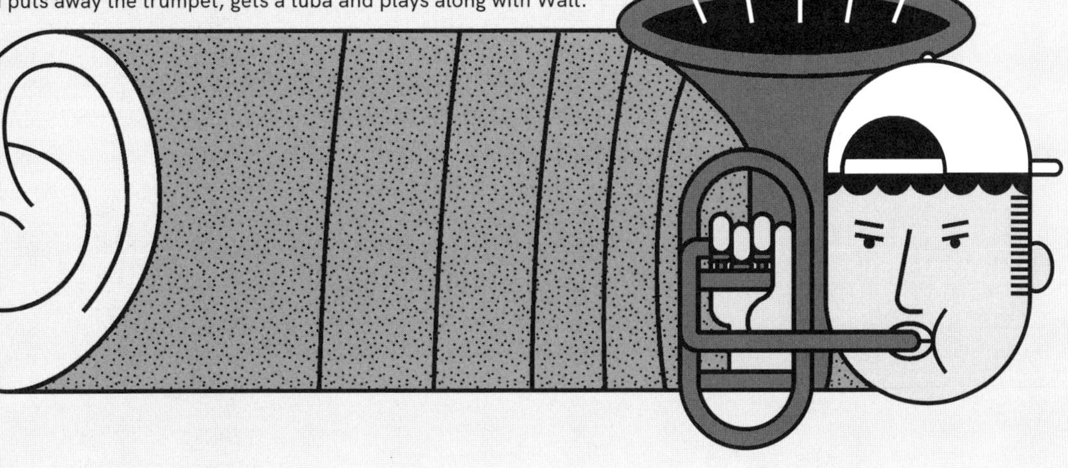

Now Zoya can hear both melodies simultaneously, even when she stays in the middle between Walt and Yoshi. This is because the two melodies have very different frequencies.

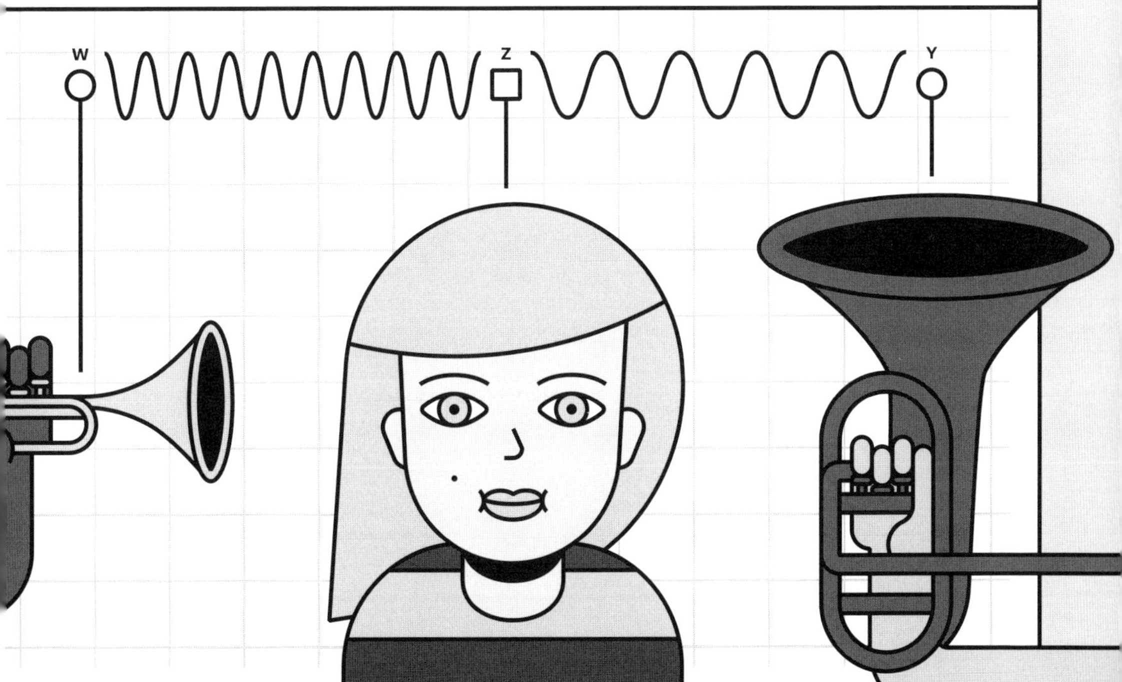

Story by Petar Popovski / Art by Peter Gregson

9

Time and Frequency in Wireless Communications

Medium access control (MAC) protocols use time as an underlying physical variable. The main entity at a MAC layer is a packet, represented by suitably modulated/coded symbols at the physical layer. On the other hand, the fundamental characterization of communication channels is based on the notion of a *channel use* and information theory is concerned with the capacity to transmit information bits per channel use. A channel use should be understood as an abstract transmission opportunity rather than something that occurs at regular time intervals. For example, the pigeon channel, discussed in Chapter 6, has a certain capacity per channel use that is determined independently of the fact that the pigeon flies once per day.

This chapter makes a connection between, on the one side, the channel uses in a communication channel and, on the other side the actual data carrying signals that occur in time. This is instrumental to create the physical basis for the notion of time used in MAC protocols and other protocols running at the higher layers. The physical signals used for communication are *analog signals and waveforms*, which are continuous functions of time. The chapter discusses how a digital communication signal is mapped onto physical signals. In this way, the problem of sending information per an abstract channel use becomes a problem of sending information per time unit. It is thus not immediately clear whether the discrete communication channels that we have described so far are relevant in such a setting. Another important concept introduced in this chapter is the one of a *frequency*, practically central to any communication system. In that sense we will discuss the notions of *signal bandwidth* and *carrier frequency*.

9.1 Reliable Communication Requires Transmission of Discrete Values

We start by establishing the fact that reliable communication is essentially a process of sending discrete rather than continuous values. Indeed, noise is a continuous waveform that, with probability one, has a non-zero value at any time instant. Hence, a transmitted waveform will almost surely be distorted by noise at the receiver. In the best case, the only thing that the receiver can state reliably is that the received waveform belongs to a certain class of waveforms. Here the term "class of waveforms" plays a role that is analogous to the one of the noise circle from Chapter 5: the transmitted waveform corresponds to the

Wireless Connectivity: An Intuitive and Fundamental Guide, First Edition. Petar Popovski.

center of the noise circle, while the noise circle itself corresponds to the set of random waveforms that can be received when the transmitted waveform is polluted by noise. We are thus interested in the largest number of M waveforms that can be reliably distinguished at the receiver. Here "reliably" means that the probability of error is below a predefined value or, ideally, zero. The transmission of one of those waveforms, selected randomly and uniformly, corresponds to sending $\log_2 M$ bits of information. However, the difference with the communication channels discussed before is the fact that now we consider physical signals that consume actual time. Instead of finding how many bits per channel use one can send, the proper question now is: *how many bits per second (bps) can be reliably communicated by using analog signals?*

A way to find the data rate in bits per second, denoted by R, is to first find how many independent channel uses there are per second, denoted by D, calculate the capacity in terms of bits per channel use (bits/c.u.), denoted by C, and then find R through:

$$R = D \cdot C \quad \text{(bps)}. \tag{9.1}$$

We have used D to denote the number of real-numbered channel uses per second. In this chapter we will refer to a channel use also as a *degree of freedom (DoF)*, such that in this case there are D real-numbered DoFs per second. As we are going to use the term DoF quite extensively in this chapter, it deserves a short elaboration.

The term DoF bridges the concepts used for abstract communication channels and the ones used to describe analog waveforms. The bridging role of DoF is related again to the idea of layering, recurring throughout this book. Namely, (9.1) assumes that we have decomposed the problem of communication with analog waveforms into two independent sub-problems:

1. Using continuous waveforms, create a certain number of independent channel uses per unit time. This *discrete* set of channel uses is obtained by a certain sampling of the continuous waveform.
2. Apply communication strategies over the discrete set of channel uses, following the strategies developed for a baseband model of a communication system, treated in the previous chapters.

We note that whenever we use the term DoFs we mean real DoFs. If we observe a single channel use in which a complex symbol can be sent, this can be seen as a complex DoF that consists of two real DoFs. Hence, the real DoF can be treated as an atomic transmission unit for analog signals that can be used to build other types of DoF.

As it is always the case in layering, this separation is only a specific instance in which the communication problem can be solved. Nevertheless, as Shannon already stated in his original paper and later Slepian proved rigorously, this strategy is optimal. The reason is that every continuous signal of interest in communication engineering can be almost perfectly synthesized as follows: take a finite set of D basis waveforms, scale each waveform with a real coefficient and add the scaled waveforms. The resulting waveform is a linear combination of the basis waveforms and its shape can be controlled by choosing the D real coefficients. Assuming that the basis waveforms are predefined and known in advance by the transmitter and the receiver, then the information in the resulting waveform is stored only in the D coefficients that determine the linear combination. This explanation provides an idea for the use of the term DoFs to describe the number of independent discrete inputs that preserve the information about a waveform. Throughout the chapter it will become clear how D is determined and the physical constraints that affect it.

9.2 Communication Through a Waveform: An Example

Let us assume that the TXmodule of Zoya is a black box that can accept a real number as an input and produce as an output a waveform that is scaled with that same real number. Taking the specific example from Figure 9.1, if the input value is 1, the output waveform is shown in Figure 9.1(a), while if the input value is -3, then the waveform in Figure 9.1(a) is multiplied by -3. Let us further assume that once the waveform is produced at the output of the TXmodule, it is transferred to the RXmodule of Yoshi without any distortion, such that in the absence of noise Yoshi receives exactly the same transmitted signal. Furthermore, assume that the RXmodule makes its decoding decision based only on the value that is sampled at the instant where the waveform achieves its largest magnitude. This instant occurs at 0.5 s at the output of the TXmodule. Note that in any physical system there is a delay τ, not depicted in the figure, that equals to the time that is needed for the signal to propagate between Zoya and Yoshi. We need to assume that Yoshi knows this τ, such that he can set the timing for making a decision to $\tau + 0.5$ s. All these assumptions are very idealized and are used to provide a pedagogical example, ignoring the difficulties Yoshi needs to go through in order to learn τ, as well as the fact that the wireless channel always introduces an attenuation and other distortions to the signal sent by Zoya.

If Zoya sends the two pulses to the input of her TXmodule in a way that the pulses are separated at least $T = 1.5$ s in time, then there is no time overlap between the waveforms. The RXmodule will output the observed values, sampled at the time instants with highest amplitudes. Hence, the system TXmodule–wireless channel–RXmodule can be treated as a black box that defines a discrete communication channel: every T s a real value is applied at the input and a noisy value is obtained at the output. Figure 9.1(b) depicts the situation in which the pulses are put $T = 1.5$ s apart and there are four possible inputs $z \in \{-3, -1, 1, 3\}$ that represent the bit pairs 00, 01, 11, 10, respectively. The output signal produced at the sampling or observation instants can be written as:

$$y = z + n \tag{9.2}$$

where n is the noise, not depicted on the figure. Once the waveform channel can be mapped into a discrete communication channel, we can use the communication methods described in the previous chapters. An important point about the example in Figure 9.1(b) is that, when the time separation is $T \geq 1.5$ s, then one can be certain that the channel uses can be independently modulated and what is transmitted/received during one channel use is independent of the other channel uses. Assuming that the capacity per channel use is C (bits/c.u.) and the pulses are separated for $T \geq 1.5$ s, the equation (9.1) can be rewritten as:

$$R = \frac{C}{T} \quad \text{(bps)}. \tag{9.3}$$

The previous equation suggests that lowering T increases the data rate. If we start to lower T below 1.5 s, then the uses of the discrete communication channel start to overlap in time

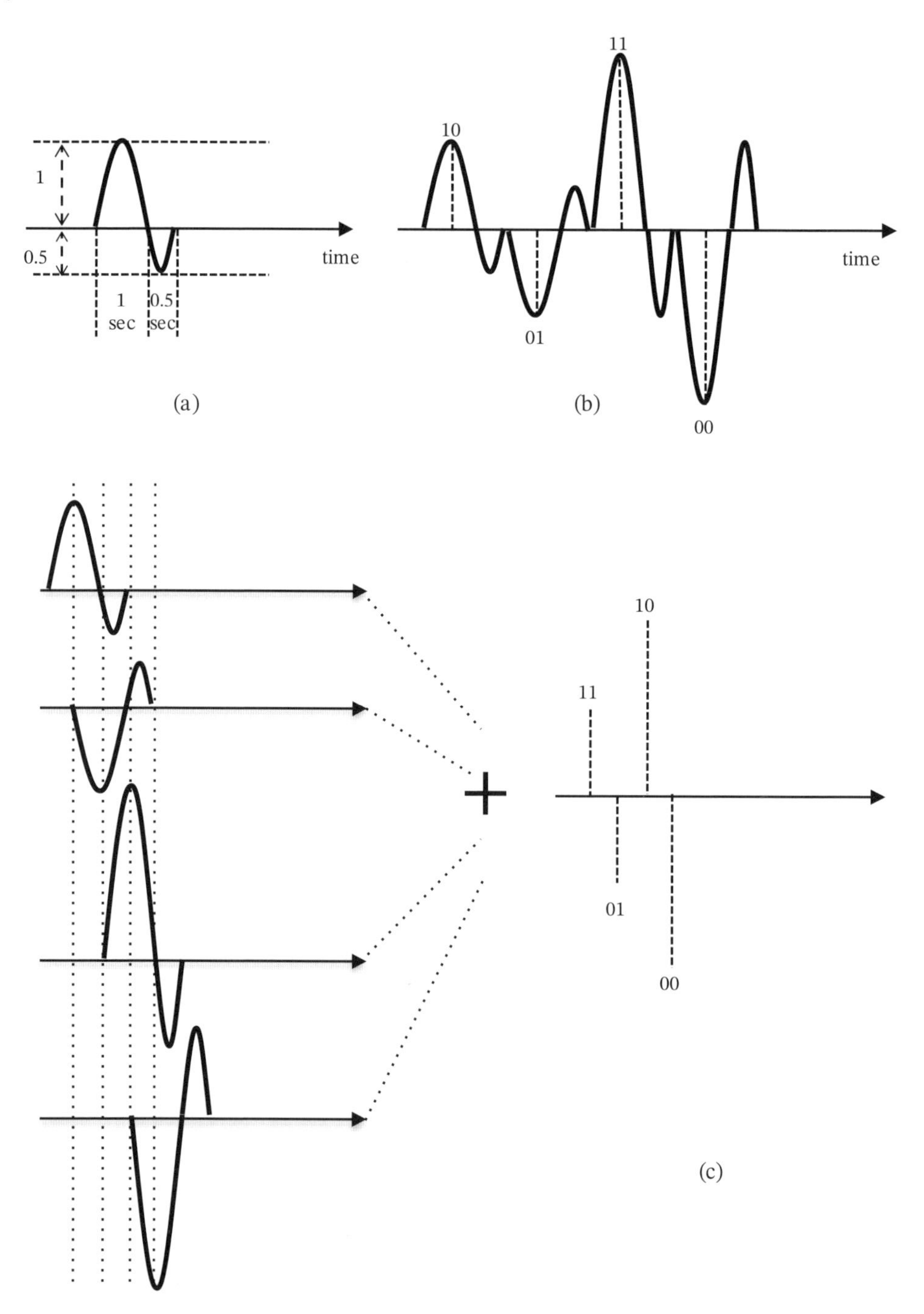

Figure 9.1 A simple example that illustrates how to create a discrete communication channel using continuous waveforms. (a) The elementary waveform. (b) The waveform can be used to carry Gray mapped quaternary symbols: 00 is sent by multiplying it by −3, 01 by −1, 11 by 1, and 11 by 3. The waveforms are sent with a minimal possible separation in time to avoid overlap and there are $D = \frac{1}{1.5} = 0.67$ (c.u/s). (c) Densely packed waveforms with $D = \frac{1}{0.5} = 2$ (c.u/s). The figure only depicts the discrete instants that define the outputs of the communication channel; a full illustration would include a sum of the continuous waveforms contributing to the overall transmitted/received signal.

and can give a rise to *intersymbol interference (ISI)*. With this, the channel uses are no longer independent.

As a simple example of ISI, not related to the example in Figure 9.1, the following baseband received signal can be considered:

$$y_k = z_k + 0.5z_{k-1} + n \tag{9.4}$$

where we see that the kth received symbol by Yoshi is dependent both on the $(k-1)$th and kth symbols transmitted by Zoya. However, even if the waveforms of different symbols overlap in time, what matters is whether there is an ISI that ultimately appears in the baseband representation, as in (9.4). As the example in Figure 9.1(c) shows, setting $T = 0.5$ s also produces discrete channel uses that are independent and free of ISI, although their respective waveforms do overlap in time. Using $T = 0.5$ instead of $T = 1.5$ results in having three times more symbols sent per seconds compared to Figure 9.1(b), which results directly in a three times larger data rate, expressed in bits per second.

Nevertheless, T cannot be decreased arbitrarily while still avoiding ISI in the resulting discrete channel. The reader can check that $T = 0.5$ s is the absolutely minimal time separation between two pulses at which one can still get a communication channel that consists of a sequence of independent channel uses, leading to a data rate of $R = 2C$ (bps). However, if we are willing to depart from the model of a communication channel without ISI and make $T < 0.5$ s, then the discrete communication channel starts to have a *memory*, as in the example (9.4), which is an effect not considered in our baseband models so far. This is in contrast to the *memoryless* communication channels, in which the output of a channel use is affected by the input of the same channel use, but not the inputs of the previous channel uses. The example (9.4) illustrates a channel with memory, where the output y_k has a statistical dependency with both y_{k-1} and y_{k+1}. We would like to emphasize that reliable communication and definition of channel capacity is possible despite the occurrence of ISI, but it requires encoding/decoding strategies that are different from the ones used for memoryless channels.

A practical wireless communication channel inherently introduces ISI, as well as other additional distortions to the signal. The common approach to address the ISI caused by the wireless channel is a layered one: at the lower layer, use *equalization* in order to remove ISI and create a memoryless communication channel; at the higher layer, apply the known coding/modulation strategies for that channel. This leads us to the conclusion that a communication channel can be created by either removing the ISI or by keeping it to remain a part of the communication channel. Once the channel definition is fixed, we can compute its capacity.

Nevertheless, that is not the ultimate answer to the question of how much information we can carry through waveforms with certain physical features since "the channel is the part of the system we are unwilling or unable to change", while each possible ISI pattern defines a new channel. Given the interference pattern, we can compute the capacity in bits per second of the respective channel; it thus remain to find the interference pattern whose channel has the highest capacity. As a fundamental question, we would like to know the information carrying properties of the set of all possible waveforms that can be produced in a given communication system.

Consider the following communication method adopted by Zoya. In transmitting information to Yoshi, Zoya may decide to depart from the conventional strategy that avoids ISI and start to send symbols to the TXmodule using arbitrarily low T. In the meantime, Yoshi samples the output of the RXmodule using that same value of T. With this Zoya tries to increase the capacity R from (9.3) by decreasing T. At the same time, she hopes that the capacity C of the channel with ISI will not decrease so much so as to eliminate the gain obtained by sending the symbols faster. Can T be really decreased to be arbitrarily low? This line of thinking needs to take into account the other physical limitations of the system and the notion of frequency characteristics of the waveforms.

9.3 Enter the Frequency

A very simple way to increase the data rate of the system based on the elementary waveform from Figure 9.1(a) would be to compress the waveform in time. For example, if the same waveform is compressed to take in total 0.15 s instead of 1.5 s, then one gets ten times more channel uses per second that, with unchanged number of bits per channel use, leads to a ten times higher data rate. Compressing the waveform to a tenth of the original duration implies that one allows ten times faster changes in the waveform and therefore ten times faster operation of the circuits and modules underlying the communication system. However, there must be physical constraints on how fast the changes in a waveform may occur; otherwise one can make the duration of the waveform infinitely short and thus get the data rate to grow to infinity. Those physical constraints are often expressed through the *frequency characteristics* of the system and the associated waveforms. There is neither space nor ambition to cover the details of the rich area of frequency analysis, Fourier transform, time-frequency diagrams, etc., but we will provide intuitive arguments to highlight the role that frequency plays in communication.

9.3.1 Infinitely Long Signals and True Frequency

Frequency is a quantity that can only be defined precisely for a signal of infinite duration. An example signal of frequency f is:

$$\tilde{z}_f(t) = |A| \cos(2\pi ft + \phi) \quad -\infty < t < \infty \tag{9.5}$$

where we put a "wave" $\tilde{\cdot}$ to denote a signal that is continuous in time[1], $|A|$ is the amplitude, and ϕ is the phase of the sinusoidal signal. Once the frequency is fixed, the amplitude and the phase represent two channel uses or two DoFs that can be used to modulate data. We note here that these two DoFs are not identical in a communication/statistical sense, since noise or other physical distortions affect the amplitude and the phase in two different ways.

A bigger conceptual problem with these two DoFs is their practical significance as channel uses for a communication system. If Zoya wants to use these two DoFs to communicate with Yoshi, then she needs to send the information only once, infinitely back in the past, and the transmitted symbols will last indefinitely. Common statements such as "change the

1 This will only be used when there is a danger of confusion.

frequency between time t_1 and t_2" do not make sense simply because, by the strict definition, a frequency cannot change in time! This paradox and the discrepancy between the "mathematical" frequency and "practical" frequency has been noted already in the early days of communication engineering, as eloquently put forward by Dennis Gabor in his seminal article "Theory of Communication". Nevertheless, for the discussion that follows we embrace the absurdity of the setup in which Zoya sends the symbols infinitely back in the past and let them last indefinitely. On the other hand, we require Yoshi to receive the symbols sent by Zoya by turning his receiver on for a finite duration of time T.

An equivalent way to represent a signal from (9.5) that has a frequency f is:

$$\tilde{z}_f(t) = z_{I,f}\cos(2\pi ft) - z_{Q,f}\sin(2\pi ft) \tag{9.6}$$

where $z_{I,f}$ is the *in-phase (I)* and $z_{Q,f}$ is the *quadrature (Q)* component at the frequency f. Given $z_{I,f}$ and $z_{Q,f}$, the values of $|A|$ and ϕ in (9.5) can be uniquely determined, and vice versa. The representation (9.6) is convenient as $z_{I,f}$ and $z_{Q,f}$ can be seen as two DoFs that are of the same statistical nature, in the sense that if noise is added to $\tilde{z}_f(t)$ it is going to affect $z_{I,f}$ and $z_{Q,f}$ in statistically identical ways. In other words, $z_{I,f}$ and $z_{Q,f}$ are two channel uses of the same kind that can be independently modulated, which will become even clearer when we take into account the impact of the noise. The information about these two DoFs can be compactly represented as a single complex symbol:

$$z_f = z_{I,f} + jz_{Q,f} \tag{9.7}$$

such that the signal $\tilde{z}_f(t)$ (9.6) can be obtained as:

$$\tilde{z}_f(t) = \mathrm{Re}\{z_f e^{j2\pi ft}\}. \tag{9.8}$$

Although $z_{I,f}$ and $z_{Q,f}$ can be seen as two independent DoFs, we note that when Zoya transmits $\tilde{z}_f(t)$, Yoshi observes interference between $z_{I,f}$ and $z_{Q,f}$ as they are both received simultaneously, while the objective is to extract each of them as a separate value. Yoshi can achieve this by using the fact that the signals $\cos(2\pi ft)$ and $\sin(2\pi ft)$ are *orthogonal* in any time interval that is an integer multiple kT where $T = \frac{1}{f}$ is the period of the signal (9.6). This means that if the signal from (9.6) is received during an interval of length kT, where k is an integer, the orthogonality property guarantees that there is a way for Yoshi to extract each of the values $z_{I,f}$ and $z_{Q,f}$ independently, without interference from the other one. This can be written as follows:

$$\begin{aligned} z_{I,f} &= \frac{2}{T}\int_{-\frac{T}{2}}^{\frac{T}{2}} \tilde{z}_f(t)\cos(2\pi ft)\mathrm{d}t \\ z_{Q,f} &= -\frac{2}{T}\int_{-\frac{T}{2}}^{\frac{T}{2}} \tilde{z}_f(t)\sin(2\pi ft)\mathrm{d}t \end{aligned} \tag{9.9}$$

where, for convenience, we have chosen to use the orthogonality on the interval $\left[-\frac{T}{2}, \frac{T}{2}\right)$, but any interval of length T would work. If the signal (9.6) is received during an interval shorter than T, then it is not possible to extract $z_{I,f}$ and $z_{Q,f}$ in such a way that they are free of each other's interference.

The idea of extracting interference-free discrete symbols from continuous waveforms that interfere is appealing and we will generalize beyond the I/Q signals at a single frequency. Let Yoshi set his receiving interval to length T and, without loss of generality, let that interval be $\left[-\frac{T}{2}, \frac{T}{2}\right)$. It is interesting to find out the following: how many independent, non-interfering channel uses (DoFs) can Yoshi extract from the signal sent by Zoya? The orthogonality property extends to signals with different frequencies. Having a fixed interval of length T, then any sinusoidal signal of frequency $\frac{k}{T} = kf$, regardless of its phase, is orthogonal to a sinusoidal signal of frequency vf as long as the integers k and v are different $k \neq v$. Using complex number notation, this can be written as:

$$\frac{1}{T}\int_{t}^{t+T} e^{j(2\pi kft+\phi_k)}e^{-j(2\pi vft+\phi_v)}dt = \delta_{k,v} = \begin{cases} 0 & k \neq l \\ 1 & k = v \end{cases} \tag{9.10}$$

where $\delta_{k,v}$ is the Kronecker symbol, while ϕ_k and ϕ_v are used to emphasize that the phase of the sinusoidal signal does not affect the orthogonality. The latter implies that the I component at the frequency kf is orthogonal to both the I and the Q component at the frequency vf. We note also that when the constant signal, having a frequency of 0, is the only "sinusoidal" signal in which one cannot differentiate the I and Q components such that it can carry only one DoF. Let us denote by $z_{I,kf}$ and $z_{Q,kf}$ the I and the Q components, respectively, sent by Zoya at the frequency kf, and observe the following signal:

$$\tilde{z}(t) = \sum_{k=0}^{\infty} \tilde{z}_{kf}(t) = \sum_{k=0}^{\infty} z_{I,kf}\cos(2\pi kft) - z_{Q,kf}\sin(2\pi kft) \tag{9.11}$$

where we set by definition $z_{Q,0} = 0$, i.e. the quadrature component at frequency 0 does not carry information. By receiving $z(t)$ over an interval of length T and using the orthogonality property, Yoshi can receive an infinite amount of symbols $\{z_{I,kf}, z_{Q,kf}\}$ that are free from interference from each other. The symbols are extracted as follows:

$$\begin{aligned} z_{I,kf} &= \frac{2}{T}\int_{-\frac{T}{2}}^{\frac{T}{2}} \tilde{z}(t)\cos(2\pi kft)dt \\ z_{Q,kf} &= -\frac{2}{T}\int_{-\frac{T}{2}}^{\frac{T}{2}} \tilde{z}(t)\sin(2\pi kft)dt \end{aligned} \tag{9.12}$$

where the "−" sign is due to the definition of a quadrature component in (9.6). We note that the interval $-\frac{T}{2} \leq t < \frac{T}{2}$ is not special in any way: Yoshi can choose any interval of length T and use the orthogonality property in that interval. This is because the signal $z(t)$ is periodic and the information about $z_{I,kf}$ and $z_{Q,kf}$ is identical anywhere; recall that the symbols are sent in infinite past and last until infinity. However, it is important to stick to the same length T, as this is determining the fundamental frequency $f = \frac{1}{T}$ and the *harmonics* kf of the orthogonal signals.

By choosing all $z_{I,kf}$ and $z_{Q,kf}$, Zoya synthesizes a periodic waveform with period T, such as the one depicted in Figure 9.2(a). On the other hand, Yoshi receives the signal only during a single, specific interval of length T, as in Figure 9.2(b). In fact, while the transmitter Zoya carries out *synthesis*, Yoshi is faced with the problem of *analysis* of a waveform: given the waveform from Figure 9.2(b), find out the coefficients $\{z_{I,kf}, z_{Q,kf}\}$ that have produced that waveform. This is known as *Fourier analysis* and the representation of the signal $\tilde{z}(t)$ in (9.11) is known as *Fourier series*. If for some k both coefficients are zero $z_{I,kf} = z_{Q,kf} = 0$, then this is often referred to as "the frequency kf is absent from the signal". Otherwise, if at least one of $z_{I,kf}$ or $z_{Q,kf}$ is not zero, it is considered that the frequency kf is present in the signal or, in other words, the signal carries energy at the frequency kf.

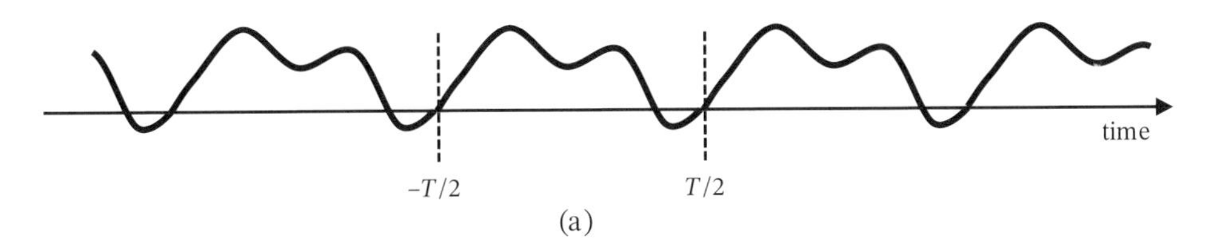

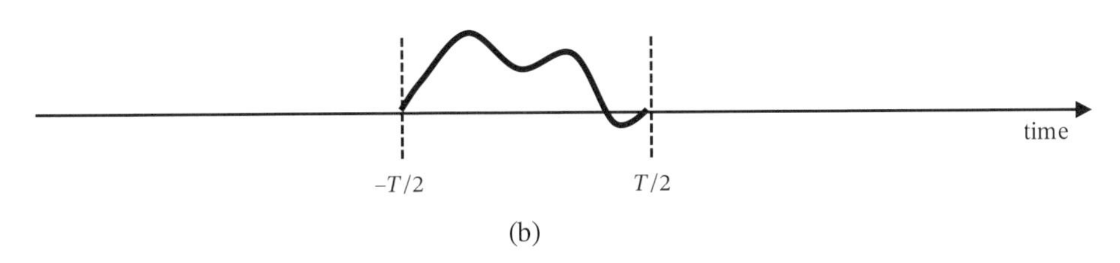

Figure 9.2 (a) The periodic waveform with a period of T sent by Zoya. (b) Yoshi observes only a fragment of the waveform of duration T, which contains all information about the waveform.

9.3.2 Bandwidth and Time-Limited Signals

From the Fourier series representation of a signal in (9.11) one can conclude that there is an infinite number of sinusoidal waveforms that are orthogonal within the finite time interval of length T. In other words, Zoya can potentially send an infinite number of interference-free symbols $\{z_{I,kf}, z_{Q,kf}\}$ to Yoshi, since k can be arbitrarily large. Then Yoshi observes infinite number of symbols within a finite time, which means that the data rate is infinite.

Nevertheless, this is physically not possible due to the following. The energy that the signal carries at the frequency kf within the limited time T is proportional to:

$$E_{kf} = (z_{I,kf}^2 + z_{Q,kf}^2)T \tag{9.13}$$

such that the total energy carried by the signal is:

$$E = \sum_{k=0}^{\infty} E_{kf}. \tag{9.14}$$

To keep the things physically sane, the energy E needs to be finite. One way to ensure this is to put the constraint that Zoya's signal is allowed to contain frequencies that belong to a limited frequency band, such that $E_{kf} = 0$ for all $k \geq k_{\mathrm{H}}$. Here $f_{\mathrm{H}} = k_{\mathrm{H}}f$ is the largest frequency contained in the signal. Alternatively, the energy per frequency can decrease to zero as the frequency grows to infinity, such that the sum (9.14) is kept finite.

One can try to increase the speed of sending new symbols by shortening the duration of the waveforms that represent the symbols. The physical limit on how short the waveform can be depends on the physical limits of the system on how fast it can process the symbols. The frequency representation provides an alternative view of the physical limits: if high frequencies kf are present in the signal, then it means that Zoya and Yoshi communicate with signals that change very quickly. Putting a limit on $k_{\mathrm{H}}f$, the highest frequency in the signal, or requiring that the energy per frequency goes to zero as k goes to infinity, means that we are limiting the speed at which the information carrying signal can change.

This brings us to the notion of a *bandlimited signal*, which has non-zero energy only at the frequencies that lie within a certain band $[f_L, f_H]$. In other words, the signal contains the frequencies within the limited band $[f_\mathrm{L}, f_\mathrm{H}]$. This is a rather simplified definition that will be revised later on, but it is sufficient for our present discussion. Although the notion of frequency can only be defined for a signal of infinite duration, we can meaningfully speak about the frequencies that the receiver can observe in a signal within a time window T, as in Figure 9.2(b). However, one should keep in mind that these are the frequencies that are defined for the signal that is a periodic extension of the observed time-limited signal, as in Figure 9.2(a).

If Zoya uses *lowpass* transmission and includes the frequency $f_\mathrm{L} = 0$, then in a *bandwidth* of size W [Hz] contains at most

$$\mathrm{DoF}_{\mathrm{lowpass}} = 2\frac{W}{f} + 1 = 2WT + 1 \tag{9.15}$$

non-zero coefficients $\{z_{I,kf}, z_{Q,kf}\}$, which can be seen as channel uses that Zoya can modulate with information. The multiplier 2 in (9.15) is due to in-phase and quadrature components for each non-zero frequency and $+1$ is the DoF of frequency 0. Alternatively, Zoya can create a *passband* signal that contains the frequencies from the range $[f_\mathrm{L}, f_\mathrm{H}]$ with $f_\mathrm{L} > 0$, such that the bandwidth of the signal is $W = f_\mathrm{H} - f_\mathrm{L}$ and the number of available DoFs is

$$\mathrm{DoF}_{\mathrm{bandpass}} = 2\frac{W}{f} = 2WT. \tag{9.16}$$

This is the point where we need to revise our assumption that Zoya communicates with periodic signals of infinite duration. Namely, Yoshi receives the signal from Zoya by observing only a limited interval of time T and what Zoya transmits outside of that interval does not affect the reception of Yoshi in any way. Therefore, Zoya can use any other time outside of that interval to send other symbols to Yoshi, which removes the annoying assumption that she should have started to send sinusoidal signal infinitely back in the past. For example, Zoya picks a frequency kf and in the interval $[-T/2, T/2)$ she modulates it with the symbols $z_{I,kf}(1)$ and $z_{Q,kf}(1)$, where "1" stands for the first modulated symbol. In the next interval $[T/2, 3T/2)$, she sends the second set of symbols $z_{I,kf}(2)$ and $z_{Q,kf}(2)$. In general, during the ith symbol interval $[(2i-3)T/2, (2i-1)T/2)$, Zoya modulates kf with the symbols $z_{I,kf}(i)$ and $z_{Q,kf}(i)$. We therefore refer to T as a *symbol duration* or *channel use duration* or *DoF duration*. The signal obtained in this way, when observed during the entire time axis, is not periodic and therefore cannot be treated as a "signal of frequency kf", understood in a strict sense. However, within each interval $[(2i-3)T/2, (2i-1)T/2)$, which corresponds to a single symbol duration, the signal looks like a portion of a periodic signal and Yoshi can use the orthogonality property to extract the symbols sent by Zoya without any *ISI*.

As long as Yoshi is perfectly synchronized with Zoya, then within each symbol interval T one can consistently speak about symbols carried at a particular frequency $kf = \frac{k}{T}$. However, if Yoshi uses a time window for reception that has a duration different from T, then the orthogonality property is lost. Even more, if Yoshi uses the same symbol duration T, but its reception window is not aligned to the timing of the symbols sent by Zoya, then the orthogonality property is also lost. This is illustrated in Figure 9.3 for a transmission that

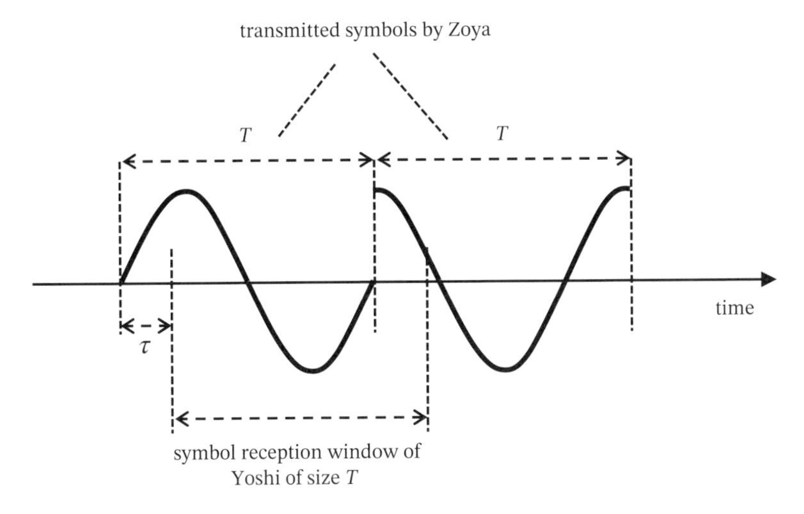

Figure 9.3 Zoya uses the frequency f to transmit the complex symbol $(0, -1)$ at time 0 and the symbol $(1, 0)$ at time T. Yoshi's reception window has the correct duration T, but is misaligned and therefore Yoshi does not observe a sinusoidal signal within his reception window.

occurs at a frequency $f = \frac{1}{T}$. In this example, Zoya and Yoshi have agreed that the symbols should be placed as $[(i-1)T, iT)$ instead of $[(2i-3)T/2, (2i-1)T/2)$. However, Yoshi has the start of his reception window misaligned for a duration of τ. Zoya transmits the symbols $z_{I,f}(0) = 0, z_{Q,f}(0) = -1$ in the interval $[0, T)$ and the symbols $z_{I,f}(1) = 1, z_{Q,f}(1) = 0$ in the next interval. Within his misaligned interval, Yoshi does not observe a sinusoid, that is, a signal that represents a "cut" from an infinite sinusoidal signal with frequency f, and therefore the orthogonality property cannot be applied[2]. Instead, the information carried in the symbol transmitted at time T affects the decoding of the symbol sent at time 0 and Yoshi experiences ISI.

9.3.3 Parallel Communication Channels

The most important quality that the concept of frequency brings to our communication models is the possibility of creating *parallel communication channels*, which can exist simultaneously at different frequencies. Once T and W are fixed, then in each time interval T, Zoya can send $2WT$ different symbols in parallel (ignoring the $+1$ for lowpass signals). This is the basis for *orthogonal frequency division multiplexing (OFDM)*, a transmission scheme that has become dominant within broadband wireless transmission technologies since the 2000s. In OFDM, each of the sinusoidal signals that is modulated with symbols is called a *subcarrier* and we adopt the same terminology to denote a certain frequency that is modulated with data. The notion of a *carrier* is reserved for the frequency that is usually the central frequency of a passband signal. In this sense, a carrier frequency can be used to

2 To be precise, if it happens that Zoya sends the same symbol consecutively, then $z_{I,f}(1) = z_{I,f}(0)$ and $z_{Q,f}(1) = z_{Q,f}(0)$, such that the signal observed by Yoshi is still sinusoidal. However, this is not true in a statistical sense, as symbols change randomly from one interval to another.

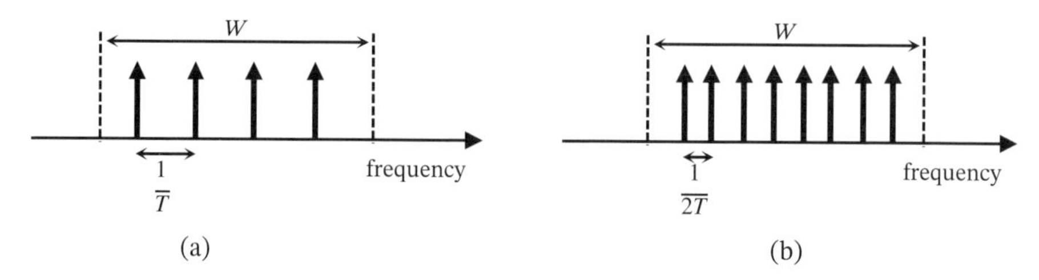

Figure 9.4 Number of channel uses in a given bandwidth W. Each vertical arrow represents a frequency subcarrier. (a) The channel use duration is T and the separation between two neighboring subcarrier frequencies is $\frac{1}{T}$. (b) The channel use duration is $2T$ and the separation between two neighboring subcarrier frequencies is $\frac{1}{2T}$.

shift a lowpass OFDM signal into an equivalent passband signal; then the data-modulated subcarriers are placed around this carrier frequency and hence the name.

Let us assume that the available bandwidth for communication is W and each subcarrier is modulated with Mary symbols, such that each symbol carries $\log_2 M$ data bits. Each channel use has a duration of T, corresponding to the time interval over which the different subcarriers with frequencies $\frac{k}{T}$ are orthogonal. Since now each channel use has a well defined duration, one can calculate the data rate in bits per seconds:

$$R = \frac{2WT\log_2(M)}{T} = 2W\log_2 M \quad \text{(bps).} \tag{9.17}$$

It is interesting to note that the data rate does not depend on the choice of the channel use duration, but it depends on the available bandwidth W. In order to see why this is the case, see the example in Figure 9.4. Each vertical arrow represents a subcarrier that be modulated with information, or, equivalently, two channel uses of DoFs (in-phase and quadrature). When the channel use duration doubles, the number of subcarriers within the bandwidth also doubles; however, this does not change the number of DoFs per second, which remains constant and equal to $2WT$.

Although the data rate in (9.17) does not explicitly depend on T, the whole discussion is based on the assumption that a certain value of T has been fixed in advance and is used as a time reference to modulate all subcarriers.

9.3.4 How Frequency Affects the Notion of Multiple Access

The parallel channels, defined on orthogonal frequency subcarriers, offer the possibility of sharing the medium among several wireless links simultaneously, without using some form of TDMA. We can take the example in which two mobile terminals, Zoya and Yoshi, communicate with a base station Basil. The total available bandwidth is W. All parties (Zoya, Yoshi, and Basil) agree in advance on the symbol period T, during which the symbols that are modulated on the sinusoidal carriers do not change. It is also agreed in advance that the frequencies that lie in the subband that contains the lower half of the frequency band W are allocated for communication between Basil and Zoya, while the ones from the upper half are allocated for communication between Basil and Yoshi. If a third user Xia joins the system to communicate with Basil, then she also needs to use the same symbol period T and Basil needs to re-allocate the bandwidth W by dividing it into three parts.

The method of channelization in which different users use different frequencies is known as *frequency division multiple access (FDMA)*. In fact, in this case, since we assume that all frequency subcarriers are orthogonal, we can speak about *orthogonal frequency division multiple access (OFDMA)*. Given the bandwidth W and using the same arguments as in Figure 9.4 it can be shown that the number of DoFs, and therefore the capacity, of TDMA and (O)FDMA is identical. In a TDMA transmission scheme, Zoya is allocated the whole bandwidth W within the symbol duration T, thus using the $2WT$ DoFs within that time. If there are M users in the system and they are served in turn using a round robin, then the average number of DoFs available to each user per unit time is $\frac{2WT}{MT} = \frac{2W}{M}$. Alternatively, each user can be served all the time, but using only a fraction of the available bandwidth $\frac{W}{M}$, which again leads to $2\frac{W}{M}$ DoFs per unit time. In the case of FDMA Zoya can apply a frequency filter and receive only the frequencies that are allocated to her. Similarly, if TDMA is used, Zoya can apply a form of time filtering and be active only during the time window allocated to her. In practice, frequency filters in FDMA (though not in OFDMA) are less flexible compared to the time filtering, such that TDMA is the preferred option when the traffic of different links varies and the resources need to be allocated dynamically.

Despite this difference, the framework introduced so far cannot really capture the true difference between TDMA and FDMA, as in both cases we require all users to be perfectly synchronized. In TDMA this is required in order to preserve the synchronization of the time windows allocated to the users, while in FDMA in order to preserve the orthogonality among the subcarriers belonging to different users. On the other hand, the relation (9.17) suggests that the frequency bandwidth W determines the data rate of a communication link, irrespective of the choice of the symbol duration T.

Perfect synchronization is not so much of an issue in a cellular communication scenario where both Zoya and Yoshi communicate with a base station Basil. This is because Basil takes care to ensure that everybody uses the same T and the same synchronization reference. However, one can think of a scenario in which the same bandwidth is used by two links that are belonging to different systems, but are still within a spatial proximity and can interfere with each other. For example, one link is Zoya–Yoshi and the other link is Xia–Walt. If these links use different technology, then there is no reason to impose that they are synchronized and use the same symbol timing T. Instead, it would be desirable to find a way to divide the bandwidth W into two portions of, say $\frac{W}{2}$ (Hz) each, allocate one portion to the link Zoya–Yoshi, the other portion to Xia–Walt, and let each link define its own symbol timing and synchronization reference. This would indeed bring a new quality to our communication models, enabling the coexistence of two or more independent wireless links that are within a spatial proximity, requiring only a pre-allocation of frequency bands, but not any temporal coordination among them. In order to understand how this can be achieved, we first need to see how noise enters into the model for waveform communication.

9.4 Noise and Interference

9.4.1 Signal Power and Gaussian White Noise

Physically, the waveform $\tilde{z}(t)$ transmitted by Zoya represents a voltage or a current such that the instantaneous power is proportional to $\tilde{z}^2(t)$, multiplied by a constant. We can conveniently choose the constant, without losing generality, and define the average power within a given interval of duration T as:

$$P_z = \frac{2}{T}\int_{-\frac{T}{2}}^{\frac{T}{2}} \tilde{z}^2(t)\mathrm{d}t. \tag{9.18}$$

This choice of the constant is motivated by the way the receiver is designed to extract the in-phase and quadrature components in (9.12). Hence, the overall signal power is given as:

$$P_z = \sum_{k=1}^{\infty} z_{I,kf}^2 + z_{Q,kf}^2 = \sum_{k=1}^{\infty} P_{z,kf} \tag{9.19}$$

where $P_{z,kf}$ is the total power at the frequency $kf = \frac{k}{T}$. The physical interpretation of (9.19) is that the total power in the signal is equal to the sum of the total power carried in all the frequencies that are integer multiples of $\frac{1}{T}$.

The waveform $\tilde{y}(t)$ received by Yoshi can be obtained by adding noise $\tilde{n}(t)$ to the transmitted one $\tilde{z}(t)$:

$$\tilde{y}(t) = \tilde{z}(t) + \tilde{n}(t). \tag{9.20}$$

The equation (9.20) is a simplified model of a communication system where the only distortion of the transmitted signal occurs due to the additive noise. It is noted that the channel coefficient h, introduced in the baseband communication model, is absent and we will bring it back to the discussion later on. Since the transmitted signal $\tilde{z}(t)$ is completely specified by the signals carried by its constituent frequencies, we will describe the noise $\tilde{n}(t)$ through its impact on the frequencies observed by the receiver.

The signal that Yoshi gets when he tries to extract, for example, the in-phase component of frequency kf is:

$$y_{I,kf} = \frac{2}{T}\int_{-\frac{T}{2}}^{\frac{T}{2}} [\tilde{z}(t) + \tilde{n}(t)]\cos(2\pi kft)\mathrm{d}t = z_{I,kf} + n_{I,kf} \tag{9.21}$$

where

$$n_{I,kf} = \frac{2}{T}\int_{-\frac{T}{2}}^{\frac{T}{2}} \tilde{n}(t)\cos(2\pi kft)\mathrm{d}t \tag{9.22}$$

is the projection of the additive noise, or a noise sample, to the in-phase component of the frequency kf. The common model used for the noise $\tilde{n}(t)$ is the *additive white Gaussian noise (AWGN)*. Here "additive" is self-explanatory, while "white" means that $\tilde{n}(t)$ affects all frequencies in identical way: $n_{I,kf}$ or $n_{Q,kf}$ is a Gaussian random variable with zero mean and a variance of

$$E[n_{I,kf}^2] = E[n_{Q,kf}^2] = \frac{P_\mathrm{N}}{2} \tag{9.23}$$

which is identical for each frequency kf. The noise components of two different frequencies kf and vf are uncorrelated, which for Gaussian random variables also means independent. Furthermore, the noise components of the I and Q signals at the same frequency are also independent.

AWGN is a theoretical construct of "the most random" noise that varies wildly and affects each frequency in the same statistical way. Since each of the I/Q component of a given frequency kf is affected by a noise with average power $\frac{P_N}{2}$, then the total noise power at the frequency kf is P_N. However, note that $f = \frac{1}{T}$ is a choice made by Zoya and Yoshi at the time when they established their communication system, such that Yoshi's receiver collects only the noise power that occurs at the frequency $\frac{k}{T}$, where k is an integer.

Noise exists in nature independently of this choice made by Zoya and Yoshi. Had Zoya and Yoshi agreed upon a different value T_1, then Yoshi's receiver would have collected noise power at the frequency $\frac{k}{T_1}$. In fact, the model of Gaussian noise assumes that it affects *all* frequencies and is white over a continuum of frequencies. However, since the power of the continuous signal is a sum of the powers that the signal carries at different frequencies, we are arriving at a physical paradox: if $\tilde{n}(t)$ is a white signal that has the same power at all frequencies, then the average power of $\tilde{n}(t)$ is infinite! This paradox is resolved by observing we are not interested in $\tilde{n}(t)$ per se but only in how $\tilde{n}(t)$ manifests itself in the "eye of the beholder", which here is the receiver of the communication system. The receiver processes a limited set of frequencies and therefore gets a limited noise power. By treating the noise samples as independent across frequencies, we are making the worst-case assumption about how our communication system is affected by external random disturbances. On the other hand, this implies that, from a communication perspective, any correlation among the noise samples that affect different frequencies can be useful. The existence of correlation means that, in principle, one can learn something about the noise at one frequency by observing the noise at another frequency.

Taking into account that the received signal of the Q-component has identical form as the one of the I-component in (9.21), we can compactly write the received signal at frequency kf by using a complex notation:

$$y_{kf} = z_{kf} + n_{kf} \tag{9.24}$$

where the complex noise is $n_{kf} = n_{I,kf} + jn_{Q,kf}$ and the baseband transmitted signal $z_{kf} = z_{I,kf} + jz_{Q,kf}$.

We now bring the channel coefficient back in the picture. While noise is a phenomenon present at the receiver, the transmitted signal $\tilde{z}(t)$ also undergoes a distortion when propagating from Zoya's transmitter to Yoshi's receiver. In the simplest case, the I/Q components at each frequency kf are scaled with their respective coefficients, such that Zoya's signal at frequency kf that enters Yoshi's receiver is given by:

$$h_{I,kf}z_{I,kf}\cos(2\pi kft) - h_{Q,kf}z_{Q,kf}\sin(2\pi kft) \tag{9.25}$$

which leads to the received signal:

$$y_{kf} = h_{kf}z_{kf} + n_{kf} \tag{9.26}$$

where $h_{kf} = h_{I,kf} + jh_{Q,kf}$ is the channel coefficient. We have thus arrived at the baseband model from Chapter 5; now we can rely on the principles of modulation and coding that are applicable to the baseband model. Taking that the total transmitted power at the frequency kf is given by $P_{z,kf}$, see (9.19), then from Chapter 5 we find the SNR at the frequency kf as:

$$\gamma_{kf} = \frac{|h_{kf}|^2 P_{z,kf}}{P_{\mathrm{N}}}. \tag{9.27}$$

The context of multiple frequencies that are independently modulated, while each frequency channel is scaled independently, sheds a new light on the notion of water filling. Namely, the average power P available at the transmitter needs to be distributed across the frequencies kf following the principles of water filling. For *frequency selective channels* the channel coefficients change from frequency to frequency, which causes the SNR to be different for a different frequency channel, although the noise power is identical for all frequencies. Specifically, the change in the SNR is due to the change of the *channel gain* $|h_{kf}|^2$, and not the phase of the channel coefficient h_{kf}.

When multiple orthogonal frequencies are independently modulated, the analog waveform that results from the superposition of these frequencies may have very large variations in the signal amplitude. This leads to the problem of *peak-to-average power ratio (PAPR)*. The transmitted signal is a sum of sinusoids, each of the frequencies kf independently modulated with data symbols. During a certain symbol time, it can happen that the combination of modulated data symbols is unlucky and the modulated frequencies superpose to a waveform that has a very high peak value. These high peaks put pressure on the power amplifier at the transmitter, regardless of the fact that the average power stays constant. The problem can be mitigated by avoiding the peaks while keeping the average power constant. However, any method for avoiding the undesired data combinations that lead to high peaks will necessarily restrict the way different frequencies (subcarriers) are modulated. This means that a frequency cannot be modulated independently of the other frequencies, which results in decrease in the available DoFs and therefore the data rate.

9.4.2 Interference between Non-Orthogonal Frequencies

Let Zoya and Yoshi agree upon a symbol time T_1 and communicate with each other using a single frequency f_1, where $f_1 = \frac{1}{T_1}$. Independently of them, let Xia and Walt agree upon a different symbol time T_2 and let they choose $f_2 = \frac{k}{T_2}$ as a frequency to communicate, where k is an integer. Let us assume that T_2 is chosen to be much longer than T_1, such that the symbol transmitted from Xia to Walt does not change while the symbol is sent from Zoya to Yoshi, see Figure 9.5(a). Other than that, we assume that T_2 is chosen independently from T_1, in a sense that $\frac{T_2}{T_1}$ is not necessarily an integer, such that the frequencies f_1 and f_2 are not necessarily orthogonal on the interval of length T_1. Finally, we assume that $f_2 > f_1$ and let $\Delta f = f_2 - f_1$.

The question we want to investigate is: how does the transmission from Xia to Walt interfere with Yoshi's reception of Zoya's signal? We start by observing that Yoshi receives the signal over a multiple access channel, with Zoya and Xia as transmitters:

$$\tilde{y}(t) = \tilde{z}(t) + \tilde{x}(t). \tag{9.28}$$

The signal $\tilde{z}(t)$ sent by Zoya will be referred to as a useful signal, while $\tilde{x}(t)$ sent by Xia as an interfering signal. In (9.28) we have assumed that the channel coefficients are 1 and, for

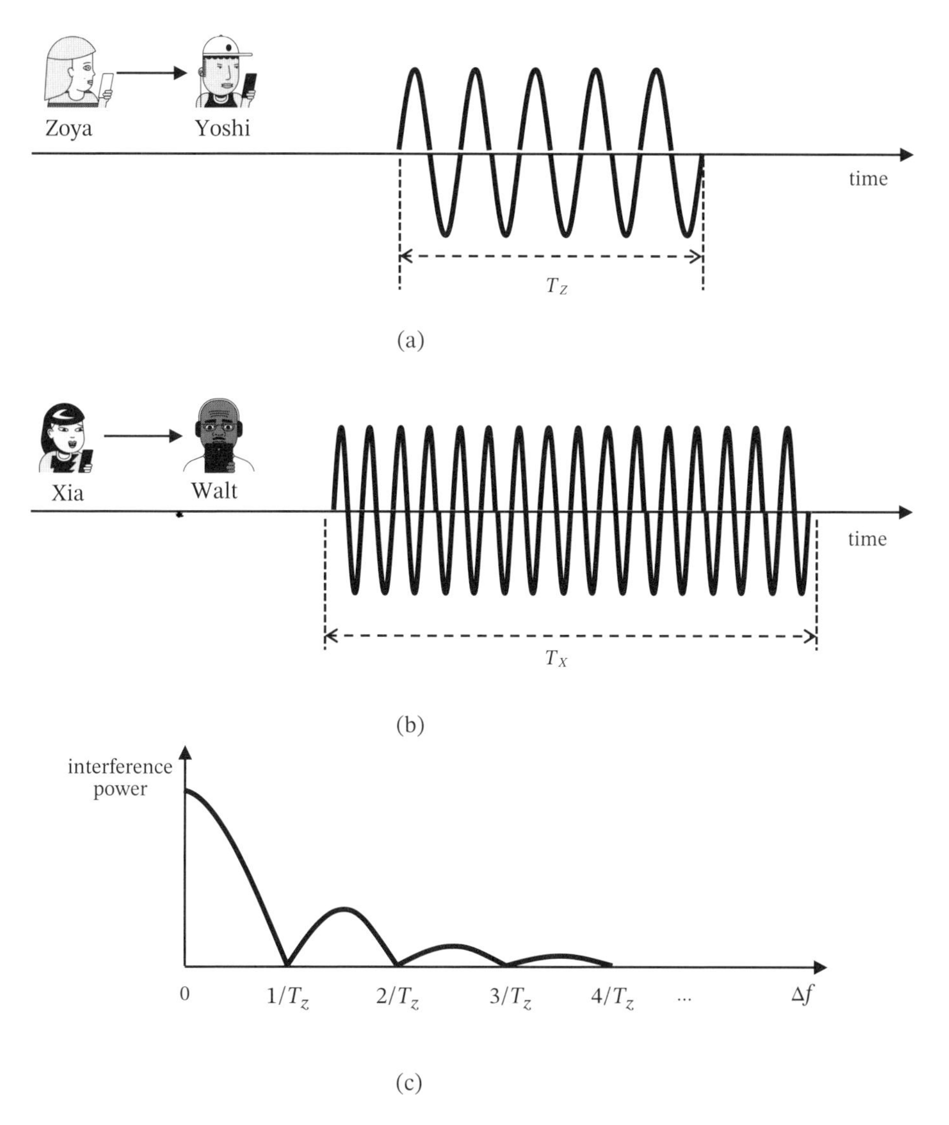

Figure 9.5 Interference between signals with different frequencies. (a) Observed signal transmitted from Zoya to Yoshi. (b) Interfering signal sent from Xia to Walt. (c) Interfering power as a function of the frequency difference.

the moment, ignore the impact of the noise. Both signals are sinusoids with amplitude 1:

$$\tilde{z}(t) = \cos(\omega_1 t) \qquad \tilde{x}(t) = \cos((\omega_1 + \Delta\omega)t + \phi) \tag{9.29}$$

where $\omega_1 = 2\pi f_1$ and $\Delta\omega = 2\pi\Delta f$. Here ϕ denotes the phase of Xia's signal at the start of the observed symbol interval in which Zoya sends to Yoshi. The phase ϕ is another element used in our discussion to indicate the lack of coordination between the links Zoya–Yoshi and Xia–Walt. Specifically, we assume that the phase ϕ of the interfering signal is chosen uniformly randomly within the interval $[0, 2\pi)$.

From the received signal Yoshi obtains:

$$y = \frac{2}{T_1} \int_0^{T_1} \tilde{y}(t) \cos(\omega_1 t) dt = 1 + x \tag{9.30}$$

where 1 is the useful symbol and x is the interference that the signal at frequency f_1 receives from the signal at frequency $f_1 + \Delta f$:

$$x = \frac{2}{T_1} \int_0^{T_1} \cos(\omega_1 t) \cos((\omega_1 + \Delta\omega)t + \phi) dt. \tag{9.31}$$

Clearly, the interfering signal x plays a role that is similar to the role of the noise, but the statistical characteristics are quite different. In our model, x draws its randomness from the random phase ϕ. Similar to the way we have treated noise, we look at the variance of the random variable $E[x^2]$, which measures the interfering power received by Yoshi. The expectation is taken with respect to the uniform random choice of ϕ. Differently from the Gaussian noise, the interfering power of x depends on the difference between the frequencies $\Delta f = f_2 - f_1$, as well as the choice of the symbol durations. More precisely, the average power of the interference depends on $\Delta f T_1$ and this dependency is plotted in Figure 9.5(b). It is seen that when $\Delta f T_1$ is an integer, then the interference power is zero; this is because in that case $f_2 = \frac{l}{T_1}$ and the two signals are orthogonal over the interval of length T_1.

Recall that the Gaussian noise is chosen such that it affects the signal in the "worst possible" way, without any correlation between different noise samples. In this sense, the behavior of the interference term is quite different. To illustrate this fact, let us denote by x_1 the interference term that affects the observed symbol from Zoya in Figure 9.5(a). Let us assume that, once that symbol is finished, Zoya sends another symbol of duration T_1 (not depicted in the figure), and let us denote the interference term that affects this subsequent symbol as x_2. In the considered example, the random phase ϕ of the interfering signal from Xia is selected at the start of the observed symbol. Hence, the same random choice in Xia's signal affects both symbols that Zoya sends to Yoshi. In other words, x_1 and x_2 are both dependent on the same value of ϕ and, considering that the values of T_1, T_2, f_1, and f_2 are fixed, we can conclude that the interference terms x_1 and x_2 are correlated. This is clearly different from the uncorrelated noise samples that affect the neighboring symbols. Hence, modeling the interference samples as independent Gaussian samples should be done with caution.

Returning to Figure 9.5(b), it can be seen that when two uncoordinated links use different frequencies, the impact of the interference decreases as the separation Δf between the frequencies of the two links increases. This decrease is not monotonous, as there are zeros and bumps, but the tendency is clear. When the frequencies become sufficiently separated, then the interference power becomes negligible compared to the noise power, such that the interference can be neglected. In this way the concept of frequency brings a qualitative novelty to the multi-user communication: *a sufficient separation in frequency makes the interference negligible and the links can operate simultaneously without synchronizing/coordinating the links.* This feature makes it possible to divide the frequency spectrum into different frequency channels/bands and allocate different frequency bands to different communication systems that are not coordinated. In order to get a full frequency characterization of an information carrying, random signal, there is a need to explain the concept of power spectrum.

9.5 Power Spectrum and Fourier Transform

The frequency representation based on Fourier series results in frequencies that are present in the signal (contain energy) and, consequently, frequencies that are absent from the signal (zero energy). Following this line of thought, Xia can cause interference to Zoya only if Xia puts energy at the frequency that Zoya uses for communication. On the other hand, we have seen from the example in Figure 9.5(b) that Xia can cause interference to Zoya even though they are not using the same frequency. In fact, the difference in the frequency Δf between the two interfering signals can be an arbitrary real number. Hence, there must be something contradictory about the frequency representation based on a Fourier series.

Let us look again into how we have arrived at the frequency representation. If a signal sent by Xia is limited to an interval of duration T_X, then we *can* choose the interval for the Fourier series to be T_X. Based on that, one may conclude that Xia's signal contains energy at the frequencies that are integer multiples of $\frac{1}{T_X}$. The key to the contradiction about frequencies present in the signal and interference caused to other signals lies in the choice of T_X. Recall that the we have defined the frequencies featured in a signal of duration T_X by assuming that the same signal is periodically repeated outside of the interval T_X. It is important to understand that, when the signal is not intrinsically periodic, then the choice of the interval length T_X is not unique. That is why it was stated above that the interval used in the Fourier series *can* be chosen to be T_X, but this is not the only choice.

Let us look closer in this issue and consider the signal of finite duration T depicted in Figure 9.6(a). One possible representation with Fourier series is to take $T_X = T$, from which it will follow that the signal contains energy at the frequencies $\frac{1}{T}$. Alternatively, we can choose $T_X = T_1 > T$, which will result in frequency representation that contains the frequencies $\frac{k}{T_1}$.

Hence, the contradiction vanishes by observing that the set of frequencies that are contained in a signal represented through Fourier series depend on the choice of T_X, which, for a finite-length signal, can be arbitrary. In the example in Figure 9.6(a) we only require that $T_X \geq T$. On the other hand, Figure 9.5(b) shows the statistical impact of a random interfering signal, seen as the distribution of the power across different frequencies. This naturally raises the question: can we find an objective statistical description of the way in which a certain signal interferes with random victim signals at different frequencies? Here by *objective* we mean that the description is not tailored to a specific victim signal and can be applied to arbitrary victim signals.

This kind of description is given by the *power spectrum*, also known as the *power spectral density* of the random, information carrying signal. In order to relate the notion of power spectrum to the example from the previous section, let us assume that Zoya transmits to Yoshi a periodic, infinitely long sinusoidal signal of frequency f. On the other hand, Xia sends to Walt a modulated signal created in the following way. Xia sends a sinusoidal signal of frequency $f + \Delta f$ and divides the time into slots of duration T_X, where each slot corresponds to the transmission of a single data symbol. At the beginning of the ith time slot, Xia modulates data by randomly choosing the phase of the sinusoidal signal according to the ith data symbol she wants to transmit. Then the average power of the disturbance that Yoshi experiences as a function of the frequency difference $f + \Delta f$ looks like Figure 9.5(b) and it represents the *power spectrum* of Xia's signal.

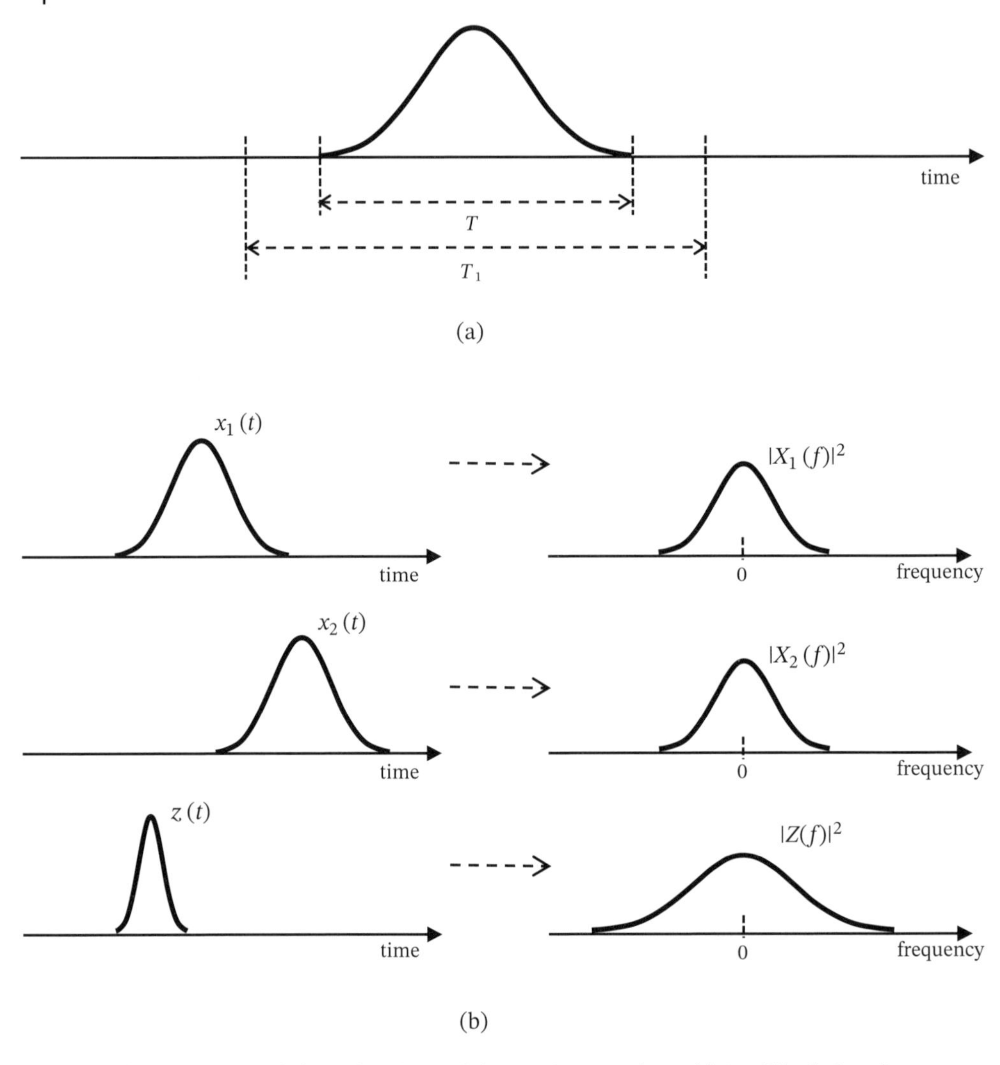

Figure 9.6 Signals with finite duration and the Fourier Transform. (a) Possible choice of two different intervals for Fourier series; in both cases the information about the useful signal is preserved. (b) Two signals $x_1(t)$ and $x_2(t)$ with identical energy spectrum, but they occur, and can cause interference at different time instances: here $z(t)$ gets interference from $x_1(t)$, but not from $x_2(t)$.

Although we have arrived at the notion of the power spectrum of a signal by treating the interference it creates to another sinusoidal signal, the power spectrum is an intrinsic property of a signal. Xia's signal is a specific type of an information carrying signal, represented by a cyclostationary statistical process, whose statistical properties vary cyclically in time. Any random signal that can be represented as a continuous function in time is also information carrying signal and one can calculate its power spectrum, provided that the random signal satisfies certain statistical properties, such as stationarity (see below). For the signals that usually occur in communication engineering, these properties are satisfied and one can safely use the power spectrum to describe how the power of a given signal is distributed across frequencies.

The precise understanding the notion of power spectrum requires introduction of the *Fourier transform*. Consider again the deterministic finite-length signal $x(t)$ of Xia in Figure 9.6 and assume that the interval T_X used in Fourier series increases its value. When T_X goes to infinity, then the interval for obtaining the Fourier coefficients approaches a correct description of the signal: it captures the fact that the signal is aperiodic and has value zero outside the interval of length T. However, the problem with the Fourier series is that infinite T_X will make all coefficients equal to zero due to the $\frac{1}{T_X}$ factor. The Fourier transform is given as:

$$X(f) = \int_{-\infty}^{\infty} x(t)e^{-j2\pi ft}dt \tag{9.32}$$

which is a (complex) continuous function of the frequency f rather than a discrete set of coefficients associated with frequencies in Fourier series. The *energy spectrum* of the deterministic signal $x(t)$ is given by $|X(f)|^2$ and the energy contained in $x(t)$ can be calculated using Parseval's theorem:

$$E_X = \int_{-\infty}^{\infty} \tilde{x}^2(t)dt = \int_{-\infty}^{\infty} |X(f)|^2 df \tag{9.33}$$

from which it is seen that the unit of $|X(f)|^2$ is joules per hertz.

Now let us assume that the signal $x(t)$ of Xia has the same finite duration as the signal in Figure 9.6(a), but the signal is random. If $x(t)$ is well behaved and has only finitely high values within that interval, then $X(f)$ exists, but it is also random and the same is valid for $|X(f)|^2$. We remind the reader that we have arrived to this point by trying to statistically characterize the interference that Xia causes to the signal that Yoshi receives from Zoya. Due to the properties of the Fourier transform, the energy spectrum $|X(f)|^2$ does not depend on the exact placement of the signal $x(t)$ on the time axis. Consider the example in Figure 9.6(b), where $x_1(t)$ and $x_2(t)$ are two possible signals that can be sent by Xia. The power spectra of $x_1(t)$ and $x_2(t)$, denoted by $|X_1(f)|^2$ and $|X_2(f)|^2$, are identical and they overlap with the power spectrum of Zoya's signal $z(t)$, denoted by $|Z(f)|^2$. However, we can see that $x_1(t)$ overlaps in time with $z(t)$ and therefore, if Xia sends $x_1(t)$, she causes interference at Yoshi's receiver. On the other hand, if Xia sends $x_2(t)$, Yoshi receives $z(t)$ free of interference. This phenomenon is not captured by the energy spectrum of the signals $x_1(t)$ and $x_2(t)$, as their energy spectra are identical.

The previous discussion leads us to conclude that, if we want to provide statistical characterization of the interference made by Xia, then we need to assume that Xia's signal has an infinite duration. In this case, regardless of where on the time axis Zoya's transmission takes place, it will experience statistically the same interference from Xia's signal. One technical requirement for this to be true is to require that Xia's random signal is stationary (in fact, wide-sense stationary). Now that we require that Xia's signal is of infinite duration, it cannot be guaranteed that the energy calculated in (9.33) is finite. On the other hand, the signal power still needs to be finite in order to put a realistic constraint on Xia's transmitter. Recall that, in defining the information-theoretic Gaussian communication channel, we have limited the power, but not the energy.

In order to arrive at the precise definition of the power spectrum, we take a cut of length T from Xia's infinite signal $x(t)$ and denote it by $x_T(t)$, such that $x_T(t) = x(t)$ for $-\frac{T}{2} \leq t \leq \frac{T}{2}$ and $x_T(t) = 0$ outside of that interval. Since $x_T(t)$ is of finite duration, its Fourier transform exists and is denoted by $X_T(f)$. The average power P of $x(t)$ can be found by averaging over an increasingly larger cut and we can establish the relation:

$$P = \lim_{T\to\infty} \frac{1}{T} \int_{-\frac{T}{2}}^{-\frac{T}{2}} x^2(t)\mathrm{d}t = \lim_{T\to\infty} \frac{1}{T} \int_{-\infty}^{\infty} x_T^2(t)\mathrm{d}t$$

$$\stackrel{(a)}{=} \lim_{T\to\infty} \frac{1}{T} \int_{-\infty}^{\infty} |X_T(f)|^2 \mathrm{d}f = \int_{-\infty}^{\infty} \lim_{T\to\infty} \frac{1}{T} |X_T(f)|^2 \mathrm{d}f = \int_{-\infty}^{\infty} S_X(f)\mathrm{d}f \tag{9.34}$$

where (a) follows from (9.33) and $S_X(f)$ is the power spectrum. The latter equation is valid for a deterministic signal, while the desired statistical characterization of the power spectrum can be given by taking the expected value as follows:

$$S_X(f) = \lim_{T\to\infty} \frac{1}{T} E[|X_T(f)|^2]. \tag{9.35}$$

Further analysis shows that $S_X(f)$ is a Fourier transform of the autocorrelation function of $x(t)$. This means, for example, if the values of $x(t)$ are weakly correlated with the nearby values $x(t+\tau)$, where τ is small, then $x(t)$ exhibits fast variations and therefore the power density $S_X(f)$ will not be zero for high values of the frequency f. When $x(t)$ is only correlated with itself, and not with any other $x(t+\tau)$, then $S_X(f)$ has equal power density across all frequencies and thus infinite power. This is the case of the Gaussian noise, which plays the role of the worst-case uncorrelated random disturbance.

An important property of the Fourier transform is that a signal of finite duration does not have a limited spectrum and there is no frequency above which the energy density of the signal is strictly zero. This would imply that two finite-duration signals that overlap in time will, highly likely, not be separable in frequency and will therefore cause interference to each other. The statistical power spectrum is calculated for a signal that is stationary. On the other hand, no finite duration signal can be considered stationary, since after some time, the finite duration signal will be identically equal to zero.

Assessing the interference requires that we observe the power spectrum and make assumptions about how the signals overlap in time. For example, take one symbol of duration T sent by Zoya and placing it randomly in time such that this symbol overlaps fully with a transmission from Xia. From the perspective of this symbol of Zoya, the signal of Xia satisfies the statistical properties of an interfering signal that has infinite duration and we can use the power spectrum of Xia to assess the power (variance) of the interference caused to Yoshi's receiver. When the power spectrum of Xia is far below the noise level at Yoshi's receiver, and therefore far below the level of Zoya's signal, then we can assume that the signals of Zoya and Xia are separated in frequency.

We have thus arrived at the answer to the question posed earlier in this chapter: *how do you separate in frequency the signals from two communication links that are not synchronized in time?* The separation in frequency is rather approximate: there will always be an interference between the signals, as a signal that is strictly bandlimited has an infinite duration.

9.6 Frequency Channels, Finally

In the first chapter we used the concept of a frequency channel and transmitters/receivers that are tuned to a given channel, but at this point we have a better picture about the physical meaning of a frequency channel. A waveform, modulated with data, is said to occupy a certain frequency band if the energy contained outside of that band, i.e. the values of its power spectrum outside of that band, are negligible. A frequency channel is band of contiguous frequencies, associated with a certain *filter mask*. The role of a filter mask is to provide bounds that limit the amount of energy radiated outside a given frequency band. In this way, the interference created towards the signals transmitted within the adjacent frequency bands is low or, at least, acceptable. Furthermore, the filter mask also limits the energy of the signal within the frequency band of interest, which is related to the limit on the maximal transmission power that can be used in a given frequency band. The analog signal that conforms to the filter mask can, for all practical purposes, be treated as a bandlimited signal.

9.6.1 Capacity of a Bandlimited Channel

Let W be the frequency band in which the communication signal should contain its energy, or at least most of it. The noise is white, having a flat spectral density of N_0 over the entire bandwidth W. Hence, the total noise power that affects the received signal is $P_\mathrm{N} = N_0 W$. Let T be a very long time interval in which we observe the communication signal. Then, assuming that the power of the received information carrying signal is P, what is the capacity of the channel in bits per seconds?

Since the signal is constrained to have most of its energy to the bandwidth of size W, the total number of real DoFs that are present in the signal observed through a duration T is $m = 2WT$. In deriving the capacity, we will assume that all m DoFs are statistically identical: the received power of the useful signal at each DoF is identical and each transmitted symbol at a given DoF is affected by an identical random process that represents the noise. The signal power per DoF can be calculated as:

$$P_\mathrm{DoF} = \frac{P}{2WT} \tag{9.36}$$

while the noise power per DoF can be calculated as

$$P_\mathrm{DoF,N} = \frac{P_\mathrm{N}}{2WT} = \frac{N_0 W}{2WT} = \frac{N_0}{2T}. \tag{9.37}$$

The m DoFs can be treated as m channel uses of the same AWGN channel in which the SNR is equal to:

$$\gamma = \frac{P_\mathrm{DoF}}{P_\mathrm{DoF,N}} = \frac{P}{N_0 W} \tag{9.38}$$

and the capacity per one real channel use (DoF) is:

$$C^\mathrm{DoF} = \frac{1}{2}\log_2(1+\gamma) = \frac{1}{2}\log_2\left(1 + \frac{P}{N_0 W}\right) \quad \text{(bit/DoF)}. \tag{9.39}$$

For fixed W, this capacity is achieved when T goes to infinity, such that n also goes to infinity and the capacity in bits per second is given as:

$$C_W = \frac{2WTC^\mathrm{DoF}}{T} = W\log_2\left(1 + \frac{P}{N_0 W}\right) \quad \text{(bps)} \tag{9.40}$$

which is the well known expression by Shannon for the capacity of bandlimited signals.

At the beginning of this chapter, we stated that it is not immediately clear why it is optimal to communicate with analog waveforms through a layered (separation) approach, which consists of modulating discrete samples and then using them to synthesize an analog waveform. The intuitive answer comes from the sampling theorem by Nyquist–Shannon–Kotelnikov, which states that a bandlimited signal can be completely described by the sampling process with a sufficiently high frequency. One concern comes from the fact that the communication signal is not strictly bandlimited, but it also contains some energy outside of its prescribed band. In fact, one may aim to squeeze more bits through the channel by modulating directly the analog waveforms while still satisfying the constraints of the filter mask. However, as rigorously proved by Slepian, the separation into sampling and synthetizing waveforms is optimal in sense of communication capacity, when the transmitted data packets are asymptotically long.

By normalizing the data rate with the bandwidth W, we get the *spectral efficiency*, denoted by ρ. The maximal spectral efficiency can be obtained by normalizing (9.40) with W:

$$C = \log_2(1+\gamma) = \log_2(1+\text{SNR}) \quad ((\text{bit/s})/\text{Hz}) \tag{9.41}$$

which corresponds to a capacity of a channel in which each channel use is a complex value. Hence, one should always choose the spectral efficiency to be $\rho \leq C$.

The energy spent per transmission of a single complex symbol can be calculated as

$$E_s = \frac{PT}{WT} = \frac{P}{W}. \tag{9.42}$$

We have divided by WT instead of $2WT$, as we want to find the energy per one complex DoF. On the other hand, the energy of the noise per one complex DoF is $\frac{N_0 WT}{WT} = N_0$, such that we get:

$$\text{SNR} = \frac{P}{N_0 W} = \frac{E_s}{N_0}. \tag{9.43}$$

For given spectral efficiency ρ, expressed in bits per symbol, one can calculate the normalized measure of the *signal-to-noise ratio per information bit*:

$$\frac{E_b}{N_0} = \frac{E_s}{\rho N_0} = \frac{\text{SNR}}{\rho}. \tag{9.44}$$

The normalized SNR is sometimes denoted by E_b/N_0. In order to get intuition about it, the inequality that needs to be satisfied by the spectral efficiency can be written as:

$$\rho \leq \log_2\left(1 + \rho\frac{E_b}{N_0}\right) \tag{9.45}$$

which means that the minimal E_b/N_0 that is required to attain a certain spectral efficiency ρ can be found from:

$$\frac{E_b}{N_0} \geq \frac{2^\rho - 1}{\rho} \tag{9.46}$$

which is termed the *Shannon limit* and is used as a reference to determine how efficient a given modulation/coding scheme is. In order to send reliably a single bit via each complex channel symbol, the minimal E_b/N_0 can be calculated to be 1 or 0 dB. An interesting case is the minimal required E_b/N_0 to support reliable communication with spectral efficiency that approaches zero $\rho \to 0$, which is $E_b/N_0 = -1.59$ dB. The interpretation of the case $\rho \to 0$

can be obtained by writing $\rho = \frac{C_W}{W}$ and letting the bandwidth of the signal W go to infinity. Hence, the minimal energy per bit is achieved by using a very wide bandwidth to transmit the data signal.

9.6.2 Capacity and OFDM Transmission

We have derived the capacity by considering bandwidth W and a duration T in which the total number of DoFs is $2WT$. Additionally, we have assumed that all DoFs are statistically identical. In Section 9.3.3 we introduced the parallel frequency channels in an OFDM transmissions, which is one possible way to use the DoFs offered by the bandwidth W. There are $m = WT$ subcarriers, each carrying 2 DoFs corresponding to the I and the Q components, respectively. The simplest analog waveform that can be used to modulate the information on a subcarrier is a rectangular pulse of duration T. This means that, at the start of a symbol, we select the I/Q components of each subcarrier according to the data that needs to be modulated and keep them constant until the next symbol, when a new selection is made.

Let us observe l consecutive symbols, each of duration T. The total number of channel uses (DoFs) during these symbols is $2lm = 2lWT$. If all subcarriers are statistically identical, such that the received power at each subcarrier is identical, then all DoFs are statistically identical and the capacity of an OFDM transmission is given by the capacity derived in the previous section. However, in frequency selective channels the received power across subcarriers differs, as explained in Section 9.4.1. In this case, the capacity of an OFDM transmission should be computed as a capacity of a water filling transmission over parallel, non-identical Gaussian channels. Note that, if all DoFs are statistically identical, then a single codeword is spread across all subcarriers and consumes all $2lm$ channel uses. On the other hand, if each subcarrier is modulated independently, then there are m different codewords, each codeword consuming l channel uses.

The power spectrum of an OFDM signal for which each subcarrier is modulated with rectangular pulses is, clearly, not confined to a bandwidth W, and part of its energy is leaked outside of that band. The shape of the power spectrum can be changed by changing the shape of the pulse in an OFDM transmission, while being careful not to destroy the orthogonality. Another way to shape the power spectrum is to depart from the OFDM paradigm and not consider a symbol-per-symbol transmission, but treat the set of $2WT$ DoFs as a general set of channel uses, not necessarily statistically identical, that can be used to create suitable codewords and send data.

9.6.3 Frequency for Multiple Access and Duplexing

One way to use the availability of different frequency channels is through FDMA, where each frequency channel is allocated to a different transmitter or different link. If the link Zoya–Yoshi uses a frequency channel that does not overlap with the frequency channel of Xia–Walt, then the interference between these two links is negligible, practically zero. The central frequency f_0 of a channel is called a *carrier*. This term reflects the traditional way in which a data-carrying signal is placed within a frequency channel. The signal is modulated in a lower band and then multiplied/mixed with a sinusoidal signal of frequency f_0 to be carried into the frequency channel centered at f_0. Finally, the signal is sent through a filter centered at f_0 in order to limit its bandwidth to the constraints set by the frequency channel. In the sequel we will refer to a frequency channel through its center frequency, for example "channel f_0".

The separation of the signals in frequency removes the need for spectrum sharing and/or time synchronization between two unrelated links, such as Zoya–Yoshi and Xia–Walt. There is still need for coordination to avoid interference, as one link should choose the channel f_0 and the other one the channel f_1. Furthermore, there is also a requirement on synchronization, as the devices need to be capable of generating the frequencies f_0 and f_1. Imperfectly tuned frequencies do not guarantee that the interference between the two signals will follow the levels that are prescribed by the filter mask. Yet, speaking in terms of interference avoidance, acceptable frequency synchronization is not as restrictive as time synchronization. Recall from the Fourier series representation that time synchronization requires that both links use identical symbol time, which is not required in FDMA.

From Chapter 1 we have assumed that *time division duplexing (TDD)* has been used: Zoya and Yoshi choose a single frequency channel and they take turns in using it as transmitters/receivers, respectively. For example, this has been used in a setting with multiple terminals connected to Basil and the use of suitable frames for downlink and uplink transmission. In the same chapter we also described the possibility of full duplexing, which is simultaneous transmission and reception over a *single* frequency channel. The availability of multiple frequency channels brings the opportunity to introduce *frequency division duplexing (FDD)*: Zoya and Yoshi agree upon two frequency channels, f_1 and f_2. Zoya uses the channel f_1 to transmit to Yoshi, such that Yoshi's receiver is tuned to f_1. Conversely, Yoshi's transmitter and Zoya's receiver are tuned to f_2. Hence, Zoya and Yoshi can simultaneously transmit to each other, without causing interference, or more precisely, *self-interference* from its own transmitted signal to the received signal.

Neither time division nor frequency division has a definitive advantage over the other. For example, one advantage of TDMA over FDMA is that the amount of time resource allocated to a user can be dynamically adjusted. In Chapter 1 we have described a TDMA system with multiple terminals and a base station, in which the latter had the flexibility to allocate resources dynamically from frame to frame. This is not the case in FDMA, since the frequency filters do not have the same flexibility and cannot be easily and dynamically adjusted to allocate a different bandwidth. On the other hand, the advantage of an FDMA system is that it can be used by multiple non-sychronized links.

In terms of duplexing, the cost in a TDD system is the turnaround time needed to switch from transmitting to receiving and vice versa. Analogously, the cost in an FDD system is the gap, unused bandwidth, which separates the band for transmission from the band for reception. Full duplexing is appealing as it seems to integrate the advantages of both TDD and FDD. However, while we treat TDD and FDD as interference free; the price paid in full duplexing is that the receiver needs to suppress the self-interference at the same frequency. Another important difference between TDD and FDD occurs when we take into account the spatial dimension and the fact that the propagation of wireless signals takes actual time, as discussed in the next chapter.

We conclude this section by noting the fundamental equivalence of TDD and FDD in terms of data rates. Let us first look at FDD, fix the time slot for a transmission to be T and let one frequency channel have a bandwidth W. The total bandwidth used in FDD is $2W$ and the total number of DoFs in the observed time-frequency chunk is $2WT$. For fair comparison with the TDD system, it has to operate with a bandwidth of $2W$, such that the total number of DoFs is again $2WT$. However, the switching between uplink and downlink happens after a time $\frac{T}{2}$, but that is irrelevant for the total data rate achieved in both directions. In this idealized comparison, we neglect the DoFs that are used due to the frequency gap in FDD and turnaround time in TDD.

9.7 Code Division and Spread Spectrum

9.7.1 Sharing Synchronized Resources with Orthogonal Codes

Time division and frequency division are examples of orthogonal multiplexing, where each transmission occupies exclusively a time slot or a frequency band and in that way interference in avoided. When the transmitters or the links are not synchronized in time, strict orthogonality in frequency is impossible, as there is always residual interference. However, this interference is negligible when the frequency bands are sufficiently separated.

We now address the question whether the two signals can occur simultaneously in time, occupying the same frequency bandwidth, but the respective receiver can still extract the desired signal with little or even no interference. In fact, we have already stated that this is possible if the two interfering links, Zoya–Yoshi and Xia–Walt are synchronized in time. Alternatively, in an uplink scenario all terminals are synchronized to Basil, a common base station. Let us treat the simpler case, in which the mobile devices MD_1, MD_2, MD_3 and MD_4 communicate with Basil in the uplink. Basil determines the symbol duration T_0 and the timing of the slot symbols, such that the four terminals are synchronized to it. Furthermore, Basil determines how to allocate the communication resources to the terminals. Since the symbols of all terminals are synchronized, they can orthogonally share the frequencies $\frac{k}{T_0}$, where k is an integer. Those frequencies appear through the Fourier series representation of the symbols. Let W_0 be the width of the frequency band that is available for communication and is shared by all terminals. Then the total number of orthogonal resources available for communication, each of them regarded as a DoF, within the duration of a single symbol is $2W_0T_0$.

To make the example more specific, we assume that $W_0 = \frac{4}{T_0}$, such that there are four available frequencies in the band W_0. One way in which Basil can decide to share the DoFs among the terminals is to use frequency division and allocate each frequency exclusively to one terminal. However, frequency division is *only one possibility* for orthogonally sharing the available $2W_0T_0 = 8$ DoFs among the four terminals. Recall that each frequency carries two DoFs, associated with cos(I-component) and sin(Q-component), respectively. Basil can thus allocate the I component of a given frequency to one terminal and the Q component to another. In fact, one can check that there are $\frac{8!}{2!2!2!2!}$ possible orthogonal allocations and each of them is structurally analogous to a time or a frequency division.

The main feature of the orthogonal allocations described so far is that each DoF is exclusively used by one user. We are now interested in a conceptually different type of division that is based on a code. Basil allocates a *code sequence* to each of the four terminals, each sequence consisting of 4 symbols. The allocation is given in Table 9.1. We will call these sequences *spreading sequences* or *spreading codes* as they will be used to spread the

Table 9.1 Orthogonal codes allocated to the terminals.

Terminal	Code			
MD_1	1	1	1	1
MD_2	1	−1	1	−1
MD_3	1	−1	−1	1
MD_4	1	1	−1	−1

same data symbol across multiple DoFs. The scalar product between the sequences of, for example, the mobile devices MD_1 and MD_3 is:

$$1 \cdot 1 + 1 \cdot (-1) + 1 \cdot (-1) + 1 \cdot 1 = 0. \tag{9.47}$$

The reader can check that the scalar product between two different sequences is always zero, while the scalar product of a code sequence with itself leads to the value 4.

The sequences are orthogonal in a sense that the scalar product of two different sequences is zero. Recall that the orthogonality of two frequencies on a given time interval enables the receiver to extract the data that is modulated onto one frequency without an interference from the data modulated onto the other frequency. Here the idea is to modulate the data onto a code instead of a frequency and use the orthogonality among the sequences to be able to extract the desired data without interference from the other data transmissions.

Let b_i denote the complex data symbol sent by the terminal MD_i, where $i = 1, 2, 3, 4$. In order to make the presentation of the main ideas easier, let us at first assume that the channel coefficient from each terminal to Basil is always equal to 1, regardless of which DoFs are used for transmission. For example, if all terminals send their symbol using the jth of the four available frequencies, then the complex signal that Basil receives is at this frequency is

$$y_j = b_1 + b_2 + b_3 + b_4 + n_j \tag{9.48}$$

where n_j is the complex noise at the jth frequency. We use a convenient notation for the codes from Table 9.1: let c_{ij} denote the jth element of the code for the terminal MD_i. Then MD_i transmits its symbol b_i at the frequency j by multiplying it by c_{ij}, such that the received signals at the four frequencies are:

$$\begin{aligned} y_1 &= b_1 + b_2 + b_3 + b_4 + n_1 \\ y_2 &= b_1 - b_2 - b_3 + b_4 + n_2 \\ y_3 &= b_1 + b_2 - b_3 - b_4 + n_3 \\ y_4 &= b_1 - b_2 + b_3 - b_4 + n_4. \end{aligned} \tag{9.49}$$

In order to decode the signal from MD_1, Basil uses the code of MD_i, multiplies y_j by c_{ij} and sums them up, which leads to:

$$r_i = \sum_{j=1}^{4} c_{ij} y_j = 4b_i + \sum_{j=1}^{4} c_{ij} n_j. \tag{9.50}$$

As already indicated above, the code orthogonality removes the interference among the users and Basil needs to make a decision on the transmitted symbol b_i based on r_i, affected only by the noise. Let P_N be the power of each of the noise samples n_j. Since $|c_{ij}|^2 = 1$ the total noise power is $4P_\mathrm{N}$. Similar to maximum ratio combining, discussed in Chapter 5, the desired signals are combined coherently, unlike the noise. If the average power of the transmitted symbol b_i is P, then the total power of the desired signal $4b_i$ is $16P$, such that the SNR at which the desired symbol is decoded is:

$$\gamma_i = \frac{16P}{4P_\mathrm{N}} = \frac{4P}{P_\mathrm{N}} = 4\gamma_T \tag{9.51}$$

where γ_T is an SNR calculated for each individual transmission of a complex symbol. In other words, the resulting SNR is identical as if only the terminal MD_i repeats its transmitted symbol b_i four times, without any transmission from the other users. This form of multiple access is called *code division multiple access (CDMA)* and it is different from TDMA and FDMA in that no DoF is allocated exclusively to a user.

In our example we have used a code to spread the data symbols across four frequencies, such that y_j is the received signal at the jth frequency. However, from the equation (9.49) it does not follow that y_j should be a DoF associated with a frequency: it can be any subset of two DoFs from the eight available ones. As an example, instead of spreading the symbols in frequency, as done above, we can spread the symbols across time. Let us look at the same setup in which the data symbol is transmitted during time T_0, using bandwidth $W_0 = \frac{4}{T_0}$, but a different way to use the eight available DoFs. Each terminal transmits four times, the duration of each transmission is $\frac{T_0}{4}$ and this will be referred to as the *chip duration*, since the symbol duration remains as T_0. Now y_j can be interpreted as the jth received chip. Note that the noise power per DoF must be invariant with respect to the way the DoFs are used, i.e. frequency or time. Hence, the noise n_j retains the same statistical characterization as in (9.49).

Here we introduce some standard terminology. The variant in which spreading is done with chips is called the *direct sequence spread spectrum*. The reason it is called a spread spectrum is that, considering that the duration of the symbol is T_0, a sufficient bandwidth to send one complex symbol is $\frac{1}{T_0}$. However, in order to repeat the same symbol four times during the same symbol interval, the bandwidth should be spread to $\frac{4}{T_0}$. The spreading factor G, in this case $G = 4$, is also called a *processing gain*, since the way the received signal is processed, see (9.51) is equivalent to maximum ratio combining in which the number of received instances of the symbol is equal to G. For simplicity, we use spread spectrum to denote any form of spreading symbols across multiple DoFs, not only spreading in time.

9.7.2 Why Go Through the Trouble of Spreading?

We could show that what spreading can offer in multiplexing four terminals is basically the same as if each user has been exclusively allocated orthogonal resources. Indeed, the total power used per symbol by each terminal is $4P$ such that the same SNR can be achieved when the symbol is transmitted only once, over exclusively allocated DoFs, with a power of $2\sqrt{P}$. Why then go through the trouble of spreading when the same thing can be achieved in a simpler way?

One reason why the advantage cannot be appreciated is that we have not taken into account the constraints that come from the hardware side. Namely, transmitting at power

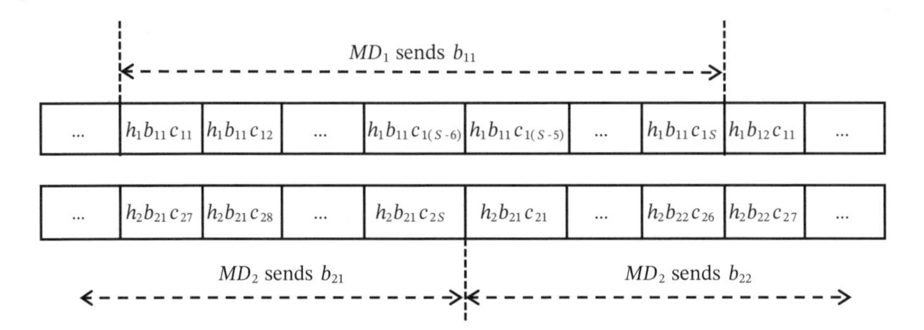

Figure 9.7 Illustration of chip synchronous, but not symbol synchronous, transmissions of two mobile device MD_1 and MD_2.

P over l channel uses is better from the perspective of a power amplifier compared to transmitting over a single channel use with power lP. Another reason why we cannot see any advantage in using a spread spectrum is that we have made the assumption that all the terminals are perfectly synchronized to Basil and all the channel coefficients are equal to 1. We now remove these assumptions and investigate the consequences.

Let the number of chips in the spreading sequence be S, but now we select S to be much larger than one. We consider spreading in time, such that each symbol spreads through S complex channel uses. The channel coefficient of the mobile device MD_i to Basil is h_i and does not change across all channel uses. In the first step we assume a light form of asynchronism (this will be briefly revised afterwards): the terminals are assumed to be chip synchronous, but not necessarily symbol synchronous. This means, for example, that the kth chip of a symbol sent by MD_1 occurs at the same time as the vth chip of a symbol sent by MD_2, but k and v are not necessarily equal. We use b_{ij} to denote the jth symbol of the ith terminal. The notation can easily get complicated, such that we use Figure 9.7 to obtain some insight.

Assume at first that Basil knows the chip at which a new symbol of MD_1 starts. He also knows the (different) chip at which a new symbol for MD_2 starts. For each transmitted chip, Basil receives a complex value that is a sum of what is sent by MD_1 and MD_2, with the appropriate channel coefficients, plus noise (not depicted in Figure 9.7). Thus, the received values that correspond to the chips associated with MD_1 transmission of b_{11} are:

$$\begin{aligned} r_{11} &= h_1 b_{11} c_{11} + h_2 b_{21} c_{27} + n_{11} \\ r_{12} &= h_1 b_{11} c_{12} + h_2 b_{21} c_{28} + n_{12} \\ &\cdots \\ r_{1C} &= h_1 b_{11} c_{1S} + h_2 b_{22} c_{26} + n_{1S}. \end{aligned} \tag{9.52}$$

In order to decode the transmission from MD_1, Basil correlates the received chips with the spreading code of MD_1:

$$\begin{aligned} r_1 &= \sum_{i=1}^{S} r_{1i} c_{1i} \\ &= S h_1 b_{11} + \left[h_2 b_{21} \sum_{i=1}^{S-6} c_{1i} c_{2(i+6)} + h_2 b_{22} \sum_{i=S-5}^{S} c_{1i} c_{2(i-S+6)} \right] + \sum_{i=1}^{S} c_{1i} n_{1i}. \end{aligned} \tag{9.53}$$

This operation is sometimes referred to as *despreading* of the signal, as it effectively collects the contributions spread over multiple DoFs into a single one. The second member of the sum in (9.53), put in brackets, is the interference experienced from MD_2 and the last member is the noise collected from all chips. The asynchronous shift between MD_1 and MD_2 for the example is arbitrarily chosen to be 5. Note that it is not possible to choose the spreading sequences such that orthogonality is preserved and interference vanishes for every possible asynchronous shift between MD_1 and MD_2.

On the other hand, assuming S is large, let us assign to MD_1 and MD_2 random spreading sequences. Before the communication starts, each chip is generated by flipping a fair coin and selecting 1 or -1, according to the flipping outcome. As such, the product of two chip values $c_{1i}c_{2j}$ is a binary random variable that gets value 1 or -1 with equal probability. When S is large and the central limit theorem is put to work, the sum in the brackets from (9.53) will converge towards a Gaussian distribution, such that the interference becomes equivalent to an additional Gaussian noise. On the other hand, since it is always $c_{1i}^2 = 1$, the contributions of the desired signal from all chips are coherently combined, leading to the processing gain. The SINR at which the signal is decoded is:

$$\text{SINR} = \frac{S^2 P}{SP_I + SP_\text{N}} = \frac{SP}{P_I + P_\text{N}} \tag{9.54}$$

where P is the received power for the desired signal from a single chip, P_N is the power of the noise affecting each chip and and P_I is the power of the interference contributed through each chip. The value of P_I is not zero and it depends on the *cross-correlation* of the spreading sequences. The cross-correlation can be controlled through the selection of a suitable set of spreading sequences. Suitable choices include *pseudorandom* or *pseudonoise (PN)* sequences, and Gold and Kasami sequences. The price for getting better cross-correlation properties is the decrease in the size of the set of available spreading sequences, which practically means decreasing the number of terminals that can stay connected to Basil.

We now remove the assumption that the transmissions of different users are chip-synchronous at Basil's receiver. Note that treating the chips as complex values in (9.52) assumes that a sampling process has already taken place and it is perfectly aligned to the transmissions of both users. In practice, the waveforms sent by different users are misaligned in arbitrary ways and the receiver needs to find the timing information about each of the transmitted spread spectrum signals. This is *timing acquisition*, which can be carried out by letting the terminals transmit *training sequences*. For these sequences, the data symbol is known a priori by the receiver and the received symbols are used to detect the start of the spreading sequences and estimate the channel. For the example from (9.52), MD_1 can obtain a training sequence by setting the symbol value to $b_1 = 1$ and using its spreading sequence, where the latter is known a priori by Basil. Then Basil applies synchronization methods to find out the timing of the chip transmission for MD_1. However, in order to be able to align to the first chip of the sequence and not to another one, the spreading sequence should posses good *autocorrelation* properties and be decorrelated with the shifted versions of itself. Finally, the interference from the MD_2 is again determined by the cross-correlation, but also by the misalignment of the chip timing for the two terminals.

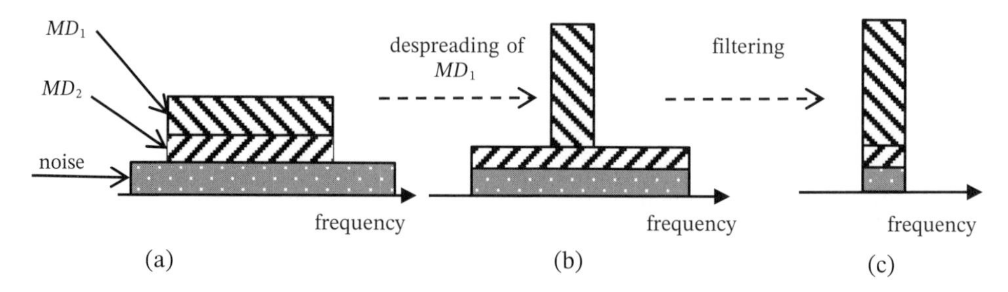

Figure 9.8 An illustration how CDMA works and why the effect of a spread spectrum signal is similar to that of a white noise. (a) The combination of the spread signals of MD_1 and MD_2 and the noise. It should be noticed that noise has a bandwidth that is wider than the bandwidth of the spread spectrum signals. (b) Despreading of the received signal by using the code sequence of MD_1. (c) Filtering out the narrowband, information carrying signal of MD_1.

9.7.3 Mimicking the Noise and Covert Communication

Figure 9.8 shows how CDMA works. The two wideband (spread) signals, see are added up, as in Figure 9.8(a), over the multiple access channel. The receiver uses the code of MD_1 to "unlock" or despread the signal from the mix of multiple signals. After despreading with the code sequence of MD_1, see Figure 9.8(b), the receiver filters out the frequencies that are not necessary to decoded the narrowband information-carrying signal from MD_1, see Figure 9.8(c). In this way, the decoding of the data sent by MD_1 is disturbed by interference power from MD_2 that is only a fraction of the total power that MD_2 contributes to the received signal. Note that here we have used the term "narrowband" information carrying signal to indicate that the bandwidth necessary to carry the signal from MD_1 before being spread is (much) lower than the bandwidth occupied after spreading.

The same figure brings more important insight. Before the spread spectrum signal is correlated and despread with the correct spreading sequence and correct timing, it appears as a signal that is similar to the noise. Indeed, the data is inherently random and the unknown spreading sequences can also be treated as random. Adding to this the fact that the chips are changing S times faster compared to the changes in the data symbol, then we arrive at a characterization similar to the one of the Gaussian noise, with wild and unpredictable changes. This feature has been noticed already from the early days of spread spectrum and it is useful to carry out *covert communication*. To elaborate, since a spread spectrum signal is mimicking the noise, a potential eavesdropper has a hard time determining that any communication is going on. In other words, spread spectrum signals can be used to achieve low *probability of interception*. Furthermore, if the eavesdropper does not know the spreading sequence of the signal he wishes to intercept, then even if he detects that there is an ongoing communication, he cannot decode it. Thus, the communication can, in principle, be kept secret and the secrecy level is proportional to the length S of the spreading sequence.

Speaking about secrecy, there is another form of spread spectrum, namely *frequency hopping*. This was originally invented by the actress Hedy Lamarr to enable secret radio communication. The main idea is to have a transmission that does not use the same frequency channel, but changes the channel at predefined time intervals. The change of the frequency channel is done according to a code that is known by the transmitter and the receiver, but not the eavesdropper. This technique can be treated as a spread spectrum, since the total spectrum that contains all the frequency channels that are hopped over is much larger than the bandwidth of the single frequency channel. Recall that, in a direct sequence spread spectrum, the information carrying signal is spread simultaneously over many DoFs. On the other hand, in frequency hopping only part of the DoFs is used at a given time. Those DoFs correspond to a frequency channel whose bandwidth is lower than the total bandwidth used for communication.

Depending on the time that the signal spends at a given frequency channel before changing to another, frequency hopping can be slow or fast. In slow frequency hopping, the frequency is changed only after a symbol, a group of symbols or a full packet is transmitted. On the other hand, in fast frequency hopping, the frequency channel may be changed multiple times even within a symbol duration.

Besides frequency hopping, there is also the concept of *time hopping*, associated with *ultrawideband (UWB)* transmissions, also called *impulse radio*. Here data is sent using very short pulses, which implies very large bandwidth. An interesting feature of UWB is that there is no carrier that is modulated, the bandwidth is determined by the pulse bandwidth itself. UWB uses time hopping in order to support coexistence/mitigate interference among multiple links.

9.7.4 Relation to Random Access

We have presented spread spectrum/CDMA as a method for non-orthogonal multiple access that enables mitigation of the interference among multiple transmissions. One can argue that random access protocols can also be seen as a class of communication mechanisms that are based on non-orthogonal access. Indeed, in the classical collision model, a random transmission choice may lead to collision or interference and, after randomly timed re-attempts, it eventually leads to a successful, non-interfered transmission. However, random access operates at a *packet level* and its objective is to arrive to a state of non-interfered transmission. On the other hand, CDMA *embraces* the interference and suppresses it through chip/symbol level processing.

A step that bridges the differences between the concept of a random access protocol and CDMA can be taken by using a model for random access that is based on a capture and successive interference cancellation (SIC). Upon receiving the set of collided/interfered signals, Basil applies correlation with all possible spreading sequences and finds the sequence, say the one of MD_1, that leads to the highest received power. Then Basil attempts to decode the signal of MD_1, treating all the remaining interference as a noise. If the decoding is successful, then a capture occurs, the signal of MD_1 can be canceled and Basil can continue to look for the spreading sequence that results in the next strongest signal. In other words, CDMA enables capture and intra-collision interference cancellation.

Random access can also be related to randomized frequency hopping. Namely, instead of choosing the randomized retransmissions only in the time dimension, one can also use the frequency dimension and, upon retransmission, also choose a different frequency channel. However, the receiver needs to know which frequency channel it needs to be tuned to in order to receive this transmission. Hence, if the transmitter needs to have a full freedom in selecting the frequency channel for retransmission, as it has in the time dimension, then the receiver needs to monitor all the frequency channels simultaneously, which brings additional costs in terms of receiver architecture and energy expenditure.

9.8 Chapter Summary

The concepts of frequency, Fourier analysis, and representation of signals are usually the prerequisites that all students need to have before going to the first course in communication engineering. Here the objective has been to provide a different perspective to those concepts, after being primed by the abstract communication models in the previous chapters. This chapter has touched upon very elementary questions, such as the following paradoxical observation: all signals encountered in practice are bandlimited, but then the Fourier transform implies that they have to have an infinite duration. It has been shown under which statistical conditions the power spectrum is defined and therefore the actual band where a certain signal resides. It has also been shown that frequency separation among different communication systems is the way to ensure absence of mutual interference without synchronization and continuous coordination among the systems. Finally, we have introduced the idea of spread spectrum and CDMA and related it to the previous discussions on random access protocols.

9.9 Further Reading

Two classical papers that treat the fundamental questions of frequency analysis and bandwidth are Gabor [1946] and Slepian [1976]. Interestingly, as pointed out in Slepian [1976], the road to the rigorous proof of the result on the capacity of channels with analog waveform was not easy and somehow missed by Shannon in his original paper Shannon [1948]. Two classical books that treat these issues in depth are Wozencraft and Jacobs [1965] and Gallager [1968]. The reader can find further discussion on multiple access (TDMA, FDMA, CDMA), OFDM and spread spectra in standard textbooks on wireless communication, such as Goldsmith [2005] and Molisch [2012].

9.10 Problems and Reflections

1. *System design for FDMA.* In Chapter 1 we presented several frames for single-channel wireless communication systems based on TDMA. Propose a similar frame based design for a system with multiple frequency channels. Focus on a downlink transmission and take into account the fact that signaling information should be received by all users. Consider the possibility for flexible allocation of resources, e.g. multiple frequency channels to a user.
2. *Co-channel interference.* Figure 9.5(c) shows an example of inter-carrier interference that decreases as the carrier frequencies become more separated. This could be used as a basis for modeling interference among different frequency channel. In that model, the communication performance of one channel is not independent of whether another frequency channel has an active transmission or not.

(a) Build a communication model that takes into account interference from other frequency channels. This is often called adjacent channel interference, but strictly speaking, the model should also take into account interference from non-adjacent channels.
(b) If these frequency channels are used as parallel channels from Zoya to Yoshi and Zoya has a limited power, investigate how the power allocation changes compared to the water filling solution.

3. *Wireless slicing with multiple frequencies.* In Chapter 4 we introduced slicing of wireless resources and illustrated it for two services, broadband and reliable low latency, respectively. The study was done in a single channel and two transmitters. Propose a similar system design, but now with more than two transmitters and with multiple available frequency channels.
4. *Frequency hopping and collisions.* Consider two wireless links that are in proximity, one from Zoya to Yoshi and the second one from Xia to Walt. Let W be the total bandwidth available for communication. The link Zoya–Yoshi divides the bandwidth W into F_1 channels, such that the bandwidth of each channel is $\frac{W}{F_1}$. The link Xia–Walt divides the bandwidth W into F_2 channels, such that the bandwidth of each channel is $\frac{W}{F_2}$. The transmissions Zoya–Yoshi are using are slots of size T_1, while the transmissions Xia–Walt are using are slots of size T_2. Both links are based on slow frequency hopping: for example, after Zoya transmits a packet at a channel $f_{1,1}$, she selects the next channel $f_{1,2}$ uniformly randomly among the F_1 channels and, during the next slot of duration T_1 she transmits a packet at the frequency $f_{1,2}$. Xia operates in a similar way, using F_2 channels. Assume a collision model, such that if the transmissions of Zoya and Xia overlap even only partially in time *and* frequency, then a collision occurs and neither Yoshi nor Walt will receive the packet.
   (a) Find the probability of collision when $F_1 = F_2$, $T_1 = T_2$ and the networks are synchronized in time, such that the slot for the link Zoya–Yoshi starts at the same time with a slot for the link Xia–Walt.
   (b) Find the probability of collision when $F_1 = F_2$, $T_1 = T_2$ and the networks are NOT synchronized in time.
   (c) Find the probability of collision when $F_1 > F_2$, $T_1 < T_2$.
5. *Spread spectrum and random access.* The base station Basil receives over a frequency band of size W from a set of K devices. The devices transmit to Basil through a random access protocol. Devise and analyze a random access protocol that is combined with spread spectrum for the following cases:
   (a) The spread spectrum is based on frequency hopping. Study the protocol both for the collision model (as in the previous problem) and for a model with capture.
   (b) The spread spectrum is based on direct sequence spread spectrum.

In a normal conversational setup there are no echoes and Walt can continuously speak out the whole sentence.

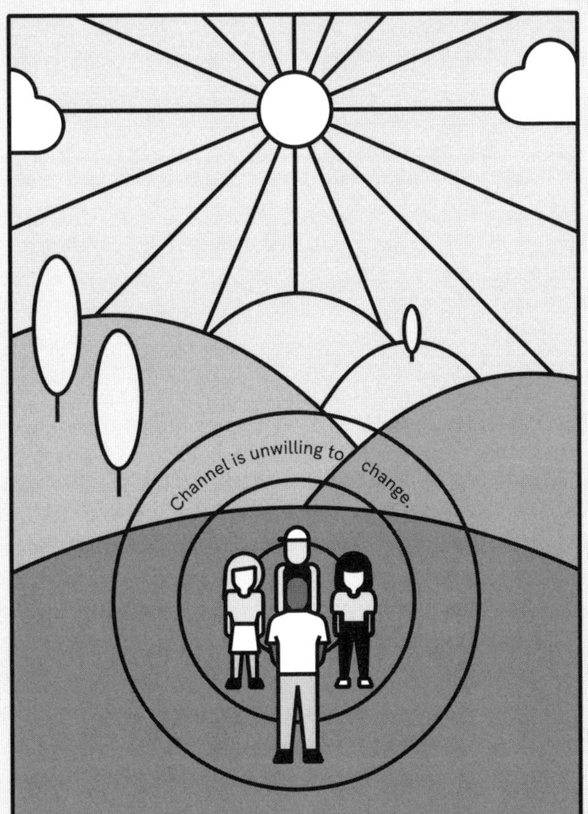

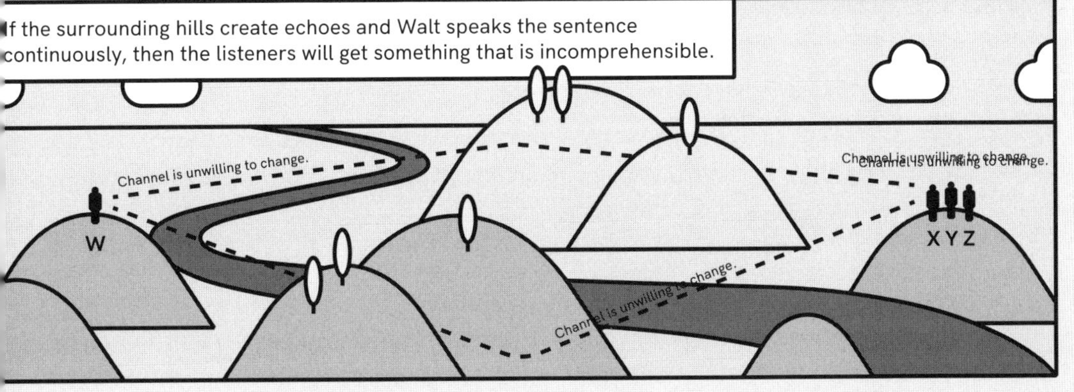

If the surrounding hills create echoes and Walt speaks the sentence continuously, then the listeners will get something that is incomprehensible.
Channel is unwilling to change.
W
XYZ
Channel is unwilling to change.

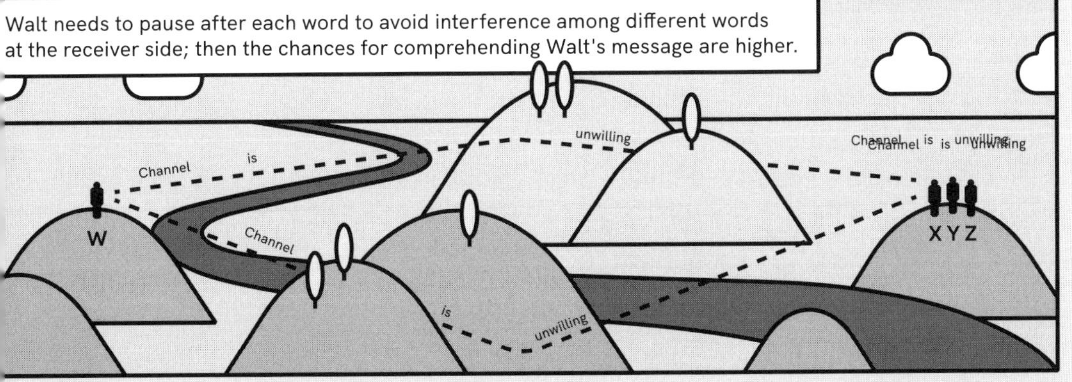

Walt needs to pause after each word to avoid interference among different words at the receiver side; then the chances for comprehending Walt's message are higher.

Story by Petar Popovski / Art by Peter Gregson

10

Space in Wireless Communications

After introducing the properties of communication signals in time, here we place the models for wireless communication into space. In this way, after bringing the temporal dimension in the previous chapter, we finalize the placement of the abstract models into the physical world through the spatial properties of wireless communications. On the one hand, this creates the basis to select realistic parameters and constraints in the models. On the other hand, understanding the underlying physics of wireless communications opens up the possibility to enrich the models; an example of this is the use of directed beams instead of omnidirectional antennas.

Consider the signal z transmitted by Zoya and the signal y received by Yoshi according to the model

$$y = hz + n. \tag{10.1}$$

Let us adopt the realistic assumption that the power of the noise n depends only on the receiver type, but not on its spatial position. The transmitted signal z depends only on the information content that Zoya sends and is thus clearly independent of the spatial positions. Hence, the only variable in (10.1) that is dependent on the spatial placement of Zoya and Yoshi is the channel coefficient h. If the distance between Zoya and Yoshi increases, then, in analogy to speech communication, we expect the amplitude of h to decrease or, in other words, the SNR of the signal received by Yoshi to decrease. However, h is not only dependent on the distance. There could be a situation in which Zoya and Yoshi are positioned close to each other, but there is an obstacle between them that additionally weakens the signal before it arrives at Yoshi. The occurrence of obstacles or other physical phenomena that affect the wireless signal can be incorporated in the model for wireless communication through a suitable modeling of h as a random variable.

This chapter is dedicated to the physical/spatial factors that determine the values of the channel coefficients and the SNR of the received signals. In an actual physical setting, the wireless communication signals need to *propagate* as waves from the sender to the receiver. We will think of these waves dominantly as electromagnetic/radio waves, but similar considerations are in place for acoustic or ultrasound waves. In order to provide a statistical evaluation of the certain propagation setting, we need to create statistical models that are suitable for representing electromagnetic propagation. The statistics of the received signal depends on the geometrical setting, and also other factors, such as random obstacles, reflectors, or the type/quality of the antenna used by the wireless devices.

Wireless Connectivity: An Intuitive and Fundamental Guide, First Edition. Petar Popovski.

10.1 Communication Range and Coverage Area

Assume that the base station Basil occupies a point in space and radiates electromagnetic energy in a *free space*, such that the radio waves propagate in an uninterrupted manner, without reflections, absorption, or scattering. This means that, if we take a sphere of an arbitrary radius d, then different points on that sphere receive the same amount of power from Basil. This is illustrated in Figure 10.1. Let us assume that Zoya's receiver has an antenna that collects electromagnetic energy from an area of size A. To simplify, one can think of the antenna area[1] as being part of a plane. Indeed, if Zoya is at a distance d from Basil and the area of the aforementioned sphere $4\pi d^2$ is much larger than A, then the plane of size A can be approximately seen to lie on the sphere surface. Furthermore, all points at that area are approximately at a distance d, such that we can consistently say that Zoya receives her signal at a distance d from Basil.

Taking this approximation further, we can state that the total power received by Zoya is a fraction of the power P_B radiated by Basil and this fraction is proportional to the fraction that A occupies from the total area of the sphere with radius d. With this representation, the power P_Z that Zoya receives is:

$$P_Z = P_B \frac{A}{4\pi d^2}. \tag{10.2}$$

Going back to the model (10.1), the power of the desired received signal and therefore the received SNR is proportional to $|h|^2$. From (10.2) it follows that, in an idealized free space propagation, the signal received by Zoya decreases with the square of the distance d from

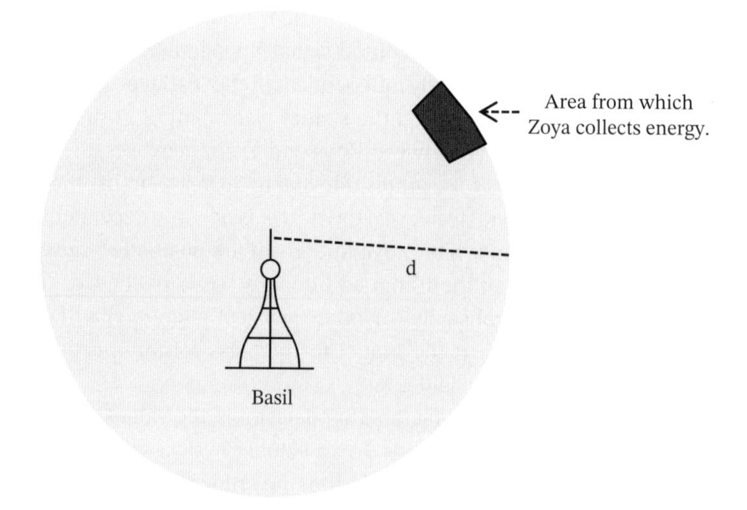

Figure 10.1 Electromagnetic radiation of the base station Basil at a distance d. If we ignore the base station body and treat Basil's antenna as a point in space, then the received energy is uniform across the sphere of radius d. Zoya collects energy from an area of size A that lies on the surface of the sphere.

1 For a large uniform aperture, the antenna area approaches its geometrical area.

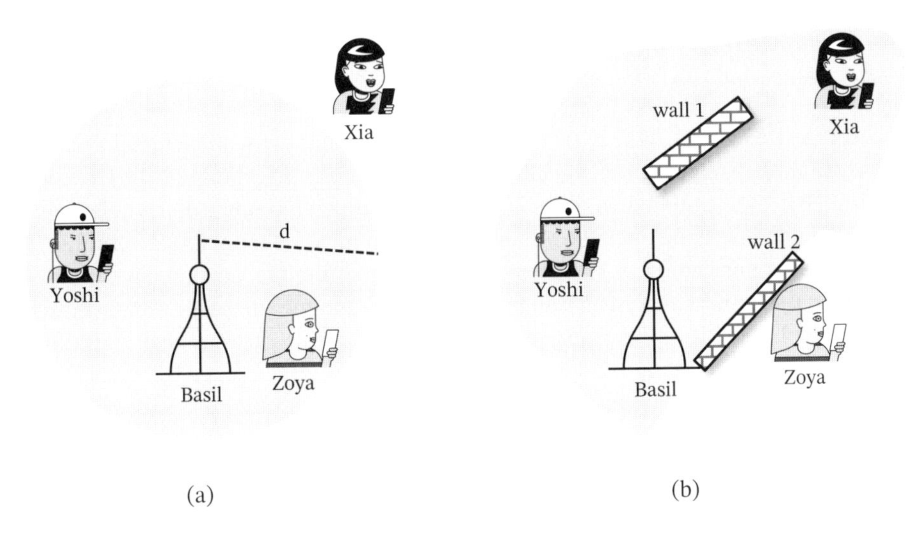

Figure 10.2 Definition of a coverage area, represented by gray shading. (a) Assuming free space propagation, the coverage area is circular, as the communication range is consistently defined by the distance. (b) If there are obstacles, reflectors, and other elements that affect the propagation, the communication range is no longer uniquely related to the distance and the coverage area gets an irregular shape.

Basil. Since the noise that affects Zoya's reception does not change with the spatial position, the SNR that Zoya experiences also decreases with the square of d.

With this, it is possible to define a *communication range* as the largest distance at which the SNR is larger than or equal to a threshold, denoted by SNR_0. Figure 10.2(a) depicts the communication range of Basil assuming free space propagation. However, when Basil and the three devices are placed on the ground, which is the usual model for a wireless communication system, the free space assumption is only valid if the ground can be treated as an ideal absorber, not reflecting any radio wave. With this assumption, the above definition of a communication range is meaningful, as it is uniquely associated with the distance between the transmitter and the receiver.

The spatial positions at which the received signal has an SNR that is larger than the threshold SNR_0 define the *coverage area*. In Figure 10.2(a), the coverage area is circular and it matches perfectly the simplest communication model used to describe access algorithms in Chapter 1. Zoya and Yoshi are within the coverage area of Basil, as their distance to Basil is less than the communication range d, while Xia is not in the coverage area.

Things start to look very different when the free space assumption is removed, as depicted in Figure 10.2(b). Between Basil and Zoya there is an obstacle, wall 2, which significantly decreases the SNR of the signal received by Zoya, such that she is not in the coverage area of Basil. On the other hand, the existence of two walls, 1 and 2, is beneficial for the propagation of the signals to Xia, as they are reflecting electromagnetic energy towards her, thus increasing the SNR of her received signal. Due to the contributions of the reflected waves, Xia is now in the coverage area of Basil.

The most important message from Figure 10.2(b) is that when the free space propagation assumption is removed the coverage area is no longer uniquely associated with the communication range. In other words, if Zoya and Yoshi are placed at the same distance from the transmitter, it can happen that Yoshi is in the coverage area, while Zoya is not, or vice versa.

The coverage area in Figure 10.2(b) is just one instance of a setup with factors that affect the propagation, such as obstacles, reflectors and user positions. The nature of the coverage area should be kept in mind when discussing the communication range of a wireless system. Indeed, in practice, one often talks about the communication range as a deterministic quantity, which is misleading. A proper characterization of the propagation environment should rely on a statistical analysis. To do that, we need to assume stochastic processes that determine the placement of the propagation factors, such as obstacles, as well the placement of the users. Having a model like that and given the distance between the transmitter and the receiver, one can treat the coverage as a random variable and thus speak about the probability that the receiver is in the coverage area of the transmitter.

Going back to our communication model, the stochastic propagation environment will result in a channel coefficient h that is a random variable. In Chapter 6 we discussed how to design communication schemes under the assumption that h is random and there is certain level of knowledge about h, called channel state information (CSI), at the transmitter or the receiver. The propagation models and the randomized mobility of users determine the statistics of h. This can provide hints on what are the realistic assumptions about the knowledge of CSI in a given communication setup. For example, if Zoya is in a fast-moving vehicle, then this will give rise to a fast dynamics of h, which would make it infeasible to obtain CSI at the transmitter, since this CSI would likely change at the time of transmission. Throughout this chapter we will touch upon different stochastic propagation factors that affect the received signal.

10.2 The Myth about Frequencies that Propagate Badly in Free Space

A radio wave is yet another electromagnetic wave, in addition to the visible light or the x-rays, whose frequencies belong to a specific part of the electromagnetic spectrum. Being an electromagnetic wave, it propagates at the speed of light c, such that the relation between the wavelength λ and the frequency f of a signal is:

$$\lambda = \frac{c}{f}. \tag{10.3}$$

In the previous chapter, we explained why one needs to be cautious when defining a frequency of an information carrying signal that has a finite duration and is not periodic. In the context of electromagnetic waves, *frequency* refers to the *carrier frequency*. The bandwidth of the information carrying signal, used to modulate the carrier frequency, is usually much smaller compared to the carrier frequency. Therefore, when talking about a frequency, we implicitly use the approximation that the frequencies that are sufficiently near to the carrier frequency have identical properties in terms of electromagnetic propagation.

The antenna of a transmitting device converts electrical current into electromagnetic waves. Conversely, the antenna of a receiving device converts electromagnetic waves into current. Wireless radio systems commonly assume that the receiver Yoshi is in the *far field* of the transmitter Zoya, which occurs when the distance d between Zoya and Yoshi is larger than several wavelengths λ. Only when the far field assumption is valid can we argue that the received power in free space propagation decreases with the square of the distance. On the other hand, when the receiver is in the *near field*, very close to the transmitter, then a different electromagnetic interaction between the transmitter and the receiver occurs, such as magnetic induction, resulting in propagation characteristics that are different from the far field. Since this book covers scenarios that are commonly associated with the far field, the relation between the received power P_r and the transmitted power P_t is given by the Friis formula:

$$P_r = P_t \frac{G_r G_t}{(4\pi d)^2} \lambda^2 = P_t \frac{G_r G_t}{(4\pi d)^2} \frac{c^2}{f^2}. \tag{10.4}$$

Here G_t is the gain of the transmitting antenna and is a quantity that indicates the antenna's efficiency in converting electric current in electromagnetic radiation. Analogously, G_r is the gain of the receiving antenna.

One important quantity that is defined to measure the impact of the distance on the received signal is the *path loss*:

$$L = 10 \log_{10} \left(\frac{(4\pi d f)^2}{c^2} \right) \quad (dB). \tag{10.5}$$

From the way it is defined, it follows that the path loss increases with the distance *and* the frequency of the signal. This, in turn, implies that the coverage area, defined through the minimal received power, decreases as the carrier frequency of the communication system increases. The conclusion would be that we need to keep frequency of the communication system as low as possible in order to have high SNR and thus an efficient communication over large distances. This is, in fact, the quick reasoning that is often used as an argument to say that the lower frequencies are more precious than the higher ones and that is why there is a constant struggle in radio spectrum regulation to get permission to use those lower frequencies.

But, is this really true? Putting the focus only on the path loss means that the impact of the antenna gains G_r and G_t is ignored, or a least they are treated as constants that are not dependent on the frequency. While it is true that the lower frequencies have more favorable properties under *practical* propagation circumstances and reach further than the higher frequencies, this phenomenon cannot be explained by looking only at the free space path loss. In fact, one needs to take into account the other propagation effects, such as diffraction and scattering; see Section 10.4 for further discussion. Furthermore, staying with the free space propagation only, it should be noted that G_r and G_t are affected by a number of factors. The relation between the antenna gain G and the *effective area* of the antenna, also called *antenna aperture*, is given by:

$$G = \frac{4\pi A}{\lambda^2}. \tag{10.6}$$

Replacing G_r and G_t in the Friis formula (10.4) with the respective values of the effective areas A_r and A_t leads to another form of the Friis formula:

$$P_r = P_t \frac{A_r A_t}{d^2} \frac{1}{\lambda^2} = P_t \frac{A_r A_t}{d^2} \frac{f^2}{c^2} \tag{10.7}$$

which, in case that the antenna apertures are constant with the frequency, would lead to an opposite conclusion: namely, that the propagation/coverage improves as the frequency increases!

However, the antennas/gains have complex dependency on the frequency, such that both forms of the Friis formula (10.4) and (10.7) should be used with caution. For example, if the gain G_t of the transmit antenna and the effective area A_r of the receiving antenna are kept fixed with frequency, then the suitable form of Friis formula is:

$$P_r = P_t \frac{A_r G_t}{4\pi d^2} \tag{10.8}$$

which would then lead to yet another conclusion: namely, the received power does not depends on the frequency (!!). The main message from these, seemingly contradictory statements, is that, when making statements about the communication range, one needs to be aware of the dependencies of the path loss and the antenna gains on the frequency.

The effective area of an antenna can be much larger than the physical area. In order to get an idea of this, one can think of a radio transmitter surrounded by a metal wire mesh that acts as a Faraday cage. With appropriate selection of the radio transmitter frequency, no signal comes out of the Faraday cage, although there are holes in the mesh, as the space around the radio transmitter is not fully covered by metal. If we think of the Faraday cage as of a receiving antenna, then it can be concluded that the effective area of the antenna also covers the holes, not only the parts physically occupied by the wires.

10.3 The World View of an Antenna

10.3.1 Antenna Directivity

The antenna gains G_t and G_r, besides being dependent on the antenna apertures, also depend on the orientation that Yoshi's antenna has with respect to Zoya's antenna. A practical antenna is almost never *istropic* in the sense that it sends/receives electromagnetic energy equally in all possible directions in three dimensions. In fact, an isotropic antenna is a theoretical construct and does not exist in practice. On the contrary, an antenna has a certain *directivity*, that is, there are spatial directions in which the antenna radiates relatively more energy. Drawing an analogy to speech communication, the quality at which Zoya listens to Yoshi's speech depends on how the heads of Zoya and Yoshi are oriented to each other, which means that that both the speaker and the listener exhibit a certain directivity.

Figure 10.3 illustrates the dependency of the radiated energy on the observed direction through the *radiation pattern* of an antenna. The value of the radiation pattern in a given direction is called *antenna gain* and it shows the ratio of the radiated energy in that direction relative to the energy radiated when the antenna is isotropic. The radiation pattern is three-dimensional and the product $G_r G_t$ in (10.4) depends on the gains of the two radiation patterns that are achieved on the line that connects Yoshi and Zoya. Figure 10.3 shows a two-dimensional projection of the radiation pattern. It can be seen that the product $G_r G_t$ can be changed by rotating the antennas within the depicted plane. The received power is maximal when the maximal gains point to each other. For the example in Figure 10.3, the

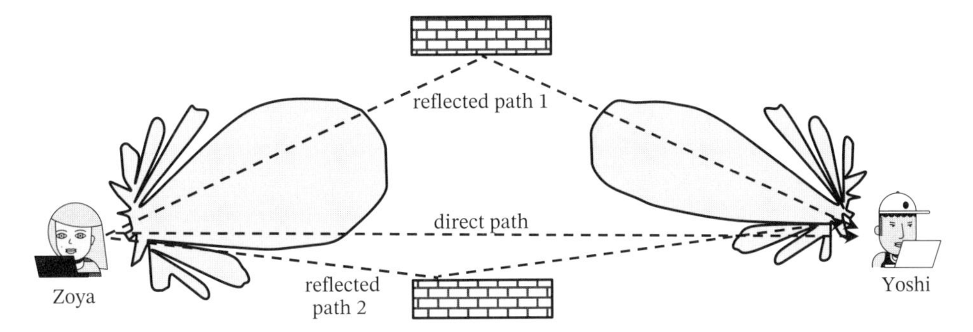

Figure 10.3 Illustration of radiation patterns of the antennas used by Zoya and Yoshi. The power of the signal that arrives from Zoya to Yoshi, and vice versa, depends on the antenna radiation patterns as well as the relative orientation of the patterns with respect to each other.

antenna gains associated with the reflected path 1 are larger than the ones associated with the direct path.

Following the explanation of the antenna gains, the Friis formula in its form (10.8) has an appealing intuitive interpretation: the received power corresponds to the fraction $\frac{A_r}{4\pi d^2}$ of the sphere area from which the receiver collects electromagnetic energy, multiplied by the gain G_t that the transmit antenna beams in the direction of the receiver. This interpretation is valid under the assumption that the gain G_t is constant for all points of A_r, which is acceptable when the distance between the sender and the receiver is large.

The direction in which the antenna radiates the maximal energy carries the *main lobe* of the beams that are radiated from the antenna. There can be weaker side lobes in the other directions, as also exemplified by the radiation patterns depicted in Figure 10.3. An important characteristic of the main lobe is the *beamwidth*, which corresponds to the angle in which the electromagnetic energy is radiated to or received from. The precise width of the beam is a matter of convention, the most common definition is that it is the angle between the two directions at which the power is halved compared to the maximal one. It can be seen that Zoya's antenna has a larger beamwidth compared to Yoshi's antenna.

Another interesting insight from Figure 10.3 is obtained by noting that no radio signal arrives from Zoya to Yoshi through the reflected path 2. Indeed, the received power is equal to zero if at least one of the antennas is oriented in a way that the gain on the line that connects Yoshi and Zoya is zero. For example, Zoya's gain for the reflected path 2 from Figure 10.3 is zero. The latter is interesting in terms of interference avoidance and cancellation. If Yoshi wants to receive a signal from Xia, while it considers the signal from Zoya to be only interference, then Yoshi can orient his antenna such that the radiation pattern is null in the direction of Zoya. However, if the radiation pattern of Yoshi is fixed, then he may end up in the unfortunate situation in which it is not possible to create a null towards Zoya, while receiving the signal from Xia at an acceptable level. This is shown in Figure 10.4. In fact, canceling the interference from Zoya is a rather heuristic approach to optimizing the communication of the link Xia–Yoshi. The optimal solution does not necessarily imply that there is a null towards Xia and the optimization problem should be defined in a more general way. For example, one formulation can be the following: given Yoshi's radiation pattern, as well as the positions and the radiation patterns of Zoya and Xia, orientate Yoshi's antenna in such a way to maximize the SINR of the signal received from Xia.

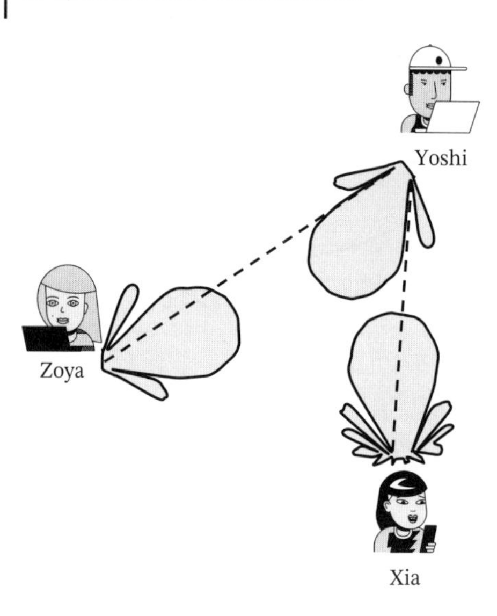

Figure 10.4 Yoshi tries to orient his antenna in a way in which the interference from Zoya is canceled. However, due to the relative position of Xia, he also cancels the desired signal from Xia.

10.3.2 Directivity Changes the Communication Models

The world view of an antenna is not isotropic and this fundamentally changes and enriches the communication models that were discussed before in this book. Namely, in all models we have used so far, the antennas were implicitly assumed to be *omnidirectional*. With omnidirectional antennas the broadcast is cheap and a single transmission reaches all the nodes in the range, as in Figure 10.5(a). This is important when the transmitter is not aware of the direction of the receiver and/or vice versa, such that a transmission does not reach the receiver, even though it is in relatively close proximity. This is of major significance in rendezvous protocols, in which the two nodes establish the connection for the first time.

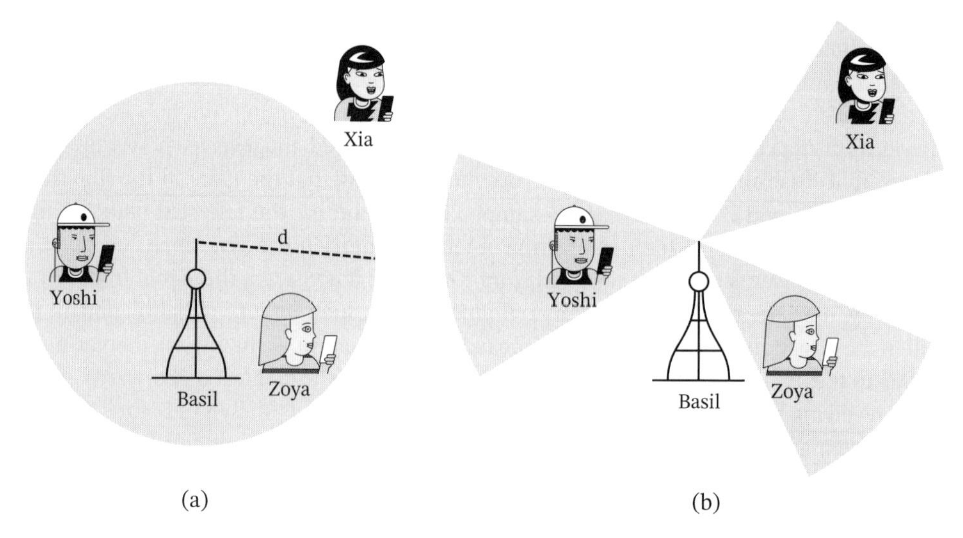

Figure 10.5 Change in the coverage area depending on the directivity of the antenna at the base station Basil. (a) Omnidirectional antenna. (b) Directed antenna.

Another situation where this occurs is a random access protocol, which may be used by a node after a prolonged period of inactivity. As a drawback, omnidirectional transmissions waste a large fraction of the radiated power in directions in which there are no receivers at all. Furthermore, a receiver with an omnidirectional antenna gets interference from all directions.

When directed antennas are used, the transmitter can send energy in a specific direction towards the receiver. On the other hand, the receiver can collect energy from a narrower spatial angle directed to the transmitter. For the example in Figure 10.5(b), Basil's antenna has a single (idealized) beam and this can be directed, in a time-division manner, towards Zoya, Yoshi, or Xia. This means that Basil deliberately changes the shape of the coverage area in time. Note that Xia is not in the coverage area of Basil if an omnidirectional antenna is used, as in Figure 10.5(a). However, when Basil uses a directed antenna in Figure 10.5(b), the electromagnetic energy is better directed and now he can communicate with Xia while using the same power as in Figure 10.5(a). The advantage of using a directed antenna is that the radio energy is beamed in relevant directions, thereby causing minimal waste. A narrower beamwidth results in a more focused transmission. Furthermore, very little, if any, interference is received from the directions that do not coincide with the beam. On the flip side, Basil and each of the connected terminals need to run a protocol in order to calibrate the beams in the right direction. This brings cost in terms of delay and overhead, which becomes even more acute if the wireless devices are moving. The delay is especially significant in rendezvous protocols, in which two devices need to discover each other and establish a link. In principle, this can be achieved by searching through space and pointing the beams in random directions until they match with each other.

From the discussion above, the reader may have arrived at an idea about hybrid operation of a node: use an omnidirectional antenna to discover devices and establish a link, and switch to a directed antenna when the actual data is transmitted.

Other factors that impact the received signal in free space propagation include: polarization, impedance matching, and the fact that the antenna parameters and the radiation pattern are dependent on the frequency. The polarization is an interesting one, since if one can keep the polarization of the transmitting and the receiving antennas perfectly matched, then the number of degrees of freedom (DoFs) for communication is doubled. Namely, there are two possible orthogonal polarizations and each of them can be treated as a separate communication channel, not interfered with by the transmissions at the other polarization. This is similar to the fact that the use of the Q component at a given frequency doubles the number of DoFs at the same frequency compared to the case in which only the I component is used. However, when the terminals are mobile, the transmitter–receiver polarizations cannot be perfectly matched and the terminal antenna needs to use polarization that offers diversity, such as a circular one, and collect at least some power, regardless of the polarization of the transmitter.

10.4 Multipath and Shadowing: Space is Rarely Free

In free space the radio signal propagates along the line that connects the transmitter and the receiver. Thus, the receiver observes the contribution from a single component, known as

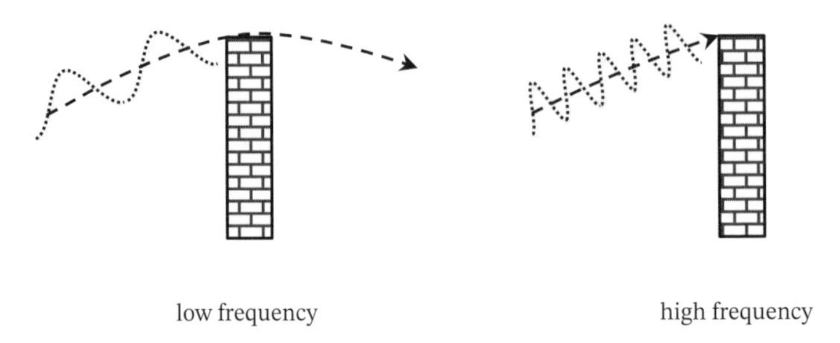

Figure 10.6 An example of an object that causes diffraction at low frequency, but acts as a blockage at higher frequency.

a line of sight (LoS) component. In a typical outdoor or indoor setting, there are a number of physical objects in the environment that affect the propagation of the radio signal from the transmitter to the receiver. The impact of the environment is exhibited through three different mechanisms: reflection, diffraction, and scattering. *Reflection* occurs when a wave hits a surface. Part of the energy is absorbed in the surface, while another part is reflected and continues to propagate. The energy carried in the reflected wave is decreased, as the reflecting surface absorbs a fraction of the incident energy. *Diffraction* is the phenomenon that describes the bending of radio waves around edges of obstacles. The occurrence of diffraction depends on the relation between the radio signal wavelength (frequency) and the obstacle. *Scattering* occurs when the radio wave hits an object of a size comparable to its wavelength and the energy is spread over multiple propagating waves.

In general, these effects are dependent on the frequency and favor the propagation of the lower frequencies, leading to a larger communication range in practical propagation scenarios. Coming back to the myth discussed in Section 10.2, we can state that lower frequencies are preferred due to the lower diffraction loss around corners as well as smaller penetration loss through walls and obstacles. For example, the wavelength of the radio waves at frequency 60 GHz is in the order of millimeters; therefore the name *millimeter wave wireless communication*. An object can act as an obstacle for the 60 GHz radio waves and result in a blockage. However, the same object can lead to a diffraction of the radio waves at frequency 6 GHz or lower, which have wavelengths in the order of centimeters. This is illustrated in Figure 10.6.

The outcome of the described propagation mechanisms is that Yoshi receives the signal sent from Zoya via multiple paths. This is known as *multipath* propagation and its analogy in speech communication is the theme of the cartoon at the beginning of this chapter. Multipath propagation in wireless communications is illustrated in Figure 10.3, where Yoshi receives the signal through the direct path and the path denoted as "reflected path 1". More precisely, this is an illustration of a two-ray propagation model, a canonical example of a multipath propagation. The signal received by Yoshi is a superposition of the two rays, which can be constructive or destructive, leading to higher or lower received power, respectively. Assuming that Zoya and Yoshi have omnidirectional antennas, unlike the illustration in Figure 10.3, the nature of the superposition of the two rays depends on the geometrical setup for Zoya and Yoshi, which is determined by their distance and the heights at which

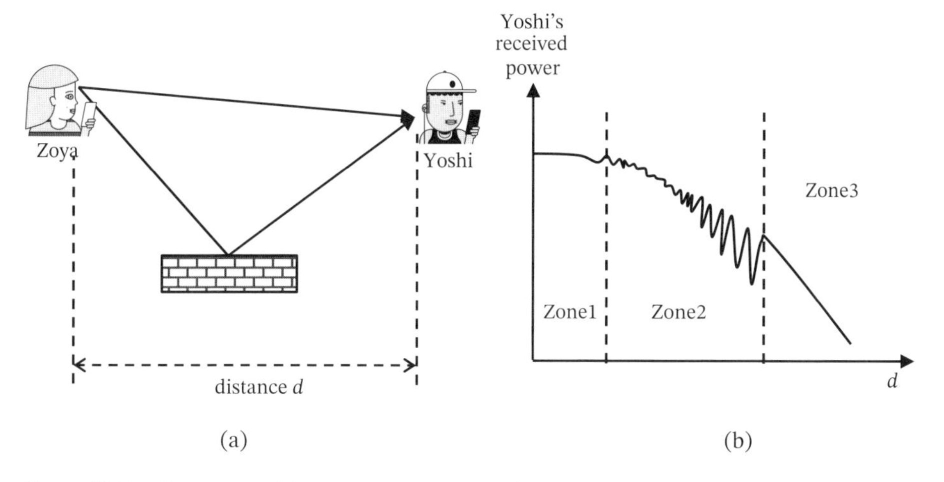

Figure 10.7 Illustration of the two-ray propagation from the sender Zoya to the receiver Yoshi. (a) The physical setup. (b) The dependency of the received power on the distance d.

their antennas are placed. Furthermore, the result of multipath propagation depends on the frequency of the radio signal and the properties of the reflecting surface.

Figure 10.7(b) shows a typical dependency of the received power on the distance between Zoya and Yoshi. As this distance increases beyond Zone1, the amount of received power starts to oscillate, but on average it decreases with the square of the distance. After a *critical distance* beyond Zone2, in Zone3 the received power starts to decrease monotonically with the fourth power of the distance, much faster than the square-of-the-distance decrease in a free space. Hence, the impact of the multipath in a two-ray model is negative at large distances. This is significant in scenarios that can give rise to two-ray propagation, such as when Zoya and Yoshi are separated by a relatively large water surface.

Zone2, in which the received power oscillates with the distance, has special significance. The separation between the two dips in the received power is in the order of the wavelength of the transmitted signal. Assume that Yoshi is placed in this region and he makes a small, unintentional move. This can result in abrupt changes in the received power, illustrating the phenomenon of *fading*. Specifically, since this fading changes within small distances that are comparable to the wavelength, it is referred to as *small-scale fading*. On the other hand, when the distance between Zoya and Yoshi is beyond the critical one, small perturbations in the distance or the frequency of the signal will result in small perturbations of the received power; in other words, Zone3 from Figure 10.7(b) does not exhibit small scale fading.

The two ray with ground reflection is a toy example of the effects of multipath. In practice, the rays from Zoya to Yoshi arrive through more than two paths. In this case, the received power that results from the superposition of multiple paths can be described with a suitable statistical model, where it is assumed that each of the reflected paths is affected by a random factor that obeys certain statistics, discussed further in Section 10.8. We remark that the use of directed antennas is diminishing the effect of the multipath, since such antennas are limiting the spatial angles from which radio waves can be received.

There is another important aspect of the multipath signals: each signal arrives at the receiver with a different propagation delay, as it follows a path of a different length. For the example of two rays in Figure 10.7(a), the reflected ray arrives τ s after the direct path:

$$\tau = \frac{\Delta x}{c} \tag{10.9}$$

where Δx is the difference between the lengths of the reflected and the direct paths, while c is the speed of light. The value τ is called a *delay spread* and it adds another dimension of complexity. Namely, when speaking about constructive or destructive interference, we have implicitly assumed that the radio waves carry the same data symbol. However, this is not necessarily true for delay spread and, at a given instant, the receiver may observe superposition of radio waves that carry two different data symbols. Due to this, we will need to deal with the problem of intersymbol interference (ISI), which we encountered in the previous chapter.

Besides the multipath, another effect that occurs when the radio waves are not propagating in a free space is *shadowing*. The diffraction effect, illustrated in Figure 10.6, introduces random attenuation for the signal that continues to propagate beyond the obstacle. These random attenuations are accumulated along the propagation path and may result in a large variation of the received power. In contrast to small-scale fading, shadowing contributes to *large-scale fading*, in the sense that the variations that it introduces are not changing significantly within small distances that are comparable to the radio wavelength.

10.5 The Final Missing Link in the Layering Model

In this section we go back to our layering model and enrich it to include the elements of the physical radio signals and propagation.

Nevertheless, at first we need a slight digression in order to relate the concept of carrier frequency to the physical features of radio communication. As discussed in the previous chapter, radio signals are modulated with a carrier frequency, such that they occupy a band that is positioned at higher frequencies. Modulating the signal with the sinusoidal carrier does not add additional information to it, it only re-positions the information carrying signal in the frequency. This means that the bandwidth occupied by the signal stays the same. One reason for doing this is the possibility to multiplex signals from different users in a frequency domain. The other reason for lifting the signals to higher frequencies is related to the physical characteristics of the antennas. Namely, the antenna size is, loosely speaking, comparable to the wavelength λ of the radio signal; for example, the classical $\lambda/2$ dipole antenna has a size of a half wavelength. Here the wavelength is calculated based on the central, carrier frequency of the signal, which is approximately correct in the common case for which the signal bandwidth is much lower compared to the carrier frequency. In order to keep the antenna size small, $\lambda = \frac{c}{f}$ also needs to be small, which implies that the carrier frequency f should be higher.

We are now ready to describe the layered model that includes the analog waveforms and propagated radio signals. This is depicted in Figure 10.8 and can be understood as an extended version of the layered model for asynchronous packet transmission from Figure 6.9, where we defined the notion of a communication channel. In Figure 10.8 we zoom into the LLChannel (low layer channel) from Figure 6.9. At the side of the transmitter Zoya, the LLChannel starts with channel coding by passing the data packet through a

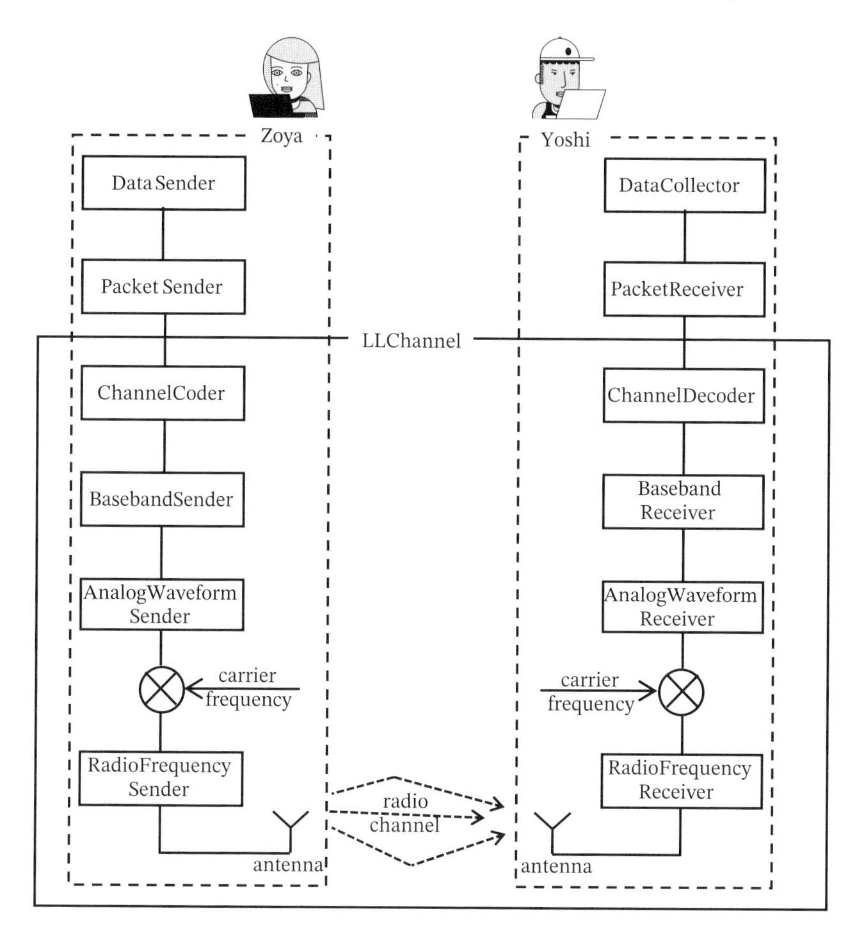

Figure 10.8 A detailed diagram of the LLCHannel (low layer channel) that shows the connection between the baseband model and the radio propagation.

ChannelCoder, which is then mapped to complex samples within the BasebandSender module[2]. As the name explains, the AnalogWaveformSender converts, or synthesizes, the samples into waveform and uses a low-pass filter to shape the waveform. This signal is then used to modulate the carrier frequency by mixing, that is, multiplication. The RadioFrequencySender amplifies the passband signal, puts it through a transmission passband filter and supplies it to the antenna that converts it into radio waves.

At the side of the receiver Yoshi, the radio signals are picked up by Yoshi's antenna. The RadioFrequencyReceiver contains a bandpass filter, which limits the spectrum of the received signal to contain most of its energy in the desired band and suppress the noise and interference coming from the other bands. Furthermore, the RadioFrequencyReceiver contains a low noise amplifier before bringing the signal to the mixer, at which the signal is downconverted to baseband analog waveform. The received waveform is noisy due to the

2 Recall that, when modulation and coding are designed jointly, as in trellis coded modulation, the Channel Coder and the BasebandSender are merged into a single module.

unwanted radiation picked up by Yoshi's antenna and the circuits in the RadioFrequencyReceiver. The block AnalogWaveformReceiver samples the analog baseband waveform and outputs digital baseband noisy samples, that is, complex numbers, to the BasebandReceiver. The signal is further processed through the ChannelDecoder, after which it exits the LLChannel and is passed on to the PacketReceiver. The channel defined from the output of the BasebandSender to the input of the BasebandReceiver is the baseband channel model, which in its most elementary form is described as in (10.1).

10.6 The Time-Frequency Dynamics of the Radio Channel

Looking at different elements in the layered model in Figure 10.8, the radio channel is the part of the overall channel that is determined by "nature" and is thus the most difficult one to control[3]. Following Massey's description of the communication channel, the radio channel is part of the system for which we are willing, but least able, to change, since it depends on the propagation environment. The information carrying analog waveform that Zoya sends can arrive at Yoshi through multiple paths, the amplitude and phase over the paths can change in time, etc. In relation to that, the basic questions we are interested in are of the following type: how is the received signal $y(t)$ distorted with respect to the transmitted signal $x(t)$? Does $y(t)$ contain energy in the same frequency band in which $x(t)$ has been properly shaped to contain its energy? In order to answer these questions, one needs to look into different factors that contribute to the dynamic behavior of the radio channel.

Considering it as a system with its transfer function, the wireless channel can be described as a linear *time-variant* system. Recall that in linear *time-invariant* systems the output $y(t)$ of the system cannot have energy at the frequencies that were not originally present in the signal input $x(t)$. This is no longer true when the system through which the signal is transferred varies in time. In order to get an idea of this, assume that the input $x(t)$ is a signal that changes very slowly, such that within the observed period it is practically constant. This implies that the frequency spectrum of the signal $x(t)$ is concentrated around frequency zero. On the other hand, assume that the channel through which $x(t)$ is sent changes very quickly and these changes are reflected in the output signal $y(t)$. Then $y(t)$ will contain energy at higher frequencies, not present in the original signal $x(t)$. However, even if the channel does not change over time, or at least over a period that is much longer than the duration of a symbol, the received signal is a distorted version of the transmitted signal, as we will see next.

10.6.1 How a Time-Invariant Channel Distorts the Received Signal

We consider a setup in which the signal $z(t)$ from Zoya to Yoshi arrives through three paths, such that the analog waveform $y(t)$ observed by Yoshi at the instant t can be described as:

$$y(t) = g_1 z(t - \tau_1) + g_2 z(t - \tau_2) + g_3 z(t - \tau_3) + n(t). \tag{10.10}$$

3 Let us not forget that we do not have full deterministic control over the other blocks, as some of the circuits introduce random noise in the communication channel and/or occasional errors in processing the signals and the data.

Treating the wireless channel as a time-variant linear system implies that, in general, the delays τ_1, τ_2, τ_3 and the channel coefficients g_1, g_2, g_3 are functions of time, but we have not denoted that explicitly in (10.10) in order to keep the notation light. We will assume that $\tau_1 \leq \tau_2 \leq \tau_3$; for example, the first path may be direct, the second reflected once and the third reflected twice, such that $\tau_1 < \tau_2 < \tau_3$. We have on purpose denoted the channel coefficients that affect the waveform by g_1, g_2, g_3 instead of h_1, h_2, h_3 in order to emphasize the fact that they may be different from the channel coefficients in the baseband model, obtained after sampling.

The model (10.10) is an approximation of the physical setup, as a large number of waves will arrive from Zoya to Yoshi after many reflections, significantly delayed beyond τ_3. However, in (10.10) it is assumed that they are all largely attenuated to a level that makes them insignificant compared to the noise.

In this section we assume that the channels are constant over many symbols. This allows us to consider a linear time-invariant system whose impulse response can be written as:

$$g(t) = g_1\delta(t - \tau_1) + g_2\delta(t - \tau_2) + g_3\delta(t - \tau_3) + n(t) \tag{10.11}$$

where $\delta(\cdot)$ is the Dirac delta function, g_1, g_2, g_3 are constant complex numbers and τ_1, τ_2, τ_3 are constant real numbers. The way we should think about the channel (10.11) is that the impulse response operates on the complex baseband representations of the analog waveforms that are defined for the channel between the AnalogWaveformSender and the AnalogWaveformReceiver from Figure 10.8. This is illustrated in Figure 10.9(a), where the three received pulses are complex numbers with an amplitude and a phase[4].

However, it is important to note that the complex numbers featured in (10.11) are not necessarily the complex coefficients used in the baseband model. In other words, the kth received symbol in the complex baseband model that corresponds to a waveform passed through the channel (10.11) is not necessarily $y_k = g_1 z_{k-k_1} + g_2 z_{k-k_2} + g_3 z_{k-k_3} + n_k$, where k_1, k_2, k_3 are discrete representations of the delays τ_1, τ_2, τ_3. This is because the complex discrete baseband model depends on the actual waveforms transmitted and the sampling process that is undertaken in the AnalogWaveformReceiver.

The difference $\tau_3 - \tau_1$ is the *delay spread* of the channel, which can be interpreted as the time duration for which the reception by Yoshi is affected by a single Dirac impulse sent by Zoya. Hence, the channel spreads the zero-duration pulse to $\tau_3 - \tau_1$, as illustrated in Figure 10.9(a). Another effect introduced by the multipath is the fact that Yoshi receives a waveform that is a distorted version of the waveform transmitted by Zoya, as illustrated in Figure 10.9(b). This occurs due to the self-interference from the replicas of the same transmitted symbol.

On the other hand, Figure 10.9(c) shows how the wireless channel can introduce ISI. Each symbol uses the same waveform, not explicitly depicted in the figure, which is subject to distortion, as in Figure 10.9(b), to which the ISI exhibits an additional impact. The separation between the first two symbols that Zoya sends is shorter than the delay spread, which results in that part of the waveform that Yoshi receives being obtained by interference between symbol 1 and symbol 2 sent by Zoya. However, after sending symbol 2, Zoya waits for a time equal to the delay spread and then sends symbol 3, whose waveform is

4 The amplitude is visible in the figure, but due to the limit of real-number representation we can only see phases 0 and π for the received pulses.

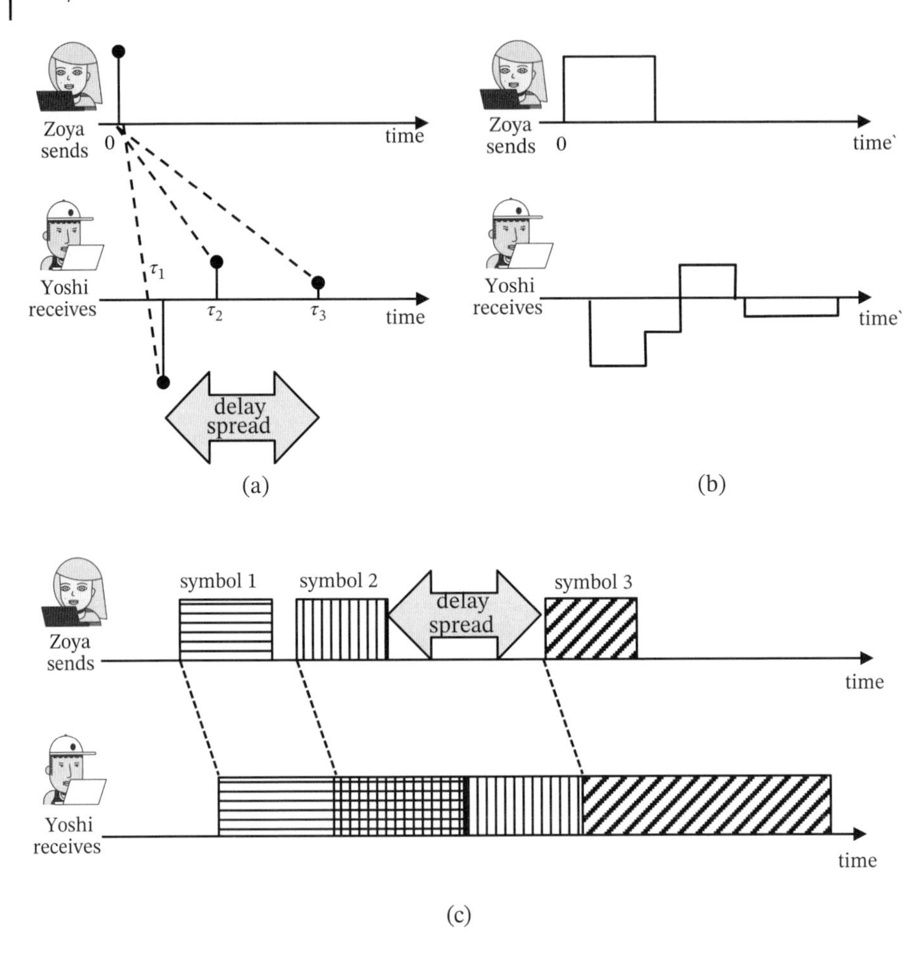

Figure 10.9 An example of a channel impulse response defined for the channel between the AnalogWaveformSender and the AnalogWaveformReceiver from Figure 10.8. (a) The channel response features three paths, which do not change over time, and the delay spread is $\tau_3 - \tau_1$. The phase of the first-arriving pulse is π, while it is zero for the other two. (b) How the channel distorts a rectangular waveform sent by Zoya. (c) Yoshi observes intersymbol interference between symbol 1 and symbol 2. However, Zoya waits sufficiently long before sending symbol 3, in this case for a time equal to the delay spread, such that it does not cause intersymbol interference with the other two symbols.

received free from interference coming from the other symbols, but still distorted due to the superposition of the arrivals through the three different paths. Thus, an obvious way to avoid ISI would be to wait for a time equal to the delay spread of the channel. Since the delay spread may vary, one can put the waiting time between the symbols to be equal to the maximal value that the delay spread could have in a given environment. However, this effectively increases the duration of a symbol and can therefore lead to a (severe) decrease in the data rate. We will discuss solutions to this problem in Section 10.7.

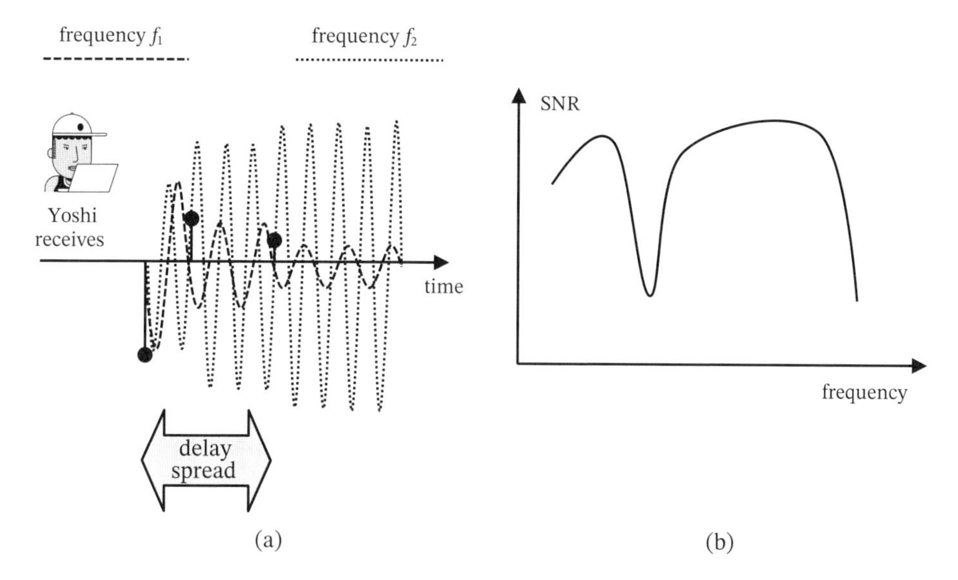

Figure 10.10 Illustration of the frequency selectivity in multipath channels. In this example Yoshi receives the signal through three paths. If Zoya sends a signal of either frequency f_1 or f_2, Yoshi observes a sinusoidal signal after the last path has arrived. In this example, frequency f_1 is attenuated more than frequency f_2, such that the overall wireless channel between Zoya and Yoshi is frequency selective.

10.6.2 Frequency Selectivity, Multiplexing, and Diversity

Figure 10.10 provides another very important perspective on the multipath propagation and delay spread. We first focus on the example in Figure 10.10(a), where the signal from Zoya to Yoshi arrives through three paths. It is assumed that Zoya can send one of the two sinusoidal signals with two different frequencies f_1 and f_2, respectively[5].

Two things are important to note from Figure 10.10(a). *First*, the signal that Yoshi observes after the third (last) path has arrived is a superposition of the contributions from the three paths and is *always sinusoidal*, with the same frequency as the signal sent by Zoya. Of course, this is true for any arbitrary number of paths and is a consequence of the general fact that sinusoidal functions are eigenfunctions of linear time-invariant systems. *Second*, two different frequencies can be affected by the channel in a different way; in this example f_1 is attenuated more compared to the f_2. In other words, the SNR for a signal that has frequency f_2 is higher than the SNR for the signal that has a frequency f_1.

This feature of the wireless channel is known as *frequency selectivity*. Figure 10.10(b) shows an example of how the SNR for the received sinusoidal signals changes with the frequency when transmitted over a multipath, frequency selective wireless channel. In the previous chapter, we introduced OFDM and the possibility of using parallel transmission across frequencies that are orthogonal over a predefined interval of duration T. If an OFDM

5 They are plotted as if occurring simultaneously to illustrate the difference in the amplitude for the resulting signal.

transmission is used in a way that the subcarriers are positioned at frequencies that are multiples of $\frac{1}{T}$, then the SNR for the signal received at each frequency can be obtained from Figure 10.10(b) by reading the SNR values that correspond to the discrete set of frequencies that represent the subcarriers or, more precisely, the centers of the subcarriers.

The way the SNR changes across frequencies depends on the structure of the multipath. Figure 10.10(b) also shows that the SNR for the two frequencies that are very close to each other is nearly identical, or, in other words, the received SNRs for two nearby frequencies are correlated. As the separation between the two frequencies increases, the received SNRs for the two frequencies become decorrelated. This leads to the notion of *coherence bandwidth*, which is the maximal separation of two frequencies such that the correlation of their received SNRs is above a certain threshold. The coherence bandwith is inversely proportional to the root mean square of the delay spread. Intuitively, when the delay spread is very low, then a large band of frequencies are superposed by the channel in approximately the same way. On the other hand, if the delay spread is large, even two very close frequencies can result in different outcomes from the superposition of the multiple paths.

The size of the coherence bandwidth is significant with respect to the overhead that is needed in order to provide the CSI at the transmitter (CSIT), which is a condition to apply water filling or other transmission strategies that are making use of the difference in SNR among the parallel channels. In other words, having CSIT, the transmitter Zoya can effectively *multiplex* multiple data streams on different subcarriers. As an example, fix the spacing of the subcarriers to $\frac{1}{T}$ and assume that the coherence bandwidth is much larger than $\frac{1}{T}$. Then a significant number of channels have the same SNR and the CSIT for them can be provided jointly, which reduces the signaling overhead. Conversely, low coherence bandwidth means that the amount of information required to describe the frequency dependency of the SNR is significant, which may lead to an explosion in the overhead required to acquire the CSIT.

However, if the CSIT cannot be made available, for example, due to costly reporting, then it is desirable that the coherence bandwidth is low. When the transmitter Zoya does not know the quality of the channel at different frequencies, then she cannot allocate the power and the rate in any non-trivial way. The best way of transmitting is to use all frequencies or subcarriers in an identical way and just hope that some of them will be sufficiently good. In a sense, Zoya diversifies her transmission risk across the frequencies and this type of transmission is often treated as a *diversity mode*. In contrast to multiplexing, where Zoya wants to maximize the number of bits sent over different streams, in diversity mode Zoya tries to send the same bits over all subcarriers, hoping to improve their reliability. If the coherence bandwidth is low, on the order of a single subcarrier, then this is the most desirable situation from the diversity viewpoint, as fading will affect each transmitted subcarrier in a random way, independent of the other subcarriers.

10.6.3 Time-Variant Channel Introduces New Frequencies

Let us now assume that there is only one path between Zoya and Yoshi. We start with the case depicted in Figure 10.11, in which there is an obstacle that blocks the direct path between Zoya and Yoshi, such that the signal arrives through a reflected path. Consider the situation in which a third actor, Victoria, has control over the reflecting surface and can

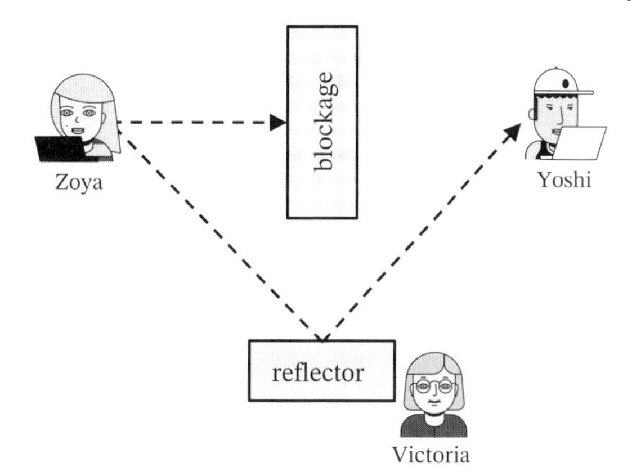

Figure 10.11 Scenario in which the direct path between Zoya and Yoshi is blocked by an obstacle, such that the signal arrives through a path that is reflected from a surface controlled by Victoria. The changes that Victoria causes to the reflecting surface, made of a meta-material, produce frequencies in the signal received by Yoshi that are not originally present in the signal sent by Zoya.

steer the way in which it reflects the incoming radio wave. We note that, with the advent of meta-materials with controllable electromagnetic properties, this situation is not only a hypothetical one. The received waveform at the AnalogWaveformReceiver, after removing the impact or *demixing* of the carrier frequency, is:

$$y(t) = g(t)z(t) + n(t). \tag{10.12}$$

Victoria effectively controls $g(t)$ and changes its phase and amplitude accordingly. Let us assume that Victoria can change $g(t)$ very quickly, within timing that is on the order of the magnitude of the timing of the symbols used by Zoya for data transmission. Then the frequency spectrum of $g(t)z(t)$ is no longer identical with the spectrum of $z(t)$, but is a result of a mixture of two signals, analogous to the case of mixing with the carrier frequency. In fact, Victoria can decide to transmit information to Yoshi and modulate $g(t)$ accordingly. In this case $g(t)$ acts as a channel for $z(t)$ and vice versa! The difference is that Zoya uses actual power to transmit, while Victoria uses power to control the reflector and free-rides on the carrier emitted by Zoya. Thus, in this example, (10.12) is in fact a multiple access channel with two transmitters, Zoya and Victoria.

The example described in Figure 10.11 illustrates the case in which the wireless channel changes due to the changes in the environment. This example reflects some of the possibilities opened by using meta-materials in wireless communications. By contrast, the traditional source channel dynamics is the mobility of the transmitter and the receiver. This is depicted in Figure 10.12. In this example, there is a single directed path between Zoya and Yoshi and the distance between them is d. For simplicity, it is assumed that Zoya transmits a sinusoidal signal of frequency f_Z. In Figure 10.12(a) both Zoya and Yoshi are standing still, such that the frequency of the signal received by Yoshi is equal to f_Z.

In Figure 10.12(b) Yoshi moves towards Zoya at a speed v. Assume that Yoshi moves within a distance $d + d_0$, where $d_0 \ll d$, such that the path loss at d and $d + d_0$ is approximately the same. However, d_0 can be comparable or even much larger than the wavelength $\lambda_Z = \frac{c}{f_Z}$ of the radio waves sent by Zoya, given by (10.3). This means that $g(t)$ changes due to the changes in its phase, leading to the *Doppler effect* or *Doppler shift*. Intuitively, by moving, Yoshi passes faster through the peaks and the valleys of the sinusoidal signal, such that the frequency he receives is higher than f_Z. The situation is opposite when Yoshi moves

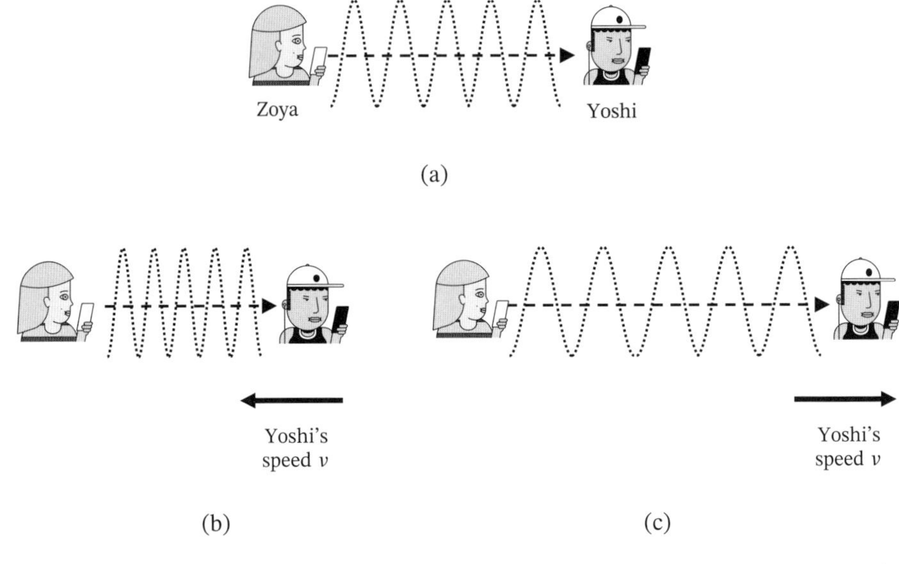

Figure 10.12 Illustration of the Doppler shift when Zoya transmits a sinusoidal signal to Yoshi at a frequency f_Z. The sinusoidal signal in all three pictures depicts the frequency f_Y received by Yoshi. (a) Zoya and Yoshi are standing still and $f_Y = f_Z$. (b) Yoshi moves towards Zoya and $f_Y > f_Z$. (c) Yoshi moves away from Zoya and $f_Y < f_Z$.

away from Zoya at a speed v, see Figure 10.12(b). Here his receiver hits the peaks and the valleys of the signal at a slower pace compared to the case at which both Zoya and Yoshi are standing still, such that the frequency received by Yoshi is lower than f_Z.

Given the speed v at which Yoshi moves towards or away from Zoya, the carrier frequency f_Y observed by Yoshi when the carrier frequency of Zoya is f_Z is given by:

$$f_Y = f_Z \pm f_Z \frac{v}{c} \tag{10.13}$$

where c is the speed of light and the "−" sign is used if Yoshi moves away from the incident wave, and "+" in the opposite case. We note that here we have illustrated the Doppler shift in the case in which the receiver moves, but the effect appears in an analogous way when Zoya moves or even both Zoya and Yoshi move. Hence, in general, v is the relative speed between Zoya and Yoshi.

Similar to the way we have defined a coherence bandwidth to capture the frequency selectivity of the channel, an important parameter that captures the time dynamics of a channel is the *coherence time*. It is defined as the minimal time between two instances at which the channel coefficients are correlated below a certain, predefined threshold value. In other words, if the interval between two instants is larger than the coherence time, then knowing the channel at one instant tells very little about the channel in the other instant. This is important if Zoya wants to adapt the transmission parameters, such as rate and power, to the instantaneous state of the channel, as in that case Zoya needs to acquire the CSI. With a short coherence time, CSI should be acquired more frequently, which leads to an increased overhead. This was discussed in Chapter 6, where we introduced the mathematical notion of a communication channel. The trade-offs put forward by the coherence time with respect to diversity and multiplexing are similar to the ones discussed in relation to the coherence bandwidth in Section 10.6.2. Namely, multiplexing requires CSIT acquisition and requires a longer coherence time in order to apply that knowledge and load the channel with the rate/power appropriately. On the other hand, transmitting without CSIT requires that the channel changes independently from one transmission instant to another, thereby offering diversity.

In the example in Figure 10.12, the Doppler spread is a deterministic quantity. In general, it is a random variable that is a result of the device mobility and changes in the environment. Similar to the relation between the coherence bandwidth and the delay spread, the coherence time is inversely proportional to the root mean square of the Doppler spread. This is even more intuitive to comprehend than the coherence bandwidth delay spread relation, since a larger Doppler spread result from a higher speed of channel change, which evidently decreases the coherence time.

Compared to the multipath, which is in general stochastic and unpredictable, there are communication scenarios in which the Doppler shift can be predicted and compensated for at the receiver, such as in inter-satellite communication.

10.6.4 Combined Time-Frequency Dynamics

We have discussed the effects of the time dynamics and frequency selectivity separately. In practice, both can take place simultaneously, leading to a complex combination of factors that distort the received signal. In order to have a glimpse into the complexity of this problem, consider the example in Figure 10.13, in which the signal from Zoya to Yoshi arrives through two paths while Yoshi moves towards Zoya. Since the moving direction of Yoshi is aligned with the direct path, the Doppler shift for the signal arriving through the direct path is determined by Yoshi's speed v. However, the projection of Yoshi's speed along the direction through which the reflected signal is received is different from v, leading to a different value of the Doppler shift for the reflected path.

Alternatively, the changes of the channel may not be continuous, but occur over a longer interval, such that the coherence time is long and the channel is quasi-stationary in time. For example, the parameters of the multipath channel, such as path delays and the coefficients associated with the paths, may change over a long interval. This will produce a

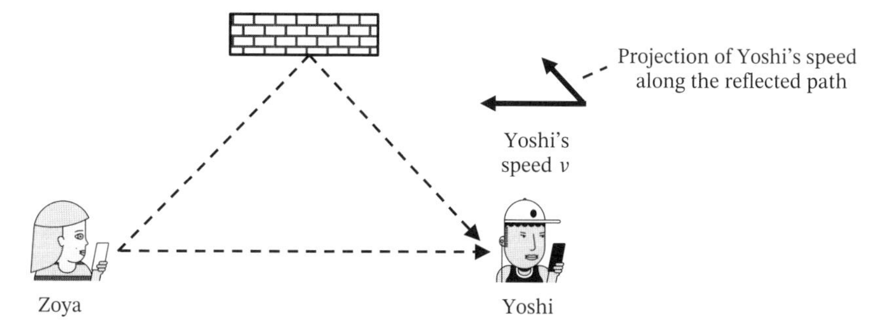

Figure 10.13 Yoshi moves towards Zoya at a speed v, but the speed along the direction of the arrival of the reflected path is smaller. This means that the Doppler shift for the direct path is different from the one for the reflected path.

changed picture of the frequency characteristics for the frequency selective channel. If Zoya wants to transmit to Yoshi by multiplexing the signals in a spectrally efficient way, then she needs to acquire CSI at time intervals that correspond to, or are shorter than, the coherence time of the channel.

10.7 Two Ideas to Deal with Multipath Propagation and Delay Spread

We have seen that multipath propagation is a major challenge in wireless communications, as it leads to distortion of the received symbols. However, multipath propagation also represents an opportunity, as it offers the receiver the possibility to harness energy arriving from multiple directions. Let us explore this opportunity.

In this section we will use the two-path channel response depicted in Figure 10.14(a). Here the delay spread $\Delta\tau = \tau_2 - \tau_1$ is the time difference between the two paths. The first idea that comes to mind is to let Zoya transmit a symbol of duration T_S that is smaller than the delay spread $T_S < \Delta\tau$, as depicted in Figure 10.14(b); this is also done at the bottom picture of the cartoon from the beginning of this chapter. Due to the small symbol (pulse) duration T_S, this wireless system is *wideband*. To facilitate the discussion, we will make one more assumption: the timings of the two paths are chosen so that they correspond to the sampling instants that define the baseband channel model. This is an approximation, because there is no reason why the delays incurred by nature are multiples of the sampling period used by the receiver.

Let h_1 and h_2 denote the baseband channel coefficients that correspond to the two paths, respectively. Since Yoshi knows that the two arrived symbol instances carry the same data, this situation is equivalent to having *two parallel* channels with coefficients h_1 and h_2. Hence, Yoshi can combine them coherently and obtain an SNR that corresponds to maximum ratio combining (MRC):

$$\gamma = \gamma_1 + \gamma_2 \tag{10.14}$$

where $\gamma_i = \frac{|h_i|^2 P}{\sigma^2}$ is the SNR of the ith, $i = 1, 2$, arrived path, P is the transmission power used by Zoya and σ^2 is the noise power.

However, this approach has a major drawback, as it relies on the assumption that no new symbol arrives from Zoya before Yoshi has received the last multipath component, as illustrated in Figure 10.14(b). This means that the period of the symbol transmission, denoted by T_R, should be larger than $\tau_2 - \tau_1$. Hence, we have $T_R > T_S$ and Zoya cannot send more than $\frac{1}{T_R}$ symbols per second, which limits the data rate. Specifically, if each symbol carries $\log_2 M$ bits, then the data rate is limited to:

$$R \leq \frac{\log_2 M}{T_R} = \frac{\log_2 M}{\tau_2 - \tau_1} \quad \text{(bps)}. \tag{10.15}$$

Since the actual values of τ_1 and τ_2 are random, T_R should be selected such as to cover the worst case, which corresponds to the largest likely value of $\tau_2 - \tau_1$ for a given environment in which the system is expected to operate. This can limit R to an unacceptably low value. What is even worse is that, with the communication method depicted in Figure 10.14(b),

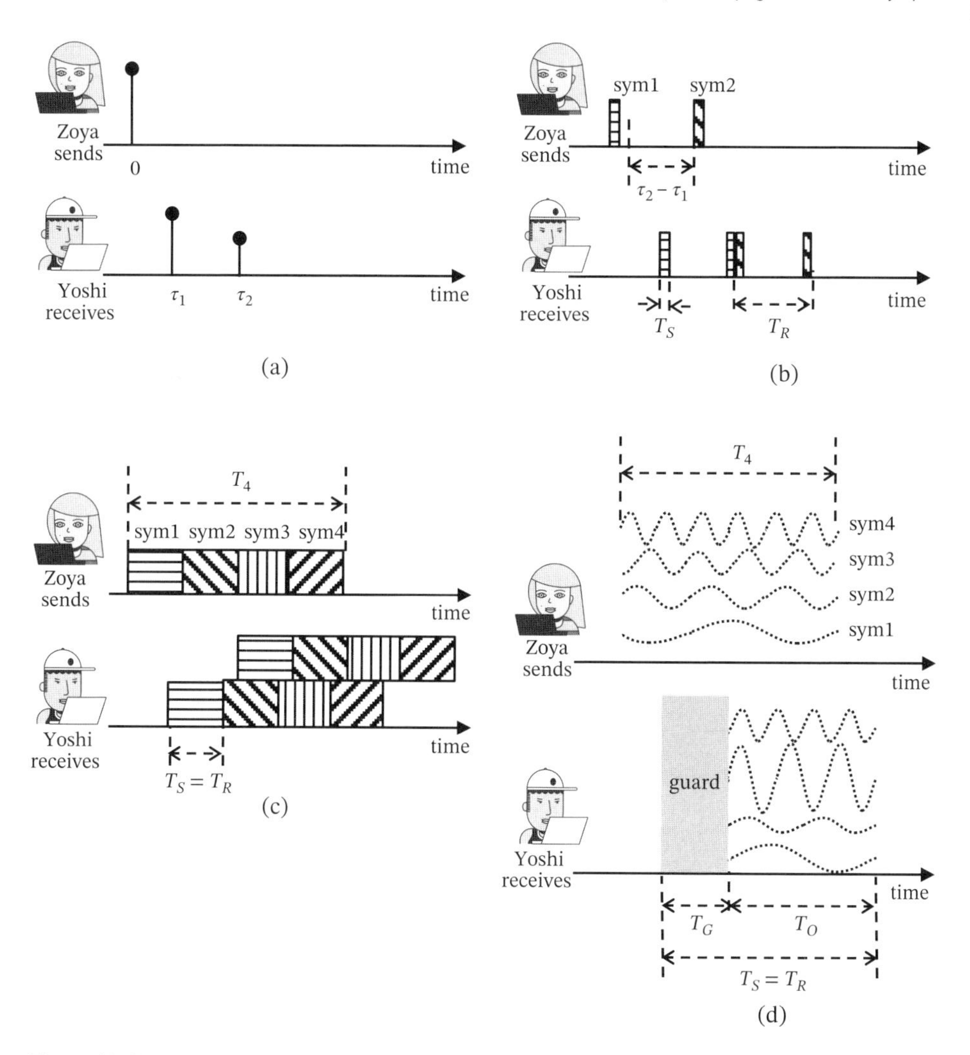

Figure 10.14 Illustration of the ideas to combat the multipath propagation. (a) The two-path channel response used in this section. (b) Conservative transmission in which Zoya extends the period of symbol transmission in order to avoid intersymbol interference. (c) Transmission based on the spread spectrum, each symbol contains G chips. At the receiver the symbols arriving through the second path are depicted on the top of the symbols arriving through the first path. (d) Narrowband transmission. The received signals during the guard interval are not depicted as they are ignored during the decoding. All sinusoidal signals are orthogonal on the interval T_O.

we do not gain by further decreasing T_S. A lower T_S means a higher signal bandwidth, such that the investment of extra bandwidth does not improve the system performance.

We are now ready to look into two different ideas that overcome this problem.

10.7.1 The Wideband Idea: Spread Spectrum and a RAKE Receiver

Let Zoya transmit each of her symbols by using a spread spectrum. Specifically, she uses a wideband direct sequence spread spectrum, such that the duration of the transmitted chip is T_c. The symbol duration is $T_S = GT_c$, where G is the number of chips per symbol, also called a processing gain. The transmission based on spread spectrum is depicted in Figure 10.14(c). As an illustration, if the wideband CDMA system uses the same bandwidth as the wideband transmission in Figure 10.14(b), then the chip duration for the CDMA signal from Figure 10.14(c) is equal to the symbol duration T_S from Figure 10.14(b). We can see from Figure 10.14(c) that Zoya transmits throughout the whole period between two symbols, such that $T_S = T_R$. We use T_4 to denote the time required to send four symbols; this will be useful in the next section, which deals with OFDM.

In order to combat the effects of the multipath propagation, the spread spectrum (CDMA[6]) scheme relies on the desirable correlation properties of the spreading sequence. Let us look at first at the preliminary operation of channel estimation, carried out before actual data is sent. For simplicity, assume that a single symbol of G chips is transmitted. This is a *pilot* symbol, its data content is known a priori by Zoya and Yoshi and is used to estimate the channel. Yoshi uses the pilot symbol by shifting it with small time steps and tries to find the time instants at which he can measure a high correlation. Due to the good autocorrelation properties of the spreading sequence, for the channel from the example in Figure 10.14, Yoshi measures two instances with high correlation, each of them corresponding to one of the arrived paths.

After the channel has been estimated by using the pilot symbol, Yoshi now knows τ_1, τ_2. Further, he knows the phases of the arrived paths and thus knows how to combine the two instances of the same symbol in order to achieve MRC, as in (10.14). However, there is a difference between the SNR that can be achieved by MRC when there is no interference among symbols, as in Figure 10.14(b), compared to the case of CDMA, as explained next.

Namely, from Figure 10.14(c) we can see that the instance of the first symbol, denoted sym1, arriving through the second path, interferes with sym2 and sym3 arriving at the same time through the first path. At this point Yoshi uses the good autocorrelation properties of the spreading sequence, as well as the fact that the data content of sym2 or sym3 is uncorrelated with the one of sym1. It is important to note that the correlation of sym1 with sym2 and sym3 is not zero; hence, the interference created by sym2 and sym3 will decrease the SINR of the second instance of sym1 compared to the SNR of this sym1 instance in the absence of sym2 and sym3. Note that the SINR of the first instance of sym1 will be also decreased due to interference with the symbols transmitted immediately before it (not depicted in the figure). To sum up, after performing MRC combining of the two received replicas, the SINR of the resulting signal will be deteriorated compared to the case of MRC combining of two paths without ISI. This is the price paid in order to reduce T_R and increase the rate compared to Figure 10.14(b).

Note that the way Yoshi should maximum-ratio combine sym2, sym3, etc. is identical to the way the instances of sym1 are combined. Of course, this is valid as long as the channel does not change. This type of CDMA receiver with maximum-ratio combing of the instances arriving through different paths is known as a *RAKE receiver*, as the receiver gathers or *rakes* different instances of the same signal.

---

6 We use CDMA here as it is commonly used in related to multipath multi-user channels; however, we emphasize that CDMA refers to use of multiple spreading sequences for multiple users, as discussed in Chapter 9, while the use of spread spectrum to address multipath propagation does not depend on the fact that there is one or there are multiple users.

10.7.2 The Narrowband Idea: OFDM and a Guard Interval

In the discussion related to Figure 10.10, it was emphasized that, when a sinusoidal signal is sent, then regardless of the structure of the multipath, the signal that is received after the last path has arrived is also a sinusoidal one and has the same frequency. Hence, the channel inherently combines the sinusoidal signals from different paths. This is, though, not MRC, but rather a superposition of the signals with amplitudes and phases determined by nature. Hence, the SNR of the resulting sinusoidal signal may turn out not to be higher than the SNR available for the signal on any of the arrived paths. In the extreme case, depending on the frequency of the sinusoidal signals and the phases of the arrived paths, the resulting signal may even be zero.

Figure 10.14(d) illustrates the idea for combating multipath with narrowband transmissions. During the time between the arrival of the first path and the last path, which is equal to the delay spread, the shape of the received signal deviates from a pure sinusoidal shape. One can adopt the strategy to ignore this non-sinusoidal part of the signal during the decoding. This part of the received signal in Figure 10.14(d) is denoted as a guard interval and has a duration of T_{G}, which in this example is $T_{\mathrm{G}} = \tau_2 - \tau_1$. In general, it is desirable that the guard interval is longer or equal to the delay spread. Note that intersymbol interference occurs only during the guard interval, which is the most important argument to throw away this part during the decoding. The total symbol duration is $T_{\mathrm{S}} = T_{\mathrm{G}} + T_{\mathrm{O}}$ and the efficiency η_T, expressed as the time fraction that is used for actual communication, is:

$$\eta_T = \frac{T_{\mathrm{O}}}{T_{\mathrm{S}}} = \frac{T_{\mathrm{O}}}{T_{\mathrm{R}}} = \frac{T_{\mathrm{O}}}{T_{\mathrm{G}} + T_{\mathrm{O}}}. \tag{10.16}$$

If the receiver throws away the part of the signal received during the guard interval, then η_T can be improved by making the useful duration of the signal T_{O} longer. However, here $T_{\mathrm{S}} = T_{\mathrm{R}}$ and this will decrease the data rate to even lower values compared to Figure 10.14(b).

One key idea here is to use multiple narrowband signals, modulated onto different subcarriers, each carrying different symbols. This already looks like OFDM, except that we have not fixed the time interval in which the subcarriers should be orthogonal. The duration of that interval in Figure 10.14(d) is denoted by T_{O} and therefore the subcarrier frequencies should be chosen as integer multiples of $\frac{1}{T_{\mathrm{O}}}$. The longer T_{O}, the more subcarriers can fit in the given bandwidth, which compensates for the rate loss. Finally, it should be noted that four symbols are transmitted during the time T_4, such that the nominal data rate in this OFDM transmission is identical to the one of CDMA from Figure 10.14(c),

In addition to the basic principle for using OFDM to deal with multipath propagation, there are a lot of details about the implementation, not all of which can be described here. Perhaps the most important among them is the choice of the guard interval T_{G}. Its value should be fixed, if not to a constant value, then at least it should be fixed for a given communication session. The value of T_{G} cannot change too fast, for example from symbol to symbol, since in that case the transmitter should find a way to inform the receiver about the actual value of T_{G}, which would be an unnecessary overhead. Furthermore, even if there can be some level of adaptation of T_{G}, it cannot be assumed that it can always adapt to the actual value of the delay spread $\Delta\tau$. One common approach is to select T_{G} to be equal to the likely maximal value $\Delta\tau_{\mathrm{max}}$ that the delay spread gets in an environment in which the wireless system typically operates. Clearly, if the actual $\Delta\tau < \Delta\tau_{\mathrm{max}}$, then the system operates sub-optimally, as too much of the signal is thrown away by the receiver. Nevertheless, this sub-optimality is the price for having a good architectural solution that scales.

Looking at Figure 10.14(d) the following observation emerges: since the signals during T_G are not used, then it is not clear why should one waste power on them. Indeed, if a transmitter sends a signal of duration T_O and does not transmit during the time T_G, then $T_S = T_O$ and $T_R = T_O + T_G$. In this case the guard interval will contain contributions from only one symbol and there will be no ISI at all. In addition, the transmitter will save power by not sending signals that are wasted anyway.

However, this is not the way OFDM is usually implemented. A standard OFDM implementation is based on a *cyclic prefix*. This is implemented by copying the last T_G s of the data symbol with duration T_O and putting them in front of the data symbol, such that $T_R = T_O + T_G$. This allows efficient implementation by using a fast Fourier transform (FFT), since the cyclic prefix is used to "trick" the circular convolution to give the same results as the linear convolution. Finally, for the example in Figure 10.14(d), all four subcarriers start with phase 0 when the data symbol of duration T_O starts. This is not a requirement, as the subcarriers remain orthogonal regardless of the phase.

To summarize, OFDM combats multipath by letting the channel do the natural combining and using orthogonality to eliminate the ISI. The penalty paid is the waste of the communication resources during the guard interval, which corresponds to the loss of DoFs in communication. On the other hand, the advantage of CDMA is that it does not waste any DoF and it uses MRC. The drawback is that it does not eliminate ISI. Furthermore, while the combining in OFDM is done by nature, though not always with a good result, the combining in CDMA results in an increased complexity of the receiver.

10.8 Statistical Modeling of Wireless Channels

The changes in the environment and in the positions of the transmitter/receiver cause fluctuations in the channel response. In order to provide statistically relevant characterization of the wireless system, these fluctuations need to be modeled as random quantities that reflect the statistical dependencies imposed by the underlying physics. Consider the channel impulse response depicted in Figure 10.9. The random quantities include the number of paths (which is three in this illustration), the arrival time of each path, as well as the amplitude and phase of each arrived path. On the other hand, the statistical model needs to cater for the dependencies between the amplitudes/phases of the different arriving paths in the channel response. Furthermore, the model needs to capture how all these random quantities are changing in time, for example from one symbol to another or, over a longer period, from one packet transmission to another.

The detailed characterization of the channel impulse response and its temporal evolution is known as *wideband channel characterization*. The wideband model based on arrival of multiple delayed paths, as in Figure 10.9, is known as the *tapped delay model*. The rationale for calling it wideband is that, in order to measure it in an ideal way, one needs to transmit a Dirac pulse, which is the ultimate wideband signal. This results in a description of the channel behavior throughout the entire frequency spectrum. The simplest case of a wideband model is the two-path model from Figure 10.14(a).

Nevertheless, we have seen that, when a narrowband sinusoidal signal is transmitted over the channel and its duration is larger than the delay spread, then the channel produces a narrowband sinusoidal signal. It is therefore of interest to provide a stochastic description of the amplitude and the phase of the resulting sinusoidal signal. This is known as *narrowband channel characterization*. Its use is not limited only to the case of sinusoidal signals. It is also applicable to the cases in which the delay spread is much lower than the symbol transmission time, such that one can use the approximation that all paths arrive simultaneously and are inseparable. With this in mind, the baseband model $y = hz + n$ that we have extensively used until now is a narrowband channel model and the physical effects of the propagation and antenna reception are summed up in the description of the channel coefficient h as a complex random variable.

10.8.1 Fading Models: Rayleigh and Some Others

One widely used narrowband model is the Rayleigh fading model. The main assumption in the Rayleigh model is that there is a large number of signals with random phases and amplitudes that arrive from different paths almost simultaneously, see Figure 10.15. The channel coefficient h is modeled as:

$$h = \sum_{k=1}^{\infty} a_k e^{j\phi_k} \tag{10.17}$$

where a_k and ϕ_k are the random amplitude and phase of the kth arrival path, respectively. This setup allows the application of the central limit theorem and the resulting channel coefficient h is Gaussian complex random variable, with zero mean and variance that

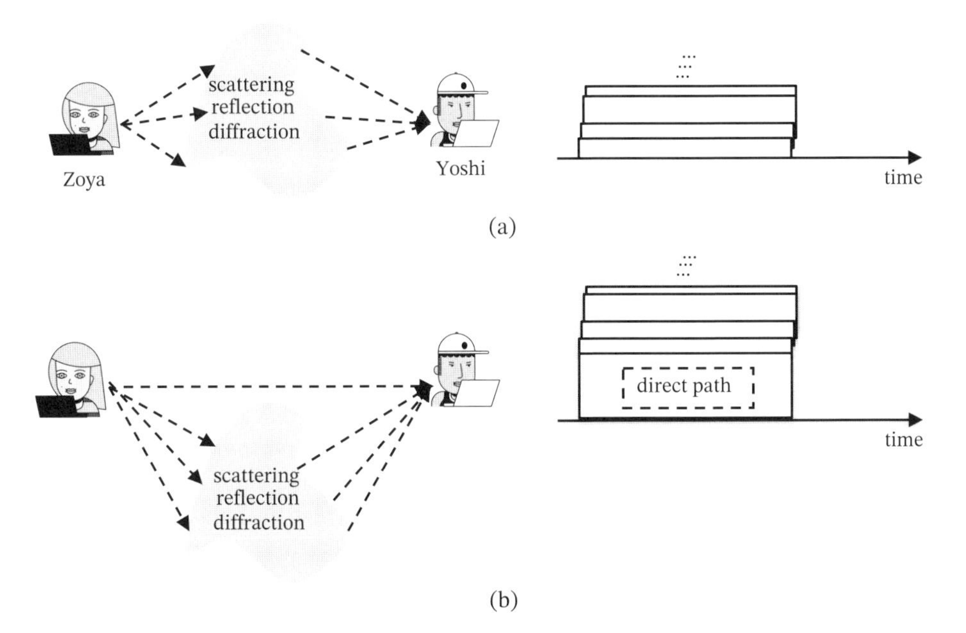

Figure 10.15 A physical setup that corresponds to: (a) Rayeigh fading, without a dominant component; (b) Ricean fading, with a single dominant component.

corresponds to the average received power. As a first approximation, the average received power can be related to the path loss. The model is termed "Rayleigh" since the distribution of the amplitude of the received signal is Rayleigh. On the other hand, the received power follows an exponential distribution and so does the received SNR. If we denote by $\overline{\gamma}$ the average received SNR, proportional to the variance of h, then the received SNR is sampled from the following exponential probability density function:

$$p(\gamma) = \frac{1}{\overline{\gamma}} e^{-\frac{\gamma}{\overline{\gamma}}}. \tag{10.18}$$

While the Rayleigh fading model is appealing to use due to its simplicity, one needs to be cautious about a specific physical setup and check whether the underlying assumptions have been met. One such setting is depicted in Figure 10.15(b). Here, in addition to the scattered components, there is a dominant component that comes through the direct path. The dominant component has practically a deterministic value which is not subject to changes due to stochastic changes in the environment. In this case the amplitude of the channel coefficient h has a Ricean distribution.

Another typical case in which Rayleigh fading is not applicable is the two-path model, corresponding to the physical setup in Figure 10.13; the movement of Yoshi in the figure is, for the moment, irrelevant. Here the number of superposing paths is insufficient to make the central limit theorem applicable. Instead, the channel coefficient can be modelled as:

$$h = a_1 + a_2 e^{j\phi_2} \tag{10.19}$$

where a_1 and a_2 are the amplitudes of the direct and the reflected wave, respectively, while ϕ_2 is the phase of the reflected wave. For fixed positions of Zoya and Yoshi, a_1 and a_2 are constant. On the other hand, ϕ_2 is uniformly distributed in $[0, 2\pi)$. We have taken $\phi_1 = 0$, but it can have an arbitrary value, as the important thing for the resulting signal is the relative phase between ϕ_1 and ϕ_2. In order to get an idea about the physical basis for the uniform distribution of $[0, 2\pi)$, consider Figure 10.7. As discussed in Section 10.4, fading occurs in Zone2, as small fluctuations in the distance result in large fluctuations of the received power. Here we may add that the plot in Figure 10.7(b) is made for a specific frequency, as the model (10.19) is narrowband, such that the picture Figure 10.7(b) changes when the frequency changes. This reveals a different source of fading: if Yoshi is placed in Zone2, but the frequency of the transmitted signal fluctuates, then the received power fluctuates as well.

Finally, a note on the correlation in the wireless channel models. In our discussion about the narrowband channel models, we have considered the fading that affects a signal of a given frequency in isolation from the other frequencies. However, the frequencies are not fading independently and, as mentioned before, the frequencies that are within the coherence bandwidth are significantly correlated, which needs to be taken into account when detailed models are made. Another issue is the temporal correlation of the channel. One simple model that is often used is the block fading model, in which the fading distribution is sampled independently at each packet transmission. Physically, this corresponds to the use of frequency hopping, in which a new packet transmission is carried out at a new frequency, uncorrelated with the frequency used previously. The model can be enriched by generating fading instances that are correlated with the previous use of the channel, rather than sampled independently.

10.8.2 Randomness in the Path Loss

We look again at the two-path model in Figure 10.7(a), but now we assume that Zoya broadcasts a packet to two receivers, Yoshi and Xia. Both receivers are placed at the same distance d from Zoya and this distance is chosen to be in Zone2. Due to the random phase in (10.19), the received power that Yoshi measures at a given time (for example, corresponding to one packet transmission) is likely different from the power measured by Zoya. However, the average power that they measure if each of them slightly perturbs the distance will be identical. In other words, the large-scale average for Zoya and Yoshi will be the same. However, this is not the case when we consider the effects of shadowing, described in Section 10.4.

A common statistical model for shadowing is using a log-normal distribution and its physical basis can be explained as follows. Each diffraction during the signal propagation attenuates the signal with a certain random multiplicative factor; let us denote this factor by L_i for the ith diffraction. Then the total attenuation of the signal is

$$L_{\text{total}} = L \prod_{i=1}^{I_{\text{D}}} L_i \tag{10.20}$$

where L is the path loss and I_{D}, which is a random variable, is the number of diffractions along the propagation. If we take the logarithm of (10.20) the result is:

$$\log L_{\text{total}} = \log L + \sum_{i=1}^{I_{\text{D}}} \log L_i. \tag{10.21}$$

Here we apply again the central limit theorem: if I_{D} is sufficiently large, then $\sum_{i=1}^{I_{\text{D}}} \log L_i$ is approximated well by a normal (Gaussian) random variable. Hence, the logarithm of the overall loss has a normal distribution, hence the name *log-normal shadowing*.

From Figure 10.7(b) it can be seen that the slope of the logarithm of the average received power is not constant, otherwise the fading oscillations would be around a line. In contrast, the slope of the logarithmic loss changes from from 2 in Zone2 to 4 in Zone3, such that the path loss is proportional to $\frac{1}{d^2}$ in Zone2 and $\frac{1}{d^4}$ in Zone3. This sets the motivation for using path loss models with *dual slope*: the path loss depends on the distance as $\frac{1}{d^\nu}$, but ν changes (increases) after a certain critical distance. More generally, the path loss exponent ν depends on the environment and typically can have a value between 2 and 6.

Finally, another variable that is subject to statistical modeling is the distance d between the transmitter and the receiver. For this purpose, one can define a suitable spatial process for generating the positions of the communicating nodes or the base stations. This is of interest to evaluate the average performance over a (large) set of links that are established on the basis of the stochastic spatial process. The mathematical methodology used is based on stochastic geometry.

10.9 Reciprocity and How to Use It

Consider[7] a situation in which Zoya and Yoshi are in a bunker and trying to communicate through a wireless device to the outside world. The device uses TDD to transmit and receive

7 The situation described here took place in an actual TV series, only the names have been changed.

in the same frequency band. After struggling for some time, Zoya and Yoshi hear a voice of an unknown person and they try to respond to him. The person cannot hear them, even if they can hear him; then Zoya proposes climbing closer to the bunker door to get better reception. After climbing, the person at the other side is able to hear them.

In the described event there is likely a factual error, explained by the feature of *electromagnetic reciprocity*, which has a very significant impact on the communication techniques and protocols. A quick description of reciprocity would go as follows: if two identical wireless devices use identical transmission power and they communicate with each other in the same frequency band by using TDD, then the SNR of the signal received by one of the devices is identical to the SNR of the signal received by the other device. In other words, the radio communication channel is reciprocal. To put it more precisely, let Zoya transmit the signal x to Yoshi, such that Yoshi receives the signal hx (we ignore the noise), where h is the channel coefficient. If Yoshi transmits the signal x to Zoya *at the same frequency*, then Zoya would receive the same signal hx. In other words, the channel h between Zoya and Yoshi is the same, regardless of the direction in which we see through it. In this description we have ignored the noise; however, since the noise power is identical for the receivers of both devices, the fact that they receive the same received power implies that they have the same received SNR.

Interpreting reciprocity in a free space and using the Friis formula implies the following: the impact that the gain of a certain antenna has on the received power is the same, regardless of whether the antenna is used for transmission or for reception. However, more remarkably, reciprocity is preserved even when the electromagnetic propagation is affected by reflection, diffraction, and scattering. On the other hand, perfect reciprocity is destroyed by random noise, since the received signal by Yoshi would be $hx + n_1$, while the received signal by Zoya would be $hx + n_2$. We can also interpret this by stating that the existence of noise makes the reciprocity imperfect and the higher the SNR, the better the approximation of the perfect reciprocity. In addition to noise, reciprocity is affected by interference, which may be different for the receiver Yoshi compared to the receiver Zoya. In this case, the higher the SINR, the closer to the situation of perfect reciprocity. It was therefore stated above that the factual error in the described event at the bunker is likely, but not certain: for example, if the person from the other side were using a higher transmit power compared to the device held by Zoya and Yoshi, then it could have happened that the person from the other side cannot hear them, while they could hear him.

Let us take the case in which Zoya wants to transmit to Yoshi and she wants to use an appropriate level of modulation/coding, selected according to the SNR of her signal as received by Yoshi. Using reciprocity, Yoshi transmits a known signal to Zoya, also known as a pilot, and using a predefined power level. By measuring the SNR of the signal received from Yoshi, Zoya knows the SNR of the signal that Yoshi would receive from her. If there were no reciprocity, then in the first step Zoya sends a pilot signal and Yoshi measures the received signal, calculating the SNR. In the second step Yoshi transmits the value of the SNR to Zoya. In the third step, Zoya adapts the modulation and coding and transmits to Yoshi. Thus, reciprocity has reduced one protocol step. The gain from reciprocity is even higher if there are more than two involved devices. Consider the scenario in which K terminals want to communicate to Basil, a base station. Basil broadcasts a pilot signal from which each of the K terminals calculates the SNR that its transmitter signal would have when received by Basil. If this is a TDMA system, then each terminal can use the appropriate modulation/coding during its transmission window.

In conclusion, reciprocity is a phenomenon rooted deeply in the physical layer and the electromagnetic features, but it has a profound effect on the wireless communication protocols.

10.10 Chapter Summary

Wireless communication, be it with radio or other types of waves, always takes place in a certain physical/spatial setup in which the communication signal propagates between the sender and the receiver. The usual approach in a textbook on wireless communication is to start from the physical environment and phenomena that guide signal transmission, propagation, and reception. Here this discussion came first in the 10th chapter. The objective of this approach was to prime the reader with the requirements of the wireless protocols and the abstract communication models, and only then carry out a kind of guided search of the physical phenomena underlying wireless communications. This chapter shed light on the notions of antenna, free space versus real propagation, the real reason why lower frequencies are more precious as well as some basic mechanism of time-frequency dynamics in wireless channels. One of the central themes in the chapter, also illustrated in the cartoon of this chapter, is multipath propagation. Two fundamentally different ideas to deal with multipath were introduced, wideband (CDMA) and narrowband (OFDM), respectively. This was followed by the ideas for statistical modeling in wireless channels. The chapter was concluded with a discussion on reciprocity, a fundamental physical property of electromagnetic propagation. This property has long-reaching consequences on the design of wireless communication protocols; a prime example of this is the systems with massive antenna arrays, discussed in the next chapter.

10.11 Further Reading

The topics of antennas and propagation are at the heart of radio communication engineering. The interested reader is directed to explore Vaughan and Andersen [2003]. Two other books on wireless communications with emphasis on the radio channels are Rappaport [2001] and Molisch [2012]. In the chapter we have touched upon the emerging topic on metamaterials; the reader is referred to Liaskos et al. [2018].

10.12 Problems and Reflections

1. *TDMA and timing advance.* In Chapter 1 we presented several frames for single-channel wireless communication systems based on TDMA. Redesign the frame to take into account the fact that signal propagation takes time and different mobile devices can be at different locations in the cell. Investigate the concept of timing advance and include it in your design. Regarding duplexing, consider two possibilities, TDD and FDD, and analyze the difference.

2. *Transmission with beams.* The objective of beamforming is to focus energy in the desired direction, but, on the flip side, the sender loses the capability for efficient broadcast/multicast. In a simple approximation, a beam-based transmission transmits within an angle θ with equal gain and does not radiate any energy outside of that angle. With this, the omnidirectional transmission is the special case $\theta = 360°$. Assume that the total transmit power is constant, regardless of the beam angle.
   (a) Build a model by which the transmission range is inversely proportional to the beam angle θ.
   (b) Assume that the base station Basil can flexibly change θ for each transmitted packet and the minimal beam angle is θ_0. Devise a transmission protocol by which Basil can efficiently broadcast information to all devices in his range. If a device cannot be reached with a beam angle θ_0, then it is considered to be outside of Basil's range.
3. *Wideband versus narrowband.* Figure 10.14 illustrates the difference between wideband and narrowband methods for combating multipath. In relation to that, the text has indicated the trade-offs: wideband uses maximum ratio combining, but has to deal with interference; narrowband removes the interference, but has to live with nature-prepared combining and the loss of the guard interval. Consider a simple model with two paths and examine these trade-offs analytically. Assume a pseudonoise spreading sequence for the wideband case.
4. *Random access with statistical channel models.* Chapter 3 introduced several enrichments of the models for random access, such as capture. Build a simulator for investigating the performance of random access protocols by assuming block fading for the following protocols:
   (a) ALOHA-type protocols with capture.
   (b) Random access with intra- and extra-collision interference cancellation.
   You can select specific codes and modulation for the simulator based on e.g. the coding and modulation used in WiFi. Note that extra-collision interference cancellation desires the channel to be constant during multiple packet transmission times.
5. *Access with physical imperfectness.* The communication models for communication over a shared wireless medium are based on a (over)simplification of the physical reality. As explained in the dark room analogy in Chapter 2, the problem in offering access to the users is the fact that the base station Basil cannot differentiate among them. However, in practice there are a plenty of factors to differentiate among the users, such as time of signal arrival or the multipath pattern that is specific for a device placed at a specific location. Furthermore, the simple, battery-powered sensor and IoT devices are often based on imperfect oscillators, such that the carrier frequency at which a device transmits is $f + \Delta f$, where Δf can be used as a fingerprint for that device. Propose methods to utilize these physical effects for designing more efficient access protocols.

Z
Z
send
Vith a Base Station that has a single antenna, Zoya and Yoshi should take turns
ı transmitting. At a given time, one of them sends only one letter of the name.
Y
Y
send
Vith a Base Station that
as a massive number of
ntennas, both Zoya and
'oshi can send whole
ames simultaneously.
Zoya
send
Yoshi
send

Story by Petar Popovski / Art by Peter Gregson

11

Using Two, More, or a Massive Number of Antennas

In this chapter we talk about the benefits, and also the challenges, when the transmission and reception are supported by more than one antenna. As the actual number of antennas will play a significant role, the antennas will be clearly illustrated in the figures, see Figure 11.1 unlike the previous chapters. The antennas can be at the receiving device Figure 11.1(a), transmitting device Figure 11.1(b), both devices Figure 11.1(c) and even distributed to other devices Figure 11.1(d). The example from Figure 11.1(d) is better known as *relaying* or *cooperative communication*. In this scenario the nodes have the following commonly used names: Zoya is a source, Yoshi is the destination and Xia is a relay or a helper device. The setup is known as cooperative, since Xia cooperates with Zoya and Yoshi to facilitate their communication.

The relay scenario in Figure 11.1(d) can be seen as a generalization of the case in which both antennas are on the same device. To see this, we need to think of a device with multiple antennas as having communication bandwidth among the antennas that is arbitrarily large, and even infinite. If the SNR between Yoshi and Xia in Figure 11.1(d) is infinite, and hence the wireless bandwidth between them is also infinite, then the relay case becomes equivalent to a multi-antenna receiver Figure 11.1(a). If the SNR between Zoya and Xia in Figure 11.1(d) is infinite, then the relay case becomes equivalent to a multi-antenna receiver.

The benefit of using multiple antennas is intuitively clear for the example with multiple receive antennas and the relay scenario. Using multiple receive antennas, as in Figure 11.1(a), Yoshi is capable of collecting electromagnetic energy from a larger area and combining the outputs of the antennas in order to maximize the SNR or cancel interference. For the relay scenario in Figure 11.1(d), it is clear that Xia helps to extend the communication range between Zoya and Yoshi by, for example, regenerating the signal of Zoya and retransmitting it to Yoshi. It can be stated that in both cases the use of multiple antennas brings benefits in terms of better reliability or higher SNR to the link Zoya–Yoshi. Nevertheless, it is less intuitively clear that similar reliability benefits are brought by having multiple transmit antennas, as in Figure 11.1(b), since there is interference among the two signals arriving at a single antenna. Precisely this lack of immediate intuition makes the transmit diversity scheme intellectually brilliant, see Section 11.2.3.

Wireless Connectivity: An Intuitive and Fundamental Guide, First Edition. Petar Popovski.

11.1 Assumptions about the Channel Model and the Antennas

Following the discussion in the previous chapter, we need to emphasize that here we will assume use of narrowband channel models. The distance between the antennas on the same device is much lower than the distance between the transmitter and the receiver, such that the signals arrive practically simultaneously. This implicitly assumes that the system applies some form of OFDM transmission and the observed communication channel can be limited to one subcarrier; the other subcarriers can be treated in an analogous way to independent communication channels.

When talking about multiple antennas at the devices, we need to differentiate between *passive* and *active* antennas. An active antenna has a radio frequency (RF) chain of active electronic components associated with it, such that in our baseband model it can produce a dedicated baseband input. Hence, if there are two active receiving antennas, as in Figure 11.1(a), then Yoshi gets two baseband outputs, one per antenna:

$$y_1 = h_1 z + n_1$$
$$y_2 = h_2 z + n_2 \tag{11.1}$$

where z is the signal sent by Zoya and n_1, n_2 are the noise samples, independent for each antenna. The channel coefficients h_1 and h_2 are, in general, different. On the other hand, if

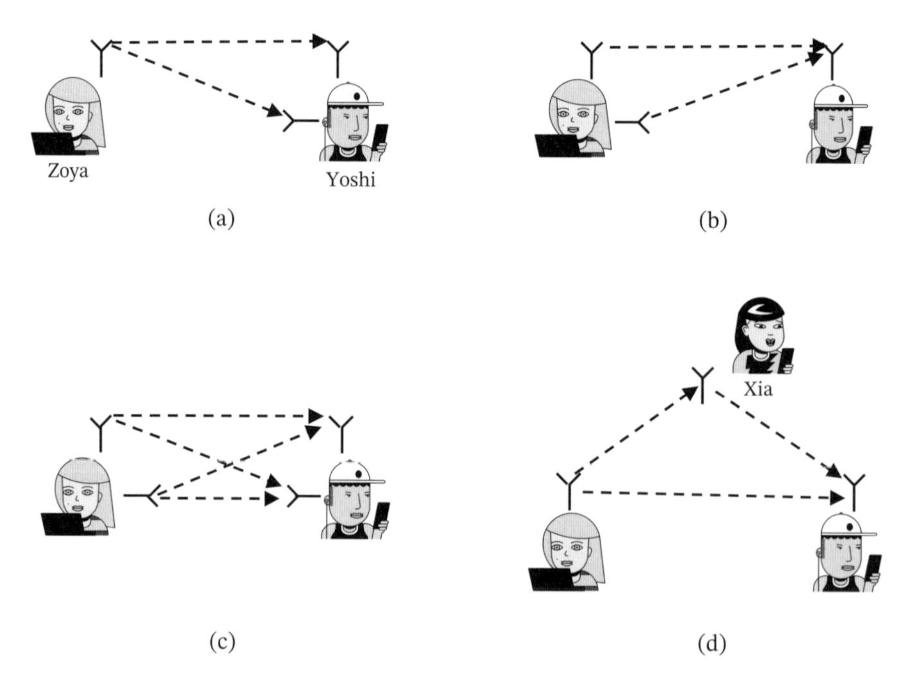

Figure 11.1 Use of multiple antennas to support the communication between Zoya and Yoshi. (a) Two receive antennas. (b) Two transmit antennas. (c) Two antennas at the transmitter and the receiver. (d) Relaying or cooperative communication.

Yoshi has two passive antennas, then there is only one RF chain associated with them and Yoshi gets a single output at the baseband:

$$y = hz + n. \tag{11.2}$$

Changes in the passive antennas can affect the value of the channel coefficient h or even distort z.

The difference between passive and active antennas is similar when we consider the antennas that have transmitting roles. If Zoya has two active antennas in Figure 11.1(b), then she can supply different baseband input to each of the antennas. Let z_i denote the symbol transmitted over the ith antenna, where $i = 1, 2$. Then the signal received by the single antenna of Yoshi is:

$$y = h_1 z_1 + h_2 z_2 + n \tag{11.3}$$

where h_i is the complex channel coefficient between the ith antenna of Zoya and Yoshi. If Zoya has two passive antennas in Figure 11.1(b), then she can send the same baseband input z to both antennas, such that the signal received by Yoshi is:

$$y = hz + n \tag{11.4}$$

where h is the composite channel between the baseband output of Yoshi and the baseband input of Zoya.

If the antennas placed on a single device are too close to each other, then the small-scale fading coefficients are correlated. This situation is not desirable, as it is intuitively clear that diversity is provided when the random variations of the quality of the channel coefficient in one antenna is independent of the random variations at the other antennas. We will reiterate this observation later on when we speak about specific techniques for multi-antenna transmission. As a rule of thumb, if the distance between two antennas is at least half a wavelength, then the small scale fading at these antennas can be considered uncorrelated. In practice, there are complex electromagnetic effects that create coupling and correlation among the antennas; even the hand grip of a device can affect this correlation.

11.2 Receiving or Transmitting with a Two-Antenna Device

An important physical distinction between the use of multiple antennas at the transmit and the receive side is that the use of an additional transmit antenna *consumes* part of the transmit power, while the use of an additional receive antenna *supplies* additional power to the receiver. We treat the two cases separately, with a longer elaboration on the use of multiple transmit antennas, as this aspect brings more novelty relative to the topics presented previously in this book.

11.2.1 Receiver with Two Antennas

As already indicated, when the two (or more) antennas are at the receiver, then the benefits are rather evident, as more antennas mean collection of more energy. A channel in which the transmitter has a single antenna, while the receiver has two or more antennas, is known as a *single input multiple output (SIMO)* channel. In the best case, the receiver Yoshi knows the channel coefficients h_1 and h_2 and, having the two antenna outputs as in (11.1), he can apply maximum ratio combining (MRC):

$$y = h_1^* y_1 + h_2^* y_2 \tag{11.5}$$

such that the resulting SNR is:

$$\gamma_{\mathrm{MRC}} = \gamma_1 + \gamma_2 \tag{11.6}$$

which is the sum of the individual SNRs of the two antennas, given as:

$$\gamma_i = \frac{|h_i|^2 P}{\sigma^2} \qquad i = 1, 2. \tag{11.7}$$

Here P is the transmit power applied by Zoya on her single antenna. Besides MRC, there are two other commonly used combining methods, *equal gain combining (EGC)* and *selection combining (SC)*. In EGC the two antenna outputs are co-phased and summed up with the same weight, such that the SNR is:

$$\gamma_{\mathrm{EGC}} = \frac{(\sqrt{\gamma_1} + \sqrt{\gamma_2})^2}{2} \tag{11.8}$$

where γ_1 and γ_2 are given by (11.7). EGC uses a weaker assumption of MRC, namely that the receiver uses the phase of h_i but not the amplitude, which may be unknown to the receiver. In SC, the receiver selects the signal from the strongest antenna, such that the resulting SNR is:

$$\gamma_{\mathrm{SC}} = \max_{1,2} \gamma_i. \tag{11.9}$$

In all three cases the resulting SNR is certainly higher than the SNR of the weakest antenna. For convenience, let us assume further on that MRC is used. The increased SNR can be used either to increase the rate or to improve the reliability. Let us consider at first the *rate improvement* and let us assume that the channel between Zoya and Yoshi changes slowly, such that there is sufficient time that Yoshi reports the value of γ_{MRC} to Zoya. Then Zoya can adapt the transmission rate to

$$R_{\mathrm{adapt}} = \log_2(1 + \gamma_{\mathrm{MRC}}). \tag{11.10}$$

If the short channel coherence time or the expensive signaling prevents Yoshi from informing Zoya about the resulting γ_{MRC}, then Zoya cannot adapt the rate and she can fix the rate to a predefined value R_{fix}. If the fading is such that

$$R_{\mathrm{fix}} > \log_2(1 + \gamma_{\mathrm{MRC}}) \tag{11.11}$$

then the selected rate R_{fix} is not supported by the channel for the given transmission and the packet is received in error, that is, an outage occurs.

In a statistical sense, the random variable γ_{MRC} has a higher value than the individual signals, such as γ_1. In other words, the probability that γ_{MRC} gets a low value is lower than the probability that γ_1 gets a low value. In this way the use of two receive antennas brings *receive diversity* and tames the fluctuation of the resulting SNR compared to a single-antenna case.

11.2.2 Using Two Antennas at a Knowledgeable Transmitter

We now turn to the case in which Zoya has two antennas and the receiver Yoshi one antenna, known as a *multiple input single output (MISO)* channel. Let us assume that Zoya has knowledge about the channels h_1 and h_2, associated with her two transmitting antennas. Each of the antennas is active and has its own RF chain, such that the transmit power P limits the total transmit power applied at both antennas. With the knowledge of h_1 and h_2, the best Zoya can do is to use *transmit MRC*, which can be described as follows. If Zoya wants to send the complex symbol b, then she uses *precoding* before sending it to a given antenna. Specifically, the symbols z_1 and z_2 sent from the first and second antenna, respectively, are:

$$
\begin{aligned}
z_1 &= \frac{h_1^*}{\sqrt{|h_1|^2 + |h_2|^2}} b \\
z_2 &= \frac{h_2^*}{\sqrt{|h_1|^2 + |h_2|^2}} b.
\end{aligned} \tag{11.12}
$$

The symbol received by Yoshi is:

$$
y = h_1 z_1 + h_2 z_2 + n = \frac{|h_1|^2}{\sqrt{|h_1|^2 + |h_2|^2}} b + \frac{|h_2|^2}{\sqrt{|h_1|^2 + |h_2|^2}} b + n. \tag{11.13}
$$

Since the power of b is P and the power of the noise n is σ^2, the SNR available to Yoshi for decoding b is γ_{MRC} as given by (11.6) and (11.7). It can be noted from (11.13) that the precoding has the effect of beaming or directing the wireless energy through the two paths in a way that allows their coherent combining at the receiver. It should also be noted that the total transmit power remains P.

Another sub-optimal way to use the two antennas when h_1 and h_2 are known is to apply *antenna selection* and use all the power only for the antenna that has a stronger channel coefficient; the SNR is given by (11.9). If Zoya only knows which antenna is stronger, but not the exact values of h_1 and h_2, then she knows which antenna should be used for transmission. However, she does not know what the resulting SNR at the receiver will be. This means that Zoya cannot adapt the data rate, but the fact that the stronger antenna is selected offers a higher reliability for a fixed-rate transmission.

The optimality of transmit MRC is somehow expected due to the reciprocity feature, but it is in fact more difficult to explain the idea behind it compared to the case of MRC at the receiver. Indeed, it may seem counterintuitive that transmit antenna selection is suboptimal and instead transmit MRC is optimal: if e.g. $|h_1|^2 \gg |h_2|^2$, why not send all power over the stronger channel? This is due to the property of the power as a quadratic constraint. Namely, assume that the channel coefficients h_1 and h_2 are real, still satisfying $|h_1|^2 \gg |h_2|^2$ and the power allocated to the first antenna is $P - \epsilon_P$, while the second antenna gets power ϵ_P. Then it can be shown that the derivative of

$$
h_1 \sqrt{P - \epsilon_P} + h_2 \sqrt{\epsilon_P} \tag{11.14}
$$

with respect to ϵ_P when $\epsilon_P = 0$ is always positive, such that regardless of how weak h_2 is, it is always beneficial to send at least some signal through that antenna. The optimality of transmit MRC is a key element of systems with a massive number of antennas, discussed in Section 11.6.

11.2.3 Transmit Diversity

In this section we stay with the MISO channel, but now assume that Zoya does not know the channel coefficients h_1 and h_2. The best she can do is to allocate half of the power to each of the antennas. However, just splitting the power into half and transmitting the same symbol from both antennas is not necessarily helpful. The received signal is equal to the one from (11.3), where we have set $z_1 = z_2 = z$, such that the resulting SNR is:

$$\gamma_{TX2} = \frac{|h_1 + h_2|^2}{\sigma^2}\frac{P}{2} \tag{11.15}$$

and is thus dependent on the relative phases between h_1 and h_2, rather than only on their amplitudes.

The solution to this conundrum is brought by transmit diversity based on the Alamouti scheme. This scheme is an example of a *space-time* code, where the intuition is built on the best traditions of communication/information theory: instead of focusing on transmission of a single symbol, augment the scheme towards the transmission of multiple symbols. Let us assume that the coherence time is of at least two symbol duration and consider two consecutive symbol transmissions. Let $z_i(j)$ be the complex symbol sent from the ith antenna during the jth symbol transmission. Zoya picks two baseband symbols that contain modulated data and are denoted by b_1 and b_2. During the first symbol transmission, she sends $z_1(1)$ from the first antenna and $z_2(1)$ from the second antenna, such that:

$$z_1(1) = b_1 \qquad z_2(1) = b_2. \tag{11.16}$$

During the second transmission she uses the same two symbols b_1, b_2 and sends:

$$z_2(2) = -b_2^* \qquad z_2(2) = b_1^*. \tag{11.17}$$

It can be seen that the symbols are modified in ways (the sign "−" and conjugation "*") that do not lose any information. The two symbols received by Yoshi are:

$$\begin{aligned} y(1) &= h_1 b_1 + h_2 b_2 + n(1) \\ y(2) &= -h_1 b_2^* + h_2 b_1^* + n(2) \end{aligned} \tag{11.18}$$

where $y(j)$, $n(j)$ are the received symbols and noise during the jth symbol reception. In order to make a decision about the symbol b_1, Yoshi combines its two antenna receptions as follows:

$$r_1 = h_1^* y(1) + h_2 y^*(2) = |h_1|^2 b_1 + |h_2|^2 b_1 + h_1^* n(1) + h_2 n^*(2) \tag{11.19}$$

where the interference from the symbol b_2 has disappeared, which means that the space-time code created an orthogonalized structure for transmission of two symbols. It can be seen from (11.19) that Yoshi achieves MRC, such that the SNR at which the symbol b_1 is received when sent with a space-time code is given by:

$$\gamma_{ST} = \frac{|h_1|^2}{\sigma^2}\frac{P}{2} + \frac{|h_2|^2}{\sigma^2}\frac{P}{2}. \tag{11.20}$$

In an analogous way, Yoshi makes the decision about the symbol b_2 by combining the outputs in the following way:

$$r_2 = h_2^* y(1) - h_1 y^*(2) = |h_1|^2 b_2 + |h_2|^2 b_2 + h_2^* n(1) - h_1 n^*(2) \tag{11.21}$$

and the SNR at which the symbol b_2 is received is identical to (11.20).

This SNR expression γ_{ST} clearly shows the nature of transmit diversity: it bets on both channels and gets the average SNR of the two channels. It can be noted that γ_{ST} is only half of the value of the SNR achieved with the optimal transmit MRC strategy; this is the price paid for not knowing h_1 and h_2 at the transmitter.

In the case of two antennas at the receiver, the schemes for receive diversity can be used in two conceptual ways: if the transmitter knows the SNR, then the rate can be adapted, while if it does not, then it uses a fixed rate, but the reliability is improved. In the case of transmit diversity without channel knowledge at the transmitter, the rate cannot be adapted as, by definition, the transmitter does not know the SNR and the scheme can be used for reliability improvement.

The usefulness of a given transmission scheme needs to be seen in the broader picture of the communication scenario and the functionalities of the other layers. This is particularly interesting for the transmit diversity scheme. A scenario in which it is clearly useful is the one of broadcast, where Zoya uses two antennas to broadcast the information that needs to be received by single-antenna devices of Yoshi, Xia, and possibly others. Each of the receiving terminals benefits from the diversity according to (11.20), but with its own channel coefficients.

Contrary to this, let us consider a case of a point-to-point link between Zoya and Yoshi for which the channel coefficients do not change during multiple packet transmissions. Furthermore, let us assume that Zoya runs a retransmission protocol based on the feedback by Yoshi. Zoya sends a packet using transmit diversity. Yoshi sends a single bit of feedback to indicate that the packet has been received correctly (ACK) or incorrectly (NACK). Assuming a time division duplexing (TDD) system, the channel is reciprocal and from the feedback sent by Yoshi, Zoya can learn the coefficients h_1 and h_2 and use this knowledge afterwards to transmit more effectively than just using a space-time code. In the simplest case, the feedback tells us which antenna is better and Zoya can use antenna selection afterwards. For completeness, we need to add that the switch from space-time code to antenna selection needs to be communicated to Yoshi, since he needs to use a different processing algorithm dependent on the transmission scheme used by Zoya.

11.3 Introducing MIMO

When both Zoya and Yoshi are equipped with two antennas as in Figure 11.1(c), the equivalent communication channel is called *multiple input multiple output (MIMO)* channel[1]. Using a combination of the techniques described for SIMO and MISO channels, one can devise methods by which both transmit and receive antennas can work together in the direction of increasing the SNR at Yoshi's receiver. However, by doing this, the resulting scheme remains qualitatively the same as in the scenario in which the transmitter and the receiver have a single antenna. This is because there is a single communication channel at the baseband, in which there is a single (complex) baseband input and a single baseband output, described by equation (11.4).

1 The term MIMO is used often, even in cases for which only one of the sides has multiple antennas.

The baseband model for a link from Zoya to Yoshi with two transmitting and two receiving antennas is:

$$y_1 = h_{11}z_1 + h_{12}z_2 + n_1$$
$$y_2 = h_{21}z_1 + h_{22}z_2 + n_2 \quad (11.22)$$

where z_i the baseband signal sent by Zoya from the transmit antenna TxAi, y_j, is the baseband signal received by Yoshi at his receiving antenna RxAj and n_j is the noise at RxAj. The propagation environment and the antennas are described by four coefficients h_{ij}, denoting the baseband channel from the TxAi to RxAj.

The key new qualitative feature brought by having multiple antennas at the transmitter and the receiver is that the link can operate with more than one communication channel. They are called *spatial channels*, since the channels result from having multiple transmitters/receivers that are placed at different spatial points. From the baseband viewpoint, the spatial channels are treated as parallel channels, independently affected by noise.

In order to get the first idea about multiple spatial channels, consider Figure 11.2. Both transmit antennas are used simultaneously, but for clarity, here we have depicted only the radio signals radiated from the antenna TxA1. It is assumed that, after undergoing various propagation effects, the signal from TxA1 arrives at each of Yoshi's receive antennas through two paths. Let us make the following very strong assumptions, which will be relaxed further. Regarding the reception at RxA1, it is assumed that the two rays (signals) that arrive from TxA1 to RxA1 interfere constructively, while the two rays that arrive from TxA2 to RxA2 interfere destructively and cancel each other. The situation is opposite at RxA2, that is the rays from TxA1 cancel each other, while the rays from TxA2 add up constructively. With these assumptions, the signals received by Yoshi are:

$$y_1 = h_{11}z_1 + n_1$$
$$y_2 = h_{22}z_2 + n_2 \quad (11.23)$$

where, due to the assumption of destructive interference, $h_{12} = h_{21} = 0$. Hence there are two parallel Gaussian channels between Zoya and Yoshi, with SNRs $\gamma_i = \frac{|h_{ii}|^2 P_i}{\sigma^2}$ where P_i is the power allocated by Zoya to the ith antenna. For easier illustration, assume that both

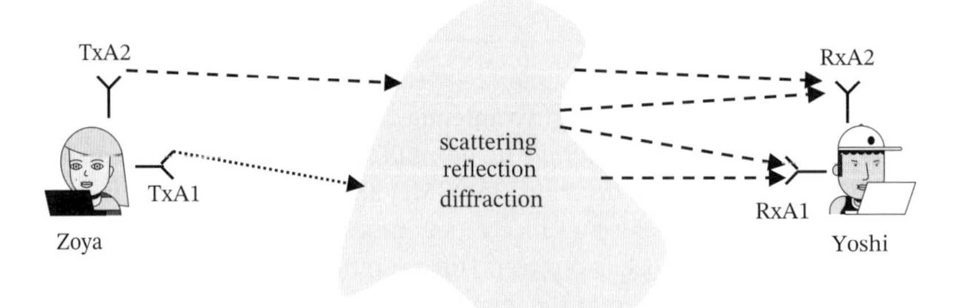

Figure 11.2 Illustration of the propagation effects when the transmitter and the receiver are equipped with two antennas. At the transmitter side, the rays from both antennas are depicted. For simplicity, at the receiver side only the rays coming from the first transmit antenna TxA1 are depicted.

channels have equal strength, such that the water filling solution will allocate equal power to both channels and the resulting SNRs are equal $\gamma_1 = \gamma_2 = \gamma$. Using normalized bandwidth $W = 1$, the capacity of (11.26) is:

$$C = \log_2(1 + \gamma_1) + \log_2(1 + \gamma_2) = 2\log_2(1 + \gamma). \tag{11.24}$$

Hence, the capacity is achieved by sending two independent data streams through each of the spatial channels. If, on the other hand, a single data stream is sent, then $z_1 = z_2 = z$ and, if Yoshi uses MRC, then the SNR of the received signal is $\gamma_1 + \gamma_2 = 2\gamma$ and the maximal rate that can be achieved is

$$R = \log_2(1 + 2\gamma) \tag{11.25}$$

for which it can easily be checked that it is lower than (11.24).

11.3.1 Spatial Multiplexing

The strong assumptions that we have made about the constructive/destructive interference at the receive antennas, may lead to the impression that the channel instance (11.26) is rather exotic. However, the possibility of creating two spatial streams is fundamental to the fact that there are two transmit and two receive antennas and spreads way beyond the exotic case (11.26). This mode of using the MIMO channel is known as *spatial multiplexing*, where multiple antennas are used to establish multiple spatial channels, i.e. simultaneous transmission of multiple data streams.

Indeed, let us assume that the destructive interference does not cancel any of the arrived signals, such that both receive antennas get a signal from TxA1 and TxA2. However, the propagation effects are such that the channel coefficients are $h_{11} = h_{21} = h_1$ and $h_{12} = -h_{22} = h_2$. It is convenient to represent the MIMO channel in a vector form[2]:

$$\mathbf{y} = \begin{bmatrix} y_1 \\ y_2 \end{bmatrix} = \begin{bmatrix} h_1 & h_2 \\ h_1 & -h_2 \end{bmatrix} \begin{bmatrix} z_1 \\ z_2 \end{bmatrix} + \begin{bmatrix} n_1 \\ n_2 \end{bmatrix}. \tag{11.26}$$

It turns out that for the specific channel (11.26), the optimal way to decode z_1 is to combine the received signals in the following way:

$$y_1 + y_2 = 2h_1 z_1 + n_1 + n_2 \tag{11.27}$$

such that, assuming that the power of z_1 is P_1, the available SNR is:

$$\gamma_1 = \frac{4|h_1|^2 P_1}{2\sigma^2} = \frac{2|h_1|^2 P_1}{\sigma^2}. \tag{11.28}$$

Similarly, the optimal way to decode z_2 is to combine the outputs as:

$$y_1 - y_2 = 2h_2 z_1 + n_1 - n_2 \tag{11.29}$$

and the SNR available to decode z_2 is:

$$\gamma_2 = \frac{2|h_2|^2 P_2}{\sigma^2}. \tag{11.30}$$

2 Vector variables represented in boldface letters, such as **x**.

Clearly, in (11.26) we have selected the channel gains carefully in order to make a point. The reader can easily check that, if the channel coefficient are such that $h_{11} = h_{21} = h_1$ and $h_{12} = h_{22} = h_2$, then it is not possible to create two spatial channels. Indeed, the received signal would be:

$$\mathbf{y} = \begin{bmatrix} y_1 \\ y_2 \end{bmatrix} = \begin{bmatrix} h_1 & h_2 \\ h_1 & h_2 \end{bmatrix} \begin{bmatrix} z_1 \\ z_2 \end{bmatrix} + \begin{bmatrix} n_1 \\ n_2 \end{bmatrix}. \tag{11.31}$$

It is worth noting that, in absence of noise $n_1 = n_2 = 0$, the two equations (11.31) are not linearly independent. This leads us to the idea that the quality of the two spatial channels is related to some degree of independency in the absence of noise, dictated by the channel coefficients. In general, this can be stated by representing the channel with two transmit and two receive antennas, referred to as a 2×2 MIMO channel, through the 2×2 matrix $\mathbf{H}$:

$$\mathbf{y} = \begin{bmatrix} y_1 \\ y_2 \end{bmatrix} = \begin{bmatrix} h_{11} & h_{12} \\ h_{21} & h_{22} \end{bmatrix} \begin{bmatrix} z_1 \\ z_2 \end{bmatrix} + \begin{bmatrix} n_1 \\ n_2 \end{bmatrix} = \mathbf{Hz} + \mathbf{n} \tag{11.32}$$

and the quality of the spatial channels depends on the properties of the matrix $\mathbf{H}$. In this sense, it is useful to represent the matrix $\mathbf{H}$ through a singular value decomposition (SVD) as follows:

$$\mathbf{H} = \mathbf{U\Lambda V}^H. \tag{11.33}$$

Here $\mathbf{U}$ and $\mathbf{V}$ are 2×2 unitary matrices, $\mathbf{V}^H$ is a conjugated and transposed version of $\mathbf{V}$ and

$$\mathbf{\Lambda} = \begin{bmatrix} \lambda_1 & 0 \\ 0 & \lambda_2 \end{bmatrix} \tag{11.34}$$

is a diagonal matrix. Let us now analyze the impact of each of these matrices.

First, note that a unitary matrix has the following feature:

$$\mathbf{UU}^H = \mathbf{U}^H\mathbf{U} = \mathbf{I} \tag{11.35}$$

where $\mathbf{I}$ is an identity matrix. This means that every unitary matrix $\mathbf{U}$ or $\mathbf{U}^H$ is invertible. Let the receiver Yoshi observe the vector $\mathbf{y}$. Then the information contained in:

$$\tilde{\mathbf{y}} = \mathbf{U}^H\mathbf{y} \tag{11.36}$$

is the same as the information contained in $\mathbf{y}$, provided that yoshi knows $\mathbf{U}^H$ perfectly, since $\mathbf{y}$ is uniquely determined by $\tilde{\mathbf{y}}$ and vice versa. Hence, Yoshi can work with the following received signal:

$$\tilde{\mathbf{y}} = \mathbf{U}^H\mathbf{y} = \mathbf{U}^H\mathbf{U\Lambda V}^H\mathbf{z} + \mathbf{U}^H\mathbf{n} = \mathbf{\Lambda V}^H\mathbf{z} + \tilde{\mathbf{n}} \tag{11.37}$$

where we have used $\mathbf{U}^H\mathbf{U} = \mathbf{I}$. Furthermore, $\tilde{\mathbf{n}} = \mathbf{U}^H\mathbf{n}$ is the transformed noise that has exactly the same statistical characteristics as the original noise vector $\mathbf{n}$: each noise sample remains Gaussian and independent of the other sample, preserving the same variance σ^2.

Finally, let us denote $\tilde{\mathbf{z}} = \mathbf{V}^H\mathbf{z}$. Now we can represent the channel between Zoya and Yoshi as:

$$\tilde{\mathbf{y}} = \mathbf{\Lambda}\tilde{\mathbf{z}} + \tilde{\mathbf{n}} \tag{11.38}$$

where $\mathbf{\Lambda}$ is a diagonal matrix and thus we have two independent spatial channels. However, the inputs to these two channels are not the antenna outputs $\mathbf{z}$, but rather the transformed antenna outputs $\tilde{\mathbf{z}}$. Again, due to the unitary property of $\mathbf{V}^H$, the total power of $\mathbf{z}$ and $\tilde{\mathbf{z}}$ are equal, such that we are not investing additional power in the system. Hence, the information carrying properties of the channel are not changed in any way, except that we have obtained a very convenient representation (11.38), for which we know how to achieve the capacity through water filling.

For the sake of clarity, we summarize the required knowledge and the transmission/reception procedures that achieve the capacity of the MIMO channel:

- At first, both Zoya and Yoshi learn $\mathbf{H}$ and based on that they calculate $\mathbf{U}, \mathbf{\Lambda}$, and $\mathbf{V}$.
- Knowing $\mathbf{\Lambda}$, Zoya applies water filling for two parallel Gaussian channels with noise variances $\frac{\sigma^2}{\lambda_1}$ and $\frac{\sigma^2}{\lambda_2}$. Based on that, Zoya determines the power level P_i of $\tilde{z}_i$, as well as the code rate/codebook for each of the two equivalent spatial channels. Since Yoshi also knows $\mathbf{\Lambda}$, he also knows the code rates/codebooks.
- Zoya creates the signal $\mathbf{V}\tilde{\mathbf{z}}$ and sends z_i through the respective antenna TxAi.
- Yoshi receives $\mathbf{y}$ and uses $\mathbf{U}^H$ to create $\tilde{\mathbf{y}}$ according to (11.36). After that Yoshi decodes each spatial channel $\tilde{y}_i = \lambda_i \tilde{z}_i + \tilde{n}_i$ separately.

The capacity of the 2×2 MIMO channel is:

$$C_{2\times 2} = \log_2\left(1 + \frac{P_1\lambda_1^2}{\sigma^2}\right) + \log_2\left(1 + \frac{P_2\lambda_2^2}{\sigma^2}\right). \tag{11.39}$$

Ideally, both eigenvalues λ_1 and λ_2 of the channel matrix $\mathbf{H}$ should be equal. However, it can happen that the channel $\mathbf{H}$ is *not well conditioned*. In other words, the channel does not provide sufficiently independent equations to Yoshi about the two unknown inputs from Zoya. In that case, the smaller eigenvalue λ_2 can be much lower than λ_1. In the extreme case, $\lambda_2 = 0$ and there is only one spatial channel. Note that the water filling takes this automatically into account, as the power that would be allocated to the spatial channel with $\lambda_2 = 0$ is zero.

In the general case of a MIMO channel, Zoya has M_T transmit antennas and Yoshi has M_R receive antennas. The maximal number of spatial channels that can be created between Zoya and Yoshi is:

$$M = \min\{M_\mathrm{T}, M_\mathrm{R}\} \tag{11.40}$$

which is intuitively clear, as this is the maximal possible number of independent inputs or outputs in the system.

11.4 Multiple Antennas for Spatial Division of Multiple Users

The electromagnetic properties of an antenna can be such that the antenna is directive and radiates within a beam that is not omnidirectional, as discussed in Section 10.3. Antenna directivity generalizes the communication model based on omnidirectional coverage by introducing coverage areas that are defined by the beams. This is illustrated in Figure 11.3(a), where Bastian achieves omnidirectional coverage by using three directed antennas, each of them having a beam that has a width of 120°. Each of the three directed antennas defines a *sector*, which is defined in a rather *geometric* way, as it shrinks the coverage to be from an angle of 360° towards a smaller angle. The use of directed antennas introduces the possibility of *space division multiple access (SDMA)*: Zoya, Yoshi, and Walt

Figure 11.3 Coverage area with directed antennas. (a) Passive directed antennas that define geometric beams and the omnidirectional coverage of Bastian is divided into three sectors. (b) Coverage defined by digital beamforming.

from Figure 11.3(a) can communicate simultaneously with Bastian (and vice versa), using the same frequency, while being orthogonal and not experiencing any interference. On the other hand, Yoshi and Xia are in the same sector and cannot be spatially separated, such that they do interfere if they simultaneously use the same frequency.

The opportunity for spatial division, offered by geometric beamforming, can be generalized further by *digital beamforming*. This is illustrated in Figure 11.3(b). Assume that Bastian has multiple antennas, while Zoya, Yoshi, Xia, and Walt have a single antenna each. If Bastian has a sufficient number of antennas, then he can create one spatial channel (beam) to each of the users. These beams are not geometric, but they still ensure orthogonality of the spatial channels of the four users. This is achieved by digital *zero forcing (ZF) precoding* and is discussed in the following section. The coverage area achieved by digital beamforming is concentrated around the target terminals, which minimizes the waste of the radiated energy. On the other hand, the fact that the coverage is extended to the nearest neighborhood of each terminal reflects the fact that the channels observed by antennas that are in close proximity are correlated.

Taking the layering view, digital beamforming is performed on top of the physical antennas, regardless of their electromagnetic properties. For example, the beamforming depicted in Figure 11.3(b) can be implemented by using omnidirectional antennas. Alternatively, the omnidirectional electromagnetic coverage can be achieved by a set of passive directed antennas; digital beamforming can still be performed on top of them. In any case, the number of RF hardware units, referred to as RF chains, is equal to the number of outputs/inputs that Bastian has in the baseband representation.

11.4.1 Digital Interference-Free Beams: Zero Forcing

We introduce the main idea behind ZF precoding through an elementary example. Let the base station Bastian have five antennas. We consider at first a downlink transmission, such that the MISO channel to each of the users is a vector with five coefficients. Let us focus only on Zoya and Yoshi and denote the respective channels by $\mathbf{h}_Z$ and $\mathbf{h}_Y$:

$$\begin{aligned}\mathbf{h}_Z &= \begin{bmatrix} h_{Z1} & h_{Z2} & h_{Z3} & h_{Z4} & h_{Z5} \end{bmatrix} \\ \mathbf{h}_Y &= \begin{bmatrix} h_{Y1} & h_{Y2} & h_{Y3} & h_{Y4} & h_{Y5} \end{bmatrix}.\end{aligned} \tag{11.41}$$

In a single transmission, Bastian wants to send a single symbol s_Z to Zoya and s_Y to Yoshi. He maps s_Z to the five transmitting antennas by using the precoding vector:

$$\mathbf{v}_Z = \begin{bmatrix} v_{Z1} \\ v_{Z2} \\ v_{Z3} \\ v_{Z4} \\ v_{Z5} \end{bmatrix}. \tag{11.42}$$

The vector $\mathbf{v}_Y$ for sending the symbol s_Y is defined in a similar manner. Bastian transmits the two symbols simultaneously:

$$\mathbf{b}_{TX} = \mathbf{v}_Z s_Z + \mathbf{v}_Y s_Y. \tag{11.43}$$

ZF precoding is implemented by a suitable choice of the precoding vectors $\mathbf{v}_Z$ and $\mathbf{v}_Y$. As in the case of superposition coding, the total power of $\mathbf{b}_{TX}$ should be limited to P_{TX}, which can be expressed as follows:

$$\|\mathbf{b}_{TX}\|^2 = \sum_{i=1}^{5} |b_i|^2 \leq P_{TX}. \tag{11.44}$$

The requirement to avoid interference to the reception of Zoya is stated as follows:

$$\mathbf{h}_Z \mathbf{v}_Y = 0 \tag{11.45}$$

while the requirement to avoid interference to Yoshi's reception is, similarly, determined as $\mathbf{h}_Y \mathbf{v}_Z = 0$. The received signal by the single antenna of Zoya is:

$$\begin{aligned} z &= \mathbf{h}_Z \mathbf{b}_{TX} + n = \mathbf{h}_Z(\mathbf{v}_Z s_Z + \mathbf{v}_Y s_Y) + n \\ &= \mathbf{h}_Z \mathbf{v}_Z s_Z + \mathbf{h}_Z \mathbf{v}_Y s_Y + n = \mathbf{h}_Z \mathbf{v}_Z s_Z + n \end{aligned} \tag{11.46}$$

where n is the noise and the last equality follows from the condition (11.45). In a similar way, Yoshi receives s_Y free of interference from s_Z. Since $\mathbf{h}_Z$ is of dimension 5, there can be many vectors $\mathbf{v}_Y$ that satisfy (11.45), or, in other words, lie in the null space of $\mathbf{h}_Z$. Bastian should choose $\mathbf{v}_Y$ such that the SNR at which Yoshi receives s_Y is maximized. On the other hand, Bastian should choose $\mathbf{v}_Z$ such that $\mathbf{h}_Y \mathbf{v}_Z = 0$, but the SNR at which Zoya receives s_Z in (11.59), which is proportional to $|\mathbf{h}_Z \mathbf{v}_Z s_Z|^2$, is maximized.

The fact that the precoding vectors $\mathbf{v}_Z$ and $\mathbf{v}_Y$ are selected regardless of the content of the actual data carrying symbols, s_Z and s_Y, puts ZF precoding in the class of *linear precoding* schemes.

Due to the channel reciprocity, the digital beams defined by $\mathbf{v}_Z$ and $\mathbf{v}_Y$ can also be used when Zoya and Yoshi transmit to Bastian in the uplink. In this case digital beamforming results in an interference-free multiple access channel. Specifically, Zoya and Yoshi send s_Z and s_Y, respectively, and Bastian receives:

$$\mathbf{b}_{RX} = \mathbf{h}_Z^T s_Z + \mathbf{h}_Y^T s_Y + \mathbf{n} \tag{11.47}$$

where $(\cdot)^T$ stands for transposed and $\mathbf{n}$ are the five noise samples from the five circuits, associated with the five receiving antennas of Bastian. This is a SIMO scenario, where, by contrast to MISO, the power constraint is put individually at each transmitter, Zoya and Yoshi. In order to extract the signal of Zoya from (11.47), Bastian applies the following ZF precoding vector of Zoya:

$$\begin{aligned} b_Z &= \mathbf{v}_Z^T \mathbf{b}_{RX} = \mathbf{v}_Z^T \mathbf{h}_Z^T s_Z + \mathbf{v}_Z^T \mathbf{h}_Y^T s_Y + \mathbf{v}_Z^T \mathbf{n} \\ &\overset{(a)}{=} (\mathbf{h}_Z \mathbf{v}_Z)^T s_Z + (\mathbf{h}_Y \mathbf{v}_Z)^T s_Z + \mathbf{v}_Z^T \mathbf{n} \overset{(b)}{=} (\mathbf{h}_Z \mathbf{v}_Z)^T s_Z + \tilde{n} \end{aligned} \tag{11.48}$$

where (a) follows from the matrix identity $(\mathbf{AB})^T = \mathbf{B}^T \mathbf{A}^T$ and (b) follows from (11.45). Here b_Z is a noisy observation of s_Z, where the resulting noise from all receive antennas is sublimed into the scalar $\tilde{n}$. In an analogous manner, Bastian extracts the signal of Yoshi from $\mathbf{b}_{RX}$ via the use of $\mathbf{v}_Y^T$.

Through the ZF precoding example above, we have shown the potential of multiple antennas to support multi-user communications in novel ways compared to the case when only a single antenna is used by the sender and the receiver. Specifically, ZF precoding achieves spatial separation of the users by processing inputs/outputs from multiple antennas in the digital, basedband domain. This is illustrated in Figure 11.4(a). However, if the channels $\mathbf{h}_Z$ and $\mathbf{h}_Y$ are far from being orthogonal, which can happen if the communication does not take place in a propagation environment without rich scattering/reflections, then ZF becomes largely sub-optimal. In the extreme case, where $\mathbf{h}_Z = \mathbf{h}_Y$, spatial separation of the users through ZF precoding is impossible.

The beamforming with passive directed antennas from Figure 11.3(a) is often termed *analog beamforming*. The three analog beams from Figure 11.3(a) do not depend on the instantaneous channels of the users and therefore cannot offer precise directivity, as in Figure 11.3(b), nor are they flexible enough to adjust to the changes of the users' channels. In contrast, the digital beams can be adapted flexibly to the users, at the expense of knowing the precise channel state information (CSI) for all antennas and all users.

When the mobile devices Zoya, Yoshi, etc. are also equipped with multiple antennas, they can also create digital beams that are focused on the desired base station (Bastian), while having null towards other, undesired base stations. If both the transmitters and the receivers are equipped with multiple antennas, then the optimization of the digital beamforming is jointly made for the transmitters and the receivers.

11.4.2 Other Schemes for Precoding and Digital Beamforming

While it is intuitively appealing to use ZF precoding and support multiple users, there are other schemes for precoding that can offer better performance. Since, by definition, ZF avoids interference among the users, SINR is not relevant and we can limit the discussion to SNR only. The other precoding schemes do not aim to cancel the interference, but can offer SNR that is larger than the SNR offered by the ZF. The signal/interference obtained from those non-ZF schemes can, in general, be illustrated as in Figure 11.4(b).

Let us stay with the MISO scenario from the previous subsection and consider a precoding based on transmit MRC, a precoding scheme that we have already encountered

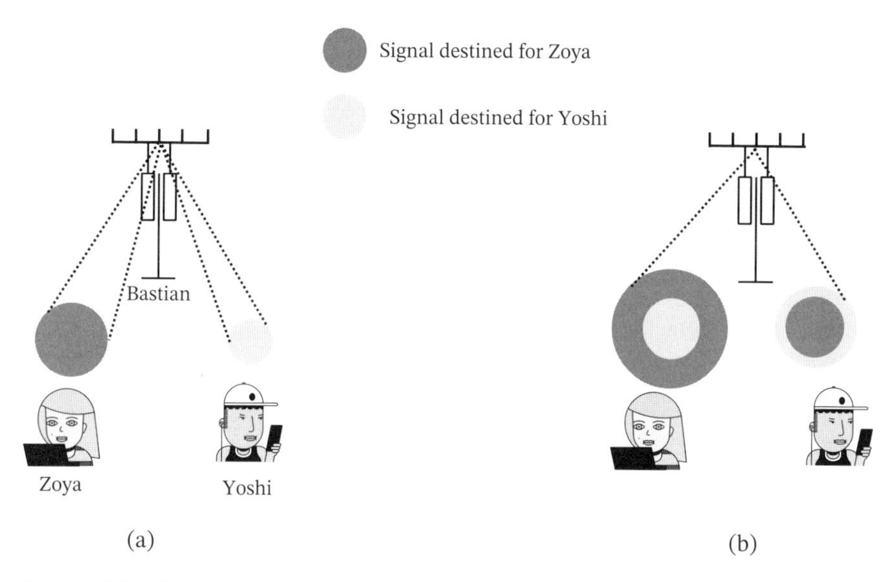

Figure 11.4 Illustration of the distribution of the signals and interference with digital beamforming. Bastian has five antennas, as in our example. The larger the circle, the higher the power of the received signals. (a) Interference-free digital beamforming using ZF precoding. (b) Digital beamforming with interference, such as transmit MRC and MMSE. The ratio between the radii of the overlapping circles indicates the SINR of the signal.

in Section 11.2.2. Bastian sends again the signal $\mathbf{b}_{\mathrm{TX}}$ as in (11.43), but now the precoding vectors $\mathbf{v}_{\mathrm{Z}}$ and $\mathbf{v}_{\mathrm{Y}}$ are chosen as follows:

$$\begin{aligned} \mathbf{v}_{\mathrm{Z}} &= \rho_{\mathrm{Z}} \mathbf{h}_{\mathrm{Z}}^{H} \\ \mathbf{v}_{\mathrm{Y}} &= \rho_{\mathrm{Y}} \mathbf{h}_{\mathrm{Y}}^{H} \end{aligned} \tag{11.49}$$

where, as before, $(\cdot)^H$ stands for conjugated and transposed. The factors ρ_{Z} and ρ_{Y} are used to regulate the transmit power in order to satisfy the power limit (11.44). This type of precoding maximizes the power of the useful signal beamed to each of the destinations, Zoya and Yoshi, but does not guarantee that there is no interference between them. The SNR performance depends on the relationship between the channels $\mathbf{h}_{\mathrm{Z}}$ and $\mathbf{h}_{\mathrm{Y}}$: the closer to zero the product $\mathbf{h}_{\mathrm{Y}}\mathbf{h}_{\mathrm{Z}}^{H}$ is, the lower the interference between the signals intended to Zoya and Yoshi, hence the higher the SINR for each of them.

A reasonable objective for a linear precoding scheme would be to maximize the SINR for both signals, decoded by Zoya and Yoshi, respectively. This scheme is called *Wiener precoding* or *minimum mean squared error (MMSE)* precoding. Intuitively, this precoder attempts to hit a trade-off between the extremes of ZF (interference avoided) and transmit MRC (power of the useful signal maximized).

The central feature of the linear precoding schemes is that the selection of the symbols transmitted to Zoya does not depend on the actual transmitted symbol to Yoshi, and vice versa. Indeed, in ZF the rate/modulation/coding of Zoya is selected knowing that there is no interference from Yoshi. For the other linear precoders that do not eliminate the

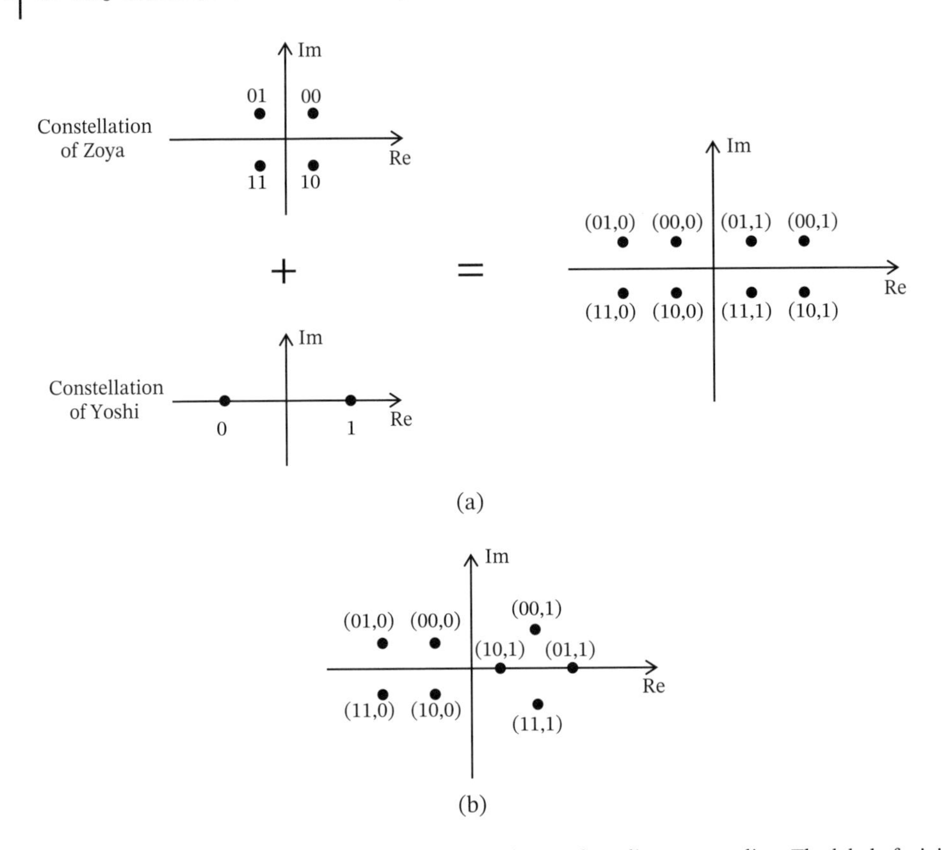

Figure 11.5 Difference between superposition coding and nonlinear precoding. The label of a joint symbol (s_Z, s_Y), such as e.g. (01, 1), means that the symbol carries the bits $s_Z = 01$ for Zoya and $s_Y = 1$ for Yoshi. (a) Superposition coding using QPSK for Zoya and BPSK for Yoshi. (b) Nonlinear joint encoding of the symbols for Zoya and Yoshi.

interference between Zoya and Yoshi, the rate/modulation/code of Zoya is selected by treating the interference from Yoshi as additional noise; and vice versa for Yoshi with respect to the interference from the signal intended to Zoya. For example, if Bastian communicates with Zoya using QPSK, then whenever he wants to send the bits 01, he sends the same complex symbol s_{01} to Zoya.

This is changed in *nonlinear precoding*, where the way 01 is encoded for Zoya depends on the data that Bastian has to send to Yoshi. The key observation is that, by selecting the outputs of his five antennas, Bastian has the possibility to simultaneously control the signals received by Zoya and Yoshi. In order to understand the essence of nonlinear precoding, it is useful to compare it to superposition coding, as illustrated in Figure 11.5. Superposition coding is shown in Figure 11.5(a), where it is assumed that Bastian communicates to Zoya using QPSK and to Yoshi using BPSK. If Zoya is able to decode and cancel the signal of Yoshi, then what is left is a signal that is encoded with the original QPSK constellation, regardless of whether Yoshi sent 0 or 1. This is no longer true for the nonlinear precoding, shown in Figure 11.5(b). The 8ary constellation in Figure 11.5(b) cannot be obtained by superposition of two independent signals. Zoya decodes the signal intended for her as $s_Z = 00$ whenever the received signal is close to either $(00, 0)$ or $(00, 1)$.

A well known form of nonlinear precoding is termed *dirty paper coding*. Looking at Figure 11.5(b), dirty paper coding for Zoya can be interpreted as follows. Bastian determines the symbol to be sent to Yoshi; think of it as a symbol that Bastian writes on paper. When encoding the symbol for Zoya, Bastian treats whatever there is already for Yoshi as "dirt" and he tries to write the symbol of Zoya around the dirt, but does not use power to cancel the dirt. However, Bastian has the freedom to decide the constellation of Zoya based on where the dirt is placed, rather than always using the same constellation, as is the case in Figure 11.5(a).

Nonlinear precoding is, in general, more complex than linear and, in practice, the preferred precoding schemes are the linear ones.

11.5 Beamforming and Spectrum Sharing

The use of multiple antennas and digital beamforming through precoding allow Bastian to apply space division multiple access (SDMA) and thereby serve multiple terminals simultaneously and in the same frequency spectrum. Recall that CDMA achieves something similar. However, CDMA relies on the spread spectrum in order to achieve user separation. In contrast to CDMA, the separation/differentiation of the users in SDMA is based on the spatial features of the users and does not require a spread spectrum. Ideally, if a separate, non-interfered, spatial channel can be defined between Bastian and each of the terminals, then a single frequency band can be simultaneously used by all users, which is very efficient in terms of spectrum requirement.

However, SDMA has its disadvantages and its implementation introduces an additional cost in terms of overhead. In CDMA, the system designer is in control of the separation between the users by the choice of the spreading sequence or the frequency hopping pattern. In SDMA, the quality of user separation depends on the relation among their channel coefficients, which are chosen by nature, such that the designer is not in control of them. The cost for implementing SDMA is seen in the fact that Bastian's precoding is critically dependent on his knowledge about the channels from all of his antennas to all the mobile devices, Zoya, Yoshi, and the others. Since these channels are changing dynamically, based on fading or user mobility, keeping the gains of SDMA may generate a significant overhead from the procedures used to acquire the CSI, which consists of the values of the channel coefficients between Bastian and all users.

The communication scenario from Figure 11.3 has Bastian in the role of coordinator and he is either source or destination for all transmissions. The use of digital beamforming is also interesting in scenarios with interfering links. Figure 11.6 depicts three interfering links, Zoya–Yoshi, Xia–Walt, and Victoria–Umer. Specifically, the figure depicts an instant at which the transmitters are Zoya, Walt, and Victoria, while the respective receivers are Yoshi, Xia, and Umer. One way to avoid interference can be to have the receiver setting up the digital beam to have null for the signals arriving from the transmitter. Alternatively, the transmitter can position the null of its digital beam at the spot at which the undesired receiver is placed. For the example in Figure 11.6, the receiver Yoshi has a null towards Walt

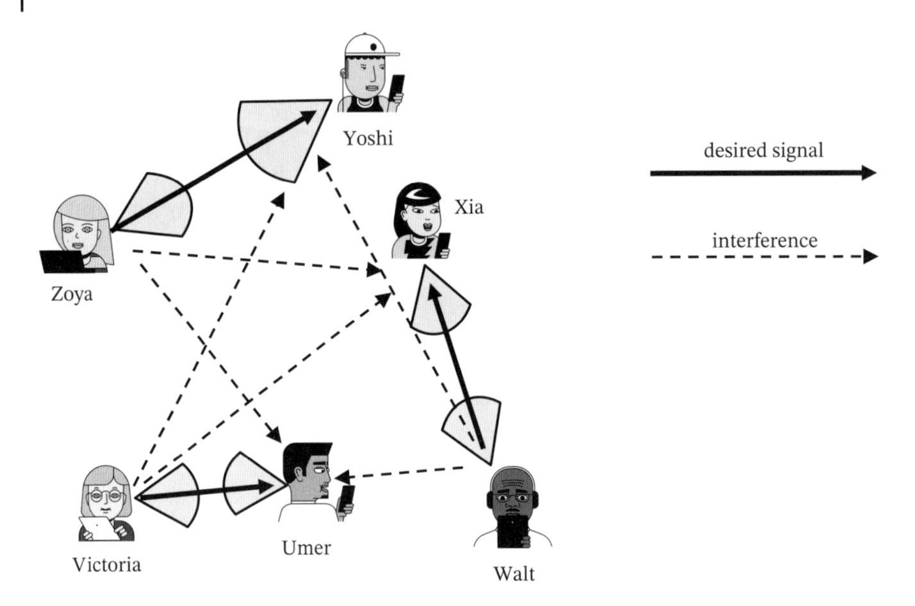

Figure 11.6 Example of spectrum sharing among three links by using digital beamforming to implement space division multiple access (SDMA).

and thus avoids interference from him. On the other hand, the interference from Victoria to Yoshi is avoided due to the fact that the digital beam of the transmitter Victoria has null towards the undesired receiver Yoshi. Similar considerations are valid for the receivers Xia and Umer. Hence, the use of multiple antennas allows the three links to operate in the same spectrum, i.e. *share* the same spectrum while not interfering with each other. We remark that the geometric form of the beams is for illustrative purposes only; as we have seen previously in the chapter, the zeros created by the digital beamforming are not necessarily associated with a geometric direction, but with a spatial point that has a specific values of its channel coefficients.

In Chapter 9 we concluded that an ideal separation in frequency is possible when the interfering links are perfectly synchronized, they share a common time interval T and agree how to share the (potentially infinite) number of subcarriers with frequencies $\frac{k}{T}$, where k is an integer. This would be an ideal spectrum sharing applicable to single-antenna devices. In practice, achieving such synchronization among heterogeneous systems is unfeasible; that is why we are resorting to frequency division in which the bands in which the system operate are sufficiently separated to make the interference negligible. Separating the users/links in SDMA does not rely on the strong assumption that the links are perfectly synchronized, but it does require knowledge of CSI, such that the transmitters/receivers can create suitable digital beams. Hence, spectrum sharing by SDMA is critically dependent on the information about the dynamically changing channel coefficients. In contrast to TDMA and SDMA, FDMA relies on dividing the spectrum in smaller bands, sufficiently separated, and allocate the links to different bands. Since FDMA does not depend on time synchronization or CSI knowledge, it appears as the most robust way of avoiding interference among independent links. At the same time, FDMA is the most wasteful in terms of bandwidth, as each link occupies a different part of the spectrum.

Instead of digital beamforming, spectrum sharing can also be achieved by analog beamforming or hybrid analog–digital beamforming. Indeed, if the nodes in Figure 11.6 have directed antennas, then these antennas can be positioned in a way that avoids the interference among the neighboring links.

11.6 What If the Number of Antennas is Scaled Massively?

In order to take advantage of digital beamforming and SDMA, multi-antenna transmitters and receivers need to acquire and/or exchange information about the channel coefficients for all of their antennas. However, even if the coefficients are known perfectly, nature may have chosen their values in an unlucky way, such that the performance offered by SDMA is not satisfactory. For example, Zoya wants to transmit to Yoshi while making a null towards the receiver Walt. However, the values of the channel coefficients are such that the use of ZF results in deterioration of the SNR for the desired receiver Yoshi. For the sake of completeness, we need to mention that a similar observation is valid when the communication nodes use antennas that are electromagnetically directed. Namely, the transmitter/receiver needs to orient the antennas to avoid interference, but the spatial position of the nodes can also be an unlucky one. For example, the desired and the undesired receiver may lie in the same line of the beam and thus prevent an efficient multi-user communication with SDMA.

Let us now consider the case in which the number of antennas used to do digital beamforming is massively scaled to hundreds, even theoretically to infinity. Then the bad-luck cases in SDMA disappear due to the statistical properties of the channel coefficients, as will readily be seen. Generally, the concept of massive antennas is applicable to base stations, since having a large number of antennas on a device is rather unfeasible. Figure 11.7 shows

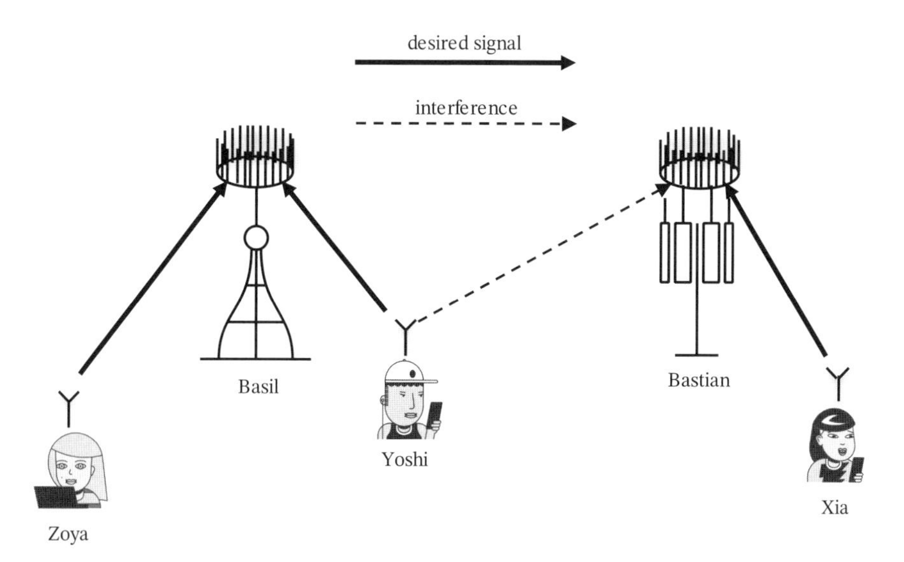

Figure 11.7 Illustration of a massive MIMO. Basil and Bastian are base stations, each with massive number of antennas. Each mobile device has a single antenna, as emphasized. Zoya and Yoshi communicate with Basil, Xia communicates with Bastian. There is also interference between the terminals and the base station to which they are not associated: the figure illustrates only the interference between Yoshi and Bastian.

such a system, termed a *massive MIMO*. Bastian has M antennas, where M is a large number. The terminals Zoya and Yoshi are assumed to have single antenna each. However, this is not limiting, as the concepts described here are also applicable when the terminals are equipped with multiple antennas. The channel coefficients from the base station to Zoya and Yoshi are denoted by $\mathbf{h}_Z$ and $\mathbf{h}_Y$, respectively. They are vectors as in (11.41), except that here they have $M \gg 1$ elements. Thus, for example $\mathbf{h}_Z$ is given as:

$$\mathbf{h}_Z = \begin{bmatrix} h_{Z1} & h_{Z2} & h_{Z3} & \cdots & h_{ZM} \end{bmatrix}. \tag{11.50}$$

11.6.1 The Base Station Knows the Channels Perfectly

Let us for a moment assume that Basil somehow knows $\mathbf{h}_Z$ and $\mathbf{h}_Y$ perfectly. At first we consider an uplink transmission, simultaneously from Zoya and Yoshi. The signal received by Basil is given by:

$$\mathbf{b}_{RX} = \mathbf{h}_Z^T s_Z + \mathbf{h}_Y^T s_Y + \mathbf{n}. \tag{11.51}$$

In order to extract the signal from Zoya, Basil creates the following decision variable:

$$\hat{b}_Z = \frac{1}{M}\mathbf{h}_Z^* \mathbf{b}_{RX} = \frac{1}{M}\mathbf{h}_Z^* \mathbf{h}_Z^T s_Z + \frac{1}{M}\mathbf{h}_Z^* \mathbf{h}_Y^T s_Y + \frac{1}{M}\mathbf{h}_Z^* \mathbf{n} \tag{11.52}$$

where for convenience we have scaled the received signal with the constant $\frac{1}{M}$. The large value of M unleashes two asymptotic statistical properties that are central to the massive MIMO performance.

The first property is related to the following quantity:

$$\frac{1}{M}\mathbf{h}_Z^* \mathbf{h}_Z^T = \frac{1}{M}\sum_{m=1}^{M} |h_{Zm}|^2 \xrightarrow{M\to\infty} \beta_Z \tag{11.53}$$

where h_{Zm} is the channel coefficient from Zoya to the mth antenna of Basil. The channel coefficients $\{h_{Zm}\}$ are random values that are sampled from a distribution where the mean square value is $E[|h_{Zm}|^2] = \beta_Z$. The property (11.53) is known as *channel hardening*: the value of the coefficient β_Z that multiplies the received symbols does not depend on the instantaneous values of the channel coefficients.

The second property is related to

$$\frac{1}{M}\mathbf{h}_Z^* \mathbf{h}_Y^T = \frac{1}{M}\sum_{m=1}^{M} h_{Zm}^* h_{Ym} \xrightarrow{M\to\infty} 0 \tag{11.54}$$

and is termed *asymptotic orthogonality*. This is the consequence of the fact that the channel coefficients of Zoya and Yoshi are independent and they are sampled from distributions with zero mean value.

Applying the properties of channel hardening and asymptotic orthogonality to (11.52) we get:

$$\hat{b}_Z \approx \beta_Z s_Z + \frac{1}{M}\mathbf{h}_Z^* \mathbf{n}. \tag{11.55}$$

Assuming that the noise has a variance of σ^2, the SNR that is available to decode the symbol of Zoya is[3]:

$$\gamma_Z = \frac{M\beta_Z}{\sigma^2} \tag{11.56}$$

where we are getting the usual processing gain from the MRC receiver. The decision variable and the correspondent SNR for Yoshi are found in a similar way.

The same statistical properties are also put to work for downlink transmissions from Basil to Zoya and Yoshi. The signal transmitted by Basil is a vector and follows (11.43) for a MISO transmitter:

$$\mathbf{b}_{\text{TX}} = \rho_Z \mathbf{h}_Z^H s_Z + \rho_Y \mathbf{h}_Y^H s_Y \tag{11.57}$$

where the scalar parameters ρ_Z and ρ_Y are used to satisfy the transmit power requirements for Basil, formulated by (11.44). The symbol received by Zoya is:

$$\mathbf{h}_Z \mathbf{b}_{\text{TX}} + n_Z \tag{11.58}$$

where n_Z is the noise at Zoya's receiver. After the normalization with the number of antennas M, the received symbol by Zoya is:

$$z = \frac{1}{M}\rho_Z \mathbf{h}_Z \mathbf{h}_Z^H s_Z + \frac{1}{M}\rho_Y \mathbf{h}_Z \mathbf{h}_Y^H s_Y + \frac{n_Z}{M} \approx \rho_Z \beta_Z s_Z + \frac{n_Z}{M} \tag{11.59}$$

and the last approximation is the result of the properties of channel hardening and the asymptotic orthogonality; note that $\frac{n_Z}{M}$ will also tend to 0 as M goes to infinity.

We can conclude that, in the downlink, the massive number of antennas allows using linear precoding that is capable of directing energy and specific data content, with a great precision, to a specific spatial point. Precoding is done in a way to take advantage of asymptotic orthogonality and avoid interference among the users receiving the downlink. Conversely, in the uplink the asymptotic orthogonality is utilized by the receiver to remove the multi-user interference almost perfectly. Hence, the massive number of antennas allows multi-user decoding without an additional spectrum investment, unlike, for example, the spread spectrum.

11.6.2 The Base Station has to Learn the Channels

The attractive features of massive MIMO are conditioned on the perfect knowledge of a massive number of channel coefficients. The fact that this knowledge cannot be perfect sets fundamental limits on the performance of practically realizable massive MIMO systems.

A feasible way to learn the channel coefficients is to have pilot transmission from the terminal, say Zoya, to the base station Basil, such that Basil simultaneously estimates all M elements of the vector $\mathbf{h}_Z$. However, due to the presence of noise the estimated value of the

3 This is only an approximation, since to assume channel hardening and asymptotic orthogonality, we have taken that $M \rightarrow \infty$, but here we are still using a finite value of M.

vector will be $\hat{\mathbf{h}}_Z$, which is only approximately equal to $\mathbf{h}_Z$. Furthermore, having obtained $\hat{\mathbf{h}}_Z$, Basil can proceed to receive uplink transmissions from Zoya. However, in order to use the same $\hat{\mathbf{h}}_Z$ for downlink transmission, it is necessary to assume that the system operates in a TDD regime and uses the feature of channel reciprocity. The fact that $\hat{\mathbf{h}}_Z \neq \mathbf{h}_Z$ means that the precoding (in the downlink) or reception (in the uplink) applied by Basil has an additional noise compared to the case when $\mathbf{h}_Z$ is known perfectly.

Massive MIMO boasts high capability for spatial multiplexing of users, such that learning of the channel coefficients for all users associated with a base station incurs significant costs in terms of training overhead. If Zoya and Yoshi want to let Basil estimate their channel vectors $\mathbf{h}_Z$ and $\mathbf{h}_Y$ simultaneously, then the pilots that they are using need to be orthogonal. In this way Basil can estimate each of them free of interference from each other, such that the estimate remains corrupted only by the noise. If Zoya transmits the training pilot by herself, then it is, in principle, possible that she uses only a single channel use for sending the pilot. However, this is not sufficient when Zoya and Yoshi need to send orthogonal pilots; in this case they need to use at least two symbols. In general, K orthogonal pilots require at least K channel uses, that is, K symbols transmitted in the uplink. This introduces a cost, as it increases the fraction of time that the system spends on overhead and thereby decreases the overall efficiency of communication.

Things are getting more complicated when we consider that there may be multiple base stations with massive number antennas, each with different set of associated users. In Figure 11.7, Zoya and Yoshi are associated with Basil, while Xia is associated with Bastian. The use large number of antennas promotes spectral efficiency and, in that spirit, we assume that Basil and Bastian operate in the same frequency band. Furthermore, note that the signal of Yoshi creates non-negligible interference at Bastian's receiver. In the classical setup for massive MIMO, it is assumed that both Basil and Bastian use the same set of K orthogonal pilot sequences. Otherwise, if one requires that all users associated in all cells use a different orthogonal sequence, the number of pilot sequences scales to a level that introduces an excessive overhead. Let us assume that Yoshi and Xia use the same pilot sequence. Then the estimate that Bastian obtains for the channel coefficients of Xia, represented by $\mathbf{h}_X$, gets contaminated by Yoshi's transmission. Indeed, let $\mathbf{g}_Y$ represent the channel coefficients between Yoshi and Bastian. Then the signal received by Bastian for each of the symbols s of the pilot sequence is:

$$\mathbf{b}_{\text{RX}} = \mathbf{h}_X^T s + \mathbf{g}_Y^T s + \mathbf{n} = (\mathbf{h}_X^T + \mathbf{g}_Y^T) s + \mathbf{n} \tag{11.60}$$

and the channel coefficients $\mathbf{g}_Y$ inevitably corrupt the estimate $\hat{\mathbf{h}}_X$.

This is known as a *pilot contamination problem*. It is a consequence of a decision regarding resource allocation, applied to the pilot sequences. One can take a different approach in order to eliminate the problem. For example, Basil and Bastian can be allocated $K/2$ orthogonal pilots each, such that no terminal associated with Basil can choose the same pilot sequence with a terminal associated with Bastian. In this case, if both base stations have more than $K/2$ associated users, then they need to dynamically allocate the pilots to the terminals. One option could be to let the terminal run a random access process in order to contend for access to a specific set of preambles. Another option is dynamic allocation of the pilot sequences based on some scheduling: for example, the terminals requiring service with low latency may be allocated a fixed pilot sequence over a longer period, while those that are tolerant to latency may contend for pilot sequences through random access. In this way, the pilot contamination problem is transformed into a protocol design problem within the domain of each of the base stations.

11.7 Chapter Summary

Multipath and other random effects of the radio propagation are traditionally seen as necessary impediments that the wireless system designer needs to deal with. However, if the transmitter and/or the receiver are equipped with multiple antennas, then these random propagation effects may be turned into an opportunity for more efficient wireless connectivity. In that sense, the use of multiple antennas in wireless communication can be seen as a giant intellectual and technological step. In this chapter we have discussed the benefits brought by the use of multiple antennas, but also the challenges associated with reaping those benefits. The chapter has first covered the ideas of transmit and receive diversity. It has then introduced the concept of spatial multiplexing or communication through multiple spatial channels, which is possible in a MIMO setup, where both the transmitter and the receiver have multiple antennas. Next, we have covered the ideas of SDMA based on directional antennas and introduced the notion of non-geometric beams that are dedicated to specific users. Finally, we have introduced the basic ideas of massive MIMO, a disruptive concept inspired by the asymptotic properties of large antenna arrays. Yet, the mechanisms of massive MIMO have already started to work with hundreds of antennas. As we will see in the next chapter, the revolutionary ideas brought by massive MIMO have potential for large architectural implications in wireless networks.

11.8 Further Reading

The topic of multi-antenna wireless communication has been a subject of many books; the reader can check Brown et al. [2012] for a pedagogical approach and a more recent book Heath Jr. and Lozano [2018]. Massive MIMO is still a very active area of research; the reader is referred to Marzetta et al. [2016] and Björnson et al. [2017] for further details. In the spirit of this book, the problem of massive MIMO has been treated as a cross-layer problem of medium access and physical layer in De Carvalho et al. [2017], where it is shown how random access redefines the pilot contamination problem.

11.9 Problems and Reflections

1. *System design for multiple antennas.* In this assignment we continue along the lines of generalizing the design of the TDMA frame (see Problem 1 from Chapter 9), but this time by bringing in the features of multiple antennas. Assume that the base station Bastian has multiple antennas, while each device connected to it has a single antenna. Propose a frame design for a TDMA based system with multi-antenna base station. Discuss separately the case of TDD and FDD.

2. *Spatial Modulation*. Problem 4 from Chapter 6 was dedicated to a communication channel that emerges when antenna activation is considered as signaling. A related technique is termed spatial modulation, where information is transmitted by means of the indices of the transmit antennas of a MIMO system.
   (a) Investigate the concept of spatial modulation from the relevant literature and relate it to the specific channel instances discussed in Problem 4 from Chapter 6.
   (b) Compare spatial modulation with the "ordinary" MIMO presented in the chapter. What are the advantages brought by spatial modulation with respect to the transmission of data and/or control information?
3. *Comparison of digital beamforming methods* Section 11.4.1 introduced ZF beamforming with an example of five transmit antennas. Figure 11.4 provided a qualitative comparison with other beamforming methods, such as MMSE and transmit MRC. Build a model that will provide a quantitative comparison of the different beamforming methods. Simulate different instances of the channel coefficients $\mathbf{h}_Z$ and $\mathbf{h}_Y$ and see how the structure of these vectors affects the results.
4. *Multiple antennas and retransmissions*. Problem 5 from Chapter 7 dealt with the new aspects that emerge when retransmission is applied to a communication signal created by superposition coding. Assume that both Zoya and Yoshi have two antennas each.
   (a) Study the possible retransmission schemes for the case in which Zoya transmits a single data stream, using transmit diversity. As discussed in the chapter, the ARQ feedback sent by Yoshi can help in the antenna selection.
   (b) Study the possible retransmission schemes for the case in which Zoya transmits two data streams, using spatial multiplexing. Each transmission of Zoya consists of two packets transmitted in parallel; note that the outcome of this transmission is that either zero, one, or two packets should be retransmitted.
   (c) Following Problem 5 from Chapter 7, add superposition coding and investigate how that changes the options for retransmission in a multi-antenna system.
5. *Multiple antennas and access protocols*. When the base station Bastian is equipped with multiple antennas, it is intuitively expected that the capability of Bastian to resolve collision from accessing devices is increased proportionally to the number of antennas. Investigate how can a random access protocol take advantage of multiple antennas for the following cases:
   (a) ALOHA-type protocols with intra-collision SIC.
   (b) Coded random access with inter-collision SIC.

   Study both the cases in which Bastian has few antennas, as well as a massive number of antennas. Pay attention to the question that channel knowledge is required from Basil as well as from the accessing devices.

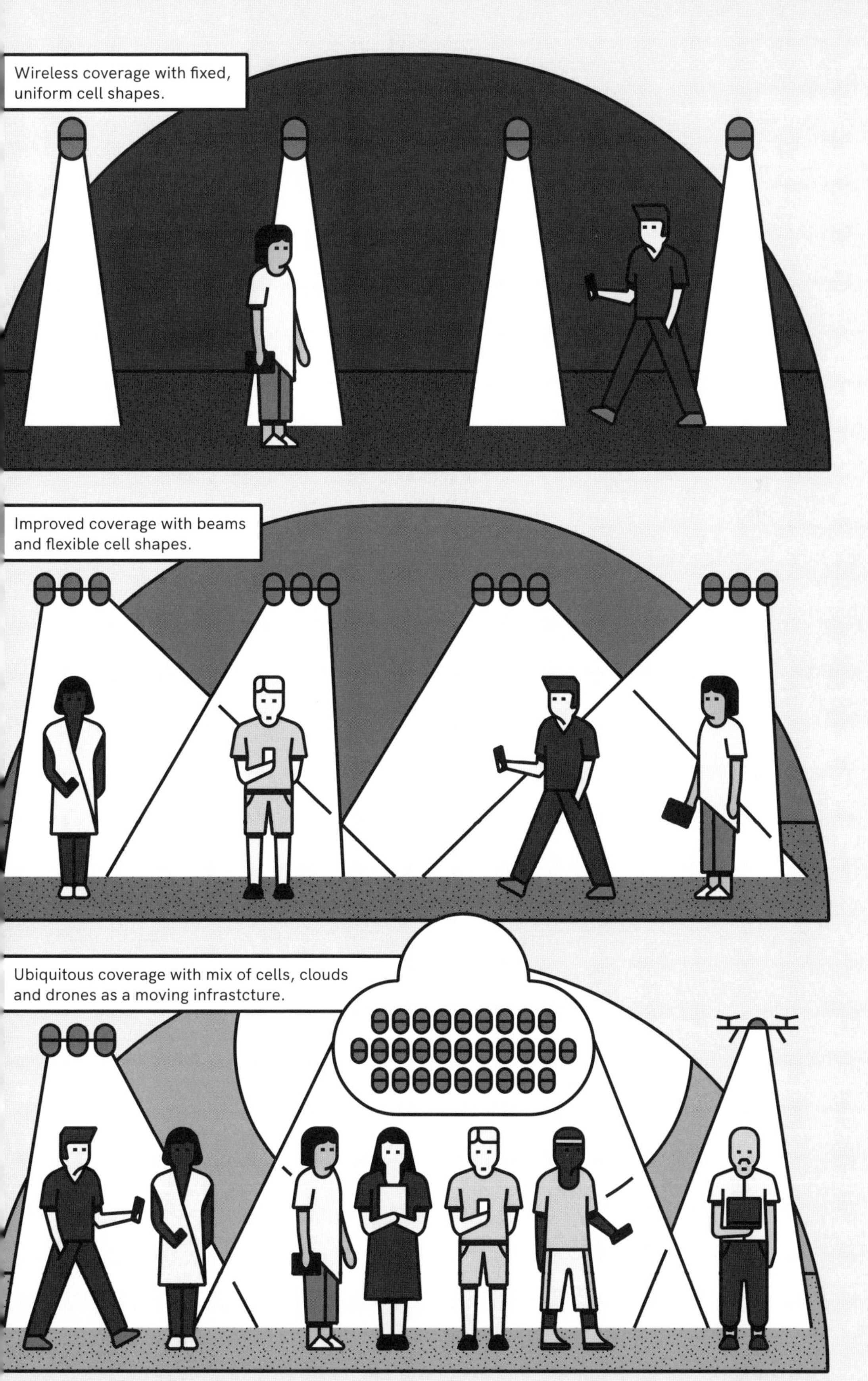

Wireless coverage with fixed, uniform cell shapes.
Improved coverage with beams and flexible cell shapes.
Ubiquitous coverage with mix of cells, clouds and drones as a moving infrastcture.

Story by Petar Popovski / Art by Peter Gregson

12

Wireless Beyond a Link: Connections and Networks

The previous chapters have been mostly limited to scenarios in which the two communicating parties, for example Zoya and Yoshi, are in close proximity and communicate either directly or through a multi-hop connection, relying on other nearby wireless devices. In the vast majority of practical cases, the communicating parties are not in physical proximity and there is a need for an elaborate communication infrastructure to establish and support the connection between them. As illustrated in Figure 12.1, the base station Basil, to which Xia communicates directly, is very often not the other communication party for Xia, but an intermediate point of the *connection* that offers Xia *access* to the infrastructure, which then carries the data through a network to Walt, the actual communication party.

This last chapter presents various aspects of the infrastructure that support connectivity between two parties, where at least one of the parties has a wireless connection. The wireless communication infrastructure should, in general, solve the following problems:

1. *End-to-end connectivity*. Offer end-to-end connection to the service or the other communicating party.
2. *Wireless coverage*. Ensure that the devices positioned within a given area (city, building, production plant, etc.) have reliable wireless connectivity.
3. *Mobility*. Wireless connectivity follows the users that are changing their position.
4. *Interference management*. Avoid or mitigate interference among different wireless connections.

In order to understand the design approach to these infrastructure related problems, we need to look at first at the different types of wireless connectivity that should be supported by the communication infrastructure.

12.1 Wireless Connections with Different Flavors

12.1.1 Coarse Classification of the Wireless Connections

The underlying assumption behind most of the connections so far has been that the communicating parties are humans. This is referred to as *human-to-human (H2H)*

Wireless Connectivity: An Intuitive and Fundamental Guide, First Edition. Petar Popovski.

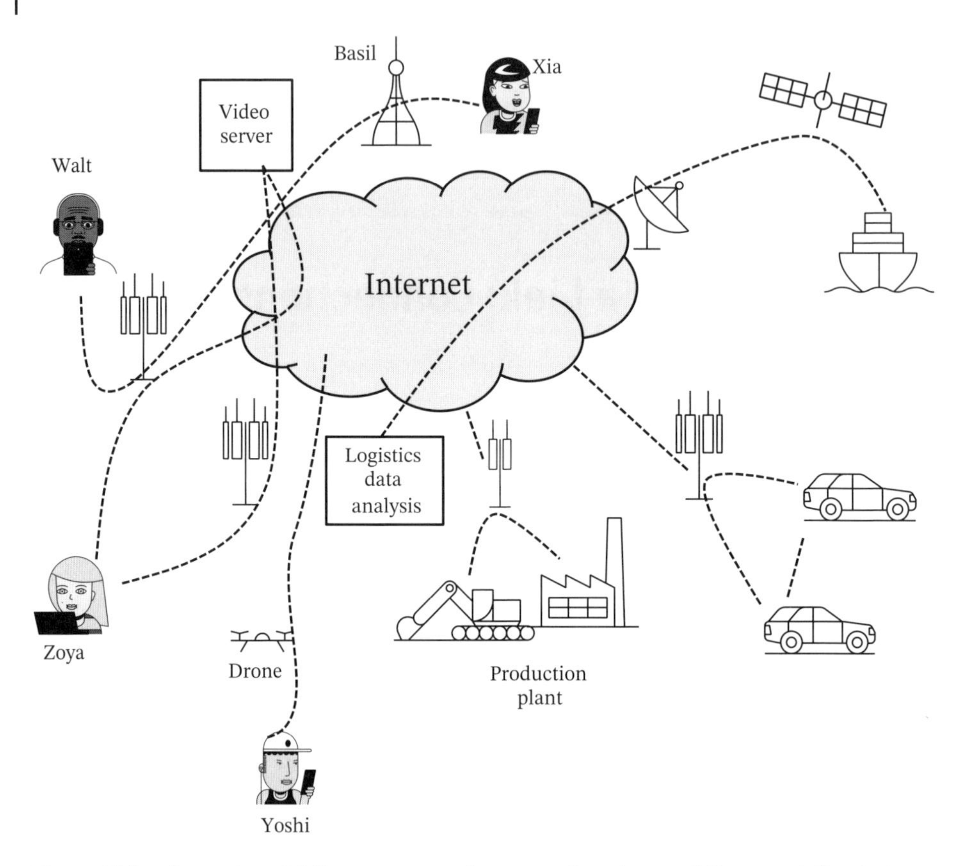

Figure 12.1 Illustration of different use cases for connections that are fully or partially wireless. Walt and Xia have voice communication. Zoya streams video using multiple paths in order to gain diversity. Yoshi gets wireless coverage through a drone. The IoT devices on a ship exchange data through the satellite with a server for logistic data analysis. The wireless connections within a production plant are supported by a base station that has edge computing capabilities. The cars exemplify entities that, despite their high mobility, need to stay connected to the infrastructure.

communication and it reflects voice communication, the most traditional service in wireless mobile networks. We note that this type of connection comes in handy when we seek to provide an analogy between wireless and speech communication. However, for a significant fraction of wireless connections, one of the parties is a human, while the other party is a server that hosts a website or streams a video. This is *human-to-machine (H2M) communication*. The third case is related to things and objects that are equipped with wireless connectivity, often related to the connectivity paradigm of the *Internet of Things (IoT)*, which is why these things and objects are referred to as *IoT devices*. The type of connection among IoT devices or between an IoT device and a computer that resides in the infrastructure is *machine-to-machine (M2M) communication*. An IoT connection may also have a human in the loop. For example, this is the case for an operator that remotely controls a moving IoT device, such as a robot or a drone. Strictly speaking, this is not M2M communication and another term to denote the case in which at least one of the parties is a machine (IoT device) is *machine-type communication (MTC)*.

The classification of connections given above is rather coarse and the variability across different types of connectivity explodes once we start to parametrize according to the use cases, mobility, requirements in terms of rate, latency, reliability, etc. Figure 12.1 illustrates connections associated with multiple use cases. Here "internet" is used in terms of "network of networks", rather than a network that runs on the internet protocol (IP), although the latter is very often the case. Walt has a voice connection to Xia and each of them is connected to his/her respective base station (BS), while the two BSs are connected over the internet. Zoya wants to stream video, she is in the range of two different BSs and she is connected to both of them. In this way the video is streamed to Zoya through multiple paths; this type of diversity can result in a lower latency and more stable, reliable video connection. The IoT devices on a ship use a satellite connection in order to exchange logistic data with a suitable server. The production plant is equipped with a small, dedicated BS that facilitates the local interconnection among the robots and other objects on the plant. This BS acts as an *edge computing node* and supports the local computations, but is also connected to the global internet for exchanging data with other actors (not depicted on the figure). The wireless infrastructure should also support cars and other entities that exhibit high mobility. Figure 12.1 illustrates this point with two cars connected to a BS, but also using *direct device-to-device (D2D)* links to communicate and coordinate with each other.

12.1.2 The Complex, Multidimensional World of Wireless Connectivity

The connections illustrated in Figure 12.1 represent a tiny subset of the set of all possible connections that are fully or partially wireless. These connections differ from each other in multiple dimensions, such as requirements in terms of data rate, latency, traffic arrival statistics, connection reliability, number of connected nodes, security requirements, etc. The overall result is a complex set of connections with widely heterogeneous requirements. The following example illustrates this heterogeneity, as well as a way to deal with it by representing the connections through a rather small set of requirements.

Figure 12.2 shows a communication scenario built around an observation of a physical phenomenon, which could take place, for example, in a mine, a production plant, or a large concert venue. There are four different types of wireless devices: three of them are IoT devices (SimpleIoT, ComplexIoT and HealthIoT) and the fourth one is the mobile device carried by each of the persons responsible for monitoring the physical phenomenon; in the illustration these are the laptops used by Zoya and Yoshi. SimpleIoT is a simple sensor with modest processing/computing capabilities, which sends regular, sporadic updates and, occasionally, event-driven updates. The ComplexIoT device has significantly higher capabilities for computing/processing and energy storage. It can send the same types of updates as SimpleIoT, but in addition, it is also capable of sending a live video directly to the persons as well as to the BS Bastian. HealthIoT is a sensor mounted on each person and reports her/his vital signs to Bastian. This scenario features multiple types of connections and we differentiate them through four dimensions:

- *Rate*. This can be in the order of kbps for SimpleIoT and HealthIoT, while it can be orders of magnitude higher for video transmission from ComplexIoT.
- *Number of devices*. The number of SimpleIoT devices can be significantly larger than the number of any other device type.

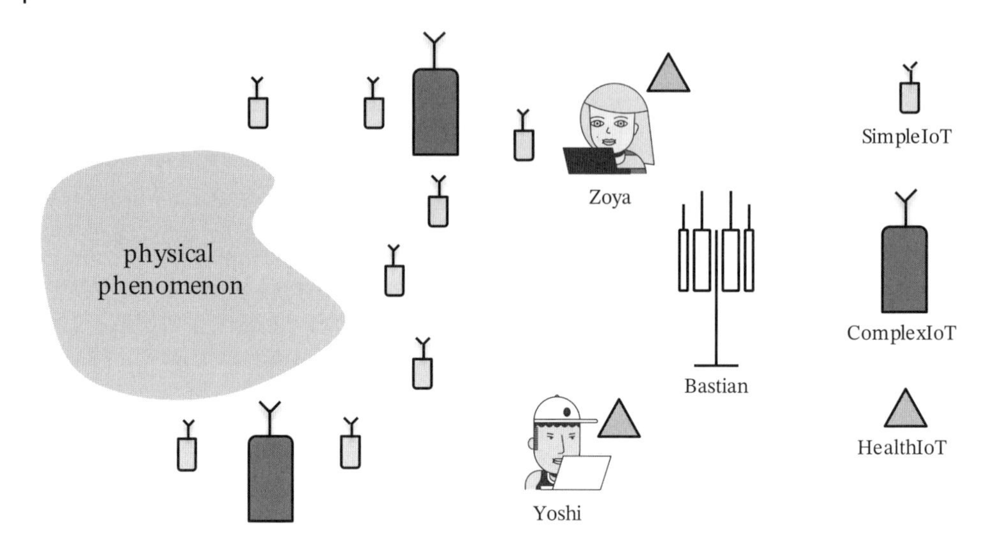

Figure 12.2 Communication scenario built around monitoring of a physical phenomenon, such as a mine, a production plant, or a large concert venue, with four different types of wireless devices: SimpleIoT, ComplexIoT, HealthIoT and a mobile device carried by each of the users.

- *Reliability*. The vital signs sent by the HealthIoT should have extremely high reliability, while the sporadic updates sent by SimpleIoT can be much less reliable. The video stream should be reliable, but not at the level applicable to the vital signs. An interesting form of reliability is required for the event-driven updates from SimpleIoT and ComplexIoT. Namely, a number of these devices will observe the same event within the physical phenomenon, such that their event-driven communication is correlated, both in terms of device activation and the correlation of the actual data transmitted. Hence, while the reliability of each of these individual updates can be low, the *collective* reliability achieved by all IoT devices reporting the event should be very high.
- *Latency*. Low latency is required by the live video stream, especially in an interactive scenario, where the users take action, such as issuing a remote control command based on the video feed. In this case the required latency is in the order of several to tens of milliseconds. In contrast, the latency of the vital sign transmission from HealthIoT can be several seconds.

The traditional way to address connection heterogeneity is a *silo approach*, in which each connection type uses a dedicated, purposefully designed wireless interface and the operation of each interface is optimized separately. With the emergence of 5G wireless systems, wireless connectivity started to be addressed through a *platform approach* to heterogeneous wireless systems. Here "platform" can mean two things. On the one hand, it can refer to one wireless standard that can be flexibly adjusted to produce wireless interfaces with different flavors, such as high-rate transmissions versus low-rate sporadic transmissions. On the other hand, it can refer to joint optimization and orchestration of heterogeneous connections that possibly use different wireless interfaces.

The platform approach needs to deal with the complexity and high variety of connection types. The way to address it is to define few connection types that correspond to a core set of of generic connectivity services. These generic connectivity types can be combined in various ways to produce different types of connectivity and support more specific services. In mathematical terms, the few generic connectivity types can be understood as "eigenvectors" of the connectivity space, capable to produce or approximate any desired type of service in that space. The International Telecommunication Union (ITU) has specified the vision for evolution of the 5G and the other wireless systems in "International Mobile Telecommunications (IMT) for 2020 and beyond" through the support of three generic types of connectivity: *enhanced mobile broadband (eMBB), massive machine-type communications (mMTC), and ultra-reliable and low-latency communications (URLLC)*. A succinct way to describe each of these services is the following:

- eMBB supports stable wireless connections with very high peak data rates, but also moderate rates for users that are not close to the BS or other access points to the infrastructure.
- mMTC supports a massive number of IoT devices, which are only sporadically active and send small data payloads.
- URLLC supports low-latency transmissions of small payloads with very high reliability from a limited set of wireless IoT and non-IoT devices, which are active according to patterns typically specified by outside events, such as alarms.

Referring to the example of Figure 12.2, it should be noted that the IoT devices associated with mMTC correspond to SimpleIoT, while those associated with URLLC would lean towards the features of the ComplexIoT device. Another important observation is that URLLC couples ultra-reliability with low latency; however, from the description of the HealthIoT device it becomes apparent that this should not always be the case. This indicates that, instead of using URLLC as a generic service, one can consider two different types of generic services, one related to ultra-reliability and the other to low latency, and combine them only in certain use cases.

12.2 Fundamental Ideas for Providing Wireless Coverage

The variety across the types of wireless connections is reflected into a variety of requirements put on the wireless infrastructure and network. Low-power sensors and other IoT devices falling into the class of mMTC connections are deployed in a wide area and require coverage from one, or possibly several BSs, in order to receive their sporadic transmissions. In contrast, high-rate transmissions and low-latency interactions intended to support a scenario with virtual reality (VR) require localized coverage with a limited range, for example indoors, offering a stable, high power received signal as well as with low interference. As another example, highly mobile users, such as cars or trains, require good mobility support and coverage within a specific extended area, such as a highway.

This and the next section are dedicated to the issue of ensuring wireless coverage over an area that is larger than coverage from a single BS, which we have referred to as Basil or Bastian throughout the book. Indeed, a single BS can cover only a limited area and support a limited number of users. The coverage is not defined to be within a fixed range around the

BS, as we have seen that it may vary based on the environment, power levels used, but also the technology applied by the BS. If, for example, the BS is equipped with massive number of antennas, then the coverage area can be much larger than the coverage of a BS equipped with a single or few antennas. The number of users that a BS can support is limited by the bandwidth allocated to it and this bandwidth is always limited, both due to the spectrum regulation (see Section 12.6), but also due to the physical limits on the wireless hardware. This leads to the conclusion that there is a need for *infrastructure for wireless access*, where many BSs and access points are interconnected in a wireless network. For example, the connection between Xia and Walt in Figure 12.1 consists of two connections to two separate BSs interconnected through a wired infrastructure.

12.2.1 Static or Moving Infrastructure

There are basically three different options for providing wireless coverage in a given geographical region: static infrastructure, moving infrastructure, and hybrid static–moving infrastructure. These are illustrated in Figure 12.3 and are discussed in the sequel.

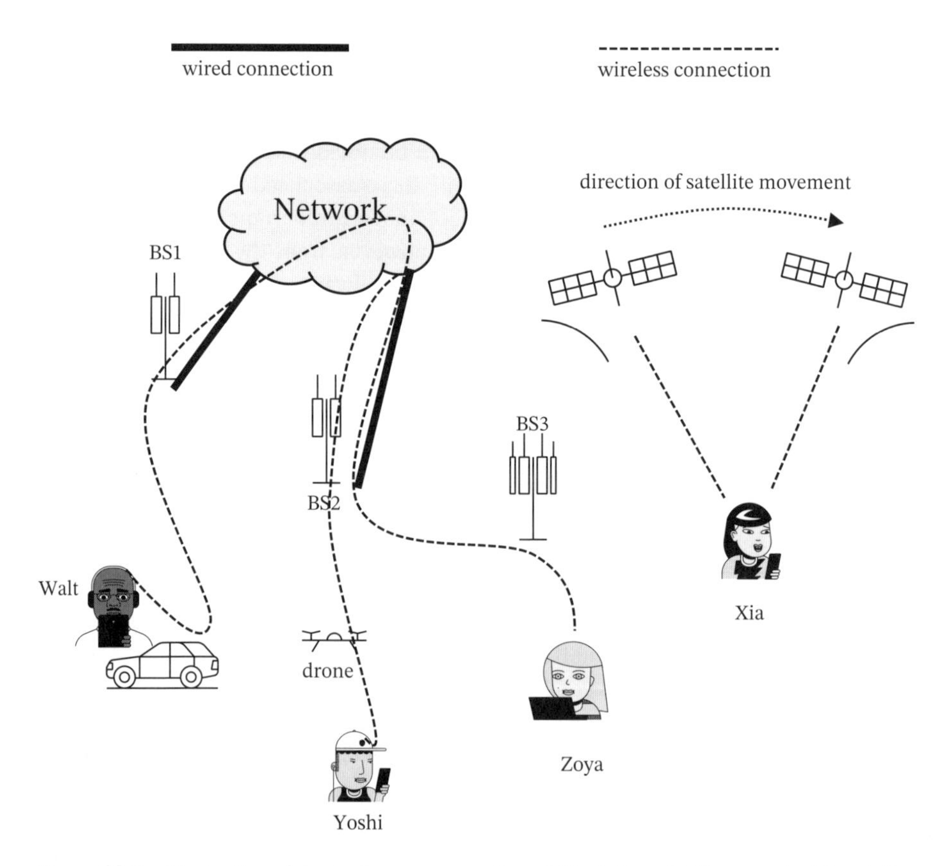

Figure 12.3 Illustration of the basic coverage options. BS1, BS2, and BS3 are part of the static infrastructure. The satellites represent a moving infrastructure. Examples of a hybrid static–moving infrastructure: Walt is connected through a car and Yoshi is connected through a drone to the fixed infrastructure.

In a *static wireless infrastructure*, the BSs and the access points are deployed at fixed locations. These locations are selected strategically, to ensure that practically any point within the considered geographical region has wireless coverage. Zoya, from Figure 12.3, uses a static infrastructure, as she is connected directly to BS3, while BS1, BS2, and BS3 are deployed statically, at fixed locations. The reason Zoya is connected directly to BS3 and not, for example, BS2, is that BS3 is the BS to which Zoya has the best wireless signal. Here the feature of "best" signal is not an absolute notion, but it changes with time and the surrounding wireless conditions. If Zoya is a mobile user, then the BS with the best signal changes as Zoya moves, such that the wireless infrastructure should hand over the connection of Zoya from BS3 to another BS. This operation is called *handover*.

Regarding *moving wireless infrastructure*, the prime example is the satellite networks, where the *low Earth orbit (LEO) satellites* are constantly moving and therefore the coverage region associated with a LEO satellite is constantly changing. Xia, in Figure 12.3, is covered by satellites that are in motion. The two satellites form a *constellation* and they cooperate in order to offer coverage to Xia. Even when she is standing still, the movement of the satellites causes handover and, at some later point, she needs to change the satellite through which she accesses the network[1]. If the two satellites are not within a sufficiently small distance, then the coverage that they provide to Xia can be intermittent, such that there are time intervals in which Xia is not covered.

The overall connection Zoya–Walt, from Figure 12.3, uses a *hybrid infrastructure*. Walt uses his mobile device to connect to the car he is sitting in, such that the car acts as a static BS from his perspective. However, the moving car is wirelessly connected to the static BSs and, as a result, the overall infrastructure to which Walt is connected is hybrid. The car acts as a wireless relay, which is the same role that BS3 has for Zoya, but the car is moving, while BS3 is static. There are other options for a hybrid infrastructure: for example, a drone can fly in a region where coverage is needed and act as a mobile BS, while it is wirelessly connected to a static infrastructure. If Xia and Zoya, from Figure 12.3, establish a connection, then they also need to use a hybrid infrastructure. This is because the satellites act as mobile BSs, wirelessly connected to the terrestrial static infrastructure, represented by fixed BSs and access points.

The following discussion will refer to static infrastructure, unless explicitly stated otherwise.

12.2.2 Cells and a Cellular Network

The area in which one BS provides wireless coverage is called a *cell*. The basic idea of a *cellular network* is to offer wireless coverage in a given large geographical area by covering the area with multiple BSs, each supporting its own cell, as illustrated in Figure 12.4(a). The cells are interconnected through servers. Any two BSs can communicate with each other through this network, which also implies that the two mobile devices, each connected to a respective BS, can communicate with each other. The longer name is *mobile cellular network*, used to emphasize the fact that a mobile user can move throughout the geographical

1 The rest of her connection from the satellites to the other communication party has not been illustrated. However, the other party could be the same constellation of satellites that provides, for example, images or sensing data from space.

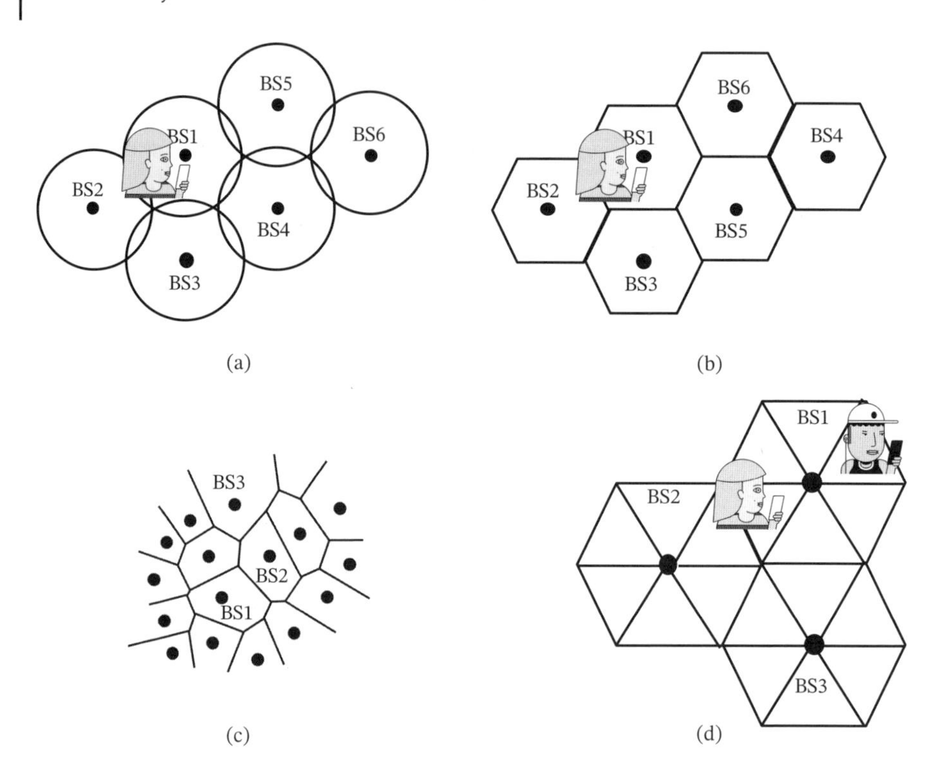

Figure 12.4 Wireless mobile coverage achieved by cellular networks. Each base station (BS) is represented by a black circle. In all example a simple, distance-only dependent propagation model is assumed. (a) Coverage of each cell defined through a minimal SNR measure in the downlink. (b) Cellular structure with regular, lattice deployment of BSs. Coverage of a BS is defined with respect to the interference from the other BSs: each cell is a Voronoi cell and each spatial point is in the coverage area of the closest BS. (c) Irregular spatial deployment of BSs and the resulting coverage with (Voronoi) cells. (d) Use of directed antennas to create sectorized cells.

area, performing a handover from one cell to another as the mobile user crosses the cell boundary.

In Chapter 10 we defined a coverage area of a cell to be the geographical area in which the signal-to-noise ratio (SNR) of the signal received by a terminal is higher than a certain threshold. Let us, for the moment, ignore the complicated propagation phenomena and assume a simple, free-space propagation model where the strength of the received signal is solely dependent on the distance from the device to the BS. Under this simplifying assumption each cell is a circle, as illustrated in Figure 12.4(a). If no point in the geographical area should remain uncovered, then the circles of different cells overlap, such that there are points that are covered by more than one BS.

In the presence of multiple BSs, Zoya should connect to the BS that offers the best signal. This means that, with the simplest free-space propagation model, Zoya connects to her closest BS. Let us assume that all BSs transmit at the same frequency and using the same transmit power. In this case, the downlink signal received by Zoya from the closest BS has the best signal-to-noise-and-interference ratio (SINR). The interference comes from all the other BSs except for the closest one. The SINR measured by Zoya for the signal received from BS1 is:

$$\mathrm{SINR}_Z = \frac{P_{Z,1}}{\sum_{i=2}^{K} P_{Z,i} + N} = \frac{\gamma_{Z,1}}{\sum_{i=2}^{K} \gamma_{Z,i} + 1} \tag{12.1}$$

where $P_{Z,i}$ is the power that Zoya receives from the BSi, N is the noise power, while $\gamma_{Z,i} = \frac{P_{Z,i}}{N}$ is the SNR that Zoya measures for the signal from BSi.

The fact that the mobile device becomes associated with the closest BS, which offers the highest SINR, calls for a redefinition of the coverage area of a given BSi. The coverage area of BSi consists of the spatial points that are closer to BSi than to any other BS. With this definition, the wireless cell becomes a *Voronoi cell* and the cellular network is represented through a Voronoi diagram. A simple illustration of a cellular network is obtained by assuming that the BSs are deployed regularly, according to a two-dimensional lattice, which results in a coverage area with hexagonal shape, as in Figure 12.4(b).

In practice, the deployment of BSs follows an irregular spatial pattern, as illustrated in Figure 12.4(c), resulting in irregular shapes of the Voronoi cells. It should be noted that the actual coverage still requires that the received signal is higher than a minimal threshold. For example, Yoshi can be in a cell that is closer to BS1 than to any other BS, but he receives a very weak radio signal from BS1 and even weaker signals from the other BSs, due to the propagation phenomena that go beyond the distance dependence. In this case Yoshi is nominally in the coverage area of BS1, but he actually does not have wireless coverage.

12.2.3 Spatial Reuse

For all examples in Figures 12.4(a)–(c) it is assumed that each BS has an omnidirectional antenna. This means that if Zoya and Yoshi are in the coverage area of the same cell, then they cause interference to each other whenever they both use the same frequency channel simultaneously. Instead of omnidirectional, the BS can use a set of directed antennas to create spatial *sectors* within its coverage area, as illustrated in Figure 12.4(d). In the simple example from Figure 12.4(d), each BS uses six directional antennas, each of them defining a sector with an angle of 60°. Zoya and Yoshi are in different sectors such that they can simultaneously *reuse* the same frequency channel and not cause interference to each other. In order to enable this reuse, each directed antenna at the BS needs to have its own hardware. Considering that some of the highest costs of the wireless infrastructure are associated with the site (tower, mast) where the BS is installed, mounting additional hardware for directed antennas takes a rather small fraction of the overall costs. Note that the directional antennas allow reusing the same spectrum within the coverage area of a BS and this should be leveled against the high costs for attaining spectrum licenses, as discussed in Section 12.6.

The idea of *spatial reuse* of frequencies, also called *frequency reuse*, is also relevant when the BSs use omnidirectional antennas. Here, the interesting question is how to distribute the frequency channels across the cells in order to mitigate the interference, while ensuring that a certain frequency channel is reused in as many cells as possible.

The hexagonal cellular structure in Figure 12.4(b) offers a pedagogical setup for explaining the frequency reuse. Assume that there are three frequency channels, denoted by

f_1,f_2, and f_3. One option is to allow all BSs from Figure 12.4(b) to use all three channels and freely allocate them to the users within their coverage area. This is called *reuse-1* channel allocation. Reuse-1 allows two neighboring BSs to use the same channel simultaneously, which results in significant interference between the respective links in the two cells. Another option is to allocate f_1 to BS1 and BS4, f_2 to BS2 and BS5, and f_3 to BS3 and BS6. This would increase the distance between two links that are allowed to use the same frequency and thus decrease the interference between them. In short, increased reuse distance decreases the inter-cell interference. The price paid for this is the flexibility/efficiency of allocation: if BS1 has three users associated with it and there are no users in the other cells, then having only one frequency channel at its disposal is inefficient. This could be addressed through dynamic allocation of channels, as well as cooperation and coordination among cells that are allocating the channels and controlling their mutual interference.

Special note should be taken of the sharing of the resources by multiple access techniques based on the spread spectrum, such as CDMA. Let us consider CDMA with pseudorandom spreading codes. The transmitters competing for resources use the same band and the interference is suppressed through the process of despreading at the decoder. The decoder applies the pseudorandom code of the desired signal and it is oblivious to the fact of whether the interference from the other spread spectrum signals comes from the same cell or the neighboring cell. It follows that CDMA is inherently suited for a reuse-1 allocation: all cells are allowed to used the same wide band while the interference from other transmitters is inherently suppressed by the despreading process. The key element that removes the need for resource allocation across cells, but also within the cell, is the use of pseudorandom sequences. This supports intermittent transmissions of the nodes and it thus carries some of the flexibility contained in the random access protocols. However, this is different from the dark room analogy, as here the receiver looks for the "name" (spreading code) of a specific set of transmitters and ignores the others. The price paid for the lack of need for resource allocation is that CDMA operates well in the case when the cells are not fully loaded. In other words, CDMA is suitable when there are not too many simultaneously active users that transmit all the time. If there are users that are active over extended periods, and the number of those users is high, then orthogonal resource allocation is, in general, better than CDMA.

Taking into account only distance based propagation effects, a cell appears to be a static structure. However, once we take into account the true propagation effects and time-variant nature of the wireless channels, the structure of the cells starts to be dynamic and this complicates the process of association with the best BS. Let us take the perspective of a mobile device, Walt, that measures the received SINR in the downlink. Assume that Walt is a static user in a scenario with realistic BS deployment, as in Figure 12.4(c). Even if he is static, he still measures changes in the SINR both due to the change of the signal power and the interference power.

At this point, the following question occurs naturally: *how does Walt know the signal strength from each individual BS, such that he can compute the SINRs with respect to the association to each of the BSs according to (*12.1*)?* This can be done, in principle, by assigning a unique pilot sequence to each BS, similar to the spreading sequences in CDMA, and they can be transmitted regularly, occupying only a part of the time-frequency grid. Clearly, a BS cannot send the pilots continuously, which would be needed to track the channel at any instant, as in that case the BS would not do anything else but send pilot sequences. Walt can periodically reassess the SINR and, in principle, change the cell association according to the best SINR.

If a cell coverage is defined according to this instantaneous SINR, then the cell shape changes in time, but also in frequency, as the cell shape for f_1 can be different from the cell shape for f_2. The latter is a result of frequency dependent propagation effects in a given physical setup that is different from free space. Frequent change of the cell association, called handover ping-pong, brings cost in terms of overhead. To cope with this, the mobile device should use a certain form of averaging/low-pass filtering of the SINR measurements in order to minimize this overhead, while still keeping acceptable performance.

12.2.4 Cells Come in Different Sizes

The coverage area of a *macrocell* is large, usually with a radius above 1 km and going up to several kms. The main benefit of a large cell size is that it can offer mobility that is not interrupted by handovers. As such, it is suitable to be deployed on highways or in rural, sparsely populated areas. Another benefit is that there is only one installation site for a large area. On the flip side, a macrocell may not use the frequency spectrum efficiently. For example, if Walt communicates with the BS using the frequency channel f_1, then this channel cannot be used anywhere else in the cell or in the same sector (if the cell is sectorized). Due to this, the macrocell has, generally, a low *area spectral efficiency*. Considering a fixed frequency band, the area spectral efficiency increases when one maximizes the data rate achieved through that band while reusing the same band in space as often as possible. Clearly, this is limited by interference among the proximate links that use the same band.

Smaller cells can have different sizes: microcell, picocell or femtocell. The same area covered by the macrocell could be covered by, say, nine small cells, as shown in Figure 12.5(a). Each of the BSs associated with those small cells could be allowed to use the channel f_1. It is, though, not immediately obvious that this is better than having a single BS serving the macrocell. In the case of a macrocell, Walt does not experience any interference from a transmitter residing within the macrocell. However, assuming that the macroBS uses fixed power, the quality of the received signal can vary in a large range due to the large number of positions that Walt can take in a macrocell. In the case of small cells, the BS is closer to Walt and thus offers a stronger received signal to his mobile device. However, now this link may get interference from eight other transmitters that reside within the area of the original macrocell, as seen in Figure 12.5(a).

The realistic propagation conditions, such as weak propagation through walls, may contribute significantly to mitigation of the interference among the cells. This is happening in urban scenarios, where it is more efficient to use small pico- and femto-cells that reuse the frequencies over shorter distances and thus increase the area spectral efficiency. This drives the idea of increasing the mobile wireless capacity in a given region through *network densification*, where the cells become smaller, the distance (and therefore the signal loss) between the user and the serving BS decreases, while fewer users compete for the same bandwidth within the same cell. All this is at the price of potentially higher inter-cell interference; yet, with the hope that the obstacles and the scatterers in the environment will mitigate the interference.

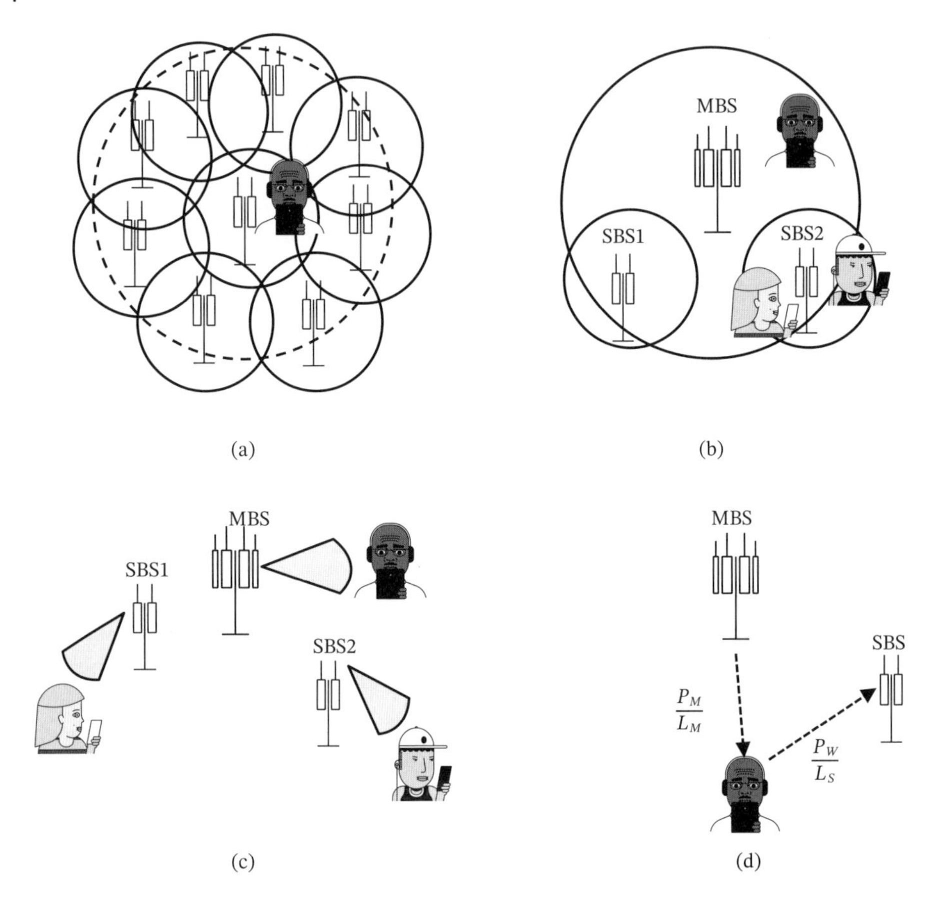

Figure 12.5 Macrocell versus small cell coverage. (a) The area of a macrocell (dashed line) is covered by nine small cells. (b) Heterogeneous network (HetNet) with a macroBS (MBS) and two small-cell BSs (SBSs). (c) Beamforming allows different proximate BSs in a HetNet to use the same frequency and stay free of interference. (d) An example of a case for decoupled downlink–uplink access.

The concept of cell-based wireless coverage can be generalized by embracing the idea that a given geographical area can be covered by cells of different size. This is illustrated in Figure 12.5(b) and is known as *heterogeneous networks (HetNets)* due to the heterogeneity in the coverage area for different cells. Another way to think about HetNets is that they cover the area with several tiers (layers), such that cells of a given size belong to the same tier. The macro BS, denoted as MBS, has a larger transmit power compared to the small-cell BS, denoted as SBS. Note that the coverage area of SBS2 largely overlaps within the coverage area of MBS. The motivation for this is the problem of non-uniform distribution of the wireless traffic demands within the large cell. For example, the part of the macrocell that is also covered by SBS2 could be an indoor lounge area where Zoya and Yoshi are downloading large files or streaming videos. This sub-area of the macrocell features users that have data demands that are much higher compared to those coming from sub-areas of the macrocell that are outdoors. In addition, the data-hungry users in a lounge do not require high mobility. The deployment of SBS2 helps to contain this localized data demand and thus *offload* traffic from the MBS.

The setup of HetNets with overlapping cells poses a significant challenge in terms of resource allocation and interference management compared to the homogeneous cellular networks discussed above. In HetNets the MBS and the SBSs may even use the same frequency channel, provided that the interference is mitigated, for example, by beamforming and use of multiple antennas as in Figure 12.5(c).

12.2.5 Two-Way Coverage and Decoupled Access

Until now, the concepts of wireless coverage and cells have been discussed exclusively with respect to the downlink transmission and the power used by the BS. Let us now take the perspective of an uplink transmission. If Walt is in a macrocell and transmits to an MBS, then his mobile terminal will, in general, use higher power compared to a case in which he is transmitting to a nearby SBS[2]. In heterogeneous networks, if we follow the rule that the mobile device should be associated according to the highest SNR/SINR, then the BS to which Walt is associated for downlink transmission may not be the same as the BS to which he is associated for an uplink transmission. This is illustrated in Figure 12.5(d). The MBS uses power P_{M} and the SBS uses power P_{S}, where $P_{\text{S}} < P_{\text{M}}$. The power used by Walt is P_{W}. The propagation loss from MBS to Walt is L_{M}, such that the power that Walt receives from MBS is $\frac{P_{\text{M}}}{L_{\text{M}}}$. Similarly, the propagation loss from SBS is L_{S} and the received power by Walt is $\frac{P_{\text{S}}}{L_{\text{S}}}$. Assume that Walt is closer to SBS, such that $L_{\text{S}} < L_{\text{M}}$. In that case the power that MBS receives in the uplink is $\frac{P_{\text{W}}}{L_{\text{M}}}$, while the power received by SBS is $\frac{P_{\text{W}}}{L_{\text{S}}}$. The uplink association of Walt is with SBS since:

$$\frac{P_{\text{W}}}{L_{\text{S}}} > \frac{P_{\text{W}}}{L_{\text{M}}}. \tag{12.2}$$

However, as $P_{\text{M}} > P_{\text{S}}$ it is possible to have

$$\frac{P_{\text{M}}}{L_{\text{M}}} > \frac{P_{\text{S}}}{L_{\text{S}}} \tag{12.3}$$

such that the downlink association of Walt is with MBS. Similar conditions for different association can be obtained by considering different interference at SBS and MBS. This type of association is termed *decoupled downlink–uplink* and it clearly breaks the boundaries of the traditional notion of cell association.

A new dimension in wireless cell coverage and resource allocation emerges when uplink and downlink are considered jointly into a problem of two-way communication. An immediate question is that of duplexing. If frequency division duplexing (FDD) is used, then the resources allocated to uplink and to downlink are fixed by the allocation of the respective bands. The consequence of this type of allocation is that a downlink signal can only be

---

2 This is important from the health perspective, since with a terminal connected to a SBS, Walt is exposed to smaller radiation from his mobile device. Thus, a denser deployment of BSs and cell densification does not necessarily cause more radiation to the wireless users. We have witnessed this tendency in different mobile wireless generations, as since 2G and towards 5G the networks have become denser while the terminal power has decreased.

interfered with by other downlink signals, but not the uplink signals, and vice versa. As the downlink interferers have fixed positions, downlink interference has a higher level of predictability, which is an advantage in terms of resource allocation. This predictability is conditioned on the use of omnidirectional antennas or fixed directional antennas (sectors) at the BSs and it disappears if the BSs use adaptive beamforming.

The major problem with FDD is the inflexibility in trading off downlink for uplink resources or vice versa, which is important for adapting to dynamic traffic patterns. This problem is addressed by time division duplexing (TDD), where the same frequency band can be allocated either for uplink or for downlink. This allows for dynamic allocation of uplink/downlink resources in order to respond to the actual traffic demand in downlink or uplink. This variant is known as *dynamic TDD*. As a disadvantage, TDD can lead to a more unpredictable interference, which calls for cooperation and coordination of the neighboring cells.

12.3 No Cell is an Island

The relationship between multiple links belonging to the same cell can be described either as *competition* for resources or as interference. Competition for resources is expressed, for example, through the problem of who gets exclusive rights to use channel f_1. In contrast, interference occurs for schemes based on non-orthogonal use of the same wireless resources, such as CDMA. Following this line of thought, the basic relationship between two links belonging to different cells is interference, unless the BSs of those cells communicate with each other in order to coordinate or cooperate to avoid/mitigate the interference. No cell operates in isolation, as the BSs need to be part of a network that coordinates them in some way in order to support handover, the fundamental operation in a cellular network. Furthermore, resource allocation and frequency reuse across cells are also relying on the interconnections among the BSs. In this section we will explore the question of interconnecting the BSs, as well as architectures for wireless coverage based on coordination/cooperation among different cells.

12.3.1 Wired and Wireless Backhaul

The notion of a *relay* in a wireless networking is commonly understood to be a *wireless relay*, that is, a node that has wireless connection to the mobile device, but also a wireless connection to the rest of the infrastructure. However, the general notion of a relay is a transceiver that can receive data from one or more nodes, create new packets or symbols based on the received data, and then transmit this newly created data further towards other nodes. Thus, in general, the physical layer of the receiver through which the data has been received is not necessarily equal to the physical layer used to send this data further. The transmission from Zoya to BS3 in Figure 12.3 is wireless and BS3 uses also a wireless transmission to relay the data from Zoya to BS2. However, BS2 uses a wired connection to relay the data from Zoya further towards the core network. The data keeps being relayed until it reaches Walt.

This is a simple illustration of the fact that relaying is a fundamental operation throughout the wireless infrastructure and network, while wireless relaying is only a specific option.

The link through which a BS is connected to the core network or has a peer connection to another BS is termed a *backhaul link*. To differentiate, we will use the term *access link* for the wireless link between a device and a BS.

Using a wired backhaul has advantages that are inherent for wired communication, such as no need for a wireless spectrum, absence of interference, and relatively static channel quality. However, wired connections from the BSs to the infrastructure also induce installation costs. Furthermore, there are scenarios in which the backhaul is necessarily wireless, such as in the case of a moving infrastructure with satellites or drones. Another advantage of the wired backhaul is that the BS can communicate with the devices over the wireless interface while simultaneously using the wired backhaul. This is also possible when the backhaul is supported by a dedicated wireless interface, different from the one used for the access link. In order to avoid interference between the backhaul and the access link, the backhaul link should, generally, use a different transceiver and a frequency band different from the one used by the access link.

The use of a dedicated interface and spectrum for wireless backhaul, termed *out-of-band relaying*, is often a luxury. From a cost, but also from an engineering viewpoint, a more interesting case is *in-band relaying*, where the same wireless interface and the same spectrum is used both for the access and for the backhaul links[3]. Let us consider the scenario in Figure 12.6(b), with an access link Zoya–SBS and a backhaul link MBS–SBS. We have used here the terminology of MBS and SBS as this scenario can occur in HetNets. Namely, an SBS is deployed in an area where larger traffic density is expected and an SBS needs a backhaul link to connect to the infrastructure. In this case the backhaul link is provided by using part of the resources of the MBS to communicate with the SBS. Nevertheless, the reader should note that this discussion on relaying has a much broader scope than the scope of a wireless backhaul in HetNets. For example, it can cover a scenario in which Zoya and Yoshi stand at different parts of the globe and communicate by using a satellite as an in-band relay.

12.3.2 Wireless One-Way Relaying and the Half-Duplex Loss

Let us at first treat only the uplink communication from Zoya as a source to the MBS, which serves as a destination. In a downlink transmission the roles of source and destination are swapped, such that it can be treated similarly. Let the transmission rate of Zoya be fixed to

$$R = \frac{D}{T} \quad \text{(bps)} \tag{12.4}$$

where D is the packet size in bits and T is the slot duration. For the moment, the reader can consider the simplest range-dependent collision model, where the rate is R if the distance is below a threshold or 0 otherwise. We use the case from Figure 12.6(a) to serve as a reference, where Zoya is connected directly to the MBS. The transmission of a packet $\mathbf{Z}_i$ in the uplink takes a single slot to reach the MBS, such that the goodput is $\text{Gp} = R$ (bps).

In Figure 12.6(b) Zoya is not in the range of the MBS and needs to communicate to the MBS through two hops, using an SBS as a relay. In this case the SBS is assumed to have a half-duplex transceiver, such that Zoya's packet $\mathbf{Z}_i$ takes two slots to get to the destination.

3 Unless explicitly stated otherwise, in the following "relaying" stands for in-band relaying.

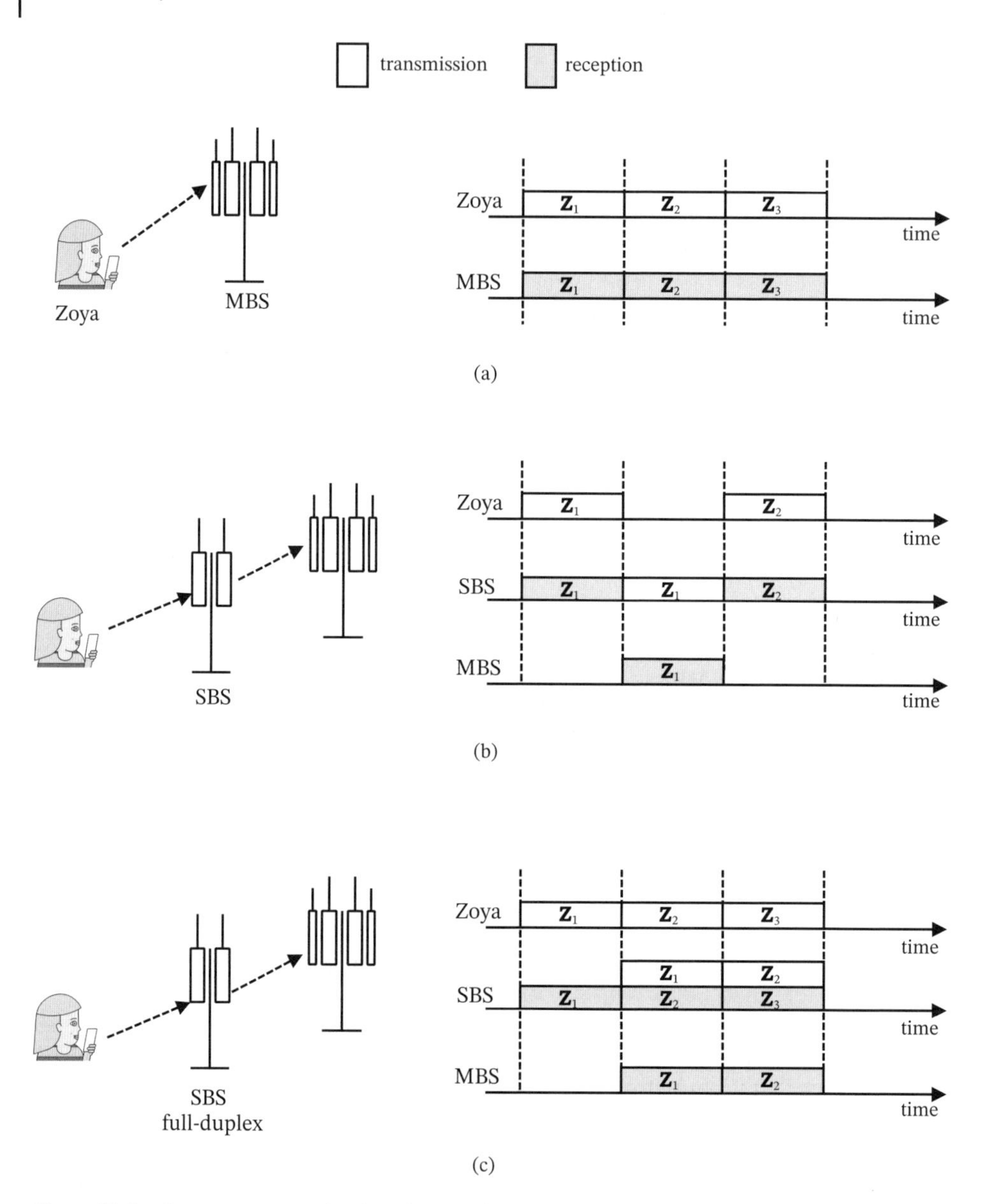

Figure 12.6 One-way communication via a wireless relay that provides a backhaul link. (a) Reference system without a relay. (b) Half-duplex transceiver at the relay SBS. (c) Full-duplex transceiver at the relay SBS.

If Zoya sends in total L packets, then the total number of slots required for MBS to get all packets is $2L$, such that the goodput is:

$$\text{Gp} = \frac{LD}{2LT} = \frac{R}{2} \quad \text{(bps)}.$$

This can be improved if the SBS is assumed to have a full duplex transceiver, such that it can simultaneously receive from Zoya and send to MBS. As illustrated in Figure 12.6(c), if Zoya transmits L packets, then MBS gets all of them after $L + 1$ slots. The goodput is:

$$\text{Gp} = \frac{LD}{(L+1)T} = \frac{LR}{L+1} \stackrel{(a)}{\approx} R \quad \text{(bps)}$$

where the approximation (a) is valid when L is very large.

From this discussion it follows that one-way wireless relaying with half-duplex receivers introduces loss in goodput and this is a major challenge for implementing in-band wireless backhaul, SBS–MBS. Using full-duplex transceiver can mitigate this loss, at least in the simple model that we have used in the example in Figure 12.6. However, the use of a more realistic model, such as non-ideal self-interference cancellation at the full-duplex SBS or distance-dependent rate, can change this conclusion. For example, if the achievable rates in the links Zoya–SBS and SBS–MBS are different, then the goodput of Figure 12.6(c) will be, asymptotically, equal to the rate which is the smaller of the two.

12.3.3 Wireless Two-Way Relaying: Reclaiming the Half-Duplex Loss

Let us now consider two-way communication between Zoya and MBS and assume that TDD is used. The ith packet sent from Zoya to MBS is denoted again by $\mathbf{Z}_i$, while the jth packet sent from MBS to Zoya is denoted by $\mathbf{M}_j$. Each of them transmits the packet at rate R. If there is a direct link between Zoya and MBS, as in Figure 12.6(a) plus an arrow from MBS to Zoya, then each of them sends in every other slot. During $2L$ slots, Zoya sends L packets, such that her goodput is:

$$\text{Gp}_Z = \frac{LD}{2LT} = \frac{R}{2} \quad \text{(bps)}$$

and the goodput achieved by the MBS is also $\text{Gp}_M = \frac{R}{2}$ (bps). The aggregate, two-way *sum goodput* is:

$$\text{Gp}_{\text{sum}} = \text{Gp}_Z + \text{Gp}_M = R \quad \text{(bps)}.$$

The case of two-way relaying with a half-duplex transceiver at SBS is shown in Figure 12.7(a). Here a full two-way transaction, in which the packet $\mathbf{Z}_i$ reaches MBS and the packet $\mathbf{M}_i$ reaches Zoya takes four slots. Hence, if the exchange contains L packets from Zoya and L packets from MBS, the total number of slots required to send them is $4L$. The sum goodput in this case is:

$$\text{Gp}_{\text{sum}} = \frac{2LD}{4LT} = \frac{R}{2} \quad \text{(bps)}$$

and, again, halved compared to the direct link. The use of full-duplex transceiver at SBS can again compensate for the loss, as shown in Figure 12.7(a). For sending L packets from

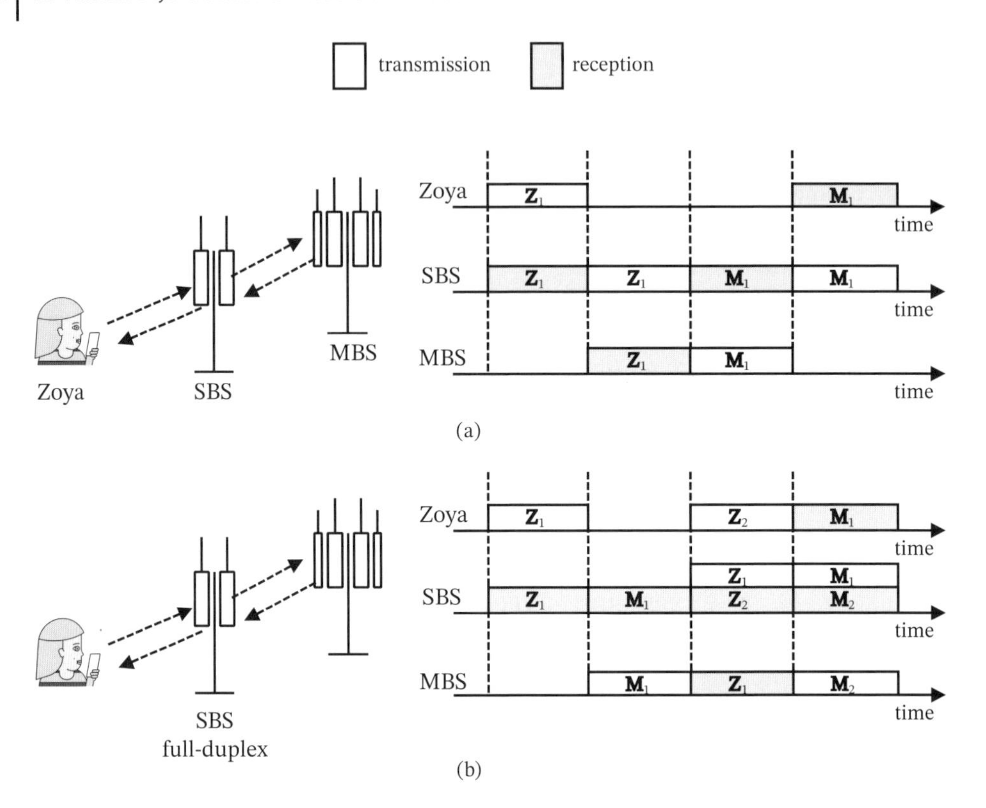

Figure 12.7 Two-way communication via a wireless relay that provides a backhaul link. (a) Half-duplex transceiver at the relay SBS. (b) Full-duplex transceiver at the relay SBS.

Zoya and L packets from the MBS, the total number of slots that need to be used is $2(L+1)$ and the sum goodput is:

$$\mathrm{Gp}_{\mathrm{sum}} = \frac{2LD}{2(L+1)T} \approx R \quad \text{(bps)}$$

where the approximation is for large L.

However, the two-way scenario gives rise to new and interesting communication modes with a half-duplex transceiver at the relay SBS. On Figure 12.8(a), after two slots in which SBS collects the packets $\mathbf{Z}_1$ from Zoya and $\mathbf{M}_1$ from the MBS, it broadcasts the packet $\mathbf{S}_1 = \mathbf{Z}_1 \oplus \mathbf{M}_1$, where $\oplus$ is the XOR operation. Since Zoya knows $\mathbf{Z}_1$, she can extract $\mathbf{M}_1$ from the received signal as follows:

$$\mathbf{Z}_1 \oplus \mathbf{S}_1 = \mathbf{Z}_1 \oplus (\mathbf{Z}_1 \oplus \mathbf{M}_1) = \mathbf{M}_1.$$

The MBS extracts $\mathbf{Z}_1$ from $\mathbf{S}_1$ in a similar manner. The total time for a two-way transaction between Zoya and MBS is three slots, such that one can calculate the sum goodput to be:

$$\mathrm{Gp}_{\mathrm{sum}} = \frac{2R}{3} \quad \text{(bps)}$$

which is an improvement compared to the case in Figure 12.7(a) without the use of a full-duplex transceiver.

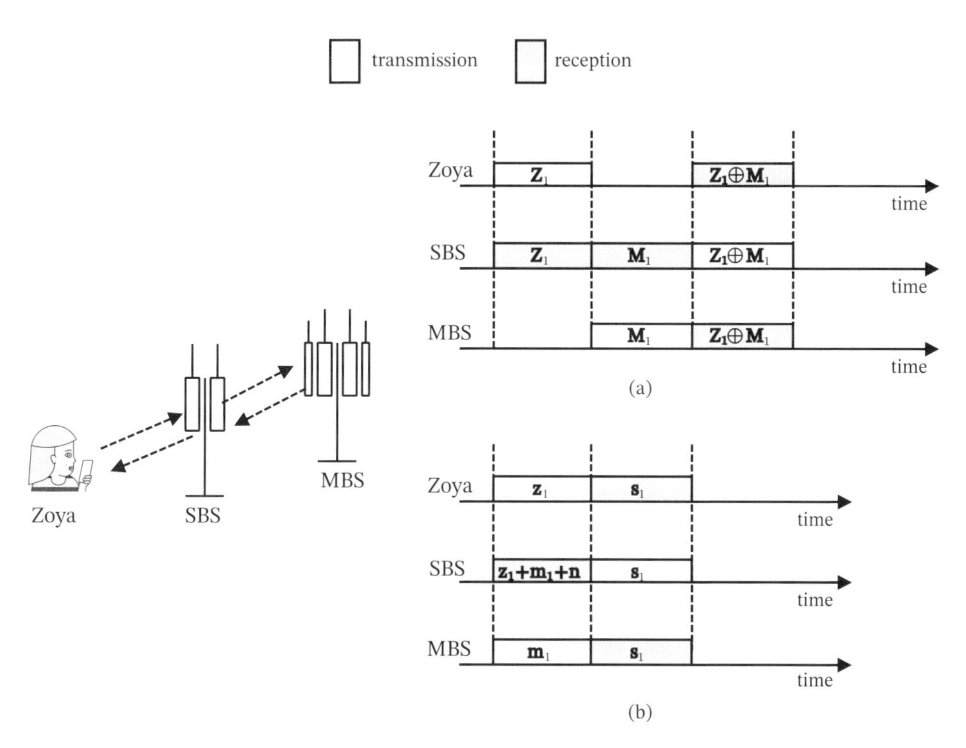

Figure 12.8 Two-way relaying that uses network coding. (a) Digital network coding. (b) Analog or physical-layer network coding.

This type of transmission is an instance of *network coding*, which is a generalization of the traditional routing function of a network node. Namely, in classical routing a packet at the input of a network node is replicated at the output (or more outputs) in order to be forwarded further to the destination (or multiple destinations). In network coding, the routing node forwards functions of the packets that it receives at the input. The SBS in Figure 12.8(a) acts as a network node and needs to route packets between the source and the destination. The network coding function in the example is represented by the simple XOR operation, after which the SBS takes advantage of the broadcast nature of the wireless channel and thus saves one slot.

The network coding operation in Figure 12.8(a) is also referred to as *digital network coding*, since the function of the input packets is computed in the digital domain. The shared nature of the wireless channel offers the possibility for network coding at the physical layer or *analog network coding*, as illustrated in Figure 12.8(b). Here we use the small letters $\mathbf{z}_1$ and $\mathbf{m}_1$ to denote the baseband (analog) representations of the packets $\mathbf{Z}_1$ and $\mathbf{M}_1$, respectively. For simplicity, we have also assumed that the channel coefficients of the links Zoya–SBS and MBS–SBS are both 1, such that the signal $\mathbf{s}_r$ received by the SBS is:

$$\mathbf{s}_r = \mathbf{z}_1 + \mathbf{m}_1 + \mathbf{n}_S \tag{12.5}$$

where $\mathbf{n}_S$ contains the noise samples that affect the reception at SBS.

The key observation here is that the SBS is not the final destination of any of the packets, such that it does not need to decode them. Instead of decoding $\mathbf{z}_1$ and $\mathbf{m}_1$ individually, the decoder of the SBS can aim to decode a function of $\mathbf{z}_1$ and $\mathbf{m}_1$. Specifically, it can try to decode $\mathbf{S}_1 = \mathbf{Z}_1 \oplus \mathbf{M}_1$ directly from $\mathbf{s}_r$ and then forward it in the next slot, after which Zoya and MBS will decode the desired packet using their own packet, as in the example with digital network coding. This method of relaying is termed *physical-layer network coding* or *compute and forward*.

Another possibility is that the SBS does not attempt to decode anything from $\mathbf{s}_r$, it just amplifies the signal $\mathbf{s}_r$ and uses the second slot from Figure 12.8(b) to broadcast $\mathbf{s}_r$ back to Zoya and the MBS. This type of relaying is termed *amplify and forward (AF)*. To simplify, assume that the amplification coefficient is 1 and the SBS[4] retransmits $\mathbf{s}_r$, such that Zoya receives:

$$\mathbf{z}_r = \mathbf{s}_r + \mathbf{n}_Z = \mathbf{z}_1 + \mathbf{m}_1 + \mathbf{n}_S + \mathbf{n}_Z$$

where $\mathbf{n}_Z$ is the noise of Zoya's receiver. Similar to the use of its own packet in digital network coding, here Zoya can subtract $\mathbf{z}_1$ from $\mathbf{z}_r$ and get:

$$\mathbf{z}_r' = \mathbf{z}_r - \mathbf{z}_1 = \mathbf{m}_1 + \mathbf{n}_S + \mathbf{n}_Z \tag{12.6}$$

and thus try to decode $\mathbf{m}_1$. Equation (12.6) looks as if there is a direct link between Zoya and the MBS, except that now there is an extra noise $\mathbf{n}_S$ inserted by the SBS. The reception at the MBS is treated similarly. Assuming that the rate R is chosen such that both Zoya and the MBS can decode the desired signals despite the extra noise from the SBS, then the total time for a two-way transaction between Zoya and the MBS is two slots and the goodput is:

$$\mathrm{Gp}_{\mathrm{sum}} = \frac{2R}{2} = R \quad \text{(bps)}$$

which is the same as if there were a direct link between Zoya and the MBS. Well, almost the same, as we have made some idealizing assumptions, such that the SBS can perfectly decode $\mathbf{S}_1 = \mathbf{Z}_1 \oplus \mathbf{M}_1$ in compute and forward or that the noise n_S is negligible in AF. We note that AF relaying is also possible for one-way relaying, where the SBS amplifies the signal received from one party and forwards it to the other party.

We conclude the discussion on relaying and backhauls by looking briefly at the layering issue. If the backhaul is wired or uses a dedicated wireless interface, then there is a strict layering separation between the physical and the link layer: the packet decoded from the wireless part is put into a digital form and then forwarded through the backhaul (and vice versa). The separation of physical and link layers remains when the same wireless interface is used, but the relaying methods are conventional, as in Figures 12.6 and 12.7. However, the separation starts to get blurred when the network coding techniques are applied in Figure 12.8. For example, in AF relaying, the SBS does not use its link layer at all; it keeps the forwarding at the physical layer, which, as known from before, is not intended to have a role in forwarding.

4 This assumption implies that the SBS has a higher transmission power from Zoya and the MBS, but we ignore the detailed analysis here.

12.4 Cooperation and Coordination

The concept of a cellular network is implicitly assuming at least some coordination and cooperation among the cells towards supporting handovers. Nevertheless, handover is perhaps the simplest idea for cooperation among different BSs and, as we will see, other elements of the wireless infrastructure. Another idea for cooperation that follows from the discussion so far is the one of mutual coordination of wireless resources in order to avoid or mitigate interference. For example, two BSs in neighboring cells that use TDD and operate in the same frequency channel can coordinate the scheduling of the mobile devices such as to minimize the interference between the links formed in the different cells.

In the following we will present other ideas and schemes for cooperation that utilize the nature of wireless propagation. Similar to the way abstract art deviates from real-world shapes, these cooperation concepts push the system architecture away from the basic concept of a cell, such that the "cell" eventually becomes difficult to identify.

12.4.1 Artificial Multipath: Treating the BS as Yet Another Antenna

Consider the two BSs in Figure 12.9(a). Yoshi is connected to BS1 and located close to the border between the two cells. As he moves and the signal strengths change, he needs to make a decision to be handed over from BS1 to BS2. This is a *hard handover* or hard handoff, as one connection is broken and another one is established.

Let us look first at a downlink transmission from the mobile network to Yoshi, but now the *radio network controller (RNC)* provides the same data to be transmitted by both BS1 and BS2. At least two replicas of the downlink signal, one from each BS, arrive at Yoshi. This can be seen as an artificial multipath propagation. In addition to these replicas, Yoshi may

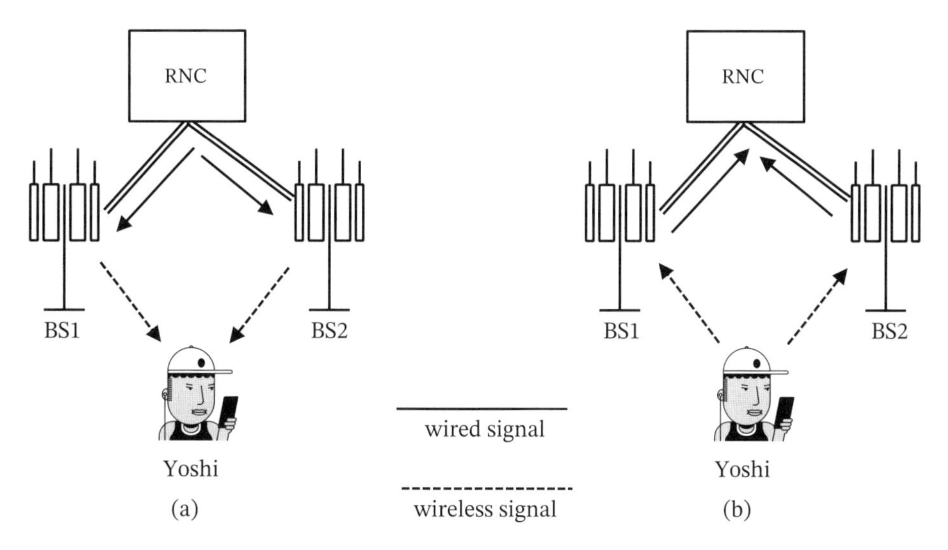

Figure 12.9 Yoshi performs a soft handover by being simultaneously served by BS1 and BS2 in (a) downlink and (b) uplink. The radio network controller (RNC) has wired links to both BSs.

receive other replicas that arrive due to natural multipath propagation. The main observation is that Yoshi can treat all replicas as being a result of multipath propagation. With this observation, the system can use the transmission methods designed to deal with the phenomenon of multipath, such as a RAKE receiver in CDMA or a cyclic prefix in OFDM, and thus decode the desired signal from the RNC. At the cell edge Yoshi ends up being served by both BSs; this is termed *soft handover*. Used in the context of CDMA, the soft handover was the driving force behind the 3G wireless cellular systems. As already mentioned, the principle of artificially created multipath works well with OFDM, under an appropriate choice of the cyclic prefix. This has been applied for a downlink transmission in single frequency networks (SFNs), used to broadcast multimedia data in a larger geographical region covered by multiple BSs.

The soft handoff mechanism for an uplink transmission works differently, as depicted in Figure 12.9(b). Here the signal transmitted by Yoshi is received by both BSs and the data should be eventually transferred to the RNC. There are several options for doing this. One is that each BS tries to decode the data of Yoshi and, if the decoding is successful, it forwards it to the RNC. This means that the RNC decodes the data correctly if at least one BS decodes it correctly. Another option is that each BS converts the received signals into bit stream and sends it to the RNC; then the RNC can run a decoding algorithm on the received bits and try to recover the data. Yet another option, which is de facto AF relaying, is that each BS takes each received wireless symbol, along with the added noise, and forwards it through the backhaul to the RNC. After receiving the noisy baseband symbols, the RNC runs a decoding algorithm. As in the wireless AF relaying, in this case the noise of the baseband RNC receiver adds to the noise of each BS.

Still treating the uplink of Yoshi to BS1 and BS2 from Figure 12.9(b), let us now consider the hypothetical situation in which the AF relaying is ideal. This means that the wired backhauls BS1–RNC and BS2–RNC have infinite capacity, while the receiver of the RNC does not add extra noise. For each baseband symbol y transmitted by Yoshi, the RNC receives two symbols r_1 and r_2 from BS1 and BS2, respectively:

$$\begin{aligned} r_1 &= h_1 y + n_1 \\ r_2 &= h_2 y + n_2 \end{aligned} \tag{12.7}$$

where h_i is the channel between Yoshi and BSi, while n_i is the noise at BSi. However, the equation (12.7) looks exactly like the reception of a device or a BS equipped with two antennas. Indeed, in this case the RNC can be treated as a two-antenna device and it can apply all reception techniques applicable to multiple antennas, such as maximum ratio combining (MRC). The difference from common two-antenna devices is that here the two antennas are spatially separated, thereby offering a *macro diversity* to the user Yoshi. Using this multi-antenna perspective, handover can be understood as a scheme for macro-diversity with antenna selection, as Yoshi connects to the BS to which he has the best signal.

The analogy with multiple antennas extends to transmission as well. For example, the RNC can use the two BSs from Figure 12.9(a) to implement the Alamouti scheme for transmit diversity, described in Section 11.2.3. This and other possible transmission schemes are generalizations of the basic scheme in which the two BSs send exactly the same signal, creating an artificial multipath. If Yoshi has multiple antennas, then the link between the RNC and Yoshi is a MIMO link, which brings us to the idea of distributed MIMO, discussed next.

12.4.2 Distributing and Networking the MIMO Concept

Thinking of multiple interconnected BSs as multiple antennas leads to more general concepts of a wireless infrastructure. The interconnected BSs can cooperate and coordinate with each other in order to provide a more efficient wireless service to the users in their coverage region. Different instances of this type of wireless infrastructure have been devised under different names, such as *coordinated multipoint (CoMP)* operation, *network MIMO* or *distributed MIMO*. With this view, the inter-cell interference can be treated, for example, similarly to the inter-antenna interference in MIMO systems and it can be embraced rather than avoided.

This principle in the operation of a CoMP is illustrated in Figure 12.10, where we have depicted only the wired backhaul that interconnects the two BSs, while the backhaul connections to the network are not depicted as they are not relevant for this discussion. Keep in mind that we are still assuming that the backhaul link that connects the two BSs has an infinite capacity. In Figure 12.10(a), BS1 and BS2 transmit in the downlink to Yoshi and Xia. If each BS has a single antenna, this can be treated as a problem in which a BS equipped with two antennas serves two users in the downlink. Furthermore, each of the BSs can have multiple antennas, such that BSi has $M_i > 1$ antennas. In that case, the CoMP acts as a single BS with $M_1 + M_2$ antennas and can serve Yoshi and Xia by using, for example, beamforming. Figure 12.10(b) shows a CoMP example for uplink transmissions from Yoshi and Xia, where the coordinated BSs can use various techniques for multi-antenna reception. Two-way communication for the users Yoshi and Xia with, for example, TDD, is implemented by switching between downlink configuration (a) and uplink configuration (b).

A non-traditional CoMP setup is shown in Figures 12.10(c) and (d), where Yoshi and Xia transmit or receive in opposite directions. Here the two-way communication with TDD between the users and the infrastructure is achieved by switching between (c) and (d). Taking the case in Figure 12.10(c), BS1 transmits to Yoshi and creates interference to BS2, which receives an uplink transmission from Xia. In addition, Xia creates interference to Yoshi's receiver. The signal $b_{r,2}$ received by BS2 is:

$$b_{r,2} = h_{1,2}b_{t,1} + h_{X,2}x + n_2 \tag{12.8}$$

where $h_{1,2}$ is the channel between BS1 and BS2, $b_{t,1}$ is the signal transmitted by BS1, $h_{X,2}$ is the channel between Xia and BS2, x is the signal sent by Xia and n_2 is the noise at the receiver of BS2. Since the BSs are interconnected with a backhaul link (still with an infinite capacity!), BS1 can forward $b_{t,1}$ to BS2 and, similar to the extra-collision interference cancellation, BS2 can cancel it from (12.9) to obtain:

$$b'_{r,2} = h_{X,2}x + n_2. \tag{12.9}$$

This cancellation requires BS2 to know h_{12}, but this is reasonable to assume, considering that the BSs are static, the channel between them changes slowly and there is sufficient time to estimate it. In contrast, the interference that Xia creates to Yoshi cannot be canceled and, if the channel between Xia and Yoshi is not weak, then this interference degrades the overall system performance.

It is not immediately obvious which configuration from Figure 12.10 is optimal and it depends on the traffic patterns of the users, rate requirements, and the actual channels.

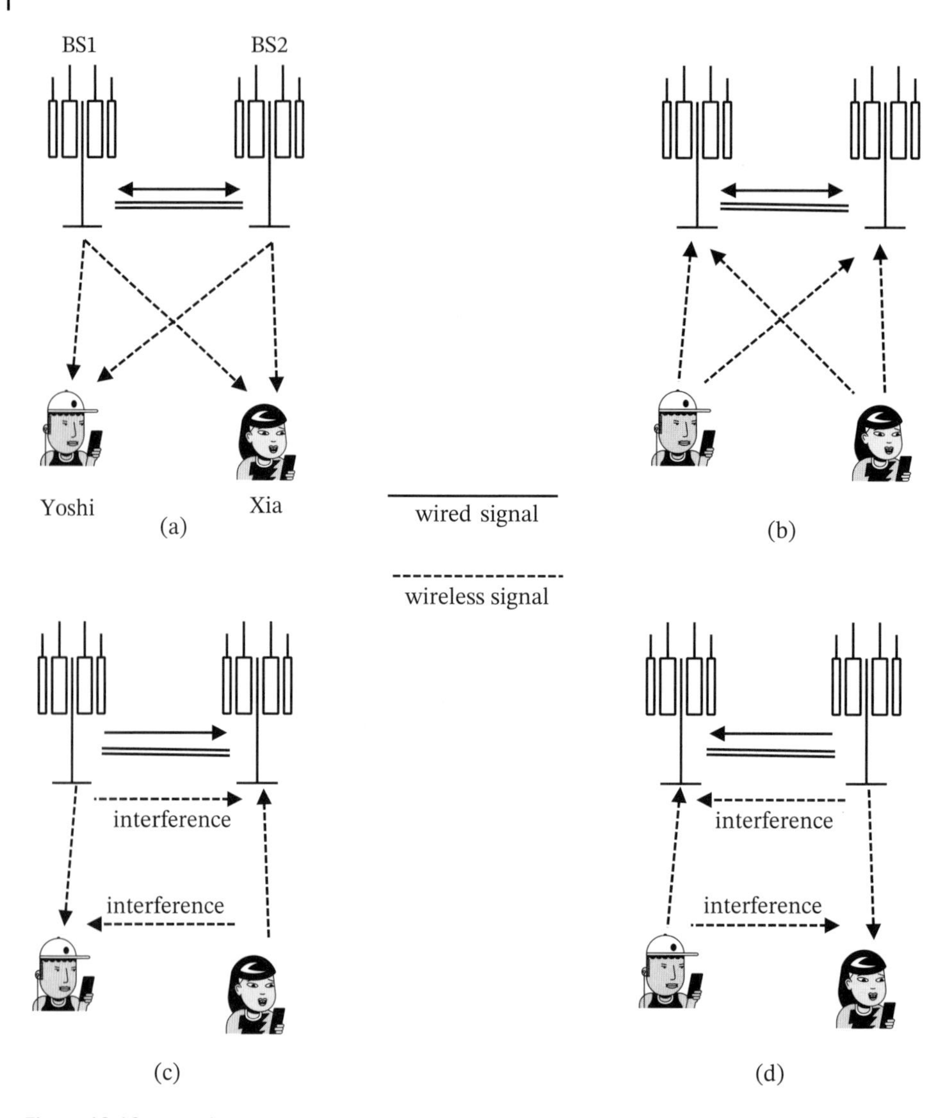

Figure 12.10 Wireless coverage with cooperation and coordination among BSs interconnected via a wired backhaul. (a) Downlink transmission. (b) Uplink transmission. (c),(d) Mixed uplink–downlink transmission. The wired signal in (a) and (b) is bidirectional to denote full cooperation between the BSs. The unidirectional wired signal on (c) and (d) shows the direction in which information is provided from the source BS to enable interference cancellation at the destination BS.

For example, in the extreme case where Yoshi has only downlink and Xia only uplink traffic, the optimal scheme should be sought between two possibilities. The first one is to use only Figure 12.10(c). The second one is to time-multiplex between Figure 12.10(a) and Figure 12.10(b), but assuming a no transmission to Xia in Figure 12.10(a) and no transmission from Yoshi in Figure 12.10(b). As a last remark, the interconnected BSs in Figure 12.10(c) emulate a distributed form of a full-duplex device.

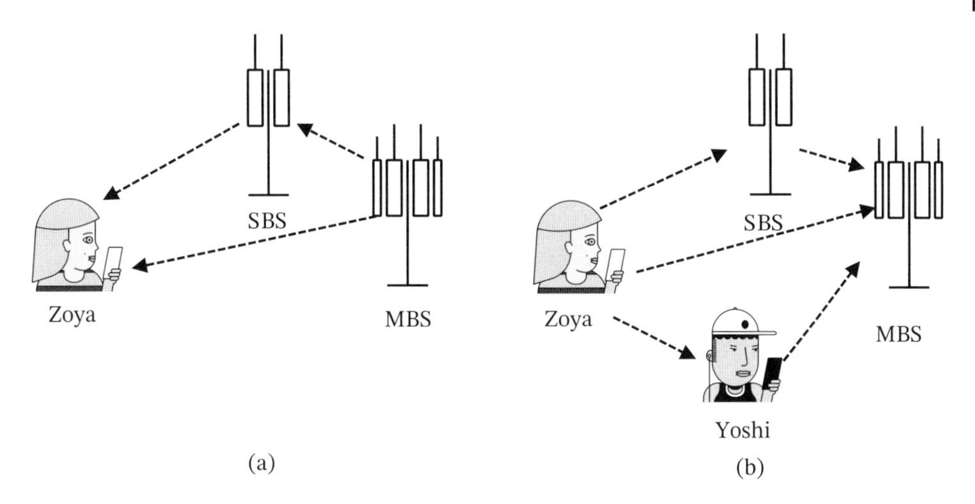

Figure 12.11 Cooperative wireless communications. (a) Cooperative downlink transmission towards Zoya with wireless backhaul between the MBS and SBS. (b) Cooperative uplink transmission from Zoya, where the SBS and Yoshi appear as helpers towards the destination MBS.

12.4.3 Cooperation Through a Wireless Backhaul

Here we address the question of establishing cooperation between BSs if the backhaul link is wireless. This brings us back to the cases of wireless relaying, however, using a viewpoint that is different from the one used in the previous section. Consider the downlink transmission from MBS to Zoya in Figure 12.11(a). The difference from the wireless relaying scenarios discussed previously is that now there is also a direct transmission from MBS to Zoya. For example, there can be an out-of-band wireless backhaul between the MBS and SBS with a dedicated wireless interface that is different from the one used to communicate with Zoya. If the wireless backhaul has a huge (infinite) capacity, then this case boils down to the cases with wired backhaul. Namely, the MBS and SBS contain the same data and can act as two transmit antennas that cooperate to send the data towards Zoya.

As before, the case that is more interesting is the one in which MBS–SBS uses the same, half-duplex wireless interface as Zoya. In the first step the MBS broadcasts its signal m_1 and the signals received by the SBS and Zoya are:

$$\begin{aligned} s_{r,1} &= h_{M,S} m_1 + n_S \\ z_{r,1} &= h_{M,Z} m_1 + n_{Z,1} \end{aligned} \tag{12.10}$$

where $s_{r,1}$ is the signal received by the SBS, $h_{M,S}$ is the channel MBS–SBS, n_S is the noise at the receiver of the SBS, $z_{r,1}$ is the signal received by Zoya in the first step, $h_{M,Z}$ is the channel MBS–Zoya, and $n_{Z,1}$ is the noise at Zoya's receiver. Let us assume that the channel MBS–SBS is much stronger than MBS–Zoya, such that $|h_{M,S}| \gg |h_{M,Z}|$. Then after the first step, the SBS has perfectly decoded the symbol m_1. In the second step the SBS sends m_1 to Zoya and she receives:

$$z_{r,2} = h_{X,Z} m_1 + n_{Z,2} \tag{12.11}$$

where $h_{X,Z}$ is the channel SBS–Zoya and $n_{Z,2}$ is the noise at Zoya's receiver in this second transmission. Now Zoya can treat $z_{r,1}$ and $z_{r,2}$ as outputs of two receive antennas and use MRC to decode m_1. The SNR available to Zoya for this decoding is:

$$\gamma_Z = \gamma_{Z,M} + \gamma_{Z,S} \tag{12.12}$$

where

$$\gamma_{Z,M} = \frac{|h_{M,Z}|^2 P_M}{N_Z} \qquad \gamma_{Z,S} = \frac{|h_{M,S}|^2 P_S}{N_Z} \tag{12.13}$$

where P_M and P_S are the transmit powers of the MBS and SBS, respectively, and N_Z is the noise power at Zoya's receiver. With this scheme, the highest achievable rate between the MBS and Zoya is:

$$R_{M,Z} = \frac{1}{2}\log_2(1 + \gamma_Z) \tag{12.14}$$

where the factor $\frac{1}{2}$ is due to the fact that each symbol sent between the MBS and Zoya consumes two channel uses. This is known as *cooperative MIMO* or *distributed MIMO*.

The described scheme is perhaps one of the simplest among the schemes for cooperative communication. As an example of a more complex scheme, in the second step both the MBS and SBS can transmit simultaneously the symbol m_1, but in such a way that their transmissions add up coherently at Zoya. This is some form of transmit beamforming with a subtle difference. Namely, when the two transmit antennas are on the same device, then the total power distributed across the antennas is constrained by a certain value. In contrast, the MBS and SBS each have an individual power constraint and they can operate at the highest transmit power. Even more complex cooperative schemes can be devised by not considering symbol-by-symbol transmission, but rather packet-by-packet transmission. It can be shown that, in these schemes, the SBS can aid the transmission of the MBS towards Zoya even if the SBS is not able to completely decode the message from the MBS.

The possibilities for cooperation increase when we consider multiple relays, sources and destinations. Figure 12.11(b) shows a scenario for uplink transmission. Zoya sends her data to the MBS through two relays, SBS and Yoshi. The connection to Yoshi is through a device-to-device (D2D) link. One can think of this D2D link as bi-directional, used by Zoya and Yoshi to help each other for uplink transmissions. In the first two steps each of them shares her own data with the other one, in the next step they transmit using transmit beamforming.

A lot of other cooperative methods are possible by allowing various level of message exchange among the nodes. As a final remark, cooperative communications require revision of the classical layering architecture from Chapter 4, as it combines functionalities of the network, link (MAC), and physical layer.

12.5 Dissolving the Cells into Clouds and Fog

12.5.1 The Unattainable Ideal Coverage

The cooperation among the BSs brings forward the idea of providing wireless coverage in a given area by abandoning the concept of a cell. Figure 12.12 shows that the area is covered

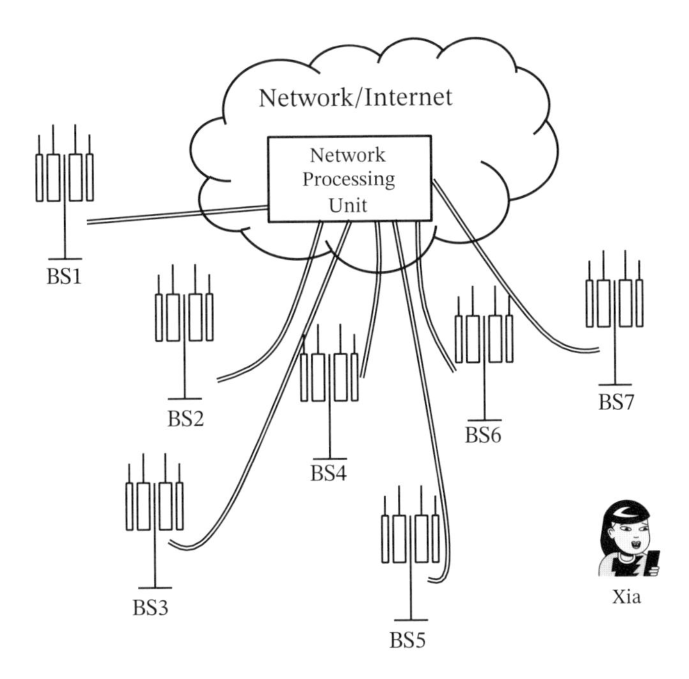

Figure 12.12 Providing wireless coverage through a system of interconnected BSs. In this example there are no direct backhaul links between two BSs. All backhaul links are connected to a central processing unit that resides in the network and treats the system of BSs as a distributed system of multiple antennas. Conceptually, Xia is connected to all BSs simultaneously.

by a set of BSs interconnected by wired backhauls, through which they coordinate and cooperate. This results in a distributed, potentially giant, multi-antenna system that offers wireless coverage. There are no cells and each user is covered by a set, sometimes a subset, of all BSs that can efficiently contribute to the transmission and reception using multi-antenna processing algorithms. This mode of operation naturally removes the need for explicit handover and interference management, as these operations are part of the overall distributed operation.

This holy grail of wireless coverage is not realizable, as it relies on several idealized assumptions about the wired backhaul links, as well as the global availability of the information about the wireless channels, both of the desired and the undesired, interfering links. In the attempt to understand why these assumptions are problematic, let us compare this setup with the case in which the user Xia communicates with only one BS Basil. The communication between Xia and Basil is based on some protocol that has a certain frame structure. The timing parameters of this frame structure are chosen in such a way to accommodate, for example, pilots that allow Basil and Xia to estimate the wireless channel between them and apply appropriate modulation and coding. If we want to preserve the same frame structure and protocol timing when Xia is served by a set of BSs, as in Figure 12.12, then the backhaul links need to have zero latency and infinite capacity. This would allow instantaneous availability of the information about all wireless channels in the network to all BSs. Furthermore, the idealized backhaul links are used to exchange the actual data that should be transmitted in the downlink, such that each BS can act as a transmit antenna that implements certain multi-antenna techniques. In the opposite, uplink direction, the ideal backhaul links are used to transfer the received baseband samples to a central unit that implements a certain multi-antenna multi-user receiver algorithm.

Let us see what happens when these ideal assumptions are removed. When the area covered by many interconnected BSs is large, the effect of the limited speed of light starts to be noticeable, as the propagation time becomes comparable to the timings that are applied in the communication protocols. On top of the propagation time, there is an additional latency due to the processing at each of the BSs. Finally, there is also a latency contribution due to the fact that the backhaul link has a finite data rate and thus it takes time greater than zero to transmit a certain portion of data.

12.5.2 The Backhaul Links Must Have a Finite Capacity

We have thus arrived to the point where we have to explain the consequences of having wireless backhaul links with finite capacity. Let us at first consider the easy case, where the finite-capacity backhaul is used to transmit data that is already in a digital form. This is the case, for example, when the network sends a data packet towards the BS for a downlink transmission. Let the backhaul link rate from the network processing unit to Basil be R_{B} (bps), while the rate on the wireless downlink transmission from Basil to Xia is R_{D} (bps). Then it can be shown that the overall downlink rate, measured from the network to Xia, must be:

$$R = \min(R_{\mathrm{B}}, R_{\mathrm{D}}). \tag{12.15}$$

The case which is more challenging is the one in which the finite-capacity link carries information that is an analog form. This happens in at least two cases:

- The BS needs to provide the network processing unit with channel state information (CSI) that it measures with respect to every user in the network. Furthermore, if the BS has multiple antennas, such information needs to be provided for each antenna.
- The baseband signals received by each BS in the uplink need to be transferred to the network processing unit, such that it can apply multi-antenna processing algorithms for receiving and decoding these signals.

In order to see the effect of the finite-capacity backhaul links, let us assume that Walt wants to send an analog signal to Victoria over a noiseless link that can carry up to R (bps). This is a problem of lossy source coding mentioned in Section 8.1. Let w denote an analog symbol observed by Walt, which he needs to quantize and transmit to Victoria. Victoria tries to reconstruct the signal w, but cannot do it perfectly, such that her reconstruction of the symbol w, denoted by v, is given by:

$$v = w + n_{\mathrm{Q}} \tag{12.16}$$

where n_{Q} is the quantization noise used to model the incorrect reconstruction of w by Victoria. As in channel coding, the optimal way to do quantization is that Walt gathers multiple symbols and applies quantization to all of them, instead of quantizing symbol-by-symbol. The variable n_{Q} can often be modeled as Gaussian noise and each noise sample is drawn independently from the same Gaussian distribution with zero mean and variance σ_Q^2. The variance σ_Q^2 decreases with the rate as 2^{-R}, since a higher rate leads to better precision of the signal recovered by Victoria.

The implication of this is that the finite capacity of the backhaul links can be modeled as a source of additional noise. This noise does not allow us to claim straightforward gains from the communication architecture in Figure 12.12. Instead, the following engineering trade-offs are at play. On the one hand, multi-antenna processing algorithms aim to improve the desired signal and mitigate/eliminate interference among different links in the covered area. On the other hand, the gains from the multi-antenna processing is counteracted by the additional noise that is infused in the system through the quantization of the analog signals.

12.5.3 Noisy Cooperation with a Finite Backhaul

The trade-offs uncovered in the previous section are easily illustrated through the simple examples from Figure 12.13, where Yoshi and Xia transmit in the uplink. Their signals are received by BS1 and BS2 and are given by, respectively:

$$\begin{aligned} b_{r,1} &= h_{Y,1}y + h_{X,1}x + n_1 \\ b_{r,2} &= h_{Y,2}y + h_{X,2}x + n_2 \end{aligned} \tag{12.17}$$

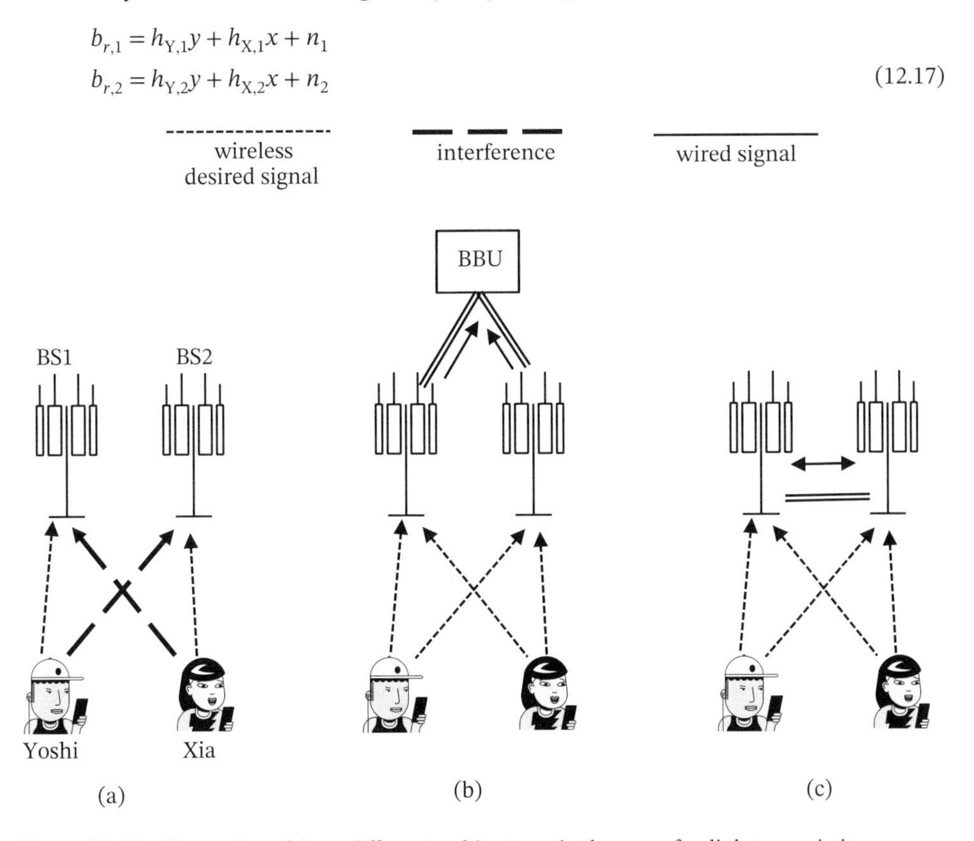

Figure 12.13 Illustration of three different architectures in the case of uplink transmission. (a) Yoshi is associated with BS1, Xia with BS2. (b) The two BSs quantize and forward the analog signals to a central baseband unit (BBU), which acts as a MIMO receiver. (c) The two BSs cooperate directly by exchanging quantized signals and each BS acts as a MIMO receiver.

where $h_{Y,i}$ is the channel between Yoshi and BSi, and $h_{X,i}$ is the channel between Xia and BSi. y and x are the signals transmitted by Yoshi and Xia, respectively. n_i is the noise at the receiver of BSi.

Figure 12.13(a) depicts the traditional cellular case in which Yoshi is associated with BS1 and Xia with BS2. For decoding of the desired signal y, BS1 treats the interfering x as additional noise, which decreases the SINR that is available to decode y. Similar considerations are valid for BS2, where y creates interference and impairs the decoding of Xia's signal x.

In contrast, in Figure 12.13(b) the two BSs quantize and forward the received analog signals to a central processing unit, often termed the *baseband processing unit (BBU)*. The quantized signals $b_{q,1}$ and $b_{q,2}$ available to the BBU are:

$$
\begin{aligned}
b_{q,1} &= b_{r,1} + n_{q,1} = h_{Y,1}y + h_{X,1}x + n_1 + n_{q,1} \\
b_{q,2} &= b_{r,2} + n_{q,2} = h_{Y,2}y + h_{X,2}x + n_2 + n_{q,2}
\end{aligned}
\tag{12.18}
$$

where $n_{q,i}$ is the quantization noise added by sending the signal of BSi through the wired backhaul. The BBU uses the signals $b_{q,1}$ to apply a certain MIMO reception algorithm and decode simultaneously x and y instead of treating any of them as an interference. However, this is done at the expense of an additional noise $n_{q,1}$ and $n_{q,2}$.

The third case, depicted in Figure 12.13(c), is somewhat intermediate. The two BSs cooperate directly through the wired backhaul and exchange the quantized versions of the received analog signals. For example, the two received baseband signals by BS1 are $b_{r,1}$, given in (12.17), and $b_{q,2}$, given in (12.18). Now BS1 can apply locally some algorithm for MIMO reception and decode both x and y. However, BS1 is not interested in x sent by Xia and it uses it just as a means to better recover the desired signal y from Yoshi. After decoding y, BS1 forwards it further to the network (not depicted in the figure). BS2 operates in an analogous way.

All these impairments may reduce the enthusiasm for the wireless communication infrastructure depicted in Figure 12.12. One should be careful in assessing the gains from these cooperative architectures and accurately account for the impairments, the latency, as well as the overhead required to operate these schemes. The exchange of the channel state information needs to be fast, otherwise the channel values associated with, for example, BS1 used by BBU or the other BSs may be outdated and lead to deteriorated performance. Nevertheless, the use of the backhaul links for cooperation and coordination opens a myriad of new possibilities for designing transmission/reception schemes, along with the associated protocols. As an example, referring to Figure 12.12, each BS may only exchange channel state information and (quantized) data with the neighboring station. In that case, BS2 exchanges information with BS1 and BS3, but not BS4, and then uses this partial information to apply a certain MIMO technique for transmission/reception.

12.5.4 Access Through Clouds and Fog

We take another look at Figures 12.13(b) and (c). From the performance viewpoint, the architecture Figure 12.13(c) seems superior, since each BS can apply the same MIMO algorithms as the BBU from Figure 12.13(c), but with a lower amount of quantization noise. However, the architecture from Figure 12.13(c) has a higher cost, since each of the BSs needs to have a unit similar to the BBU. This means that the architecture from Figure 12.13(c) cannot easily scale to larger systems with multiple BSs. On the other hand, the BSs from Figure 12.13(b) can have a significantly reduced functionality, as they do not need to decode the data or run the whole protocol, but only act as a remotely deployed antenna for the BBU.

This brings us to the concept of a *cloud radio access network (C-RAN)*. A C-RAN can be represented by Figure 12.12 and Figure 12.13(b), with slightly different terminology. The BSs from Figure 12.13(b) are named *remote radio heads (RRHs)* in order to emphasize the fact that they are not fully functional BSs, in the sense in which they are used in cellular networks. Furthermore, the wired (but also possibly wireless) links from Figure 12.13(b) are called *fronthaul links*, rather than backhaul links, in order to emphasize the fact that they carry information situated at a layer that is lower than the one usually carried by a backhaul.

We have already explained the main functionality brought by the cloud radio architecture, which is joint processing of interfering wireless signals within a given coverage area. Compared to the expensive BSs, RRHs can be deployed with a higher density, thus bringing the wireless infrastructure closer to the wireless users. Another dividend from the cloud radio architecture is that the infrastructure can orchestrate the computing and communication resources to adapt to the actual density of the users. In order to see this, consider Figure 12.13(a) and take the case in which there are no users associated with BS1, while there are many users associated in BS2. Then the resources of BS1 stay idle, while BS2 is overloaded. In contrast, Figure 12.13(b) will just direct all processing resources to BS2 (which should now be named RRH2).

Once the conceptual quantum leap from cellular BSs to C-RAN is made, we enter a new avenue for designing communication algorithms and protocols. For example, instead of outsourcing all processing to the central BBU, each remote radio head can have enhanced functionality and run locally some physical and link-layer procedures. This raises the question of *functionality split* between the central unit and the remote radio heads. There are multiple ways in which the functionality split can be implemented. In the spirit of the (eternal) discussion about where the division between the hardware and software should be, one can initiate discussion between the functionality split in the cloud radio architecture. For example, the remote radio heads may carry out part of the physical layer operations and handle some of the MAC layer signals, but the error control decoding may be left to the BBU.

Offering more functionality to the remote radio heads spreads the functionality of the cloud towards the end points of the infrastructure. This concept is often referred to as a *fog radio access network*. In a fog radio access network the concept of splitting the workload between the central cloud and the remote radio heads is not limited to functionality splitting only. In fact, the remote radio heads regain their status as BSs and completely handle some types of traffic. With this, the BS implements *mobile edge computing (MEC)*, intended to serve traffic for services and applications that require very low latency.

In general, the fog access network entails the following trade-off. On the one hand, the traffic that is handled locally achieves low latency, but cannot benefit from global interference management/cancellation that is achieved by processing the signals in the cloud. On the other hand, the cloud offers more elaborate processing of the signals at the expense of higher latency, both due to the finite-capacity backhauls, but also due to the other processing tasks that arrive at the cloud.

As a final remark, the concepts of cloud/fog access can also be combined with the benefits brought by massive MIMO. This concept is referred to as *cell-free massive MIMO*, although we need to remark that the cell has already disappeared with the concepts of cooperation, cloud, and fog. Recall that the benefit of massive MIMO is that the large antenna array allows the BS to simultaneously create multiple beams to the users and, due to the asymptotic properties of the antenna array, eliminate the inter-user interference, while maximizing the combining effect for each of the individual users. Interconnecting multiple BSs equipped with massive MIMO, as in Figure 12.12, allows each user to be served by all BSs. However, the distinctive point is that each BS creates its own beam towards the user, without coordinating the beam with the other interconnected BSs. This removes the need to exchange gigantic channel state information through the backhauls, while still allowing the BSs to cooperate through a central processing unit in transmitting downlink and decoding uplink signals.

12.6 Coping with External Interference and Other Questions about the Radio Spectrum

12.6.1 Oblivious Rather Than Selfish

There is one tacit assumption that lies beneath all the concepts and architectures presented in this chapter. Namely, we have assumed that a single network operator has full control over the radio resources and channels and this operator can decide how to allocate them to different cells, devices, cloud radio access elements, etc. Take, for example, the cells depicted in Figure 12.4: when talking about frequency/channel allocation to different cells, we assume that each cell obeys the rules prescribed by a central coordinator. These rules are in favor of the single operator that owns the cellular network. In addition, the cells may be owned by different operators, but these operators agree to cooperate according to predefined rules, which makes the allocation problem similar to the case with a single operator.

A diametrically opposite alternative would be the one in which each cell acts selfishly and accesses the radio resources without coordination with the other cells. The selfish access would lead to interference whenever two or more neighboring cells happen to use the same frequency band simultaneously. This problem is not limited to cellular systems only. Take, for example, two or more links that are in close proximity, such as the ones in Figure 12.14(a). Here each of the links aims to operate undisturbed by the transmissions taking place by the devices associated with the other links. In general, each of these links may be using a different wireless technology and in this context the term "selfish" might be rather misleading. Namely, "selfish" could be understood as a set of deliberate actions taken by, say Zoya and Yoshi, to use the shared wireless medium at the expense of the other links, Xia–Walt and Basil–Victoria. In the ideal case, Zoya and Yoshi want to be *oblivious* to the existence of any external transmitter in their proximity and thus they do not experience any interference.

One may note that random access protocols represent a paradigm in which selfish users compete to use the same frequency channel or radio resource. This can occur in an uplink transmission for the scenario depicted in Figure 12.14(b). However, in a random access setting, where the devices compete to transmit to the same receiver, they do use the same, or at least compatible, communication technology and belong to the same system.

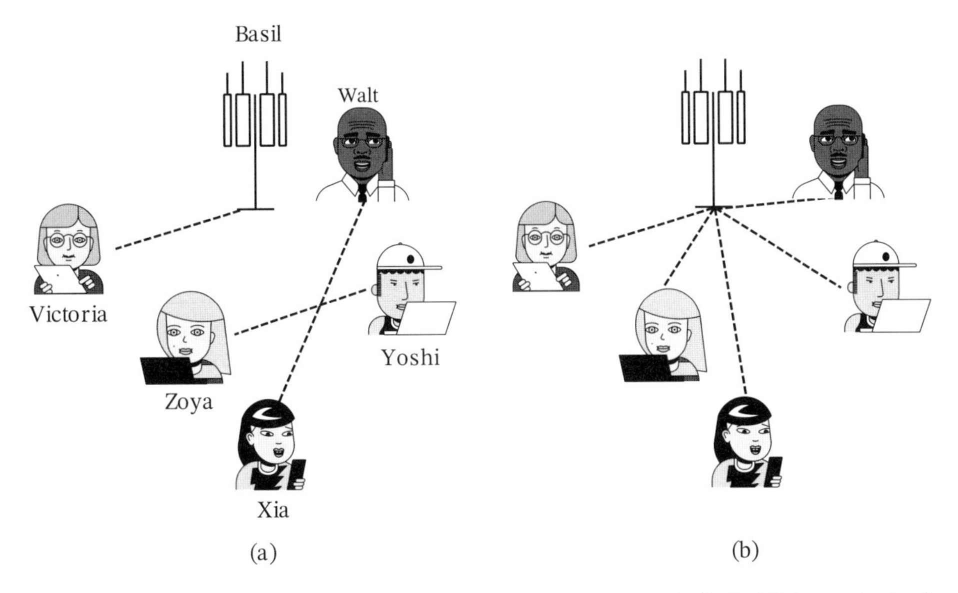

Figure 12.14 Competition for wireless communication resources. Only desired links are depicted. (a) Three links that are unrelated to each other and possibly use three different wireless technologies. (b) Five competing links that use the same wireless technology. Basil acts as a common receiver in the uplink and common downlink transmitter.

This technology can, in general, be used to coordinate the competing devices and ensure that they transmit without conflicts by using, for example, time division. However, as discussed in Chapter 2, random access is basically used due to the low efficiency of the time-division scheduling in the case when the device activity is sporadic and random.

The question is: how can we achieve a regime in which unrelated links do not interfere with each other and are even oblivious to each other? In Chapter 9 we introduced the concept of a frequency channel and why it is possible to treat the transmissions at two different frequency channels as non-interfering. In order to avoid interference between two or more uncoordinated/uncooperative wireless systems that operate in close proximity, we simply allocate a different band of the frequency spectrum to each system. This would create an (almost) ideal situation for sharing the spectrum, in which each wireless system remains oblivious to the presence and operation of the other systems.

However, at this point it is valid to ask who allocates these frequency channels to different links. Furthermore, it is not reasonable to assume that each wireless technology can operate in a dedicated frequency band, such that two unrelated wireless technologies never cause interference to each other. These concerns bring us to the problems of management of the frequency spectrum, consisting of the rules and mechanisms for allocation and sharing of frequency bands.

12.6.2 License to Control Interference

We take a step back and look into a very simple model of a single wireless link. Zoya transmits the signal z to Yoshi, such that he receives:

$$y = hz + n + i \tag{12.19}$$

where h is the channel coefficient, n is noise and i is interference. Note that all four variables h, z, n, i on the right-hand side are random. Yoshi is only interested in learning the data z. In previous chapters we have discussed the impact of h and z and, to some extent, the interference i. As elaborated in Chapter 8, from the viewpoint of wireless channel capacity, it is beneficial for Yoshi to estimate h such that Zoya can use modulation and coding schemes that assume that h is known. The noise term n represents the "unknowable" disturbance, often modeled by the most pessimistic case of a Gaussian noise. Its statistics cannot be controlled and the way to deal with it is the use the methods for reliable transmission, such as error-control coding.

However, the nature of the interference i and its statistics are fundamentally different from the noise n. Interference, as discussed in this context, is created by valid wireless communication signals and it is directly influenced by the communication protocols followed by the interfering nodes. In that sense, interference can exhibit correlation, unlike white Gaussian noise: think, for example, of the case in which the interferer uses a protocol that repeats the same symbol twice.

By buying a *spectrum license*, a connectivity provider or operator gains the right to control the statistics of the interference within a given frequency band. Within this frequency band, the operator can define frequency channels, define a common frame structure, allocate resources to cells or radio access clouds, etc.

For the sake of illustration, let us assume, for the moment, that interference is also a Gaussian random variable and that the protocols used by the interferers control only its power, that is, the variance of the random variable i. Then the relevant measure becomes the SINR:

$$\text{SINR} = \frac{S}{N + I} \tag{12.20}$$

where S is the signal power, N is the noise power and I is the interference power. The SINR is directly related to the transmission reliability that can be achieved for given rate. Then the owner of a spectrum can control the power of I or even, by doing a proper allocation of the resources in a cellular network, make $I \approx 0$. In contrast, if a connectivity provider does not own a license for a given frequency spectrum, then the power of I can vary arbitrarily and it becomes difficult, if not impossible, to give any statistical performance guarantee for the communication links affected by that interference.

The area of spectrum management is complex and multifaceted, involving aspects of technology, politics, and economy, and it is impossible to capture all these aspects in this short section. Nevertheless, several things related to spectrum licensing need to be emphasized. A connectivity provider is interested in obtaining a spectrum license within a given geographical area where this provider plans to offer wireless coverage. In a common example, this area is a territory of a country and there is a country spectrum regulation authority that allocates the frequency band to a given operator. Next, the authority that allocates the frequency band needs to have means of enforcing spectrum policy and ensuring that the spectrum license owner does not get unlawful interference from other transmitters. This is, in particular, essential for the frequency bands allocated for safety-critical purposes, such as wireless systems owned by the military or law enforcement.

Finally, spectrum is a *scarce* resource. Not every wireless technology can have its own licensed band. Due to the more favorable propagation conditions, the lower frequency bands are, in general, more desirable and hence have a higher economic value. Such are the frequencies below 1 GHz. As the carrier frequency increases above 6 GHz towards the millimeter wave frequencies, the bands that can be allocated for a specific use become, generally, larger. For example, allocating a 1 GHz band centered around 60 GHz is feasible, while doing the same at a carrier frequency below 1 GHz is hardly so. Hence, not all lower frequency bands can be offered for use through licensing. Some of the frequency bands are designated for sharing among different systems, not under the control of the same owner, and even for different technologies. These are called *unlicensed bands* and are discussed next.

12.6.3 Spectrum Sharing and Caring

In general, an unlicensed band offers open access to wireless technologies, but the price paid is the unpredictable interference that can be, potentially, arbitrarily high. In order to mitigate this interference, wireless technology and the devices that use it to operate in an unlicensed band need to follow certain rules prescribed for that band. The rules for spectrum usage have been put forward to prevent unfair use, even abuse, of the unlicensed band by a single wireless technology or system. In other words, wireless technologies should share the unlicensed spectrum, but simultaneously care about not causing too much interference to each other.

This type of wireless system operation by caring for the others is said to follow a *spectrum regulation*, which implements the principles of a certain *spectrum etiquette*. The technology requirements for the rules specified within a spectrum etiquette should be minimal in order to permit even very simple devices to follow the rules. An obvious candidate for such a rule is a limit on the transmitted power. This is a prime example of a technology neutral rule, since the transmitted power level is a parameter that is present in every wireless technology. However, this rule still does prevent the occurrence of excessive interference: the power limitation is imposed on each device, while the number of devices deployed at a given location can be arbitrarily high.

Consider the scenario in Figure 12.15(a), where a malicious user deploys a number of devices $U_1, U_2, U_3, \ldots$ in the proximity of the link Zoya–Basil. The collective interference caused by these devices is excessive, but the malicious user is, strictly speaking, not guilty of anything, since each of his devices respects the power limitation. Figure 12.15(b) shows the case in which the devices U_1, U_2, and U_3 transmit unconstrained at the same frequency channel and depicts the resulting aggregate interference. Although here we look into the interference as observed by the link Zoya–Basil, it is clear that the devices $U_1, U_2, U_3, \ldots$ create interference to each other as well. For example, the device U_1 in Figure 12.15(b) has two transmissions. The first transmission is interfered by U_2 and U_3, the intended receiver of the transmission by U_1 does not receive the data and the second transmission in Figure 12.15(b) is, in fact, retransmission. This is a vicious circle, as retransmissions cause more interference, which causes more retransmissions, etc. Thus, unconstrained transmissions can easily result in a *tragedy of the commons*, where none of the devices using the unlicensed band can derive communication value from it.

An idea for avoiding excessive interference would be to have certain coordination and cooperation among the devices. This could impose power limitation on the whole group of

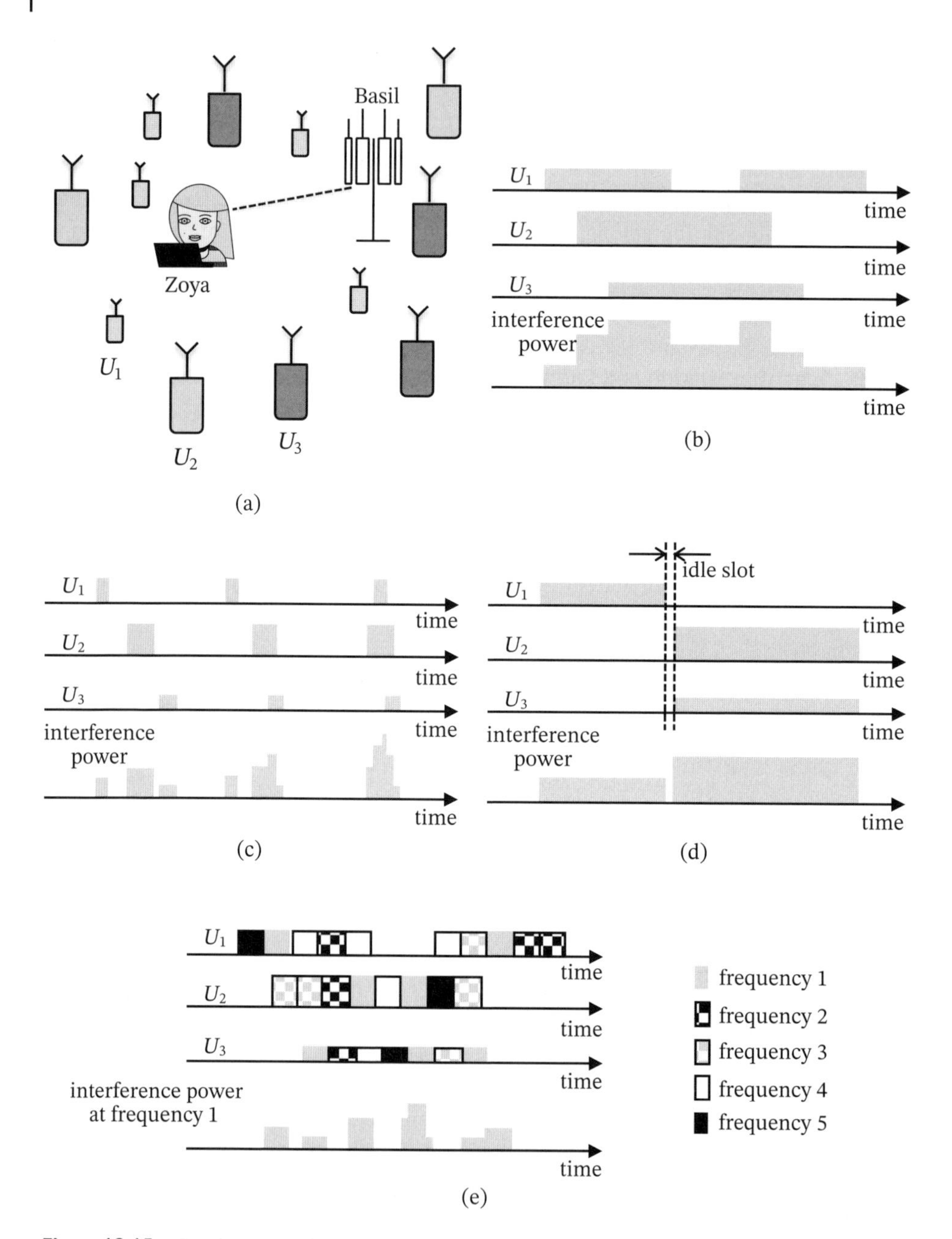

Figure 12.15 Coexistence and spectrum sharing in unlicensed frequency bands. (a) The link Zoya–Basil can get interference from an arbitrary number of unlicensed band devices $U_1, U_2, U_3, \ldots$ deployed in the proximity of Basil and Zoya. Figures (b)–(e) show the interference caused by the group of transmitters U_1, U_2, U_3. In Figures (b)–(d) all devices use a single frequency channel while the specific cases shown are: (b) unconstrained transmissions; (c) transmissions constrained by a duty cycle; (d) spectrum sharing with listen-before-talk (LBT), i.e. carrier sensing. Figure (e) illustrates operation with frequency hopping over five frequency channels. Only the interference at frequency 1 is depicted.

devices $U_1, U_2, \ldots$ as they will coordinate in a way that only few, or at best one, transmits at a given time. Such operation would result in lower interference for Zoya and Basil. However, this deviates from the requirement for a minimal, technology neutral specification, as the devices would need to implement a wireless protocol that allows them to communicate with each other. The reader should keep in mind that the devices $U_1, U_2, \ldots$ may use different, incompatible wireless technologies, such that any wireless interface for coordination and cooperation can be seen as an extra burden on the device.

12.6.4 Duty Cycling, Sensing, and Hopping

By excluding the possibility for explicit coordination and cooperation, an unlicensed device is left with two options: (1) self-restraint on the transmission activity; and (2) implicit coordination and cooperation through the interference that the different devices and systems cause to each other. One straightforward mechanism for self-restraint is the specification of a *spectrum mask*, which specifies how the power can be limited within different sub-channels of the unlicensed frequency band, as well as how it affects the neighboring frequency bands.

A representative example of operation with a self-restraint is *duty cycling*, illustrated in Figure 12.15(c). The idea of duty cycling is that a device limits its activity on a given frequency channel such that it does not occupy the channel with transmissions for more than a certain percentage of time. When specifying the duty cycle, it is also important to specify the observation time interval T in which the activity percentage is measured. For example, if $T = 20$ s and the duty cycle is 1%, then the unlicensed device should have a mechanism to suppress its transmission in a way that ensures that it does not have more than 200 ms of transmission in any interval of 20 s. It should be noted that the observation interval T imposed by the regulation of the unlicensed spectrum cannot be too small. If, for example, $T = 1$ ms, that means that a packet transmission cannot be longer than 200 ms. This would make the time of a single symbol transmission very short, which would require a transmission bandwidth that is in contradiction with the bandwidth of the allocated frequency channel. Furthermore, the requirement for short symbols somewhat removes the technology neutrality, as it imposes high performance requirements on the unlicensed devices.

Another method for sharing an unlicensed frequency channel is based on carrier sensing and is depicted in Figure 12.15(d). This type of spectrum sharing has already been introduced in the chapter on random access, Section 2.3. The term used for carrier sensing in the context of an unlicensed spectrum is LBT, which emphasizes the fact that it implements a certain spectrum etiquette. Using carrier sensing, incompatible wireless technologies can implicitly coordinate with each other through the interference caused to each other. It should be noted that carrier sensing is less technology neutral compared to duty cycling, as it requires that the wireless transmitted also has a receiver. This may not be desirable for very simple sensing devices, powered by batteries, which transmit very sporadically, and any use of receiving capability decreases their lifetime. These devices should preferably choose duty-cycled operation.

Yet another method for spectrum sharing is based on *frequency hopping*, as depicted in Figure 12.15(e). As frequency hopping is a form of spread spectrum, it entails graceful degradation of the communication performance as the number of proximate interfering links increases. From the perspective of a single frequency channel, a transmitter behaves similarly to duty cycling. Let us take again the example from above, with an observation period of $T = 20$ (s). Assume that there are 10 frequency channels, no channel can be used for more than 2 s and frequency hopping for each device follows a (pseudo)random sequence. For each specific frequency channel, a transmitter that follows this rule for frequency hopping creates interference that corresponds, on average, to a duty cycle of 10%, as also illustrated in Figure 12.15(e). We note that frequency hopping is also less neutral compared to the duty cycling, as it imposes that the unlicensed system is capable of operating over multiple frequency channels.

In addition to specifying the regulatory rules for the unlicensed bands, the regulatory authority also needs to verify that the devices follow the rules. This should prevent a malicious user from not following the duty cycle, using a shorter idle slot in LBT or occupying only one or very few "good" frequencies in frequency hopping.

12.6.5 Beyond the Licensed and Unlicensed and Some Final Words

Licensed and unlicensed spectrum operation represent rather two extremes in the way spectrum is used. A frequency band is licensed to a given operator for countrywide usage, but there can be parts of the territory where the operator does not need this channel as there are no wireless users attached to that operator. Alternatively, a frequency channel may be used in some area only during the morning or another part of the day, while it is unused for the rest of the day. These observations, as well as a number of related scenarios, lay down the motivation for *dynamic spectrum access*: abandon the static licensing and allow agile access to the spectrum resources, where a transmitter can identify the opportunities for using a given frequency band and carry out the communication in that band. Wireless systems with such behavior have been studied under the name of *cognitive radio*.

The main premise of the cognitive radio is that there is an incumbent, primary user that has a license to use the frequency band in question and the cognitive radio should find a way how to use the same spectrum as a secondary user, without disturbing the communication of the devices associated with the primary user. Keep in mind that the primary user has paid for a very expensive license in order to be able to control the interference, such that any external interference decreases the utility that the primary user can derive from the spectrum. Furthermore, a common assumption in cognitive radio has been that the primary user cannot, or does not want to, change anything in its operation or the associated transceivers in order to allow or facilitate secondary spectrum usage.

One can see cognitive radio as being a more capable spectrum user that arrives in a more advanced technological age and is capable of operating "around" the primary user. To do that, a cognitive radio should be capable of identifying precisely the conditions under which it can transmit without degrading the performance of the primary user. Identification of spectrum opportunities can happen in different ways. One is the use of a database, which contains information about the opportunity to use a frequency at given space-time slot. The database reflects the usage of the spectrum over a longer period and is not suitable for identifying intermittent spectrum opportunities. The latter can be done by using *spectrum sensing*, by which a cognitive radio device detects whether there is a primary transmitter using the band and, if there is not, transmits its own data. The fundamental problem with spectrum sensing is that it identifies the primary transmitter, while the interference created by the cognitive radio device happens at a primary receiver. Let us reuse the example from Figure 12.15(a), assuming that the link Basil–Zoya represents the primary spectrum user and U_2 is a cognitive, secondary spectrum user. Assume that U_2 can detect transmissions from Zoya, but the path loss between Basil and U_2 is too large (for example, due to an obstacle) to offer a reliably detectable signal. In that case, U_2 may identify a spectrum opportunity exactly at the time when Zoya is receiving from Basil and thus interfere with her reception.

Despite these challenges, the ideas behind dynamic spectrum access and cognitive radio should be seen as a step in the right direction, as they are challenging the established order of wireless radio communication. The rules applied in spectrum regulation follow principles that involve minimal specification of the technology requirements and expectations put on the devices. We emphasized this in the previous section when we discussed technology neutrality in relation to the rules for sharing of an unlicensed spectrum. In fact, one can see the principles and rules applied in spectrum regulation as *technology axioms*: minimal assumptions upon which the whole wireless ecosystem is built, that is, the entire population of various wireless technologies and their associated transceivers.

Differently from mathematical axioms, these technology axioms can change over a longer time, reflecting the change in the minimal technology level of the wireless systems. Although infeasible, it is useful to engage in the speculative exercise:*How would we design the rules for using and sharing the spectrum if we are able to start from a blank slate (no incumbent wireless devices), but having the present technological level?*

This speculation may lead us new ways of using the spectrum. We can then add the restrictions of the incumbents and see how they constraint the freedom of the blank slate rules. Here are a couple of examples of potential spectrum usage scenarios, some of them highly speculative:

- Instead of having a primary user that does not accept any interference, the primary user can extract economic value from its licensed spectrum by selling the right to use this spectrum to secondary, and even tertiary, users at a given space and within a given time. This type of operation can leverage the technologies that support decentralized trustful transactions among digital actors, such as blockchain or other distributed ledgers.
- Instead of allocating a frequency band to an administrative territory (country), the allocation can happen in a way that utilizes advanced wireless technologies for beamforming and spatial separation and allocate the frequency channel more precisely in space. Regarding spectrum sharing, devices that use narrower beams and/or lower power can be awarded larger duty cycles.
- A secondary spectrum user that does not operate "around" the primary user, but rather in symbiosis with the primary user. In our example from Figure 12.15(a), U_2 can offer to act as a form of relay to Zoya and Basil and facilitate their communication, but as a remuneration get the right to use the same spectrum. This means that the technology used by Zoya and Basil should have some level of flexibility to allow such facilitation from a device that comes as a spectrum user at a later point in technological development.

The quest for enabling future dynamic spectrum usage puts forward the following question: *How do you specify the spectrum usage rules and design the associated technologies such that spectrum users behave in a good and flexible way in the future, when another technological advancement is in place?* Here the technological advancement refers not only to wireless technologies, but also localization, precise synchronization, radio sensing, distributed ledgers, artificial intelligence (AI) and, last but not least, quantum communication.

12.7 Chapter Summary

The usual pedagogical starting point in wireless mobile networking is the concept of a cell and a cellular network. In this book these concepts came into the last chapter and there are two reasons for this. First, the cellular concept is not fundamental in the same way in which information theory (Chapter 6–8) or physics of wireless communications (Chapter 9 and 10) are. Second, a cellular network is only one way to achieve coverage in a given geographical area and there are other architectures that can achieve the same thing. In this chapter we have attempted to provide a perspective on the evolution of the ideas for wireless coverage, starting from cells, through heterogeneous networks, cooperative communication, and up to the contemporary concepts of cloud/fog radio access and cell-free architectures. Finally, we have tackled perhaps the most important techno-economic question in wireless communications, that of spectrum regulation and spectrum sharing. Within the short space allocated in this book, we have discussed the basic principles of licensed and unlicensed spectra. It is the belief of this author that spectrum regulation is based on certain technological axioms that determine what can be considered to be technologically neutral from the viewpoint of the requirements that are put forward for the devices that use and share certain spectra. As technology progresses, these axioms need to be changed in order to pave the way for more flexibility in spectrum use and support advanced interactions among the devices that use the same spectrum, while belonging to different networks and operators.

12.8 Further Reading

Classification of different types of connectivity has been one of the central points in defining 5G wireless systems, see Osseiran et al. [2016]. The cellular network concept has been treated in standard textbooks for wireless communication, for example Goldsmith [2005] and Molisch [2012]. Network MIMO has been treated in Gesbert et al. [2010]. An insightful study of cooperative communication is given in Lozano et al. [2013], while further information on two-way relaying can be found in Popovski and Koike-Akino [2009]. Cloud radio access is covered in Quek et al. [2017], and the concept of fog radio access in Tandon and Simeone [2016]. Essentials of spectrum management can be found in Cave et al. [2007], and cognitive radio and dynamic spectrum access in Hossain et al. [2009] and Biglieri et al. [2013].

12.9 Problems and Reflections

1. *System design for wireless networks*. Frame design has been treated in the text, as well as in the problems in the previous chapters, in increasing order of complexity (TDMA, FDMA, multiple antennas). This design becomes even more complex in a wireless network, in which there is no single node (BS) defining, synchronizing, and coordinating the frame and the use of communication resources. Discuss the possible system design based on frame and the associated trade-offs for cellular networks and cloud/fog based networks. Specifically, consider the signaling provisions that need to be made in order to integrate wireless relays in the system.
2. *Two-way communication with coordinated BSs*. Figure 12.10 shows four possible transmission modes. All devices in Figure 12.10 are assumed to be half-duplex. If both Xia and Yoshi have two-way traffic patterns to/from the infrastructure, then these traffic patterns can be supported by a combination of these four transmission modes.
   (a) Build a model to analyze the conditions in which certain transmission modes are preferred (e.g., due to traffic asymmetry or excessive interference).
   (b) Compare the system with two coordinated BSs to the system with a single full-duplex BS.
3. *Cloud/fog with wireless relaying*. The architectures based on cloud radio access or fog radio access are usually based on wired/optical connections between the remote radio heads (RRHs) and the central processing unit. In contrast, in this chapter we have seen that there are methods for cooperation and relaying and, in particular, two-way relaying, that have the potential to support wirelessly connected RRHs.
   (a) What are the trade-offs involved when considering wireless versus wired connections to the RRHs?
   (b) Investigate possible scenarios where wirelessly connected RRHs can be useful, such as distributed computation tasks based on IoT/sensor data.
4. *Frequency hopping versus carrier sensing*. A system designer wants to use an unlicensed band for a sensor network. The network should provide real-time information about the status of a critical infrastructure and occasionally send alarm messages that are event-driven. The real-time status information does not require excessive reliability, but it is desirable that it is received with low latency. The alarm message should be received with extremely high reliability. The designer has an option to use the unlicensed spectrum with carrier sensing (LBT) or with frequency hopping. Provide an analysis that will lead to the choice of the spectrum sharing method. Build a simple simulation to facilitate this analysis. Consider two different cases:
   (a) Uplink-dominated traffic, where the uplink data from the sensors is larger than the downlink data/commands that come from the infrastructure/cloud.
   (b) Symmetric traffic, in which the sensors carry out two-way interactive communication with the cloud.

5. *Clean slate spectrum sharing*. This last problem is the speculative exercise already posed in the text: how would we design the rules for using and sharing the spectrum if we were able to start from a blank slate (no incumbent wireless devices), but still having the present technological level? For example, if it is mandatory that all devices have precise synchronization and localization (e.g. provided by GPS), which modes of spectrum sharing are possible?

Bibliography

ETSI EN 300 744 V1. 5.1 (2004-11). *Digital Video Broadcasting (DVB) Framing structure, channel coding, and Modulation for Digital Terrestrial Television*. European Broadcasting Union, 2004.

Norman Abramson. The aloha system: Another alternative for computer communications. In *Proceedings of the November 17–19, 1970, Fall Joint Computer conference*, pages 281–285. New York: ACM, 1970.

Venkat Anantharam and Sergio Verdu. Bits through queues. *IEEE Transactions on Information Theory*, 42(1):4–18, 1996.

Erdal Arıkan. Channel polarization: a method for constructing capacity-achieving codes for symmetric binary-input memoryless channels. *IEEE Transactions on Information Theory*, 55(7):3051, 2009.

Dimitri P. Bertsekas and Robert G. Gallager. *Data Networks*, volume 2. New Jersey: Prentice-Hall International, 1992.

Giuseppe Bianchi. Performance analysis of the ieee 802.11 distributed coordination function. *IEEE Journal on Selected Areas in Communications*, 18(3):535–547, 2000.

Ezio Biglieri, Andrea J. Goldsmith, Larry J. Greenstein, Narayan B. Mandayam, and H. Vincent Poor. *Principles of Cognitive Radio*. Cambridge University Press, 2013.

Emil Björnson, Jakob Hoydis, Luca Sanguinetti, et al. Massive mimo networks: spectral, energy, and hardware efficiency. *Foundations and Trends® in Signal Processing*, 11(3-4):154–655, 2017.

Tim Brown, Persefoni Kyritsi, and Elizabeth De Carvalho. *Practical Guide to MIMO Radio Channel: With MATLAB Examples*. John Wiley & Sons, 2012.

John I Capetanakis. Tree algorithms for packet broadcast channels. *IEEE Transactions on Information Theory*, 25:505–515, 1979.

Enrico Casini, Riccardo De Gaudenzi, and Oscar Del Rio Herrero. Contention resolution diversity slotted aloha (crdsa): an enhanced random access schemefor satellite access packet networks. *IEEE Transactions on Wireless Communications*, 6(4):1408–1419, 2007.

Martin Cave, Christopher Doyle, and William Webb. *Essentials of Modern Spectrum Management*. Cambridge University Press, 2007.

Xu Chen, Tsung-Yi Chen, and Dongning Guo. Capacity of gaussian many-access channels. *IEEE Transactions on Information Theory*, 63(6): 3516–3539, 2017.

Thomas Cover. Broadcast channels. *IEEE Transactions on Information Theory*, 18(1):2–14, 1972.

Wireless Connectivity: An Intuitive and Fundamental Guide, First Edition. Petar Popovski.

Thomas M. Cover and Joy A. Thomas. *Elements of Information Theory*. John Wiley & Sons, 2012.

Erik Dahlman, Stefan Parkvall, and Johan Skold. *4G: LTE/LTE-advanced for Mobile Broadband*. Academic Press, 2013.

Elisabeth De Carvalho, Emil Bjornson, Jesper H. Sorensen, Petar Popovski, and Erik G. Larsson. Random access protocols for massive mimo. *IEEE Communications Magazine*, 55(5):216–222, 2017.

Abbas El Gamal and Young-Han Kim. *Network Information Theory*. Cambridge University Press, 2011.

Anthony Ephremides and Bruce Hajek. Information theory and communication networks: an unconsummated union. *IEEE Transactions on Information Theory*, 44(6):2416–2434, 1998.

Dennis Gabor. Theory of communication. part 1: The analysis of information. *Journal of the Institution of Electrical Engineers—Part III: Radio and Communication Engineering*, 93(26):429–441, 1946.

R.G. Gallager. *Principles of Digital Communication*. Cambridge University Press, 2008. ISBN 9781139468602. URL https://books.google.mk/books?id=5W0aYFU02igC.

Robert G Gallager. *Information Theory and Reliable Communication*, volume 2. Springer, 1968.

David Gesbert, Stephen Hanly, Howard Huang, Shlomo Shamai Shitz, Osvaldo Simeone, and Wei Yu. Multi-cell mimo cooperative networks: a new look at interference. *IEEE Journal on Selected Areas in Communications*, 28(9): 1380–1408, 2010.

Andrea Goldsmith. *Wireless Communications*. Cambridge University Press, 2005.

Ji Hayes. An adaptive technique for local distribution. *IEEE Transactions on Communications*, 26(8):1178–1186, 1978.

Robert W. Heath Jr. and Angel Lozano. *Foundations of MIMO Communication*. Cambridge University Press, 2018. doi: 10.1017/9781139049276.

Ekram Hossain, Dusit Niyato, and Zhu Han. *Dynamic Spectrum Access and Management in Cognitive Radio Networks*. Cambridge University Press, 2009. doi: 10.1017/CBO9780511609909.

Vikas Kawadia and Panganamala Ramana Kumar. A cautionary perspective on cross-layer design. *IEEE Wireless Communications*, 12(1):3–11, 2005.

Christos Liaskos, Shuai Nie, Ageliki Tsioliaridou, Andreas Pitsillides, Sotiris Ioannidis, and Ian Akyildiz. A new wireless communication paradigm through software-controlled metasurfaces. *IEEE Communications Magazine*, 56(9): 162–169, 2018.

Angel Lozano, Robert W. Heath, and Jeffrey G. Andrews. Fundamental limits of cooperation. *IEEE Transactions on Information Theory*, 59(9):5213–5226, 2013.

Thomas L. Marzetta, Erik G. Larsson, Hong Yang, and Hien Quoc Ngo. *Fundamentals of Massive MIMO*. Cambridge University Press, 2016.

James L. Massey and Peter Mathys. The collision channel without feedback. *IEEE Transactions on Information Theory*, 31(2):192–204, 1985.

James L. Massey. Applied digital information theory. *Lecture Notes, ETH Zurich.*, 1980. URL http://www.isiweb.ee.ethz.ch/archive/massey_scr/adit1.pdf.

Andreas F. Molisch. *Wireless Communications*, volume 34. John Wiley & Sons, 2012.

Tadashi Nakano, Michael J. Moore, Fang Wei, Athanasios V Vasilakos, and Jianwei Shuai. Molecular communication and networking: Opportunities and challenges. *IEEE Transactions on Nanobioscience*, 11(2):135–148, 2012.

Afif Osseiran, Jose F. Monserrat, and Patrick Marsch. *5G Mobile and Wireless Communications Technology*. New York, NY: Cambridge University Press, New York, NY, USA, 1 edition, 2016. ISBN 1107130093, 9781107130098.

Enrico Paolini, Cedomir Stefanovic, Gianluigi Liva, and Petar Popovski. Coded random access: applying codes on graphs to design random access protocols. *IEEE Communications Magazine*, 53(6):144–150, 2015.

Yury Polyanskiy. A perspective on massive random-access. In *2017 IEEE International Symposium on Information Theory (ISIT)*, pages 2523–2527. IEEE, 2017.

Petar Popovski and Toshiaki Koike-Akino. *Coded bidirectional relaying in wireless networks*. In *New Directions in Wireless Communications Research*, pages 291–316. Springer, 2009.

Petar Popovski, Kasper Fløe Trillingsgaard, Osvaldo Simeone, and Giuseppe Durisi. 5G wireless network slicing for eMBB, URLLC and mMTC: A communication-theoretic view. *IEEE Access*, 6:55765–55779, 2018.

J.G. Proakis and M. Salehi. *Digital Communications*. McGraw-Hill International Edition. McGraw-Hill, 2008. ISBN 9780071263788.

Tony Q.S. Quek, Mugen Peng, Osvaldo Simeone, and Wei Yu. *Cloud Radio Access Networks: Principles, Technologies, and Applications*. Cambridge University Press, 2017.

Theodore Rappaport. *Wireless Communications: Principles and Practice*. Upper Saddle River, NJ: Prentice Hall, Upper Saddle River, NJ, USA, 2 edition, 2001. ISBN 0130422320.

Tom Richardson and Ruediger Urbanke. *Modern Coding Theory*. Cambridge University Press, 2008.

Lawrence G. Roberts. Aloha packet system with and without slots and capture. *ACM SIGCOMM Computer Communication Review*, 5(2):28–42, 1975.

Raphael Rom and Moshe Sidi. *Multiple Access Protocols: Performance and Analysis*. Springer Science & Business Media, 2012.

Yuya Saito, Yoshihisa Kishiyama, Anass Benjebbour, Takehiro Nakamura, Anxin Li, and Kenichi Higuchi. Non-orthogonal multiple access (noma) for cellular future radio access. In *2013 IEEE 77th Vehicular Technology Conference (VTC Spring)*, pages 1–5. IEEE, 2013.

Anna Scaglione, Dennis L Goeckel, and J Nicholas Laneman. Cooperative communications in mobile ad hoc networks. *IEEE Signal Processing Magazine*, 23(5):18–29, 2006.

Claude Elwood Shannon. A mathematical theory of communication. *Bell System Technical Journal*, 27(3):379–423, 1948.

Lin Shu and Daniel J. Costello. Error control coding. 2004.

David Slepian. On bandwidth. *Proceedings of the IEEE*, 64(3):292–300, 1976.

IEEE Wi-Fi Standards. IEEE standard for information technology–telecommunications and information exchange between systems local and metropolitan area networks–specific requirements—part 11: Wireless lan medium access control (mac) and physical layer (phy) specifications. *IEEE Std 802.11-2016 (Revision of IEEE Std 802.11-2012)*, pages 1–3534, Dec 2016. doi: 10.1109/IEEESTD.2016.7786995.

Ravi Tandon and Osvaldo Simeone. Harnessing cloud and edge synergies: toward an information theory of fog radio access networks. *IEEE Communications Magazine*, 54(8):44–50, 2016.

Kasper Fløe Trillingsgaard and Petar Popovski. Generalized HARQ protocols with delayed channel state information and average latency constraints. *IEEE Transactions on Information Theory*, 64(2):1262–1280, 2017.

David Tse and Pramod Viswanath. *Fundamentals of Wireless Communication*. Cambridge University Press, 2005.

Boris Solomonovich Tsybakov and Viktor Alexandrovich Mikhailov. Free synchronous packet access in a broadcast channel with feedback. *Problemy Peredachi Informatsii*, 14(4):32–59, 1978.

Gottfried Ungerboeck. Trellis-coded modulation with redundant signal sets part i: Introduction. *IEEE Communications Magazine*, 25(2):5–11, 1987.

Rodney Vaughan and J. Bach Andersen. *Channels, Propagation and Antennas for Mobile Communications*. IET, 2003.

David J. Wetherall and Andrew S. Tanenbaum. *Computer Networks*. Pearson Education, 2013. ISBN 1292024224, 9781292024226.

J.M. Wozencraft and Irwin Mark Jacobs. Principles of communication engineering. 1965.

Yingqun Yu and Georgios B. Giannakis. High-throughput random access using successive interference cancellation in a tree algorithm. *IEEE Transactions on Information Theory*, 53(12):4628–4639, 2007.

Index

Wireless Connectivity: An Intuitive and Fundamental Guide, First Edition. Petar Popovski.